SI EDITION

D1243866

SUSTAINABLE ENERGY

SECOND EDITION

Richard A. Dunlap

Dalhousie University

 CENGAGE

Australia • Brazil • Mexico • Singapore • United Kingdom • United States

CENGAGE

Sustainable Energy, **Second Edition, SI Edition**

Richard A. Dunlap

Product Director, Global Engineering:
   Timothy L. Anderson

Senior Content Developer: Mona Zeftel

Product Assistant: Teresa Versaggi

Marketing Manager: Kristin Stine

Senior Content Project Manager:
   Michael Lepera

Senior Art Director: Michelle Kunkler

Cover and Internal Designer:
   Lou Ann Thesing

Cover Image: Richard A. Dunlap

Production Service: RPK Editorial Services,
   Inc.

Compositor: MPS Limited

Intellectual Property
   Analyst: Christine Myaskovsky
   Project Manager: Sarah Shainwald

Text and Image Permissions Researcher:
   Kristiina Paul

Manufacturing Planner: Doug Wilke

For product information and technology assistance, contact us at
Cengage Customer & Sales Support, 1-800-354-9706.

For permission to use material from this text or product,
submit all requests online at **www.cengage.com/permissions**.
Further permissions questions can be emailed to
permissionrequest@cengage.com.

Library of Congress Control Number: 2017956525

ISBN: 978-1-337-55167-0

Cengage
20 Channel Center Street
Boston, MA 02210
USA

Cengage is a leading provider of customized learning solutions with employees residing in nearly 40 different countries and sales in more than 125 countries around the world. Find your local representative at **www.cengage.com.**

Cengage products are represented in Canada by Nelson Education Ltd.

To learn more about Cengage platforms and services, visit **www.cengage.com.**

To register or access your online learning solution or purchase materials for your course, visit **www.cengagebrain.com.**

Printed in the United States of America
Print Number: 01   Print Year: 2017

In Memory of my Father
Robert Bennett Dunlap

CONTENTS

PREFACE

Our society uses substantial quantities of energy. This energy use amounts to about 6.1×10^{20} J per year worldwide, or an average of 8.1×10^{10} J per person. Between 80 and 85% of the world's energy comes from fossil fuels, which are preferred because they are inexpensive (relatively speaking), are readily available (at least at present), and have a high energy density. As a result, an enormous infrastructure has been established for the location, production, and use of fossil fuels. The fuel of choice is oil because it is convenient, and the gasoline and diesel fuel it produces are portable and constitute our major source of fuel for transportation.

For the purpose of planning for methods to meet our future energy needs, it is important to begin by asking two questions: How long will our fossil fuel reserves last? Is it wise, from an environmental perspective, to continue to use fossil fuels?

The answers to both questions are not simple. The answer to the first question can be several tens of years or several hundreds of years depending on the conditions that are put on our fossil fuel use. Will fossil fuels continue to supply 80 to 85% of our energy needs? Will a fossil fuel–derived product be required to fulfill our needs for a portable transportation fuel? Perhaps most importantly, how much are we willing to pay for fuel? There is certainly some limit to how much we, as individuals, are willing or able to pay for the gasoline for our automobiles or for the oil or natural gas to heat our homes. However, it is important to realize that the cost of fuel is not only a financial cost. Producing fossil fuels in a form that is suitable for our needs requires energy input in order to undertake exploration to locate new fuel reserves, the extraction of the fuel from those reserves, and the subsequent processing of the fuel. If the energy needed to produce a liter of fuel is greater than the energy we obtain from burning it, then the process is not only economically unattractive but is ultimately not energy productive. If only the use of oil in the traditional sense, from known and economically recoverable reserves, is considered, then the longevity of fossil fuels will certainly be at the low end of the timescale. If coal and less traditional oil reserves are also considered, then the answer can be near the upper end of the timescale. This will be especially true if alternative sources are used to supply a substantial fraction of our energy needs.

The answer to the second question is also not straightforward. There is overwhelming evidence that the emission of greenhouse gases that results from the burning of fossil fuels has a severe impact on the environment. The magnitude and the timescale of this impact are not fully understood. If the use of fossil fuels continues for an extended period of time, then our willingness or even our ability to take steps to mitigate the effects on the environment are also unclear.

To ensure an adequate supply of energy in the future and to avoid causing a negative impact on our environment, it is important to understand how energy is utilized at present, our future energy needs, and the options for fulfilling these needs. Designing an appropriate energy structure for the future requires, not only a consideration of appropriate energy sources, but the implementation of suitable strategies to minimize energy requirements through conservation efforts.

In terms of our reliance on fossil fuels, two extreme approaches can be taken: to stop using fossil fuels now or to stop using fossil fuels when our supply is exhausted. The first approach would certainly minimize the environmental impact of fossil fuel use but would be impossible to implement because of our lack of infrastructure for the use of other energy sources. The latter approach would maximize the environmental effects and would best make use of the resources available. Whatever the final course of events, it is essential that steps toward eliminating our dependence on fossil fuel be taken immediately by developing and implementing alternative energy sources so that the environmental impact of our fossil fuel use is minimized. The latter would involve the reduction in greenhouse gas emissions by not only a reduction in fossil fuel use but also by processes such as carbon sequestration.

To put the magnitude of this task (however it is approached) into perspective, it is necessary to consider the current world power requirement of about 1.9×10^{13} W. In 50 years (roughly the time scale set by the recent Paris Agreement for a carbon neutral society), the world power requirement might be more than twice the current amount (primarily as a result of increased energy needs in developing countries). This is a rough goal that should be kept in mind when assessing the viability of any energy policy. These power requirements can be related to the output of a typical large electric generating station. These stations most commonly use fossil fuels (mostly coal and natural gas) to produce electricity and might have a typical output of about 10^9 W. The conversion to a nonfossil fuel energy economy on a timescale of about 50 years will require the construction of about $(4 \times 10^{13}$ W$)/(10^9$ W$) = 40,000$ large replacement facilities (or a corresponding number of smaller facilities). These might be large nuclear power plants, large hydroelectric stations, or equivalent-capacity facilities utilizing solar energy, wave energy, wind energy, and other sources. This amounts to the construction of more than two major nonfossil fuel power stations every day for the next five decades. Clearly, this task requires a substantial commitment.

Sustainable Energy's Purpose

The textbook *Sustainable Energy* considers in detail our present and future energy needs, options for continued use of fossil fuels, and options for establishing an alternative energy economy. This text was developed out of a course entitled "Energy and the Environment" that has been taught in the Department of Physics and Atmospheric Science at Dalhousie University since 2003. This one-semester introductory course is aimed at undergraduate science and engineering students and is taught at the sophomore level. Most students have taken freshman-level chemistry and physics, and most have had some introductory calculus. These prior courses make a suitable prerequisite for a course taught from the present text. The course at Dalhousie is taught as a follow-up to a course on climate change to give students the overall picture of how humanity interacts with the environment, although this previous course is not a necessary prerequisite for a course taught from *Sustainable Energy*. Although such a course is intended to be introductory, there is enough technical detail that upper-level science or engineering will find it a useful and informative elective.

The purpose of this textbook is to fill a niche between a significant number of texts on similar topics at a very descriptive level intended for freshmen survey courses and a few advanced (and often fairly specialized) texts aimed at senior undergraduate or beginning graduate students in engineering. The textbook is useful for science and engineering students with an interest in energy-related matters, particularly those looking to pursue a professional career in a related field, to take an introductory energy course with some reasonable technical content. In addition to filling a mostly unfilled gap in the field, the present text also provides an up-to-date introduction to a fairly rapidly changing field.

Organization

This text begins with an overview of the basic science needed for the remainder of the book, as well as a summary of our past, present, and anticipated future energy needs. The technologies currently in use to meet our energy needs are described, and the need for the development of new energy technologies on the basis of future resource availability and environmental concerns is emphasized. The text includes a separate chapter on every future renewable energy technology that could be viewed as a viable option for the production of a significant portion of our energy needs. How these developing technologies can be integrated efficiently with existing technologies is discussed, as well as approaches to conserving available energy resources. Finally, the text considers options for perhaps our greatest energy-related challenge: transportation. The viability of any alternative energy technologies is determined by its ability to fulfill various criteria. The important criteria are described in this text by the acronym CURVE, for *c*lean, *u*nlimited, *r*enewable, *v*ersatile, and *e*conomical. This acronym makes it easy for students to appreciate how different technologies may, or may not, play an important role in our future energy production. The final chapter of the book summarizes the various alternative energy sources that have been presented and analyzes how these different technologies succeed or fail in satisfying the various CURVE criteria.

Throughout the text, the complexity of energy issues is emphasized, as is the need for a multidisciplinary approach to solving our energy problems. This approach provides students with an appreciation for the real problems that are encountered in the understanding of how we produce and use energy, as well as the realization that, while exact calculations are important and necessary, a broadly based analysis is often most appropriate. The text also stresses the fact that solutions to our energy problems, both now and in the future, are not straightforward and do not have simple, well-defined solutions, and that the way ahead is far from certain. The book contains enough material for a typical one-semester (12- to 14-week) course with about 20% excess material to allow the instructor some flexibility in course design. This coverage of material allows about 2–3 hours of lecture, on average, per chapter. Instructors may also focus on specific topics to provide a more in-depth picture of certain aspects of energy. This approach may include a more detailed and probing look at some of the topics presented in the Energy Extra boxes and may require the omission of other components of the text. Some chapters from the text can be covered in less detail and/or even eliminated. Chapters 7, 12, 13, 14, and 20 can be skipped with minimal effect on continuity. Certain approaches to sustainable energy may be more or less relevant from some national and/or regional perspectives and may warrant more or less detailed course coverage.

Finally, Chapter 21 acts as a summary of the ideas presented in the text and shows how they can be integrated into our approach to future energy production. This chapter includes a number of research and design projects that provide the student with the challenge of integrating information presented throughout the text to the solution of practical problem related to energy production and use. These projects give the student the opportunity to assess information and to make decisions about the most reasonable approach to energy production and use. Such decisions often involve a consideration of scientific, technological, environmental, and economic factors and illustrate not only the complexity but the multidisciplinary nature of sustainable energy.

Chapter Pedagogical Elements

- **Learning Objectives**. Each chapter starts with a bulleted list of learning objectives, making it very clear to both instructors and students what is covered in the chapter.
- **Examples**: The text includes numerous worked examples to provide the student with the basic approach to deal with end-of-chapter problems
- **Energy Extra Boxes**. Energy Extra boxes are included in nearly all chapters. These boxes provide insight into details of specific aspects of energy and often emphasize the complex nature of the decisions required to plan for our future energy needs. They also stress that ostensibly advantageous approaches to energy are often not as beneficial as they seem and that a critical analysis is necessary to understand all aspects of the topic.
- **End-of-chapter Problems**. The end-of-chapter problems are predominantly quantitative in nature. However, most are not straightforward calculations based on substituting values from the chapter into the appropriate formulas. The problems are designed to require the students to analyze information, to make use of material from previous chapters, to correlate data from various sources (not only from the textbook itself but from library, Internet, or other sources), and in many cases to estimate quantities based on interpretation of graphical data, interpolation of values, and sometimes just plain common sense.

New to This Edition

- Updated data tables and graphs with the most current information and developments.
- Doubled the number of end-of-chapter exercises for each chapter.
- Developed more than 30 new examples throughout the text.
- Added new Energy Extra boxes.
- Expanded coverage of alternative energy methods and feasibility analysis.

New additions to each chapter include:

Chapter 1:   New content on diesel generators.
Chapter 2:   Energy Extra box on rare earth elements.
Chapter 3:   Added operation of an oil well, transport methods for oil and natural gas, and expanded application of the Hubbert model.
Chapter 4:   Added discussion on acid rain and ocean acidification; added Energy Extra box on natural vs anthropogenic climate change, added new sections on methanol production from CO_2 and international climate change initiatives.
Chapter 6:   Added Energy Extra boxes on Watts Bar Reactor, and thorium reactors and the Indian reactor program, expanded sections on fast breeder reactors.
Chapter 7:   Expanded section on design of inertial confinement fusion reactors.
Chapter 8:   Added Energy Extra box on evacuated tube solar collectors; and added section on transpired solar collectors.

Chapter 9:   Expanded discussion of parabolic trough collectors; expanded section on central receivers; added section on solar ponds; added Energy Extra box on solar updraft towers.

Chapter 10:   Added Energy Extra box on wind turbine safety.

Chapter 12:   Added discussion on integrated wind/wave generation.

Chapter 13:   Added discussion on Sihwa Lake Tidal Station.

Chapter 14:   Added section on details of physical principles of ocean thermal energy conversion.

Chapter 16:   Added section on biogas.

Chapter 17:   Added Energy Extra boxes on cogeneration in Iceland and thermoelectric generators; expanded section on hybrid vehicles.

Chapter 19:   Expanded section on commercial availability of battery electric vehicles.

Chapter 20:   Added Energy Extra box on hydrogen storage in fullerenes; Expanded section on commercial availability of fuel cell vehicles.

Ancillaries

A variety of ancillaries are available to accompany this book to supplement your course. These supplements include:

- An *Instructor's Solution Manual*.
- *Annotated Lecture Note PowerPoint Slides*, which include suggestions for teaching the material in the book.
- Sample test items for instructors.
- Additional practice problems for students.
- *Image Bank* of figures and tables from the book.

Custom Options for Sustainable Energy

Would you prefer to easily create your own personalized text, choosing the elements that best serve your course's unique learning objectives?

Cengage's Compose platform provides the full range of Cengage content, allowing you to create exactly the textbook you need. The Compose website lets you quickly review materials to select everything you need for your text. You can even seamlessly add your own materials, like exercises, notes, and handouts! Easily assemble a new print or eBook and then preview it on our site.

Cengage offers the easiest and fastest way to create custom learning materials that are as unique as your course is. To learn more about customizing your book with Compose, visit compose.cengage.com, or contact your Cengage Learning Consultant.

MindTap Online Course

Sustainable Energy, 2e is also available with **MindTap,** Cengage's digital learning experience. The textbook's carefully-crafted pedagogy and exercises are made even more effective by an interactive, customizable eBook accompanied by automatically graded assessments and a full suite of study tools.

> **CHAPTER 9: ELECTRICITY FROM SOLAR ENERGY**
>
> Chapter 9: Electricity from Solar Energy
> Introduction - Solar Electric Generation - Photovoltaic Devices - Application of Ph
> Global Use of Photovoltaics - Chapter Review - Summary
>
> **Chapter 9: Videos**
>
> Chapter 9: Quiz
> After you've read Chapter 9, answer the questions in the quiz.
> COUNTS TOWARD GRADE
>
> Chapter 9: Drop Box
> Use this drop box to submit any other assignments your instructor has assigned
> PRACTICE

MindTap gives you complete control of your course—to provide engaging content, to challenge every individual, and to prepare students for professional practice. Adopting MindTap cuts your prep time and lets you teach more effectively with videos, assessments, and more. Built-in metrics provide insight into engagement, identify topics needing extra instruction, and let you instantly communicate with your students. Finally, every MindTap adoption includes support from our dedicated, personalized team. We'll help you set up your course, tailor it to your specifications, and stand by to offer support to you and your students whenever you need us.

How Does MindTap Benefit Instructors?

- Customize and personalize your course by integrating your own content into the **MindTap Reader** (like lecture notes, audio and video recordings) or pull from sources such as RSS feeds, YouTube videos, websites, and more.
- Save grading time by leveraging MindTap's **automatically graded assignments and quizzes.** These problems include immediate, specific feedback, so students know exactly where they need more practice.
- The **Message Center** helps you to quickly and easily contact students directly from MindTap. Messages are communicated immediately by email, social media, or even text message.

> Oil Drilling and Refinery
> The purpose of an oil refinery is to turn crude oil into products that are fit for end-use, in the quantities that are required by the market. Watch this video and animation to follow the transformation process, from crude oil to end product.
>
> Oil Drilling and Refinery - Transcript
> You can read or download the transcript file.
>
> Oil Platform in Rough Sea
> Things to think about while watching this video: This is actually the accommodation platform for the workers, as there is no drilling rig. It is being filmed from the platform with the drilling rig. How are these platforms secured? What happens to the drilling rod on the drilling platform when the platform moves in rough seas?
>
> Oil Sands
> Things to think about while watching this video: How has oil sands production been affected by recent oil price fluctuations? What affect does recent policies concerning the Keystone XL pipeline have on oil sands? How would changes in the oil sands production affect U.S. oil imports and domestic Canadian oil use?
>
> Oil Shale Advertisement
> Things to think about while watching this video: This is an advertisement from 2008. How has oil shale development followed the discussion in the video since that time? How do the production quantities discussed here compare with the oil sands production in Canada?

- **StudyHub** is an all-in-one studying destination that allows you to deliver important information and empowers your students to personalize their experience. Instructors can choose to annotate the text with **notes** and **highlights**, share content from the MindTap Reader, and create custom **flashcards** to help their students focus and succeed.
- The **Progress App** lets you know exactly how your students are doing (and where they are struggling) with live analytics. You can see overall class

engagement levels and drill down into individual student performance, enabling you to identify topics needing extra instruction and instantly communicate with struggling students to speed progress.

How Does MindTap Benefit Your Students?

The **MindTap Mobile App** includes the entire eBook accompanied by flashcards, quizzes, and course alerts to help students understand core concepts, achieve better grades, and prepare for their future courses.

- **Flashcards** are pre-populated to provide a jump start on studying, and students and instructors can also create customized cards as they move through the course.
- The **Progress App** allows students to monitor their individual grades, as well as their level compared to the class average. This not only helps them stay on track in the course but also motivates them to do more, and ultimately to do better.
- The **StudyHub** is a single-destination studying tool that empowers students to personalize their experience. They can quickly and easily access all notes and highlights marked in the MindTap Reader, locate bookmarked pages, review notes and Flashcards shared by their instructor, and create custom study guides.
- The **MindTap Reader** includes the abilities to have the content read aloud, to print from the digital textbook, and to take notes and highlights directly in the text while also capturing them within the linked **StudyHub App**.

For more information about MindTap for Engineering, or to schedule a demonstration, please call (800) 354-9706 or email higheredcs@cengage.com. For instructors outside the United States, please visit http://www.cengage.com/contact/ to locate your regional office.

Acknowledgments

I am grateful for the assistance of many individuals during the development of this text. First, I am indebted to the students who have taken courses and whom I have taught at Dalhousie University on Sustainable Energy. They have served as the inspiration for this textbook and have provided feedback on the course material. I would also like to thank Jeff Dahn for numerous discussions over the years on energy related matters and Harm Rotermund for his continued encouragement and comments during the writing of the manuscript. I am also grateful to Ewa Dunlap for assistance, support, and advice throughout this project, and to German Rojas Orozco for checking the accuracy of the Examples and the Solutions to the Problems. I would like to thank the Global Engineering team at Cengage Learning for their dedication to this new edition: Timothy Anderson, Product Director: Mona Zeftel, Senior Content Developer; Teresa Versaggi,

Product Assistant; and Rose Kernan of RPK Production. Finally, I would like to thank the following reviewers who provided invaluable comments on the manuscript:

- Julie Albertson, *University of Colorado, Colorado Springs*
- Prabhakar Bandaru, *University of California, San Diego*
- Ronald Besser, *Stevens Institute of Technology*
- Christopher Bull, *Brown University*
- Larry Caretto, *California State University, Northridge*
- Kip Carrico, *New Mexico Institute of Mining and Technology*
- Gerald Cecil, *University of North Carolina at Chapel Hill*
- Chih-hung Chang, *Oregon State University*
- Timothy J. Cochran, *Alfred State College*
- Tim Healy, *Santa Clara University*
- Jin Jiang, *University of Western Ontario*
- Charles Knisely, *Bucknell University*
- David Marx, *Illinois State University*
- Chiang Shih, *Florida A&M University and Florida State University*
- Robert J. Stevens, *Rochester Institute of Technology*
- Wencong Su, *University of Michigan—Dearborn*
- Eric Stuve, *University of Washington*
- Thomas Ortmeyer, *Clarkson University*
- Songgang Qiu, *Temple University*

R. A. Dunlap
Halifax, Nova Scotia

PREFACE TO THE SI EDITION

This edition of *Sustainable Energy,* Second Edition has been adapted to incorporate the International System of Units (*Le Système International d'Unités* or SI) throughout the book.

Le Système International d'Unités

The United States Customary System (USCS) of units uses FPS (foot–pound–second) units (also called English or Imperial units). SI units are primarily the units of the MKS (meter–kilogram–second) system. However, CGS (centimeter–gram–second) units are often accepted as SI units, especially in textbooks.

Using SI Units in This Book

In this book, we have used both MKS and CGS units. USCS (U.S. Customary Units) or FPS (foot-pound-second) units used in the US Edition of the book have been converted to SI units throughout the text and problems. However, in case of data sourced from handbooks, government standards, and product manuals, it is not only extremely difficult to convert all values to SI, it also encroaches upon the intellectual property of the source. Some data in figures, tables, and references, therefore, remains in FPS units.

To solve problems that require the use of sourced data, the sourced values can be converted from FPS units to SI units just before they are to be used in a calculation. To obtain standardized quantities and manufacturers' data in SI units, readers may contact the appropriate government agencies or authorities in their regions.

Instructor Resources

The Instructors' Solution Manual in SI units is available through your Sales Representative or online through the book's website at http://login.cengage.com. A digital version of the ISM, Lecture Note PowerPoint slides for the SI text, as well as other resources are available for instructors registering on the book's website.

Feedback from users of this SI Edition will be greatly appreciated and will help us improve subsequent editions.

Cengage Learning

ABOUT THE AUTHOR

Richard A. Dunlap is a research professor in the Department of Physics and Atmospheric Science at Dalhousie University. He received a B.S. in Physics from Worcester Polytechnic Institute (1974), an A.M. in Physics from Dartmouth College (1976), and a Ph.D. in Physics from Clark University (1981). Since 1981 he has been on the faculty at Dalhousie University. From 2001 to 2006 he was Killam Research Professor of Physics, and since 2009 to 2015 he was the Director of the Dalhousie University Institute for Research in Materials. Prof. Dunlap is author of three previous textbooks, *Experimental Physics: Modern Methods* (Oxford 1988), *The Golden Ratio and Fibonacci Numbers* (World Scientific 1997), and *An Introduction to the Physics of Nuclei and Particles* (Brooks/Cole 2004). Over the years his research interests have included critical phenomena, magnetic materials, amorphous materials, quasicrystals, hydrogen storage, superconductivity, and materials for advanced rechargeable batteries. He has published more than 300 refereed research papers.

PART I

Background

Energy is an essential component of our daily lives. Throughout human history, our energy use has increased, and we now depend on a complex energy infrastructure to meet our needs for heating, lighting, transportation, and the production and distribution of all manufactured materials. Our increased energy needs have put increasing demands on the earth's resources and have had increasingly adverse effects on our environment. We are now at a stage of human development where our energy use must be critically analyzed to determine suitable future approaches to the production and use of this vital component of our lives.

Chapter 1 of this text begins with an overview of the basic scientific principles related to energy and a description of the quantitative scientific tools needed to analyze our energy use. This overview includes a summary of the various forms of energy and a quantitative description of the processes by which energy can be converted from one form to another. Also included is a survey of fundamental thermodynamics and a description of the basic principles of electricity distribution.

An overview of energy use throughout history is presented in Chapter 2. The chapter also provides the mathematical basis needed to assess future energy needs and a summary of the factors that need to be evaluated when considering possible future energy production methods.

The photograph at the beginning of this part of the text shows the Gordon Dam in Tasmania. This high head hydroelectric dam is 192 m long and 140 m high and has a maximum capacity of 432 MW_e. It became operational in 1978 and was one of the last major hydroelectric facilities to be constructed during an era of hydroelectric power development in Tasmania that began in the 1950s and continued until the 1980s. This trend in major hydroelectric development is paralleled in many other parts of the world. ■

Tim Collins/Shutterstock.com

Energy Basics

Tim Collins/Shutterstock.com

Learning Objectives: After reading the material in Chapter 1, you should understand:

- The relationship between energy and power.
- The forms of energy.
- The laws of thermodynamics.

- Heat engines and their Carnot efficiency.
- Heat pumps and their coefficient of performance.
- How electricity is generated and distributed.

1.1 Introduction

Energy may be categorized in different ways. One approach is to classify energy as either kinetic or potential. From a classical point of view, *kinetic* energy is merely the energy associated with the motion of a body. *Potential* energy may be described in terms of the nature of the interactions in a system. For example, gravitational potential energy arises from the gravitational interaction between two masses. The potential energy of water in an elevated reservoir may be converted into kinetic energy, which can be utilized to turn a turbine.

From a practical standpoint, however, it is convenient to describe the forms of energy according to how they are produced and utilized. It is also crucial to understand how one form of energy can be converted into another. In fact, when one says that energy is "produced" (e.g., by a nuclear reactor), one refers to the conversion of one form of energy that is not suitable for our needs (i.e., nuclear binding energy) into another form (e.g., electrical energy) that is more readily utilized. Energy that is extracted from our environment is *primary* energy; for example, chemical energy in the form of fossil fuels, kinetic energy in the wind, potential energy of water in a reservoir, or incident solar energy. To utilize energy, it is nearly always necessary to convert primary energy into a form that suits our needs. In this chapter, some of the basic physics of energy are explained, as well as the characteristics of some forms of energy and their conversions.

1.2 Work, Energy, and Power

Energy, E, is defined as the ability to do work, *W. Work* is the consequence of the expenditure of energy and is defined as the product of a force, *F*, acting on an object times the distance, *d*, that the object moves. This relationship can be written as

$$W = Fd. \tag{1.1}$$

This expression assumes that the force acting on the object is a constant over the time during which the object moves a distance, *d*, and that the force is acting in the same direction as the displacement. The units of work are the same as the units of energy. In the metric system, the standard unit of energy is the *joule* (J) when the force is expressed in *newtons* (N) and the displacement in *meters* (m). In terms of fundamental metric units, the *joule* is equal to $kg \cdot m^2 \cdot s^{-2}$.

It is perhaps convenient to think of the concept expressed in equation (1.1) in terms of a mechanical system. An object with a mass, *m*, lying at rest on the floor exerts a force (the gravitational force), *mg* (here *g* is the gravitational acceleration), downward on the floor. The floor exerts a force, *mg*, upward on the mass (the normal force) that cancels out the gravitational force. The net vertical force on the object is zero, and, from Newton's law,

$$F = ma, \tag{1.2}$$

the acceleration is zero, and the object does not move. The work done is zero because the distance that the object travels is zero. If an external vertical force that is equal to (or greater than) *mg* is exerted on the object, then the object can be lifted from the floor. If the object is lifted to a height *h*, then the work done, from equation (1.1), is the force times the distance, or

$$W = mgh. \tag{1.3}$$

According to the law of conservation of energy, this work is converted into gravitational potential energy, also equal to *mgh*. The work done is independent of how long the process takes or of the path taken to reach height *h*. Because of this latter property, the gravitational force is said to be *conservative*.

It is sometimes convenient to deal with power rather than with energy or work. *Power, P,* is the rate at which work is done (or the rate at which energy is expended). Power is measured in *watts* (W), and the watt is defined as 1 joule per second. Assuming that power is a constant in time, *t*, then the total energy utilized is

$$E = Pt. \tag{1.4}$$

This definition shows that $1 \ W \cdot s = 1 \ J$. Total energy is the power integrated over time so that producing (or using) 1000 W of power for 1 second represents the same amount of energy as producing (or using) 1 W of power for 1000 seconds. Equation (1.4) provides the basis for an alternative unit for the measurement of energy, the *kilowatt-hour* (kWh). The kWh is defined as the energy corresponding to a power of 1 kilowatt (1000 W) over a period of 1 hour (3600 s) so that $1 \ kWh = (1000 \ W) \times (3600 \ s) = 3.6 \times 10^6 \ J$.

Example 1.1

If a system produces 10^6 J (i.e. 1 MJ) of energy every minute, what is the power produced in watts?

Solution

If 10^6 J is released over a period of 1 hour (or 3600 s), then the energy per unit time in joules per second, which is equivalent to watts, is

$$P = E/t = (10^6 \text{ J})/(3600 \text{ s}) = 278 \text{ W}.$$

1.3 Forms of Energy

Energy can take on many forms:

- *Kinetic energy* (e.g., of a moving automobile).
- *Gravitational potential energy* (e.g., of water in a reservoir).
- *Thermal energy* (e.g., in a pot of boiling water).
- *Chemical energy* (e.g., stored in a liter of gasoline).
- *Nuclear energy* (e.g., stored in a gram of uranium).
- *Electrical energy* (e.g., used by a light bulb).
- *Electromagnetic energy* (e.g., that associated with a beam of sunlight).

As explained, these categories of energy are merely convenient ways of describing energy from different sources. They are not necessarily unique or mutually exclusive, nor is the list necessarily comprehensive. For example, thermal energy might be thought of as the microscopic kinetic energy of the molecules of a material. Both chemical energy and nuclear energy can be viewed as manifestations of the mass-energy associated with bonds in a material. However, these seven categories are a convenient way of defining the forms of energy from a practical standpoint.

To make use of energy, it is generally necessary to convert energy from the form in which it is obtained to a form that is compatible with our needs. For example, the stored chemical energy in a liter of gasoline can be converted to heat and then into mechanical energy to move a vehicle. Energy conversions are an important aspect of the utilization of any energy source, and the efficiency of these conversions is crucial to the viable utilization of the energy source. In any process, energy is always conserved. (In nuclear physics, the conservation of mass-energy, rather than the conservation of energy itself, is employed because there is an equivalence between these two quantities.) However, in any energy conversion process, all of the energy does not end up in the form needed. Each of these forms will now be discussed briefly.

1.3a Kinetic Energy

Kinetic energy is most obviously associated with moving objects. For an object of mass, m, moving at a velocity, v, the kinetic energy is

$$E = \frac{1}{2} mv^2. \tag{1.5}$$

In the metric system, m is in kilograms (kg), v is in meters per second (m/s), and the resulting energy is in joules ($\text{kg·m}^2\text{·s}^{-2}$).

Example 1.2

What is the kinetic energy associated with a 1500-kg automobile traveling at 100 km/h?

Solution

The velocity converted to m/s is (100 km/h) $\times$ (1000 m/km)/(3600 s/h) = 27.8 m/s. Using equation (1.5), the energy is given as

$$E = \frac{1}{2}mv^2 = (0.5) \times (1500 \text{ kg}) \times (27.8 \text{ m/s})^2 = 5.8 \times 10^5 \text{ kg·m}^2/\text{s}^2 = 5.8 \times 10^5 \text{ J}.$$

Kinetic energy is also associated with the rotational motion of rotating objects. The energy is given as

$$E = \frac{1}{2}I\omega^2, \tag{1.6}$$

where I is the moment of inertia of the object, and ω is its angular velocity. (The moment of inertia of an object and further details of rotational motion will be discussed in Chapter 18.) The moment of inertia is given in units of kg·m^2, and the angular velocity is given in units of s^{-1}. As before, the energy is measured in joules. Objects that have both translational motion and rotational motion have both translational kinetic energy, as given by equation (1.5), and rotational kinetic energy, as given by equation (1.6).

Example 1.3

A wheel in the form of a solid disk with a mass of m = 400 kg, a diameter of d = 0.85 m and a moment of inertia of $I = md^2/8 = mr^2/2$ rolls without slipping. The velocity of its center of mass is 30 m/s. This is a rough approximation of a wheel on a freight train. Compare the wheel's translational kinetic energy to its rotational energy.

Solution

From equation (1.5), its translational kinetic energy is

$$E_{\text{kinetic}} = \frac{1}{2}mv^2 = (0.5) \times (400 \text{ kg}) \times (30 \text{ m/s})^2 = 1.8 \times 10^5 \text{ J}.$$

If the wheel rolls without slipping, then its angular velocity is related to the velocity of its center of mass, v, and its radius, r, because $\omega = v/r$. Substituting for ω and I in equation (1.6) gives

$$E_{\text{rotational}} = \frac{1}{2}\left(\frac{mr^2}{2}\right)\left(\frac{v}{r}\right)^2 = \frac{1}{4}mv^2.$$

Substituting these values,

$$E_{\text{rotational}} = \frac{1}{4}mv^2 = (0.25) \times (400 \text{ kg}) \times (30 \text{ m/s})^2 = 9.0 \times 10^4 \text{ J}.$$

Note that the rotational energy is independent of the wheel diameter and is exactly one-half of the translational kinetic energy. These features are basic characteristics of a solid disk that rolls without slipping.

1.3b Potential Energy

Potential energy is most conveniently thought of in terms of gravitational potential, as explained. The concept of potential energy also applies to other situations, such as the energy contained in a compressed spring. In the case of gravitational potential energy, an object of mass, m, at a height h has potential energy given by

$$E = mgh. \tag{1.7}$$

This potential energy can be converted into kinetic energy by allowing the object to fall through the distance h (assuming there are no drag forces), yielding

$$E = \frac{1}{2} mv^2 = mgh. \tag{1.8}$$

The velocity of the object may thus be calculated to be

$$v = \sqrt{2gh}. \tag{1.9}$$

Example 1.4

A 75-kg person walks up a flight of stairs with a vertical height of 3 m. What is the change in that person's potential energy?

Solution

From equation (1.7),

$$E = mgh = (75 \text{ kg}) \times (9.8 \text{ m/s}^2) \times (3 \text{ m}) = 2.2 \times 10^3 \text{ J}.$$

1.3c Thermal Energy

The *thermal energy* of a gas results from the kinetic energy of the microscopic movement of the molecules. Each molecule of gas has a kinetic energy associated with it that is given by equation (1.5), where m is the mass of the molecule, and v is its average velocity. It can be shown by applying ideal gas theory that the right-hand side of equation (1.5) can be expressed in terms of the temperature of the gas as

$$\frac{1}{2} mv^2 = \frac{3}{2} k_B T. \tag{1.10}$$

Here k_B is *Boltzmann's constant* with a value of 1.3806×10^{-23} J/K, and T is the absolute temperature in Kelvin (K) (more on this in Section 1.4). The total internal energy of a collection of gas molecules is obtained from equation (1.10) by multiplying the right-hand side by the number of gas molecules present. From a practical standpoint, it is convenient to deal with macroscopic quantities such as the number of moles of gas. Thus

$$E = \frac{3}{2} nRT, \tag{1.11}$$

where n is the number of moles of gas, and R is the *universal gas constant*; $R = N_A k_B = 8.315$ J/(mol·K). *NA is Avogadro's number* (6.022×10^{23} mol^{-1}).

It is sometimes convenient (particularly for solids and liquids) to describe changes in the macroscopic thermal energy of the material in terms of the *specific heat, C*, of the material. If a quantity of energy, Q, is supplied to a piece of material of mass, m, then its temperature will increase by an amount, ΔT, given by

$$\Delta T = \frac{Q}{mC}. \tag{1.12}$$

Materials with a large specific heat require a large amount of energy per unit mass to raise their temperature by a given amount. On the other hand, these materials are able to store large amounts of thermal energy per unit mass when its temperature is raised by a relatively small amount. (The utilization of these principles is discussed in detail in Chapter 8.) If a solid is heated to its melting point, then additional energy must be provided to melt it. This energy is used to break the chemical bonds holding the solid together and is referred to as the *latent heat of fusion*. The term *latent heat* is used to distinguish it from *sensible heat* because latent heat does not change the temperature of a solid. When a liquid is heated to its boiling point, then additional energy, the *latent heat of vaporization*, is needed to cause the material to undergo a phase transition and become a gas.

Example 1.5

The specific heat of water is 4180 J/(kg·°C). Calculate the energy required to heat 500 g of water from 20°C to 80°C.

Solution
Rearranging equation (1.12) to solve for the heat gives

$$Q = mC\Delta T.$$

Using $m = 0.5$ kg, $C = 4180$ J/(kg·°C), and $\Delta T = (80°C - 20°C) = 60°C$, then

$$Q = (0.5 \text{ kg}) \times [4180 \text{ J/(kg·°C)}] \times (60°C) \doteq 1.25 \times 105 \text{ J}.$$

1.3d Chemical Energy

Chemical energy is the energy associated with chemical bonds, that is, the interaction energy between atomic electrons in a material. Energy can be absorbed or released during a chemical reaction as a result of changes in the bonds between the atoms. If a process requires energy to be input for the reaction to occur, then the process is referred to as *endothermic*. In general, these types of processes are not useful in the production of energy, although they can be useful in the storage of it. The dissociation of water (into hydrogen and oxygen) is of interest in this respect and will be discussed in more detail in Chapter 20.

Processes that release energy are referred to as *exothermic* and are of interest in this discussion. In general, oxidation reactions (i.e., the burning of materials) fall into this category. Some of the most relevant for the production of energy are reactions that involve the oxidation of carbon. The simplest of these is the oxidation of pure carbon (using oxygen from the atmosphere) and the production of carbon dioxide. This process is given by the formula

$$C + O_2 \rightarrow CO_2 + 32.8 \text{ MJ/kg.} \tag{1.13}$$

The equation indicates the amount of energy released by the combustion of 1 kg of pure carbon, that is the *heat of combustion* (in MJ/kg) of carbon, which corresponds to a release of 4.09 eV (eV = 1.6×10^{-19} J) of energy from the oxidation of one atom of carbon. Note that the energy in electron volts (eV) per atom (or molecule) for a substance of *molecular mass*, M (given in g/mol), is

$$E(\text{eV}) = \frac{(10^6 \text{J/MJ}) \cdot M \cdot (Q/m)}{(10^3 \text{ g/kg}) \cdot (6.022 \times 10^{23} \text{ mol}^{-1}) \cdot (1.602 \times 10^{-19} \text{J/eV})}, \tag{1.14}$$

where the energy released per unit mass, Q/m, is expressed in MJ/kg. The oxidation of pure carbon is sometimes a suitable approximation of the burning of coal. The burning of other fossil fuels [or other organic materials such as wood, ethanol, or municipal waste (Chapter 16)] generally involves the oxidation of hydrocarbons. Some of the simple reactions of this type are the burning of *methane* (the major component of natural gas),

$$CH_4 + 2O_2 \rightarrow CO_2 + 2H_2O + 55.5 \text{ MJ/kg;} \tag{1.15}$$

the burning of *ethanol* (a common biofuel),

$$C_2H_6O + 3O_2 \rightarrow 2CO_2 + 3H_2O + 29.8 \text{ MJ/kg;} \tag{1.16}$$

and the burning of *octane* (an important component of gasoline),

$$2C_8H_{18} + 25O_2 \rightarrow 16CO_2 + 18H_2O + 46.8 \text{ MJ/kg.} \tag{1.17}$$

Note that combustion reactions of hydrocarbons produce steam (not liquid water) as a by-product. The energies given above are referred to as higher heating values (HHV) and include the latent heat of vaporization of the steam. This energy can only be recovered if the steam produced is condensed.

An important reaction for the production of energy by animals is the combustion of *glucose*,

$$C_6H_{12}O_6 + 6O_2 \rightarrow 6CO_2 + 6H_2O + 16.0 \text{ MJ/kg.} \tag{1.18}$$

In Chapter 20, the energy released by the oxidation of hydrogen, that is,

$$2H_2 + O_2 \rightarrow 2H_2O + 142 \text{ MJ/kg,} \tag{1.19}$$

will be considered in detail.

1.3e Nuclear Energy

Nuclear energy is similar to chemical energy in that it is the energy associated with the bonds between particles. The relevant energy scale of nuclear energy relates to the bonds between the neutrons and protons within the nucleus rather than to the bonds between atoms that involve the atomic electrons. Nuclear bonds represent much larger amounts of energy than chemical bonds, typically many megaelectron volts (MeV) compared with a few electron volts. As a result, the amount of energy that can be released in nuclear reactions is many orders of magnitude larger than the energy that can be released in chemical reactions. Through the equivalence of mass and energy as given by Einstein's relation,

$$E = mc^2, \tag{1.20}$$

where c is the speed of light, nuclear energy can be related to changes in the nuclear mass (discussed in much more detail in Chapter 5). The energy released in an exothermic nuclear reaction is given in terms of the change in nuclear mass as

$$E_{exo} = \Delta mc^2. \tag{1.21}$$

Although this is also true for chemical energy, the energy associated with chemical bonds and hence the changes in mass associated with chemical reactions are very much smaller than in the nuclear case. As discussed in detail in Chapters 6 and 7, the release of nuclear energy can accompany the fission (breaking up) of heavy nuclei like those in uranium and plutonium or the fusion (bonding) of light nuclei like the isotopes of hydrogen.

1.3f Electrical Energy

Electrical energy is associated with electrons in a conductor. It is convenient to deal with the macroscopic representation of electrical energy in terms of *voltages* and *currents* without the need to be concerned with the microscopic description of the electrons. If a current, I, flows through a circuit with a resistance, R, then there is a voltage decrease, V, across the resistance given by

$$V = IR, \tag{1.22}$$

where V is given in *volts* (V) if I is in *amperes* (A) and R is in *ohms* (Ω). The power dissipated through the resistance is given by

$$P = VI. \tag{1.23}$$

When V and I are in volts and amperes, respectively, then P is in watts. From equation (1.22), this may be written as

$$P = I^2R = \frac{V^2}{R}. \tag{1.24}$$

From equation (1.4), electrical energy (in joules) is power (in watts) multiplied by time (in seconds). This is most often expressed in kilowatt-hours by dividing the energy in joules by the conversion factor 3.6×10^6 J/kWh. Correspondingly, the conversion factor between kilowatt-hours and megajoules (MJ) is 0.278 kWh/MJ.

1.3g Electromagnetic Energy

Electromagnetic radiation may be thought of in terms of associated electric and magnetic fields that form waves (such as light waves). It may also be thought of, in the quantum mechanical sense, as a collection of particles called *photons*. This radiation covers a wide range of wavelengths, as illustrated in Figure 1.1. Different wavelength regimes are, for example, X-rays, ultraviolet radiation, visible light, infrared radiation radio waves, and so on. Electromagnetic radiation from the sun (which falls largely in the visible region of the spectrum) is one of our most important sources of energy because it is the basic source responsible for most other sources of energy, such as fossil fuels, wind, solar radiation, and biomass energy. Figure 1.1 illustrates that, as the wavelength decreases, the energy content of the radiation increases. For any wave, the wavelength, λ, is related to the frequency, f, in terms of the velocity (in this case, the speed of light, $c \approx 3 \times 10^8$ m/s) by

$$\lambda = \frac{c}{f}, \tag{1.25}$$

where the wavelength is given in meters when the velocity is given in meters per second (m/s) and the frequency in Hertz (s^{-1}).

Example 1.6

Estimate the number of wavelengths of yellow light that span the distance between a computer monitor and a user.

Solution

From Figure 1.1, the wavelength of yellow light is about 600 nm, or 6.0×10^{-7} m. If a typical user sits 0.5 m from a computer monitor, then the number of wavelengths of yellow light that fit into that distance is

$$N = \frac{0.5 \text{ m}}{6.0 \times 10^{-7} \text{ m}} = 8.3 \times 10^5 \text{ wavelengths.}$$

If the electromagnetic energy is considered in terms of quanta of energy (i.e., photons), then the energy associated with each photon (E) is related to the frequency of the electromagnetic radiation (f) as

$$E = hf. \tag{1.26}$$

where h is *Planck's constant* (Appendix II). Long-wavelength radio waves may be produced artificially by electronic devices (i.e., radio transmitters). Radiation in the infrared to ultraviolet and X-ray regions of the spectrum are most commonly produced (either naturally or artificially) by electrons undergoing transitions between atomic energy levels. Short-wavelength (i.e., high energy) radiation most commonly comes from transitions involving excited states of nuclei. The energy in equation (1.26) is in joules when Planck's constant is 6.626×10^{-34} J·s. It is often convenient (as in Chapter 9) to express the energy per photon in electron volts. In this case, Planck's constant is given by 4.136×10^{-15} eV·s.

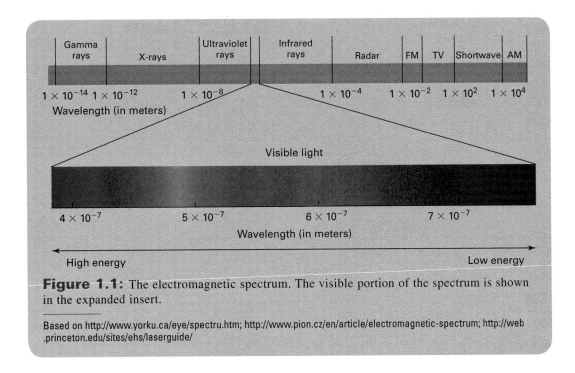

Figure 1.1: The electromagnetic spectrum. The visible portion of the spectrum is shown in the expanded insert.

Based on http://www.yorku.ca/eye/spectru.htm; http://www.pion.cz/en/article/electromagnetic-spectrum; http://web.princeton.edu/sites/ehs/laserguide/

1.4 Some Basic Thermodynamics

The thermodynamic behavior of systems is described by the *laws of thermodynamics*. A number of physical laws influence thermodynamic behavior, but four are of importance for the present discussion, and these are traditionally numbered 0 to 3:

0. Two systems that are both in thermodynamic equilibrium with a third system are in equilibrium with each other.
1. Energy is conserved.
2. A closed system will move toward equilibrium.
3. It is impossible to attain absolute zero temperature.

These are each discussed briefly below.

1.4a The Zeroth Law of Thermodynamics

The zeroth law is a generalization of the definition of thermal equilibrium. Two systems are in thermal equilibrium if they are able to transfer heat between each other but do not. This law implies that the thermodynamic state of a system can be defined by a single parameter, the *temperature*. Two systems in thermal equilibrium are defined to be at the same temperature.

Temperature may be defined in terms of the *ideal gas law*,

$$PV = Nk_{\mathrm{B}}T, \tag{1.27}$$

where P is pressure, V is volume, N is the number of gas molecules, k_{B} is Boltzmann's constant, and T is the temperature in Kelvin. No matter what system is in use (Celsius, Kelvin, etc.), two parameters define a temperature scale: the location of zero and the

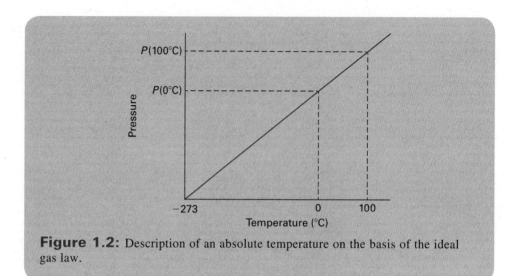

Figure 1.2: Description of an absolute temperature on the basis of the ideal gas law.

size of the degree. Reference points that define a temperature scale are often based on the properties of water, as in the case of the Celsius scales where the zero point is the freezing point of water. This zero is not absolute zero temperature, and this means that the temperature scale is not an absolute scale. To describe an absolute temperature scale using the size of the degree as defined on the Celsius scale, the ideal gas law can be used. Because a given volume of an ideal gas has a pressure that is linearly related to the absolute temperature by equation (1.27), when the temperature goes to absolute zero, then the pressure goes to zero. Figure 1.2 shows a plot of pressure as a function of temperature measured in Celsius for an ideal gas. Extrapolating the temperature until the pressure goes to zero gives the value of absolute zero on the Celsius scale. This is the basis for defining the temperature scale in Kelvin, as shown in the figure where K = °C + 273.

1.4b The First Law of Thermodynamics

Energy conservation, as it applies to thermodynamic systems, may be viewed in terms of Figure 1.3. Gas is contained in a cylinder that is sealed with a piston. If energy is supplied in the form of heat to the system by, for example, a flame, then two possible situations may be considered;

1. If the piston is held fixed, then the internal energy of the gas is increased by the addition of energy, and this is manifested by an increase in temperature.
2. If the piston is allowed to move, then some of the energy may be used to lift the piston, thereby doing work against the force of gravity.

In general,

$$Q = \Delta U + W, \tag{1.28}$$

where Q is the energy input in the form of heat, ΔU is the change in the internal energy of the gas as indicated by the change in temperature, and W is the mechanical work that

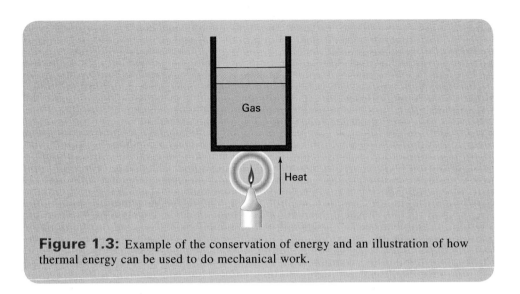

Figure 1.3: Example of the conservation of energy and an illustration of how thermal energy can be used to do mechanical work.

is done. Equation (1.28) forms the basis for understanding energy conversion processes involving heat in closed systems.

1.4c The Second Law of Thermodynamics

There are many ways of stating the second law of thermodynamics. In addition to the statement given at the beginning of this section (a closed system will move toward equilibrium), the second law may also be stated as follows:

> Heat naturally flows from a hot place to a cold place.

or

> The entropy of the universe always increases.

To see that these three statements are just alternate ways of looking at the same phenomenon, consider the behavior of a hot piece of metal and a cold piece of metal that are brought into thermal contact. This system will attain thermal equilibrium by transferring heat from the hot metal to the cold metal until the two pieces of metal are at the same temperature. When the two pieces of metal are at the same temperature, they are in thermal equilibrium, as defined by the zeroth law. Entropy is a measure of disorder. Two pieces of metal at different temperature represent a state of higher order (i.e., the hot metal atoms are separated from the cold metal atoms) than if they were in thermal equilibrium (i.e., at the same temperature). Thus, attaining equilibrium increases the overall entropy of the universe.

Thermal energy can be used to do work only if heat flows from hot to cold. An analogy is the conversion of potential energy to kinetic energy. Consider, for example, water in a reservoir at some height above a hydroelectric generating station. As long as the water remains in the reservoir, no work is done, and no electricity is generated. When the water is allowed to run downhill and through the station, electricity is generated. In this way, gravitational potential energy is converted into kinetic energy and subsequently into electrical energy. In the same way, the thermal energy contained in a piece of hot material can be converted into other forms of energy if it flows toward something that is at a lower temperature. This is the principle of operation in a heat engine, as described in the next section.

1.4d The Third Law of Thermodynamics

The third law is important for an understanding of the efficiency of thermodynamic processes. Part of the third law is the fact that a temperature of absolute zero cannot be attained. The details of the third law and its origins are not relevant and will not be discussed further.

1.5 Heat Engines and Heat Pumps

If heat moves from a hot reservoir to a cold reservoir, some of the thermal energy can be extracted to do mechanical work. A device that does this is called a *heat engine*, such as, for example, a steam turbine, a gasoline automobile engine, or a jet engine. Figure 1.4 shows a schematic of a heat engine. Heat is removed from a hot reservoir (at temperature T_h). Some of this heat is deposited in a cold reservoir (at temperature T_c), and some is used to do useful mechanical work, W. If the heat removed from the hot reservoir is Q_h, and the heat deposited in the cold reservoir is Q_c, then

$$Q_h = Q_c + W. \tag{1.29}$$

This expression follows along the lines of equation (1.28) and is a direct result of the conservation of energy as stated by the first law of thermodynamics. From a practical standpoint, Q_h can be the energy produced by burning gasoline as in an automobile engine. In this case, Q_c would be the excess heat transferred into the atmosphere, and W is the mechanical work that is done. The efficiency of this process, η, is the ratio of the useful work done, W, to the total input energy, Q_h. In percent, this is written as

$$\eta = 100 \, \frac{W}{Q_h}. \tag{1.30}$$

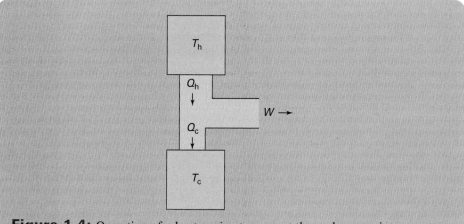

Figure 1.4: Operation of a heat engine to convert thermal energy into mechanical energy.

From equation (1.29), it is seen that

$$W = Q_h - Q_c, \tag{1.31}$$

and the efficiency becomes

$$\eta = 100\left(1 - \frac{Q_c}{Q_h}\right). \tag{1.32}$$

A consequence of the work of the French engineer Sadi Carnot is that the ratio of Q_c to Q_h can be expressed as the ratio of the reservoir temperatures:

$$\frac{Q_c}{Q_h} = \frac{T_c}{T_h}. \tag{1.33}$$

This expression allows equation (1.32) to be written as

$$\eta = 100\left(1 - \frac{T_c}{T_h}\right). \tag{1.34}$$

This form is convenient because T_c and T_h are more easily measured quantities than Q_c and Q_h. It is essential in equation (1.33) that the temperatures be measured on an absolute temperature scale (typically Kelvin). Measuring temperatures in Celsius in this equation will yield incorrect results. The efficiency as stated by equation (1.34) is known as the *ideal Carnot efficiency* and is the maximum efficiency attainable by a heat engine utilizing hot and cold reservoirs with temperatures T_h and T_c, respectively. It is seen that 100% efficiency is achieved only if T_c is zero degrees (on an absolute scale) which cannot be achieved. It is also seen in Figure 1.4 that Q_c cannot be larger than Q_h. From Carnot's relationship, $Q_c > Q_h$ would imply $T_c > T_h$ and would be inconsistent with the definition of hot and cold. Although the Carnot efficiency of a heat engine is the maximum theoretical efficiency, real heat engines typically operate at efficiencies that can be much less than the Carnot efficiency.

A *heat pump* is basically just the opposite of a heat engine; it uses mechanical energy to move heat from a cold reservoir to a hot reservoir. A schematic of a heat pump is shown in Figure 1.5 on the next page. Conservation of energy requires that

$$W + Q_c = Q_h. \tag{1.35}$$

The ratio of heat deposited into the hot reservoir to the amount of work done can be defined as

$$COP = \frac{Q_h}{W}. \tag{1.36}$$

This quantity is referred to as the *coefficient of performance (COP)* of the heat pump, and it is obvious from Figure 1.5 that this quantity will be greater than 1 (or, in percent, >100%). Using equation (1.31), equation (1.36) can be rewritten to give the ideal Carnot coefficient of performance as

$$COP = \frac{1}{1 - (Q_c/Q_h)}. \tag{1.37}$$

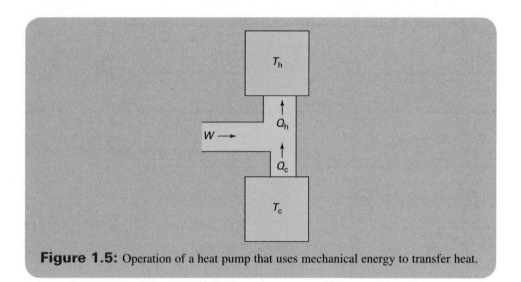

Figure 1.5: Operation of a heat pump that uses mechanical energy to transfer heat.

Using equation (1.33), this equation becomes

$$COP = \frac{1}{1 - (T_c/T_h)}. \qquad \textbf{(1.38)}$$

Heat pumps have practical applications for the transfer of heat from a cold reservoir to a hot reservoir. Air conditioners and refrigerators are examples of such devices. A heat pump can also heat a house on a cold day by moving heat from the outside to the inside. In this case, the relatively cold air outside is the cold reservoir, and the relatively warm air inside is the hot reservoir. Heat pumps specially designed for this purpose are in fairly common use in many parts of North America where winter temperatures are not extreme. Although heat pumps can be economically attractive for heating purposes, a careful consideration of capital costs, maintenance costs, local climate, and other factors is necessary to assess their viability. Applications of heat pumps are discussed further in Section 17.6.

Example 1.7

Consider an ideal heat pump operating with a cold reservoir temperature of $-20°C$ (e.g., outside on a cold winter day) and a hot reservoir at $+20°C$ (e.g., inside a home). Calculate the ideal (Carnot) coefficient of performance for this heat pump.

Solution
Noting that it is essential to use absolute temperatures in these expressions, we convert from °C to K as

$$T_c = -20°C + 273 = 253 \text{ K}$$

and

$$T_h = 20°C + 273 = 293 \text{ K}$$

Continued page 19

Example 1.7 continued

The coefficient of performance (*COP*) is found from equation (1.38) to be

$$COP = \frac{1}{1 - (T_c/T_h)} = \frac{1}{1 - (253 \text{ K}/293 \text{ K})} = 7.325.$$

This may be viewed in the context of Figure 1.5 and shows that the heat deposited in the hot reservoir (the room) is about seven times the mechanical work done.

1.6 Electricity Generation

Much of the energy used by society is in the form of electricity. The current section provides a brief overview of how electricity is presently generated and distributed. Figure 1.6 shows the distribution of energy sources used to generate electricity worldwide. It can be seen that the majority of electricity comes from fossil fuels, nuclear energy, and hydroelectricity. A very small component, about 1%, comes from other sources (solar, wind, tidal, biofuels, etc.). Fossil fuels and nuclear are similar in that they produce electricity from heat by first converting it into mechanical energy (using, for example, a turbine) and then into electricity (using a generator). Hydroelectricity is produced by converting the gravitational potential energy associated with water at higher elevations into mechanical energy and then into electricity. The details of other electricity production methods are quite variable and are discussed in detail throughout the remainder of the text. The present section concentrates on methods (specifically fossil fuels and nuclear) that convert heat into mechanical energy and then into electricity.

The conversion of heat into mechanical energy has been discussed in some detail in the previous section. Any device that works in this manner is a heat engine, and its ultimate efficiency is limited by the temperatures involved, as described by Carnot. The goal in achieving a high efficiency is to maintain the hot reservoir at as high a temperature as possible and to maintain the low temperature reservoir at as low a temperature as possible.

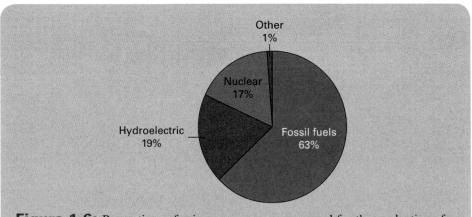

Figure 1.6: Proportions of primary energy sources used for the production of electricity worldwide.

Two approaches are commonly used for the generation of electricity from fossil fuels: thermal generation and combustion turbines. In *thermal generation*, the fuel is burned to produce heat, which is used to produce steam, which, in turn drives a turbine. These facilities most commonly use coal but can, in principle, run on any combustible fuel. *Combustion turbines* use the hot gas produced by the combustion itself to drive the turbine. Combustion turbines commonly use natural gas as a fuel but can use any highly volatile liquid fuel (i.e., gasoline). Nuclear energy is used to produce electricity exclusively by the thermal method and, aside from differences in how the heat is produced, follows closely along the lines of thermal generation from coal. In the present section, the technology of thermal generation and combustion turbines is considered.

1.6a Thermal Electricity Generation

One of the basic requirements of a thermal generation facility is a sustainable cold reservoir. Typically, this is the ocean, a large lake or river, or the atmosphere (by means of cooling towers). The general diagram of a generating station that makes use of a body of water as the cold reservoir is illustrated in Figure 1.7, and a photograph of a medium-sized generating facility is shown in Figure 1.8. The hot reservoir is steam in the boiler, which is produced by burning a fossil fuel or by a nuclear reaction. The steam at high temperature and high pressure does work as it passes through the turbine, causing the rotor assembly to rotate. The energy extracted as the steam passes through the turbine results in a reduction of both the temperature and the pressure of the steam. The typical design of the rotor of a large turbine is shown in Figure 1.9.

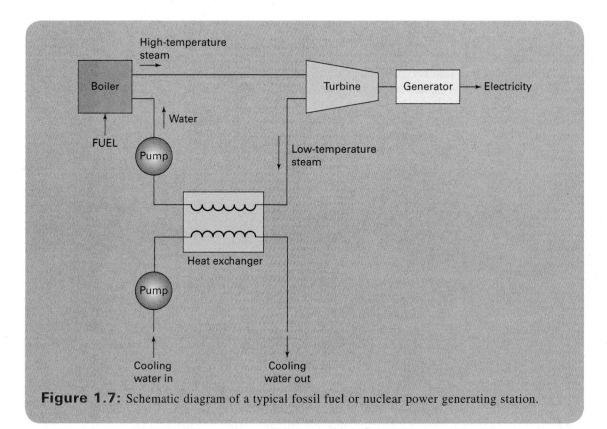

Figure 1.7: Schematic diagram of a typical fossil fuel or nuclear power generating station.

Figure 1.8: Tuft's Cove generating station in Dartmouth, Nova Scotia. This is a 350-MW$_e$ facility that uses oil or natural gas as fuel.

Figure 1.9: Rotor assembly of a turbine showing multiple stages of turbine blades.

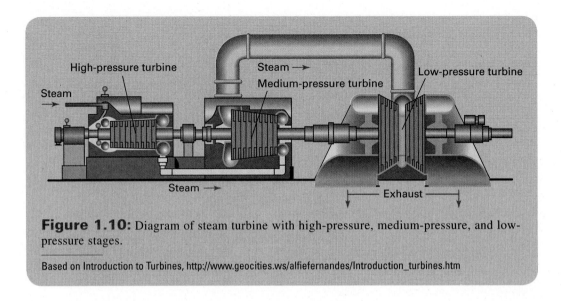

Figure 1.10: Diagram of steam turbine with high-pressure, medium-pressure, and low-pressure stages.

Based on Introduction to Turbines, http://www.geocities.ws/alfiefernandes/Introduction_turbines.htm

In many cases turbines consist of a series of stages operating at decreasing pressure as illustrated in Figure 1.10. Excess heat is transferred into the cold reservoir through a heat exchanger. The mechanical energy is converted into electricity by means of a generator. The output of these facilities is measured in watts electrical (W_e), where the subscript e stands for "electrical". The efficiency of converting thermal energy into electricity is constrained by the Carnot efficiency of converting heat into mechanical energy. (The conversion of mechanical energy into electricity is very high, typically ~90% or so). If water is used as the cold reservoir, there is a clear lower temperature limit of 0°C for T_c. Cooling towers, as illustrated in Figure 1.11, use the atmosphere as

Figure 1.11: Cooling towers typically used for thermal generating stations.

the cold reservoir. Although the value of T_c for cooling towers is not necessarily limited to 0°C, it is typically somewhat more variable than for water cooling. Typically, net efficiencies around 35–45% are achieved in the production of electricity from fossil fuels and around 30% for nuclear fuel. It is important to realize that the energy content of the fossil or nuclear fuel must be roughly three times the output in watts electrical because of the efficiency of the heat engine.

1.6b Combustion Turbines

Combustion turbines are sometimes used for the production of electricity from liquid or gaseous fossil fuels (often natural gas). The hot (burning) gas is fed directly into a turbine (much like a jet engine) and turns the turbine as it decreases in temperature and pressure. These facilities tend to be smaller than thermal (coal-fired) fossil fuel stations, and the cost of natural gas, gasoline, or oil is typically higher (per joule) than it is for coal. Combustion turbines, as shown in Figure 1.12, are often used to supplement coal-fired or nuclear electricity production during periods of high demand (Chapter 18) because they have relatively short start-up times compared to thermal generating facilities.

1.6c Diesel Generators

In areas where the local grid is isolated from a more extensive distribution system, such as inhabited islands that are not sufficiently close to the mainland to conveniently run a distribution cable or communities in remote regions, diesel generators are a suitable method of electricity production. An example of such a facility is illustrated in Figure 1.13 on the next page. This station consists of six Wartsila Sulzer ZA40 diesel generators, each rated at about 11 MW$_e$ for a total capacity of 67 MW$_e$, and serves an archipelago with a population of about 13,000. The exhaust stacks and fuel tanks are clearly seen in the photograph. The generators can burn regular diesel fuel or a variety of similar density fuel oils, such as Bunker C.

Robert Shantz/Alamy Stock Photo

Figure 1.12: Natural gas-fired combustion turbines at the Pyramid Generating Station operated by Tri-State G&T Inc. near Lordsburg, New Mexico.

Richard A. Dunlap

Figure 1.13: Diesel generating facility on an island that is not connected to the main grid in Cap aux Meules, Îles de la Madeleine, Québec, Canada.

1.6d **Distribution of Electricity**

Electricity is distributed from the generating station to the end user by means of a system of electrically conducting wires. Because all wires (other than superconductors) have resistance, some electrical energy is lost and converted into heat during transmission. From the previous discussion of electrical energy, it is easy to understand the approach to minimizing resistive losses during power transmission. For a given wire resistance, R, the power loss from equation (1.24) is

$$P_{\text{loss}} = I^2 R, \tag{1.39}$$

where I is the current in the wire. The resistance of the wire is determined by the material from which the wire is made (usually aluminum), the diameter of the wire, and its length. If a certain amount of power needs to be transmitted, P_{trans}, over some distance, then from equation (1.23),

$$P_{\text{trans}} = VI, \tag{1.40}$$

where V is the voltage. From this expression, the current is given by

$$I = \frac{P_{\text{trans}}}{V}, \tag{1.41}$$

and substituting into equation (1.39) gives

$$P_{\text{loss}} = \frac{P_{\text{trans}}{}^2 R}{V^2}. \tag{1.42}$$

Thus, it is clear that the power loss is inversely proportional to the square of the voltage. Transmitting electric power through a transmission cable is the most efficient when the voltage is as high as practical. This is the approach taken to electric power distribution: The longer the distance that the power needs to be transmitted (and the greater the resistance), then (generally) the higher the voltage. Because the voltage for residential use is typically 220 V, the voltage must be stepped down using a power distribution transformer. Figures 1.14 and 1.15 on the next page show power distribution transformers used for changing the voltage (up or down) for distribution over large distances or for residential use.

Figure 1.16 on the next page shows a typical power distribution system with high-voltage transmission lines and step-down transformers for providing appropriate voltages to industrial and residential customers. The smart grid is discussed in Section 17.4.

Richard A. Dunlap

Figure 1.14: Small power distribution transformer on a utility pole in a residential area.

Richard A. Dunlap

Figure 1.15: Transformers at an electric power distribution facility.

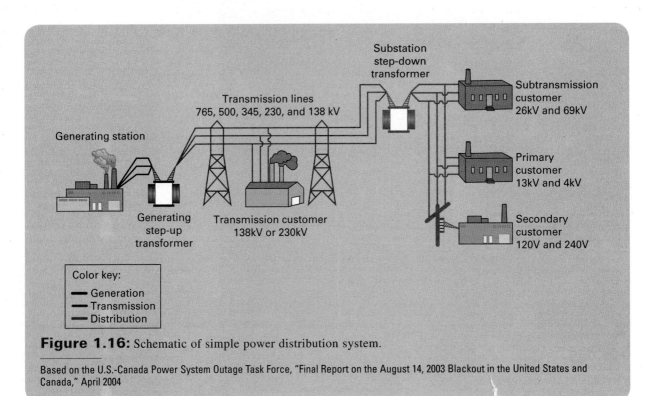

Figure 1.16: Schematic of simple power distribution system.

Based on the U.S.-Canada Power System Outage Task Force, "Final Report on the August 14, 2003 Blackout in the United States and Canada," April 2004

1.7 Summary

This chapter has shown that power is a measure of the energy produced or utilized per unit of time. Energy can exist in a variety of forms. It is convenient, from a practical standpoint, to think of these forms as kinetic energy associated with a moving object, gravitational potential energy associated with a mass that is vertically displaced, thermal energy associated with an object at an elevated temperature, chemical energy associated with the electronic bonds between atoms, nuclear energy resulting from the bonding of neutrons and protons in a nucleus, electrical energy corresponding to a flow of electric current, and electromagnetic energy associated with photons.

This chapter has also provided an introduction to the basic principles of thermodynamics: the definition of temperature, the conservation of energy, the tendency of all systems to achieve thermal equilibrium, and the inability to reach absolute zero temperature. These principles were applied to a thermodynamic system to describe the behavior of a heat engine in terms of its Carnot efficiency and a heat pump in terms of its coefficient of performance.

Some of the ways in which primary energy sources can be transformed into forms that are suitable for applications have been discussed. Examples are the conversion of the chemical energy in fossil fuels into heat energy by combustion or the conversion of mechanical energy in the wind into electricity. These conversion processes rely on the development of appropriate technologies but are also governed by fundamental physical laws. For example, more efficient automobile engines can be developed, but ultimately the limiting factors for efficiency are the laws of thermodynamics.

Finally, the chapter overviewed the basic principles for the production and distribution of electricity. The thermal generation of electricity by fossil fuels or nuclear energy is limited by the Carnot efficiency of a heat engine. On the basis of Ohm's law, the chapter has shown that losses are minimized when electricity is transmitted at high voltage.

Problems

1-1 Compare the energy scales of mechanical, chemical, and mass energy by calculating (a) the energy associated with dropping 1 kg of coal through a vertical distance of 100 m, (b) burning 1 kg of coal, and (c) converting the mass of 1 kg of coal into energy.

1-2 A simple way of looking at the energy associated with the combustion of methane [as shown in equation (1.15)] is to view the oxidation of the carbon by equation (1.13) and the oxidation of hydrogen by equation (1.19). Based on the energies involved in these processes, discuss the validity of this approach.

1-3 One cubic meter of water is poured off a 100-m-high bridge. If the change in gravitational potential energy is converted into electricity with an efficiency of 90%, how long can this energy illuminate a standard 60-W light bulb?

1-4 One gram of methane is burned, and the heat is used to raise the temperature of 1 kg of water. If the initial temperature of the water is 25°C, what is the final temperature?

1-5 (a) Estimate the energy required to raise the temperature of a cup of coffee from room temperature to 60°C. (b) Estimate the mass energy contained in a cup of hot coffee.

1-6 What is the energy [in electron volts (eV)] of a green photon?

1-7 An average person can produce considerable power output for short periods of time. However, over an entire 8-hour workday, a person might be able to average 100 W output of physical power. If energy is valued at $0.12/kWh (about the average cost of electricity from a public utility), what is the monetary value of a person's physical work per day?

1-8 Calculate the ideal Carnot efficiency of a steam turbine that utilizes steam at a temperature of 575°C and ejects water to the environment at a temperature of 35°C.

1-9 Water flows through a 1-m-diameter pipe at a rate of 0.5 m per second. If the energy associated with this flowing water is converted into electricity with an efficiency of 80% and the current flows through a 100 Ω resistor, what is the current through the resistor?

1-10 The temperature of 1 m^3 of water is decreased by 10°C. If this thermal energy is used to lift the water vertically against gravity, what is the change in height of the center of mass?

1-11 A heat pump with a coefficient of performance of 8.5 is used to heat a house. If the outside temperature is -15°C, what is the temperature inside the house?

1-12 A power transmission line is used to transmit 100 kW of power at a voltage of 10 kV with a loss of 1 kW. If the voltage is increased to 200 kV, what is power loss to transmit the same amount of power?

1-13 Steam at a temperature of 350°C is used to run a turbine. If the exhaust is at a temperature of 70°C, what is the Carnot efficiency?

1-14 (a) If 15 kW of power from a heat reservoir at 500 K is input into a heat engine with an efficiency of 37%, what is the power output?
(b) What is the temperature of the cold reservoir?

1-15 A rotating solid disk with a diameter of 0.25 m and a mass of 10 kg is used to store 1 kJ of energy. What is its rotational speed in revolutions per minute?

1-16 The lower heating value (LHV) is defined as the heat that is available from the combustion of a fuel excluding the latent heat of vaporization of the steam produced. From equation (1.16) determine the LHV of ethanol.

1-17 A local FM radio station broadcasts at 104.3 MHz. What is the wavelength of the radio waves?

1-18 One homeowner uses electricity to heat a home (with 100% efficiency) at an annual cost of $2600. A second homeowner uses a heat pump with a coefficient of performance of 7 to heat an identical home. What is the second homeowner's annual heating cost?

1-19 A coal-fired power plant operates at an efficiency of 38%. If the facility produces 6 TWh of electricity per year, what is the rate of coal consumption in kg/s. Approximate coal as pure carbon.

1-20 A 500 W electric heater is used to heat a 100 L tank of water with an efficiency of 100% and no heat loss. How long will it take to raise the temperature of the water by 20°C?

Bibliography

F. S. Aschner. *Planning Fundamentals of Thermal Power Plants.* Wiley, New York (1978).

G. J. Aubrecht II. *Energy: Physical, Environmental, and Social Impact* (3rd ed.). Pearson Prentice Hall, Upper Saddle River, NJ (2006).

R. A. Dunlap. *Experimental Physics—Modern Methods.* Oxford University Press, New York (1988).

E. L. McFarland, J. L. Hunt, and J. L. Campbell. *Energy, Physics and the Environment* (3rd ed.). Cengage, Stamford, CT (2007).

K. Wark. *Thermodynamics* (4th ed.). McGraw-Hill, New York (1983).

Past, Present, and Future World Energy Use

Learning Objectives: After reading the material in Chapter 2, you should understand:

- The energy needs of humanity throughout history.
- The current energy distribution and relationship to economic, geographical, climate, and industrial factors.
- The principles of exponential growth.
- The Hubbert model of resource utilization.
- Resource limitations to energy production and use.
- Limits of technology on energy production and use.
- Economic factors that limit energy use.
- Social factors affecting energy production.
- Environmental aspects of energy use.
- Political factors affecting energy use.
- The integration of new energy technologies with existing technology.

2.1 Introduction

Before the use of fire, humans relied only on the energy that their bodies produced to do the work needed to live their lives. This energy came from the food that each person consumed. If an average daily food consumption was 8.6 MJ, then this translated into an average continuous power utilization of $(8.6 \times 10^6 \text{ J})/(86400 \text{ s/d}) \approx 100$ W, about the heat and light output of a large light bulb. As humans implemented energy from other sources to improve their lives—first burning wood and then domesticating animals, followed by burning fossil fuels—the power utilization of each human grew. Total world energy use grew both because of the needs of the individual and because of a growing population. Throughout most of the past, humans used sources of energy that came from nature without any awareness (at least on a global scale) of any limitations to the availability of energy or any adverse environmental consequences of its use. In recent years, an awareness of both these factors arose and prompted the investigation of mechanisms for conserving energy resources, utilizing alternative sources, and mitigating the undesirable consequences of energy production and use. Planning for the future requires an analysis of past energy use, an evaluation of available resources, and a projection of future needs. The present chapter overviews the ways in which past energy use can be understood and the future can be anticipated.

Tim Collins/Shutterstock.com

2.2 Past and Present Energy Use

The average rates of energy use per person at different stages of human development are summarized in Table 2.1. This table illustrates the significance of technological advances on humanity's need for energy. It shows that the current per-capita power in the United States, 11.7 kW per person (continuous average), is more than 100 times the average power produced by a person's body. The utilization of energy sources from nature (i.e., the burning of fossil fuels) supplements our own body's energy production by this amount.

Figure 2.1 on the next page illustrates a detailed historical breakdown of energy use in the United States. The relative importance of the different energy sources can be readily seen by plotting the percentage of the total energy that comes from each source (Figure 2.2) on the next page. This graph clearly shows the transition from wood to coal that occurred in the late 19th century and from coal to oil that occurred in the first half of the 20th century. The relative importance of the different energy sources (Figures 2.1 and 2.2) is fairly typical of most industrialized nations, although some variability exists, as future chapters dealing with specific sources of energy discuss in detail. A comparison of the current energy use in the United States and that worldwide is shown in Figures 2.3 and 2.4 on page 33.

Nations vary substantially in the amount of energy they use. Figure 2.5 on page 34 shows the relationship between per-capita energy use and per-capita gross domestic product (GDP). Overall, the relationship between per-capita energy use and per-capita GDP is, as expected, a direct consequence of the degree of industrialization. However, the figure shows some minor variations that can be fairly well understood.

Table 2.1: Estimated average power used per person as a function of the stage of human development (worldwide) or as a function of year (for the United States, 1850–2000). The power values are calculated from the primary energy use.

society	power consumption (W)
hunter	100
use of fire	200
domestication of animals	500
Renaissance	1160
1850	4880
1900	5340
1950	7300
1960	8180
1970	11,000
1980	11,250
1990	11,000
2000	11,730

Based on data from G.J. Aubrecht II, Energy: Physical, Environmental and Social Impact (3rd ed.) Pearson Prentice Hall, Upper Saddle River, NJ (2006).

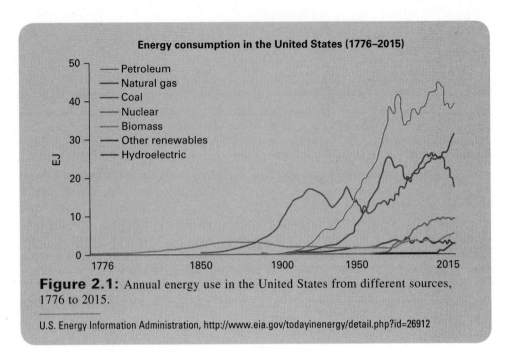

Figure 2.1: Annual energy use in the United States from different sources, 1776 to 2015.

U.S. Energy Information Administration, http://www.eia.gov/todayinenergy/detail.php?id=26912

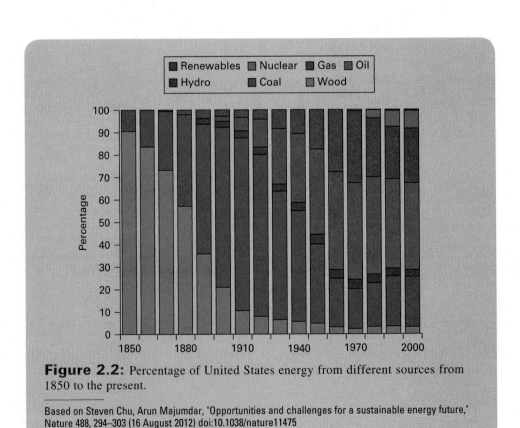

Figure 2.2: Percentage of United States energy from different sources from 1850 to the present.

Based on Steven Chu, Arun Majumdar, "Opportunities and challenges for a sustainable energy future," Nature 488, 294–303 (16 August 2012) doi:10.1038/nature11475

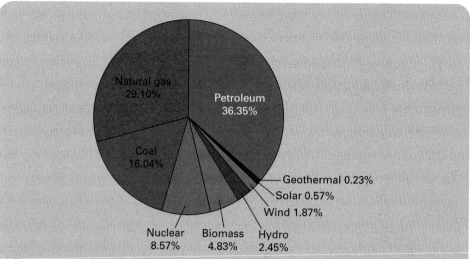

Figure 2.3: Distribution of primary energy sources in the United States in 2015.

Data from United States Energy Information Administration; http://www.eia.gov/totalenergy/data/monthly/pdf/sec1_7.pdf

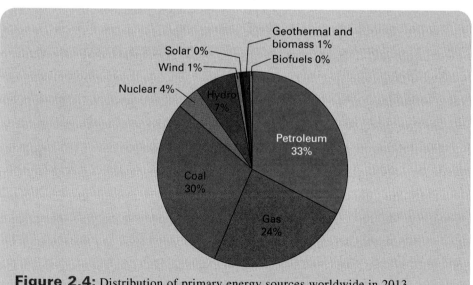

Figure 2.4: Distribution of primary energy sources worldwide in 2013.

Data are from http://euanmearns.com/global-energy-trends-bp-statistical-review-2014/

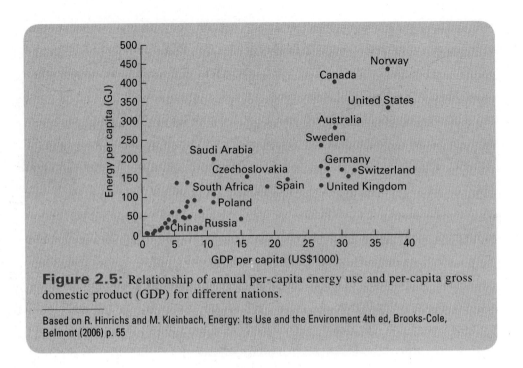

Figure 2.5: Relationship of annual per-capita energy use and per-capita gross domestic product (GDP) for different nations.

Based on R. Hinrichs and M. Kleinbach, Energy: Its Use and the Environment 4th ed, Brooks-Cole, Belmont (2006) p. 55

For example, among industrialized nations, Switzerland and Spain fall below the average line (less energy use relative to the GDP), while Canada and Norway fall above the line (more energy use relative to the GDP). Some relevant factors that have an effect on energy use are:

1. Climate.
2. Population density.
3. Type of industries.

Canada, for example, has a cool climate (resulting in greater heating costs), a low population density (resulting in greater transportation costs), and a fairly large fraction of heavy manufacturing (resulting in greater energy needs relative to product value). Norway, as well, has a severe climate. By comparison, Spain has a more moderate climate, while Switzerland has a higher population density and a greater prevalence of less energy-intensive industries, such as banking and watch making.

Energy use is sometimes given in terms of primary energy use. *Primary energy* refers to the quantity of energy as it is extracted from nature. This might be the energy content of coal that can be extracted by burning the coal or the energy content of solar radiation incident upon the earth. In most cases, the form of energy as it is obtained from the environment is not the same as the form required for an application, which is the energy after it is converted to a useful form and put to use by the end user. Converting energy from one form to another inevitably involves losses because the conversion process can never be 100% efficient. It is not that energy is not conserved; it is that it ends up in a form that is different from what is needed, often as residual unconverted energy or as excess heat. If, for example, oil is converted to heat (by burning) to heat a home, this is a fairly efficient conversion process (typically 85% or better). Similarly, converting the energy content of falling water to electricity in a hydroelectric facility is quite efficient (typically 90% or better). On the other hand, converting the

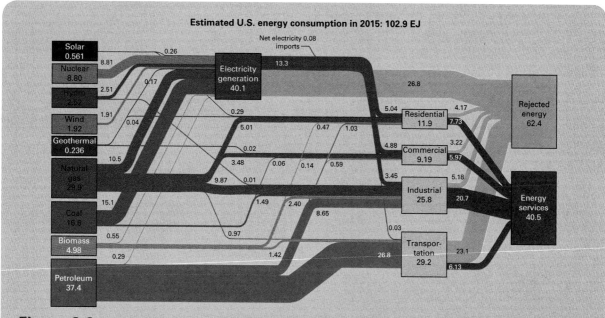

Figure 2.6: Annual energy use (in 2015) in the United States, showing sources, end use, and conversion losses. Units are quads EJ.

Lawrence Livermore National Laboratory, https://flowcharts.llnl.gov/content/assets/images/energy/us/Energy_US_2015.png

energy content of coal to electricity by burning the coal and using the heat to boil water to run a generator is an intrinsically inefficient process. Conversion efficiencies will be discussed throughout the book. The relationship of primary to end user energy, that is, the energy that is actually used by the consumer, in the United States is illustrated in Figure 2.6 and shows the significance of losses that occur during the conversion process of changing energy from its primary form into the form needed by the end user.

Clearly, humans use considerable energy, particularly those who live in industrialized nations. On average, a person living in North America might use about 12 kW of power (continuously). Humans obviously consume much more energy than they could produce on their own. To a large extent, this energy use can be accounted for by a simple analysis. Consider, as an example, a person who lives in a temperate region of North America (e.g., Boston) in a single-family, oil-heated home (with one other person) and who commutes to work by automobile (with one passenger). It is easy to account for a significant fraction of this person's energy use in terms of the energy purchased. Three sources are obvious: gasoline for transportation, oil for residential heating, and electricity for residential use. The typical annual use for these two people might be

$$2810 \text{ L gasoline at } 3.48 \times 10^7 \text{ J/L} = 9.8 \times 10^{10} \text{ J,}$$
$$2700 \text{ L heating oil at } 3.85 \times 10^7 \text{ J/L} = 1.04 \times 10^{11} \text{ J, and}$$
$$1.2 \times 10^4 \text{ kWh electricity} = 4.32 \times 10^{10} \text{ J.}$$

In a similar climate, the net energy use for heating would be similar for homes heated electrically or with natural gas. The energy content of gasoline or heating oil is close to that of the primary energy source (crude oil) because production costs (in terms of energy) are fairly minimal. The energy content of the electricity may (or may not) be close to that of the primary source. If the electricity comes from hydro-electric power, then the efficiency is high. Most electricity in North America comes (at present) from burning coal with an efficiency of around 35–40%. This means that the 4.32×10^{10} J of electrical energy actually represents about 1.1×10^{11} J of primary energy. Thus, for this person, transportation, heating, and electricity repre-sent fairly similar amounts of energy use. This, of course, would not be the same for someone living in, say, Manitoba as it would be for someone living in Florida. Nor would it be the same for a person living in an urban apartment and a person living in a rural single-family home. However, this example gives a rough estimate of the average per capita power consumption in North America as

$$0.5 \times \frac{0.98 \times 10^{11} \text{ J/y} + 1.04 \times 10^{11} \text{ J/y} + 1.1 \times 10^{11} \text{ J/y}}{3.15 \times 10^7 \text{ s/y}} = 5.0 \text{ kW}. \quad \textbf{(2.1)}$$

This is less than half of the expected value of about 12 kW. A person uses other energy, such as at a place of employment. This might be electricity for lights and computers for office workers or more substantial amounts of energy for equipment for workers in manufacturing industries. Energy is used to manufacture goods that a person uses and to produce food that a person consumes. In general, it is difficult to attribute specific energy use to specific individuals, which is why societal averages are more meaningful.

It may also be obvious from this discussion that everything that humans produce has a cost in energy. For example, if a person buys a loaf of bread at a grocery store, it might be customary to think of the "value" of the bread as, say, $1.49. The $1.49 goes, in part, to pay a farmer for growing the wheat, to pay a trucker for delivering the grain to the bakery, to pay the bakery for making the bread, to pay another trucker for deliver-ing the bread to the store, and finally to pay the store for operating expenses (as well as some profit for everyone). But the bread has a value in energy as well. The farmer used energy (in the form of gasoline or diesel fuel) to run the equipment needed to cultivate the land. The trucker used fuel to transport the grain and bread, the bakery used energy to bake the bread, and the store used energy for lights and heat. If you think about things in depth, you can see that a lot of other hidden energy costs come into play in actually get-ting the bread on your dinner table. It is also obvious that it becomes difficult to account for energy use on a per-person basis. It would be interesting to ask how the energy value of the bread (i.e., how much energy is needed to produce and distribute it) compares to the amount of energy that the bread provides to a person who consumes it (in terms of its caloric content). In this situation, the overall efficiency of the process (i.e., is a person getting as much energy from eating the bread as is used to produce it?) is less important than the goal of providing food for humankind. (Actually, in North America it requires about eight times as much energy to produce food as the food provides to the consumer.)

This kind of analysis is much more important when dealing with, for example, the energy content of a kilogram of coal. It is customary to think of the cost of energy in terms of dollars per joule, but a far more fundamental consideration is the cost of energy in joules per joule; that is, how much energy is consumed in mining, processing, transporting, and otherwise delivering 1 kg of coal compared to the energy gained by burning it and converting the heat energy into the required form of energy? Although financial considerations generally outweigh energy considerations, the bottom line is that, if it costs more (in energy) to produce the energy that is gained, then there is a net energy loss in the process. This kind of analysis is important in considering the viability of any type of alternative energy source.

This brings up another important consideration: the environmental impact of the things we utilize. We might not think of a loaf of bread (for example) as having any adverse environmental impact. However, energy is utilized to produce bread, and if that energy comes from, say, fossil fuels, then there is an environmental impact. To fully analyze the environmental consequences of every object we utilize, it is necessary to consider all stages of its production, use, and ultimate disposal. This kind of approach is referred to as *life cycle analysis* and will be considered in more detail in Chapter 4.

It is also important to understand the future implications of the information in Figure 2.1. The increase in energy use as a function of time may be understood by means of the detailed analysis of growth mechanisms. This topic is considered in the next section.

2.3 Exponential Growth

Energy use is clearly increasing. This is a result of two factors: population increases and an increase in per-capita energy use. In highly industrialized countries, the per-capita energy use is fairly constant (Table 2.1). In this case, the increase in energy use is largely governed by population changes. In the case of developing countries, both population growth and an increase in per-capita energy use are important factors. In fact, the largest component of the increase in world energy use is in rapidly developing countries with large populations (e.g., China and India). In an idealized future, all nations might achieve a similar level of industrialization, and the average per-capita use of energy would be relatively constant worldwide. In this case, population growth would be the dominant factor governing energy use. For an assessment of future energy needs, it is important to be able to predict overall changes in energy use in the short term and farther into the future. A general mathematical analysis of growth is a good starting point.

The simplest type of growth is linear growth. For a quantity, N, the time dependence is given by a constant dN/dt as

$$N(t) = N_0 + \frac{dN}{dt} t, \qquad (2.2)$$

where N_0 is the value at time $t = 0$, as shown in Figure 2.7. On a linear graph, this is described by a straight line with a constant slope (dN/dt).

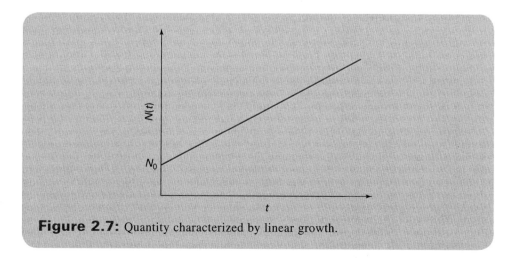

Figure 2.7: Quantity characterized by linear growth.

Another type of growth is exponential growth. The rate of change of the quantity, N, is related to the magnitude of that quantity by the expression

$$\frac{dN}{dt} = aN. \tag{2.3}$$

This may be integrated to give

$$N(t) = N_0 \exp(at). \tag{2.4}$$

A graph of this relationship is shown in Figure 2.8. It is most easy to identify true exponential growth by plotting data on a semilogarithmic scale. Taking the natural logarithm of both sides of equation (2.4) gives

$$\ln[N(t)] = \ln(N_0) + at. \tag{2.5}$$

Converting the natural logarithm to a base 10 logarithm (for ease of data presentation) gives

$$\log[N(t)] = \log(N_0) + at \log(e). \tag{2.6}$$

Thus a semilog plot of $N(t)$ as a function of t is a straight line with a slope of $a\log(e)$, as shown in Figure 2.9. The time required for the quantity N to double (called the doubling time) is found by setting $N(t) = 2N_0$ in equation (2.4) and solving for t_D as

$$t_D = \frac{\ln(2)}{a}. \tag{2.7}$$

It is also convenient to define a growth rate, R (in percent per unit time; e.g., % per year). This is given from equation (2.4) as

$$R = 100 \cdot [\exp(a) - 1]. \tag{2.8}$$

In the case where R is small (typically less than 10% per unit time), this expression may be approximated in terms of the doubling time as

$$R = 100 \frac{\ln(2)}{t_D}. \tag{2.9}$$

Example 2.1

A particular quantity has a doubling time of 20 years. If the quantity has a value of 10^6 at time $t = 0$, what is its value after 5 years?

Solution

Solving equation (2.7) for the constant a gives

$$a = \frac{\ln(2)}{t_D} = \frac{0.693}{20 \text{ y}} = 0.0347 \text{ y}^{-1}.$$

From equation (2.4), the quantity at $t = 5$ y is

$$N(t = 5 \text{ y}) = 10^6 \times \exp(0.0347 \text{ y}^{-1} \times 5 \text{ y}) = 1.189 \times 10^6.$$

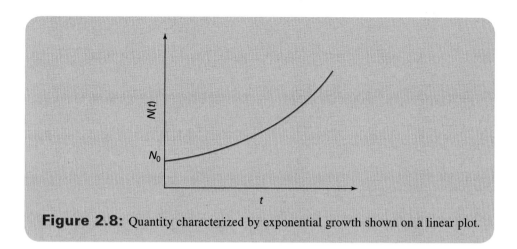

Figure 2.8: Quantity characterized by exponential growth shown on a linear plot.

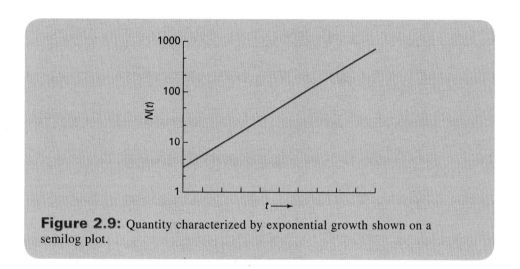

Figure 2.9: Quantity characterized by exponential growth shown on a semilog plot.

Table 2.2 gives the relationship of doubling times (in years) compared to the growth rate (in percent per year) for true exponential growth.

It is now possible to apply this information to actual data. The world population during the past millennium is shown in Figure 2.10 on page 41, where the estimated total population as of 2016 is approximately 7.44×10^9. On a linear scale, these data certainly look fairly exponential. In fact, in recent years they are reasonably well described by an exponential with an average annual growth rate of about 1.6%, corresponding to a doubling time of roughly 43 years. If the world population continues to double every 43 years, some predictions about the future can be made. Table 2.3 shows the predicted population and the total mass of humans (assuming a mass of 70 kg per person) for various years in the future.

Because the total mass of the earth is about 6×10^{24} kg, these predictions are clearly impossible to make very far into the future; that is, the population of the earth cannot increase exponentially with a growth rate of 1.6% per year for the next 2000 years. Even 1.6% annual growth over the next 650 years would result in a population of approximately 1.5×10^{14}, which corresponds to 1 human per square meter of

Example 2.2

A quantity grows at a rate of 1% per year. When will it reach 10 times its current value?

Solution

Solving equation (2.8) for a as a function of R gives

$$a = \ln\left(1 + \frac{R}{100}\right).$$

Substituting in a value of $R = 1$ gives

$$a = \ln(1 + 0.01) = 9.95 \times 10^{-3} \text{ y}^{-1}.$$

From equation (2.4), the quantity as a function of time (relative to the present value) is

$$\frac{N(t)}{N_0} = \exp(at).$$

Solving for t gives

$$t = \frac{1}{a}\ln\left(\frac{N(t)}{N_0}\right),$$

so using $N(t)/N_0 = 10$ and this value for a and solving for t gives

$$t = \frac{\ln 10}{9.95 \times 10^{-3}\text{y}^{-1}} = 231 \text{ y}.$$

Table 2.2: Annual growth rates and corresponding doubling times.

% growth per year	t_D (y)
1	69.7
2	35.0
3	23.4
4	17.7
5	14.2
6	11.9
7	10.2
8	9.0
9	8.0
10	7.3
20	3.8
50	1.7
100	1.0

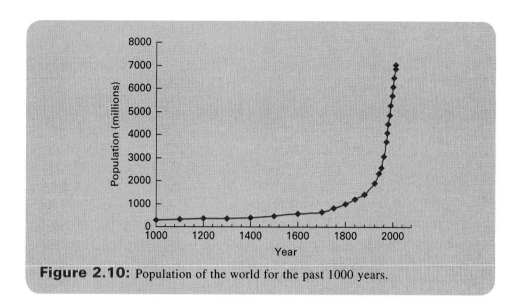

Figure 2.10: Population of the world for the past 1000 years.

the earth's land area. Certainly, the exponential growth of the human population will be limited long before this happens. A careful analysis of limits to human population must be based on the ability of the planet to produce enough food to feed its population. Some estimates have suggested a limit to the sustainable human population of about 5 billion (less than the current value). Much depends on assumptions concerning food production technology and the politics of distribution, but uncertainty in the future climate is also a significant factor. Most studies indicate that a world population of much more than 10 billion would place an unmanageable strain on food production capabilities. In Figure 2.11 on the next page, the trends in world population are seen from the plot of annual growth rate as a function of year. Clearly, the trend toward a slower growth rate is evidenced, and this analysis suggests that an equilibrium value will be approached at some point in the not too distant future.

Table 2.3: World population and mass of humanity as a function of year for 1.6% annual growth.

year	population	mass (kg)
2016	7.43×10^9	5.22×10^{11}
2200	1.41×10^{11}	9.88×10^{12}
2400	3.47×10^{12}	2.42×10^{14}
2600	8.50×10^{13}	5.96×10^{15}
2800	2.08×10^{15}	1.45×10^{17}
3000	5.11×10^{16}	3.58×10^{18}
3200	1.25×10^{18}	8.78×10^{19}
3400	3.08×10^{19}	2.15×10^{21}
3600	7.55×10^{20}	5.29×10^{22}
3800	1.85×10^{22}	1.30×10^{24}
4000	4.54×10^{23}	3.18×10^{25}

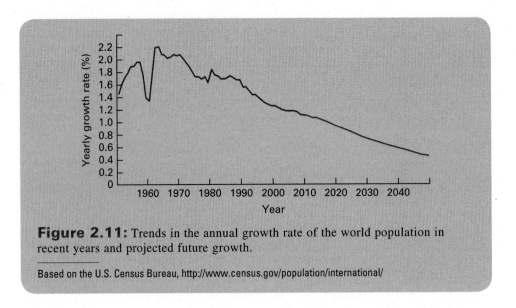

Figure 2.11: Trends in the annual growth rate of the world population in recent years and projected future growth.

Based on the U.S. Census Bureau, http://www.census.gov/population/international/

The overall trends in historical and predicted future energy use for some portions of the world's population are shown in Figure 2.12. The significance of increasing population and degree of industrialization in developing countries (e.g. China) is clearly evident in the figure. This figure suggests that industrialized nations are in a region of linear growth in their energy but that the rate of growth differs substantially among nations. One prediction for overall world energy use is illustrated in Figure 2.13. Such predictions suggest a global energy requirement of 8 to 10×10^{20} J per year by the year 2050. Ultimately, total energy use will be limited by limitations to human population and to the availability of energy resources. Although this type of prediction serves as

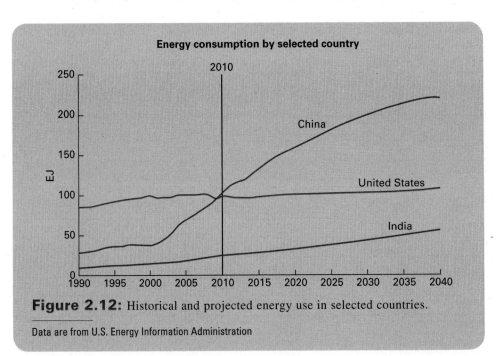

Figure 2.12: Historical and projected energy use in selected countries.

Data are from U.S. Energy Information Administration

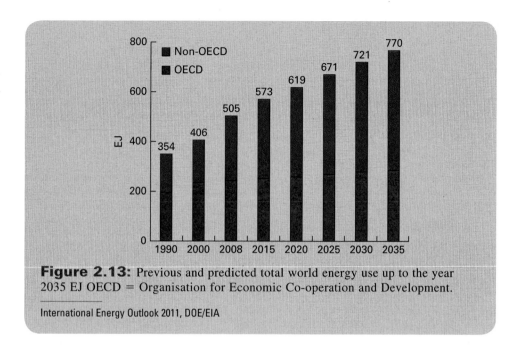

Figure 2.13: Previous and predicted total world energy use up to the year 2035 EJ OECD = Organisation for Economic Co-operation and Development.

International Energy Outlook 2011, DOE/EIA

a basis for understanding the energy needs of the future, clearly many unknowns and a great deal of uncertainty are involved in the analysis. This will be discussed further in Chapter 21.

2.4 The Hubbert Model of Resource Utilization

It is clear from the previous section that making predictions for the future is not necessarily straightforward or simple. Also obvious is that even minor variations in parameters such as annual percentage of growth can lead to enormously different future predictions. However, the previous discussion provides some general guidance for understanding our future energy needs and the utilization of available resources.

A general model that deals with resource utilization was developed by the American geophysicist, M. King Hubbert, in 1956. The basic assumptions of this model are as follows:

- When it is first realized that a resource is useful, the utilization of that resource begins slowly. This is because efficient procedures for utilizing the resource and an appropriate infrastructure need to be developed.
- Once the appropriate infrastructure has been developed, resource utilization increases.
- When the resource becomes scarce, utilization decreases and eventually stops.

This model can be readily applied to the use of oil. When the usefulness of oil (as a heating fuel and later as a transportation fuel) was realized, the locations of oil deposits had to be determined, oil wells needed to be constructed, and a system for the distribution and use of oil had to be put in place. Once these resources were established, the

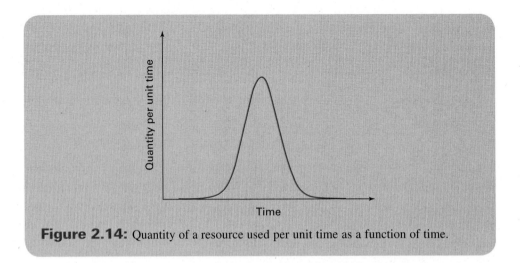

Figure 2.14: Quantity of a resource used per unit time as a function of time.

use of oil increased. It is now getting to a point where it is becoming more difficult to locate new oil resources, and it is beginning (or will soon begin) to be more difficult and expensive to make use of those resources. This will be discussed in much more detail in the next chapter.

Given all that, a graph of resource utilization (for oil this would be the quantity of oil used per year) as a function of time is illustrated in Figure 2.14. The amount of the resource used per unit time can be expressed as the derivative of the total quantity used, Q, as dQ/dt. If the curve in Figure 2.14 is integrated over time, then the total amount of the resource used up to that point would be obtained; that is,

$$Q(t) = \int_0^t \frac{dQ}{dt} dt. \tag{2.10}$$

A graph of $Q(t)$ in Figure 2.15 shows the typical sigmoidal behavior expected for resource utilization. As time increases, the curve asymptotically approaches a value Q_∞. This is the total amount of the resource available (or at least economically viable). The application of this model to actual data requires knowledge of the width of the

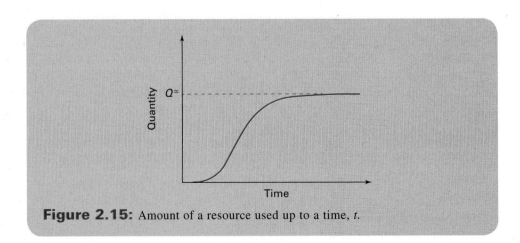

Figure 2.15: Amount of a resource used up to a time, t.

ENERGY EXTRA 2.1
Limits to growth

Predicting future conditions on earth is, at best, difficult. How humanity interacts with nature will determine the environment of the future. Understanding the consequences of our actions is not straightforward because the ecosystem is very complex. Specifically, with regard to energy utilization, it is important to understand the environmental impact of our actions and the availability and need for resources in the future. Although a carefully considered analysis of our resource utilization can provide substantial insight, there is really no consensus on how best to ensure the best future conditions for humanity.

One of the most influential approaches to modeling the impact of humanity on the environment was published in the 1972 book, *Limits to Growth*, by authors Donella H. Meadows, Dennis L. Meadows, Jørgen Randers, and William W. Behrens III. A revised edition, *Limits to Growth: The 30-Year Update*, by Donnella Meadows, Jørgen Randers, and Dennis Meadows was published in 2004. The basic hypothesis of the authors is that variables such as world population and resource depletion change exponentially in time, whereas technology's ability to increase resources is linear. This work has pointed out the importance of exponential growth in the following way: If the quantity of a resource is R and it is used at a constant rate, C, then the duration of its availability, t_0, is merely

$$t_0 = R/C.$$

If, however, use grows exponentially with a growth rate of, r, then the total resource can be written as the integral over the exponentially increasing use rate (beginning as C at $t = 0$) as

$$R = \int_0^{t_0} Ce^{rt}dt = \frac{C}{r}(e^{rt_0} - 1)$$

and solving for t_0 in terms of R and C gives

$$t_0 = \frac{1}{r}\left[\ln\left(\frac{rR}{C} + 1\right)\right].$$

Consider, as an example, an initial quantity of a resource of 10^{12} kg, which is being used at a constant rate of 10^9 kg per year. Clearly the resource will last for $t_0 = 1000$ y. If however, use increases at an annual rate of 5% or ($r = 0.05$ y^{-1}) then the lifetime of the resource will be

$$t_0 = \frac{1}{0.05\ \text{y}^{-1}}\left[\ln\left(\frac{0.05\ \text{y}^{-1} \times 1000\ \text{y}}{1} + 1\right)\right] = 78\ \text{y},$$

substantially less than the estimate based on constant use. It is clear from this example that making reliable estimates very far into the future is extremely difficult. Minor uncertainties in resource availability or growth rate can yield vastly different results.

Although the approach of the authors of *Limits to Growth* has gained some acceptance, it has also been criticized. These differing opinions only serve to emphasize the complexity of the problem and the difficulty in making accurate predictions. The work, however, has been instrumental in emphasizing the need for the careful scientific analysis of this problem and the interrelationship of the diverse factors affecting long-term ecological changes.

Topic for Discussion

Consider the world population as a function of time, as shown in Figure 2.10. In the context of limited resources, sketch a graph of expectations for the world population over the next 200 to 500 years. What analogies can be drawn with the Hubbert model, as illustrated in Figure 2.15?

distribution in Figure 2.14 (which can be gained from historical data on use rates) and knowledge of Q_∞ (which can be gained from an analysis of known and speculated unused resources). This approach will be discussed in more detail with regard to the use of fossil fuels in the following chapters.

2.5 Challenges for Sustainable Energy Development

The most reasonable approach to long-term sustainable energy production is not obvious. If a clear solution to the world's future energy needs existed, we would have a well-defined road to achieving our energy goals. Many options need to be considered, and, as discussed in this text, trade-offs must be considered in the development of any new technology. Any viable sustainable energy option must not only make a positive impact on our energy requirements for the present and near future, but it must also have a positive influence on the quality of life for future generations. In general, factors that will influence our choices for new energy technologies to pursue include the:

- Availability of the necessary resources.
- Availability of the necessary technology.
- Consideration of economic factors.
- Consideration of social factors.
- Environmental impact.
- Consideration of political factors.
- Ability to integrate new technology with existing technology.

These factors will now be discussed.

2.5a Resource Limitations

The power that is available from alternative energy resources depends on the nature and extent of the resource, as well as the existence of a viable technology to utilize the energy source. Although nonrenewable energy resources, such as fossil fuels or nuclear, are limited in terms of the total energy available, truly renewable resources may be expected to be virtually unlimited in their total availability. However, the power available, even from renewable resources, is limited. Obviously some of the renewable energy technologies discussed in this text, such as solar and wind, are prevalent in most parts of the world, but others, such as geothermal or tidal, are more limited in their distribution. The power available from these resources follows similar patterns. These points are discussed in detail throughout the book and, in particular, in Chapter 21.

It is important to realize, however, that the power available from a particular resource may be extensive, but our ability to harness that energy may be limited by other factors. The utilization of solar energy for the production of electricity is a good example. Solar energy is not infinite because the sun is a finite object, and the earth intercepts only a very small fraction of the energy that the sun produces (see Chapter 8). However, even this small fraction of the sun's energy is very much more than is needed to satisfy all of humanity's needs. To utilize the sun's energy, appropriate devices, such as photovoltaic cells (Chapter 9), are needed to convert electromagnetic energy into electrical energy. One of the common photovoltaic cells for the harvesting of significant solar energy is the CIGS (copper-indium-gallium-selenide) cell. Indium, gallium, and selenium are all relatively rare (and expensive) elements, and they are at least one of the reasons for the economic limitations to solar energy

ENERGY EXTRA 2.2
Rare earth elements—Are they really rare?

The rare earth elements (often known as the lanthanides) are those elements with atomic numbers between 57 and 71. Scandium (atomic number 21) and yttrium (atomic number 39) are generally also classified with the rare earths, as they tend to occur in the same geological deposits and have similar chemical properties as lanthanum. Compared with other elements in the earth's crust, most of the rare earths are about as abundant as many other elements that we think of as common, such as tin, lead, and tungsten. They are much more abundant than silver, gold, or platinum. There are two factors that contribute to their name "rare earth": they tend to occur together in geological deposits and are relatively difficult to separate from one another because of their similar properties, and they tend to be fairly dispersed geographically rather than occurring in concentrated deposits. Both of these features arise because of their particular chemical properties and make their commercial utilization more difficult than their actual abundance would suggest. As well, rare earth deposits often contain trace amounts of uranium and thorium, leading to the additional difficulty of having to deal with radioactive contamination from mine tailings.

The rare earth elements have a number of interesting properties (including magnetic, electrical, and optical properties) that result from their electronic configuration. The table shows the possible commercial applications of the rare earths, including many such as magnets, lasers, superconductors, and fossil fuel cracking catalysts that the present text discusses. In fact, rare earth-containing permanent magnets are an essential component of wind turbine generators and electric vehicle motors.

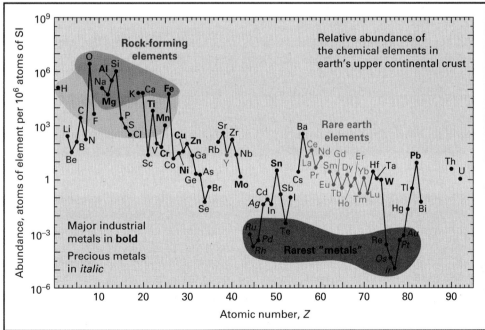

Abundances of various elements in the earth's crust plotted as a function of their atomic number.

U.S. Geological Survey, Fact Sheet 087-02, Rare Earth Elements—Critical Resources for High Technology. https://pubs.usgs.gov/fs/2002/fs087-02/

Continued on page 48

Energy Extra 2.2 continued

Applications of the rare earth elements.

atomic number	symbol	name	applications
21	Sc	Scandium	Aeronautical alloys, gas discharge lamps
39	Y	Yttrium	Lasers, superconductors
57	La	Lanthanum	High refractive index glass, hydrogen storage
58	Ce	Cerium	Polishing powder, oil cracking catalyst
59	Pr	Praseodymium	Magnets, lasers
60	Nd	Neodymium	Magnets, lasers
61	Pm	Promethium	Unstable
62	Sm	Samarium	Magnets, lasers
63	Eu	European	Lasers, fluorescent lamps
64	Gd	Gadolinium	Lasers, high refractive index glass
65	Tb	Terbium	Magnets, magnetic materials, lasers, fluorescent lamps
66	Dy	Dysprosium	Magnets, magnetic materials, lasers
67	Ho	Holmium	Lasers
68	Er	Erbium	Lasers, fiber optics, steel alloys
69	Tm	Thulium	Lasers, gas discharge lamps
70	Yb	Ytterbium	Lasers, steel alloys
71	Lu	Lutetium	High refractive index glass

In view of the geographical and geological distribution of the rare earth elements along with their role in industry, particularly alternative energy technologies, the mining and processing of these materials are of utmost importance. It is interesting to look at the availability of rare earth elements from a historical perspective.

The first rare earth elements were discovered in the late 18th century. Promethium (atomic number 61) was the last rare earth element to be discovered. Its existence had been predicted in the early 20th century, but because it has no stable isotopes it was not produced until 1945, when nuclear reactors became available. It was first isolated in 1963, and the longest-lived isotope, ^{145}Pm, has a half life of 17.7 years.

Up until 1950 rare earth production was fairly small and came primarily from deposits in India and Brazil. In the early 1950s, applications of rare earths grew, and newly utilized deposits in South Africa produced the majority of material. This period is known as the Monazite Placer Era and continued until around 1965. At that point, deposits in Mountain Pass, California became the major source of production. By the mid-1980s, deposits in China began producing rare earths and became the major source of these elements by the mid-1990s. Much of this develop-ment was at the expense of production elsewhere,

Continued on page 49

Energy Extra 2.2 continued

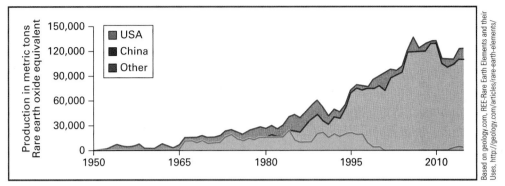

Rare earth oxide production.

particularly in the United States, where production virtually halted because it could not compete economically. By 2010 China produced about 95% of the rare earths worldwide, although it was estimated that it had only 23 to 37% of the known resources.

In late 2010 China cut back on its export of rare earths to other countries. While Chinese officials claimed that this was motivated by environmental concerns, the growing internal market for rare earths in China may have also been a factor. These export restrictions caused a dramatic increase in the price of rare earth elements worldwide. They also resulted in increased production outside of China. The graph shows that production in the

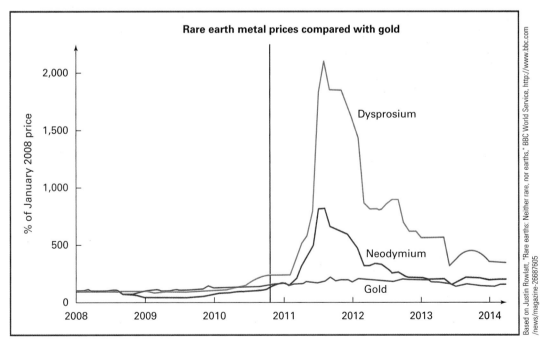

Market price of neodymium and dysprosium compared with gold. Prices are normalized to the price in January 2008.

Continued on page 50

United States (in Mountain Pass, California) resumed after being essentially dormant for the previous decade. In addition, other sites where rare earths occur are being considered. Most notable are resources in Australia and Canada. In early 2015 China eased rare earth export restrictions, perhaps due to declining market prices.

Despite the difficulty in mining and processing rare earth metals and the environmental consequences of such production, they are not really rare. The events of the past two decades or so have shown that the availability of rare earth metals is influenced more by political and economic factors rather than their actual abundance in nature.

Topic for Discussion

The most common permanent magnet material used in electric vehicle motors is $Nd_2Fe_{14}B$. Neodymium accounts for about 15% of all rare earth metal production worldwide. Locate information about the mass of the permanent magnets used in a typical electric vehicle. If 10% of all gasoline vehicles worldwide were retired and replaced with electric vehicles each year, how would that affect current rare earth production and utilization?

utilization. Table 2.4 shows the quantity of these elements that would be needed to produce photovoltaic cells that could provide all the power necessary for human society. It is clear that, at current production rates, several thousand years would be needed to develop an all–solar energy infrastructure based on CIGS cells. Although production of these elements can be increased, there are ultimately limits to their total availability. Other possible photovoltaic technologies exist, although many of the most promising, from certain standpoints such as efficiency, also suffer from similar resource availability difficulties. This simple example illustrates the materials challenges that often accompany the implementation of alternative energy technologies.

2.5b **Technological Limitations**

In some instances, for example wind energy, a suitable technology exists that enables us to effectively make use of the resource, although this, in itself, does not ensure that its development will be viable economically and environmentally. Sometimes technological barriers inhibit the development of a resource. In some cases, the lack of a suitable technological infrastructure is the result of a lack of basic scientific knowledge. In some cases, appropriate mechanisms for applying scientific knowledge in a way that is both practical and economical need to be developed. Fusion energy is

Table 2.4: In, Ga, and Se needed to produce copper-indium-gallium-selenide photovoltaic cells required to generate 1 GW electricity and power sufficient to meet global needs (about 18 TW).

element	to produce 1 GW (t)	to fulfill world power needs (t)	annual production (t)
In	90	3.8×10^6	755
Ga	30	4.2×10^5	435
Se	180	7.5×10^6	2170

Data adapted from "Byproduct Mineral Commodities Used for the Production of Photovoltaic Cells" U.S. Survey, Circual 1365.

one example where further basic research is necessary to understand fully the ways in which this energy source can be made viable. Photovoltaics is an example of a field where functioning devices are in common use, but further research is needed to improve efficiency, make them more economically competitive, and make their extensive use feasible from a materials availability standpoint.

The development of renewable energy touches on a very wide variety of diverse fields, ranging from, say, biochemistry (for biofuel synthesis), to semiconductor physics (for photovoltaics), to plasma and nuclear physics (for fusion energy), and to surface science and materials research (for tidal energy). The development of high-temperature superconductors for possible energy-related applications is an example of how scientific and technological advances can be combined to make progress toward improved energy systems. Some of the ways in which superconductors can contribute to energy systems include low-loss power transmission cables, superconducting magnet energy storage, and light-weight, high-output generators for wind energy applications. High-temperature superconductors were first discovered in the lanthanum-barium-copper-oxide system in 1986 by Johannes Georg Bednorz and Karl Alexander Müller. They observed a superconducting transition temperature of 35 K. Less than a year later, researchers discovered superconductivity in the yttrium-barium-copper-oxide system at 92 K. Because this pushed the superconducting transition temperature above the temperature of liquid nitrogen, it opened up the possibility of the simple, inexpensive applications of superconducting materials. Unfortunately, high-temperature superconducting materials are ceramics, and fabricating them in the form of flexible wires from which power lines, magnetic coils, or generator windings can be prepared is not straightforward. In addition, boundaries that typically form between the grains of superconducting material block the flow of current. First-generation high-temperature superconducting "wires" were made from bismuth-strontium-calcium-copper-oxide–based materials. These were expensive and difficult to fabricate. Second-generation high-temperature superconducting "wires" have been developed in recent years and have resolved many of the problems that affected earlier versions. These "wires" are made from yttrium-barium-copper-oxide and are actually thin films of superconducting material deposited onto a metal ribbon substrate, as shown in Figure 2.16. The technology that has been

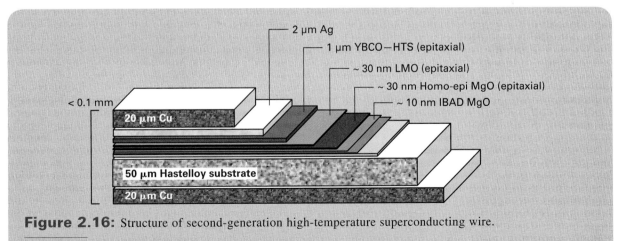

Figure 2.16: Structure of second-generation high-temperature superconducting wire.

Based on V. Selvamanickam, "Coated Conductors: From R&D to Manufacturing to Commercial Applications," EUCAS | ISEC | ICMC Superconductivity Centennial Conference, September 19-23, 2011, Den Haag, The Netherlands, http://www.superpower-inc.com/system /files/2011_0225+Barcelona+Wind+Seminar_Selva.pdf and http://www.superpower-inc.com/content/2g-hts-wire

Courtesy Brookhaven National Laboratory

Figure 2.17: The world's first superconducting power transmission line in New York in 2008 (connection from distribution station to underground cables).

developed for the preparation of these ribbons eliminates difficulties with boundaries between grains and provides a suitable substrate with good mechanical properties for the preparation of long robust conductors. The first commercial test of this technology for power transmission occurred in New York in 2008 (Figure 2.17) and has led the way to possible future developments that will improve the world's energy systems. The general properties of superconductors and their application for energy storage devices will be discussed further in Chapter 18.

In a number of areas, fundamental research (rather than engineering design) is needed to make advances in the development of alternative, renewable energy. Some notable examples covered in this text that present significant research challenges are:

- New organic photovoltaic cells that are reasonably efficient and are much more cost-effective than conventional semiconductor based materials (Chapter 9).
- Suitable economical membranes for the exploitation of salinity gradient or osmotic energy (Chapter 14).
- Methods for production of cellulosic ethanol (Chapter 16).
- Efficient non-lithium-based secondary batteries that will provide a cost-effective basis for widespread electric vehicle development (Chapter 19).
- Economical and efficient methods for direct hydrogen production (e.g., solar hydrogen, Chapter 20).

2.5c Economic Factors

The ultimate commercialization of any energy technology must consider economics as a major factor in determining its viability. This is particularly a concern for large-scale energy producers and distributors, such as public utilities, that ultimately provide electrical energy to customers. The present overview of economics focuses on this area. The development and construction of new installations for the production of electricity require long-term financial viability. The cost to produce electricity may be modeled in the following way. The cost per kilowatt-hour of electricity generated is given as

$$C = C_{\text{fuel}} + C_{\text{operating}} + \frac{I \cdot CRF}{Rf(8760 \text{ h/y})}, \tag{2.11}$$

where C_{fuel} is the cost per kilowatt-hour for fuel, $C_{\text{operating}}$ is the operating and maintenance cost per kilowatt-hour, I is the total capital installation cost, R is the total maximum capacity (in kW), f is the capacity factor, and CRF is the capital recovery factor. The last term on the right-hand side of equation (2.11) gives the contribution to the cost per kilowatt-hour of generated energy that comes from the capital investment costs amortized over the payback period for the facility. The capacity factor is the fraction of the total theoretical capacity that is actually achieved. The capital recovery factor takes into account the accrued interest on the capital investment and is given by

$$CRF = \frac{i(1 + i)^T}{[(1 + i)^T - 1]}, \tag{2.12}$$

where i is the annual interest rate expressed as a fraction (i.e., 5.1% would be 0.051) and T is the payback period. The choice of payback period for a particular facility must obviously be shorter than its life expectancy, and 15 years is a common expectation for the payback period of many facilities. Higher-risk (i.e., untested) technologies necessitate shorter payback periods, whereas well-established and reliable technologies, such as coal-fired or nuclear facilities, may tolerate longer payback periods.

In general, fuel costs are important for generating technologies such as coal, natural gas, or nuclear thermal plants, or combustion turbines. They are typically not of relevance for many renewable technologies, such as solar and wind energy. Contributing factors that may be relevant in equation (2.11) for some systems include both positive and negative factors on the right-hand side of the equation. Positive factors on the right-hand side of equation (2.11) (i.e., those that increase the cost per kilowatt-hour of electricity) include decommissioning costs at the end-of-life cycle. These are probably most notable for nuclear power plants, where radioactive waste disposal must be considered. Negative factors on the right-hand side of equation (2.11) (i.e., those that decrease the cost per kilowatt-hour of electricity) include waste heat recovery sales and end-of-life cycle salvage recoveries. The former may be of relevance, for example, for coal-fired cogeneration plants (Chapter 17), where excess heat from burning coal (i.e., hot steam or water exhausted from the turbines) is sold for heating buildings. In the latter case, for example, rare (and valuable) elements such as indium, gallium, or selenium may be recovered from photovoltaic cells at the end of their life.

The implementation of financial models of energy production depends on a number of factors. A simple approach to the use of equations (2.11) and (2.12) might include an evaluation of fuel costs (for appropriate technologies) based on an assessment of energy content of the fuels and an analysis of conversion efficiencies. Operating costs depend on a number of factors, including facility design and local labor and materials costs. Infrastructure costs, of course, depend on the energy resource being utilized but can also vary considerably depending on the design of the facility. Some typical examples of capital construction costs per kilowatt of rated capacity for different energy technologies are shown in Table 2.5 on the next page. The effect of rated capacity on the cost of a wind turbine is illustrated in Figure 2.18 on page 55. The data show that, for small turbines, there is a clear relationship between the cost per rated

Example 2.3

Calculate the effects of capital recovery costs on the price per kilowatt-hour of electricity produced by a 1.5-MW wind turbine running at a 35% capacity factor. The total installation cost was $2,300,000, and the interest rate is 6.2% over a payback period of 15 years.

Solution

From equations (2.11) and (2.12), the contribution to the cost per kilowatt-hour of generated electricity from the capital recovery term is

$$\frac{I}{Rf(8760 \text{ h/y})} \cdot \frac{i(1 + i)^T}{[(1 + i)^T - 1]},$$

where I = $2,300,000, R = 1500 kW, f = 0.35 y^{-1}, i = 0.062 and T = 15 years. This equation gives the contribution to the cost of electricity as

$$\frac{\$2,300,000}{(1500 \text{ kW}) \times (0.35 \text{ y}^{-1}) \times (8760 \text{ h/y})} \times \frac{0.062(1 + 0.062)^{15}}{[(1 + 0.062)^{15} - 1]} = \$0.052/\text{kWh}.$$

This represents the most significant component to the cost of producing electricity for a wind turbine.

Table 2.5: Typical capital system costs in the United States in dollars per kilowatt of installed capacity (US$/kW) for utility scale generating stations using different technologies.

energy source	infrastructure cost (US$/kW capacity)
natural gas (thermal)	917
natural gas (combustion turbine)	973
wind (on-shore)	2213
hydroelectric	2936
coal	3246
solar (photovoltaic)	3873
geothermal	4362
solar (thermal)	5067
nuclear (thermal neutron fission)	5530

Adapted from data in "Updated Capital Cost Estimates for Utility Scale Electricity Generating Plants-April 2013." U.S. Energy Information Administration. Fossil fuel facilities do not include carbon capture and storage

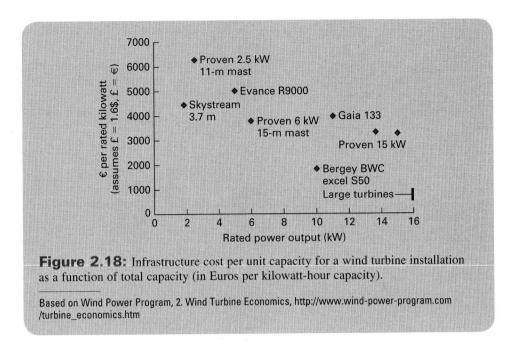

Figure 2.18: Infrastructure cost per unit capacity for a wind turbine installation as a function of total capacity (in Euros per kilowatt-hour capacity).

Based on Wind Power Program, 2. Wind Turbine Economics, http://www.wind-power-program.com /turbine_economics.htm

capacity and that larger turbines are clearly favorable economically. The net cost per kilowatt-hour generated for large facilities can be estimated on basis of this discussion and is indicated for different energy technologies in Table 2.6. A comparison of these data with those in Table 2.5 illustrates the importance of fuel costs, operating costs, and particularly capacity factor in determining the net cost of electricity.

Variability in the average cost to generate electricity among countries can result from different approaches to electricity generation and differences in national economics. The cost to the consumer can also vary considerably within a country due to

Table 2.6: Typical average cost of electricity per kWh (US$/kWh) in the United States generated by different technologies.

energy source	electricity cost (US$/kWh)
coal	0.040*
wind (on-shore)	0.060
natural gas (thermal)	0.065
hydroelectric	0.070**
geothermal	0.075**
nuclear (thermal neutron fission)	0.11
solar (photovoltaic)	0.11
solar (thermal)	0.155
natural gas (combustion turbine)	0.17

Adapted from data compiled by the U.S. Department of Energy and the U.S. National Renewable Energy Laboratory as reported by en.openei.org

geographical location, type of customer, amount used, and time of day. For example, in the United States residential electricity rates vary from about US$0.08 per kWh in Louisiana to about US$0.37 per kWh in Hawaii. Some average residential electricity costs are given in Table 2.7.

Energy producers, such as public utilities, who operate energy production facilities such as coal-fired stations, nuclear plants, or wind farms, depend on equipment suppliers to provide components for these facilities and/or construct them. For alternative energy technologies, much of the development begins as basic scientific research in university or government laboratories funded by government or industry. Commercial viability is the final goal of such a process, and all aspects of the development of new energy resources must work in concert to achieve this goal. Government subsidies and incentives are therefore beneficial during the development stages. Traditionally, the corporate bottom line was always financial profit, and technologies that did not have a foreseeable profit on a suitable timescale were not considered. New developments had to have potential financial benefits that were commensurate with the risk. However, more recently, in many countries a different approach has been implemented; the *triple bottom line* (TLB). The TBL considers three aspects of possible benefits resulting from corporate actions, sometimes referred to as *people*, *planet*, and *profit*. These benefits come from the realization that good business practices for long-term sustainability need to consider more than just short-term economic gain.

The *people* aspect of the TBL concerns the fair treatment of workers and maintaining morale by providing fair salaries and benefits and by providing a safe working environment. The people benefits of TBL corporate practices are

Table 2.7: Average cost of residential electricity in various countries in US$/kwh for 2014–2015 (except as noted).

country	average cost (US$/kWh)
Serbia	0.066[1]
India	0.07[2]
China (Beijing) [for 2012]	0.078[3]
United States	0.12[4]
Canada	0.158[5]
France	0.171[1]
Brazil	0.185[6]
Netherlands	0.197[1]
United Kingdom	0.207[1]
Sweden	0.212[1]
Japan [for 2012]	0.262[7]
Italy	0.264[1]
Germany	0.322[1]
Denmark	0.329[1]

Sources:
[1]http://strom-report.de/
[2]http://cea.nic.in/reports/others/enc/fsa/tariff_2014.pdf
[3]http://english.sz.gov.cn/ln/201205/t20120517_1914423.htm
[4]http://www.eia.gov/electricity/data/browser/#/topic/7?agg=2,0,1&geo=g&freq=M
[5]http://www.huffingtonpost.ca/2016/07/21/ontario-hydro-rates_n_11107590.html
[6]http://www.light.com.br/para-residencias/Sua-Conta/composicao-da-tarifa.aspx
[7]http://thisbluemarble.com/showthread.php?t=36342

somewhat difficult to categorize but may include ensuring that suppliers and contractors also have fair worker policies or providing community benefits through educational programs.

The *planet* aspect of TBL is concerned with maintaining environmental quality. A thorough life cycle analysis (Chapter 4) of materials and practices is essential to a complete understanding of the environmental impact of business practices.

The *profit* component of TBL is probably the easiest to quantify. It is based on a traditional corporate accounting approach but may also include an assessment of the economic impact on the community in addition to internal corporate financial benefits.

Profits are essential to the survival of any business. The implementation of policies that promote the people and planet sides of TBL practices are often legal responsibilities of a business, but it should be recognized that being proactive in these areas can be ultimately beneficial to both business and community.

2.5d Social Factors

The public perception of energy is influenced by a number of factors, including economics, comfort, safety, environmental factors, and even aesthetics. There was relatively little public interest in alternative energy sources prior to the energy crisis in the early 1970s, except, perhaps, in the controversy over the risks associated with nuclear energy. Since its beginnings, nuclear energy has been controversial. This debate precedes any significant development of most alternative renewable energy resources (hydroelectric and geothermal are exceptions). Public approval of nuclear energy has generally increased in the United States over the past 40 years, as illustrated in Figure 2.19. However, approval typically decreases after a nuclear incident, as

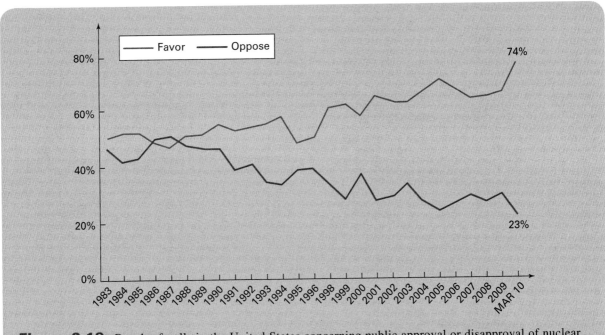

Figure 2.19: Result of polls in the United States concerning public approval or disapproval of nuclear energy.

Based on NEI, http://lh6.ggpht.com/markpf21/SLW9KFPMCwI/AAAAAAAAAQY/k_ocIDGEGEM/popo_chart%5B4%5D.jpg

evidenced by the anomaly around 1986 when the Chernobyl incident occurred. After a nuclear incident, nuclear energy regulations typically become stricter, and it is likely that the chance of further incidents (at least in the short term) decreases. Only public perception of the risks associated with nuclear energy affects the approval rating. The nuclear incident at the Fukushima reactor facility in Japan illustrates this point. Japan has, in the past, been very pro–nuclear energy. This is not unexpected because it is a highly industrialized, densely populated country with minimal indigenous energy resources. Figure 2.20 shows the results of polls in Japan following the Fukushima incident. Not surprisingly, there was a substantial drop in support for nuclear energy in the ensuing three months. Certainly both public opinion, as well as government policy (in many countries), over the implementation of nuclear energy will continue to be a topic for debate.

In recent years, public concern over energy-related issues has grown. The points of greatest interest to the general public are summarized in the results of a recent poll in the European Union, as shown in Figure 2.21. Certainly the greatest concern is for energy prices and availability. Environmental concerns are lower on the list, as is interest in conservation matters. Another view of the relative importance of economy and environment is illustrated by the results of another survey, as shown in Table 2.8. Residents in the United States, Canada, and the United Kingdom were asked which of the following approaches they favored: one in which the environment was protected at the risk of adversely affecting economic growth and the other in which economic growth was ensured at the risk of environmental damage. In this limited study, at least, environmental concerns seem to be important in North America and less so in the United Kingdom, although a significant minority favor economy over environment.

Opinions concerning the need to develop alternative energy technologies (from the same survey as illustrated in Figure 2.19) are shown in Figure 2.22 on page 60. In the European Union, there is very strong support for the development of solar energy (despite its high cost and the general concern over future energy prices). There is also significant opposition to the use of nuclear energy, compared

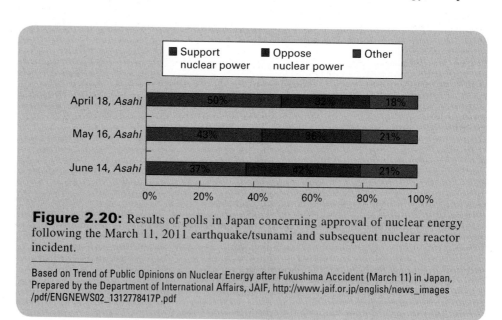

Figure 2.20: Results of polls in Japan concerning approval of nuclear energy following the March 11, 2011 earthquake/tsunami and subsequent nuclear reactor incident.

Based on Trend of Public Opinions on Nuclear Energy after Fukushima Accident (March 11) in Japan, Prepared by the Department of International Affairs, JAIF, http://www.jaif.or.jp/english/news_images /pdf/ENGNEWS02_1312778417P.pdf

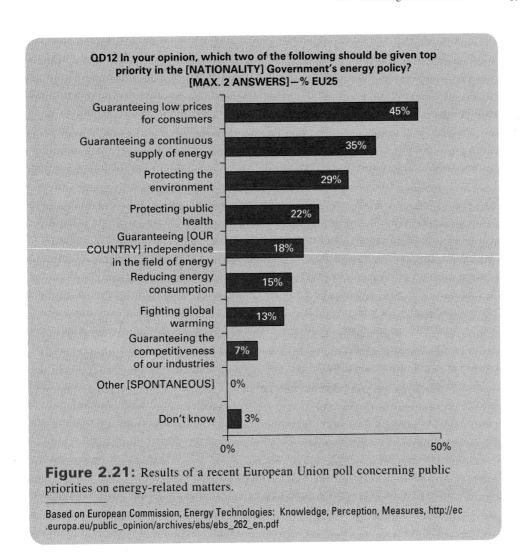

Figure 2.21: Results of a recent European Union poll concerning public priorities on energy-related matters.

Based on European Commission, Energy Technologies: Knowledge, Perception, Measures, http://ec .europa.eu/public_opinion/archives/ebs/ebs_262_en.pdf

to other energy sources. This is even the case in France, where the vast majority of electricity comes from nuclear energy (21% in favor, 44% neutral, and 33% opposed). The favorable view of solar energy is also accompanied by an optimism that it will provide the majority of energy within the next 30 years and that fossil

Table 2.8: Results of a recent public opinion poll on relative importance of environmental protection and economic growth.

nation	percent favoring environmental protection	percent favoring economic growth
Canada	55	22
United States	47	26
United Kingdom	40	33

Based on data from Angus Reid Global, http://www.angus-reid.com/issue/global-warming

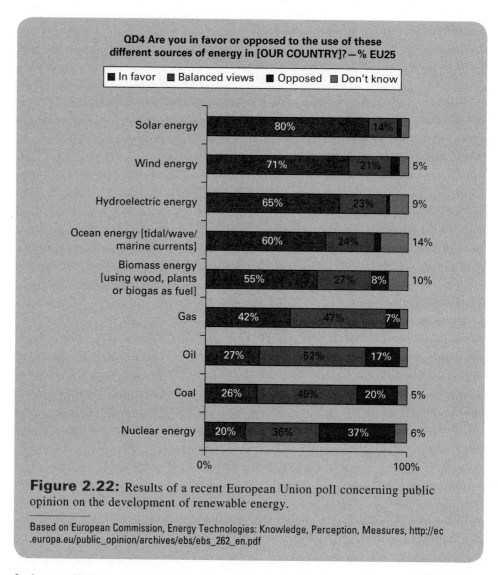

Figure 2.22: Results of a recent European Union poll concerning public opinion on the development of renewable energy.

Based on European Commission, Energy Technologies: Knowledge, Perception, Measures, http://ec .europa.eu/public_opinion/archives/ebs/ebs_262_en.pdf

fuel use will be greatly diminished. This opinion is illustrated by the data shown in Figure 2.23.

While public opinion seems to greatly favor the development of renewable energy sources and there is a positive feeling that this is possibly on an unrealistically short timescale, there is also a reluctance to accept any adverse economic consequences that could result from moving away from inexpensive fossil fuels.

2.5e Environmental Factors

Renewable sources of energy are generally considered to have less environmental impact than fossil fuels. A quantitative assessment of environmental impact is often difficult because many environmental factors do not have a direct quantitative metric. One aspect of the environmental impact of renewable energy that can be expressed quantitatively is greenhouse gas emissions (specifically CO_2). Such an assessment would include greenhouse gas emissions that occurred not only

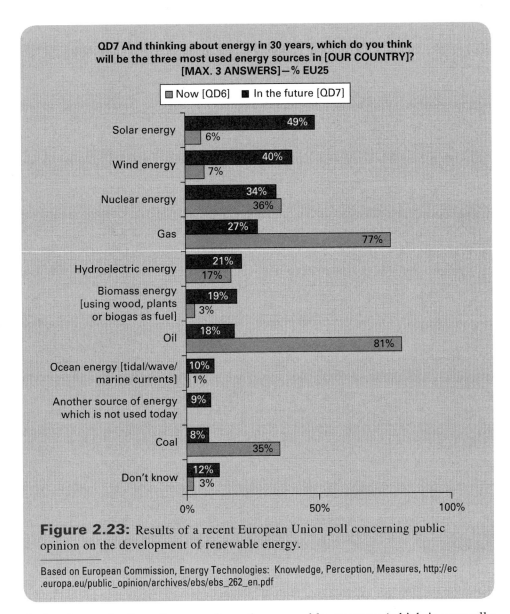

Figure 2.23: Results of a recent European Union poll concerning public opinion on the development of renewable energy.

Based on European Commission, Energy Technologies: Knowledge, Perception, Measures, http://ec .europa.eu/public_opinion/archives/ebs/ebs_262_en.pdf

during the production of energy from the renewable resource (which is generally quite small) but also during the acquisition of materials, production of necessary equipment, transportation of components, maintenance of the facility, and ultimate disposal of equipment. This kind of cradle-to-grave life cycle is discussed further in Chapter 4, and an assessment of CO_2 emissions for alternative transportation technologies is presented in Chapter 20.

Table 2.9 shows the results of a greenhouse gas life cycle analysis for several renewable energy source. The table shows the quantity of CO_2 produced per unit energy generated, averaged over the lifetime of the facility. Results for renewable energy sources are compared with those from fossil fuels. For fossil fuels, the large amount of CO_2 produced is expected, and this comes primarily from the combustion process that oxidizes the carbon in the fuel. The table gives a range of values for renewable energy sources because the actual amount of CO_2 depends on the details of the life cycle of the

Table 2.9: Greenhouse gas emissions (CO_2) per unit electrical energy generated for some fossil fuels and renewable resources.

resource	CO_2 (kg/MWh)
coal	955
natural gas	430
solar (photovoltaic)	98–167
wind	7–9
geothermal	7–9
hydroelectric (high head)	3.6–11.6

Based on data from International Energy Agency "Benign energy? The environmental implications of renewables," Paris: OECD (1998)

facility, and this can vary considerably from one facility to another. Although all renewable energy sources produce less CO_2 per unit electricity generated, none is actually carbon free. It is perhaps particularly surprising that the worst renewable energy source in this respect is solar photovoltaics. In fact, as discussed in the previous subsection, public perception of solar energy is very positive. It is generally the renewable energy source that most people see as the best hope for a sustainable future and the one that they would most like to see promoted. The reason that solar photovoltaics produces the quantity of CO_2 that it does (in fact, under some circumstances, nearly a third of that produced by natural gas) is that it is a very materials-intensive technology that extracts energy from a very low-density energy source. In other words, some of the materials used in the production of photovoltaics are very energy intensive to produce. This is typically the case, for example, for relatively rare materials where it is necessary to mine a large quantity of raw material in order to extract a small quantity of the material of interest. Added to this are the facts that photovoltaic materials require high-technology manufacturing processes and that the low density of solar energy means that a lot of photovoltaic cells need to be manufactured to generate a relatively small amount of electricity. This type of analysis indicates that the environmental impact of utilizing a renewable energy resource, from the greenhouse gas emissions standpoint, is not always obvious.

Other aspects of the environmental impact of renewable energy might be more difficult to quantify but can become apparent by means of careful analysis. As with solar energy, the low energy density of many renewables (compared with fossil fuels) is an important contributing factor in their environmental impact. The extensive land area associated with the generation of electricity by solar photovoltaics, wind, or even hydroelectricity can result in the deforestation or displacement of resources from agricultural development with a resulting impact on the environment and the quality of human life.

Biofuels are a clear example of the complexity of an overall assessment of environmental impact. As discussed in detail in Chapter 16, a careful analysis of all aspects of the impact of biofuel production and use does not necessarily provide a positive incentive for their extensive development.

The low energy density of renewables is also a contributing factor to their relatively high (in some cases) human risk factor, as discussed in detail in Section 6.9. The large quantity of material needed for many renewable technologies means risk associated with the required mining operations, material transportation, manufacturing processes, and related operations, and these typically outweigh the direct risk associated with the energy generation itself. The perception of risk or environmental impact is often an important factor in the formation of public opinion. Although wind energy is one of the least invasive of energy technologies and is generally favored as a viable new energy source, the uncertainty of local effects often results in the resistance of residents to developments in their own neighborhoods. This is sometimes referred to by the acronym *NIMBY* (not in my backyard).

Example 2.4

If a 1 GW_e coal-fired generating station that operates at a 70% capacity factor is replaced with an equivalent capacity of wind turbines, what would be the net annual savings in CO_2 emissions?

Solution

The total energy generated by the coal-fired station in one year will be

$$(0.7) \times (1000 \text{ MW}_e) \times (8760 \text{ h/y}) = 6.13 \times 10^6 \text{ MWh per year.}$$

From Table 2.9 the total CO_2 emissions will be

$$(6.13 \times 10^6 \text{ MWh}) \times (955 \text{ kg/MWh}) = 5.856 \times 10^9 \text{ kg.}$$

The CO_2 emissions from the wind turbines (assuming 8 kg/MWh) will be

$$(6.13 \times 10^6 \text{ MWh}) \times (8 \text{ kg/MWh}) = 4.904 \times 10^7 \text{ kg.}$$

This represents a savings of

$$5.856 \times 10^9 \text{ kg} - 4.904 \times 10^7 \text{ kg} = 5.807 \times 10^9 \text{ kg.}$$

2.5f Political Factors

Political decisions on energy-related matters generally follow from a nation's energy policy, as well as the efforts of individual politicians to defend the energy interests of their constituents. Energy policy provides guidelines for energy-related laws and treaties with other countries and directives for government agencies dealing with energy issues. The following items are common components of national energy policies:

- A description of national policies concerning energy generation, transmission and use
- The establishment of energy efficiency and environmental standards related to energy use
- The specification of energy-related fiscal policies, including subsidies, incentives, tax exemptions, and the like, to promote improved energy utilization
- The participation in funding programs for energy-related research and development
- The development of energy-related treaties and agreements with other countries

Specific goals that energy policies typically address include:

- To what extent is energy self-sufficiency expected?
- What future energy production methods are most appropriate?
- How will good energy practices be promoted?

- What environmental impact of energy use is acceptable?
- What is an acceptable energy technology for a transportation system?

The most appropriate means of dealing with these points can vary greatly from one country to another and depend on a number of factors such as:

- Economy.
- Climate.
- Geography.
- Natural resources.
- Population.

Although it may be reasonable for a country with a large land area, low population density, substantial natural resources, and a good economy to strive for energy independence, that goal is not reasonable for a nation that has a small land area, a large population density, and few resources.

While energy policies attempt to establish goals to ensure energy security, the specific approach to achieving these goals needs to be defined in order to implement such policies. The success of an energy policy lies not only in the suitability of its goals but also in the ability of society to overcome the necessary obstacles in order to achieve those goals. For example, the practicality of some U.S. automobile emission standards have been challenged in court by automobile manufacturers who claim that they are technologically and economically unrealistic.

General guidelines for the implementation of U.S. energy policy have been outlined in the Department of Energy's 2014 Strategic Plan. They have adopted what they refer to as an "all of the above" energy strategy, meaning that all viable energy resources will be investigated in the context of their objectives which include

- doubling renewable energy generation from wind, solar, and geothermal sources between 2012 and 2020,
- reducing cumulative carbon pollution by 3×10^9 t by 2030, and
- contributing to international efforts to address global climate change.

Specific initiatives that are part of the 2014–2018 strategic plan include the development of

- modular nuclear reactors (see Section 6.11),
- advance geothermal systems (see Chapter 15),
- offshore wind (see Section 10.4),
- commercial scale cellulosic ethanol production (see Section 16.3),
- cost-effective fuel cells (see Section 20.6),
- cost-effective plug-in vehicle technology (see Sections 17.8b and 19.4),
- cost-effective photovoltaic systems (see Section 9.4), and
- improved carbon capture and storage capabilities (see Section 4.6).

These are all areas where the development and implementation of new technologies is essential to the future production of energy that is both sustainable and environmentally sound.

Many aspects of energy policy are most appropriately considered at the regional, state/provincial, or municipal level (more in Chapter 17). These factors include, for instance, energy conservation, building codes, and other measures that are influenced by local geography and climate.

ENERGY EXTRA 2.3
Government emission control standards

Concerns over the effects of anthropogenic emissions during energy production are not new. In 1306, King Edward I of England banned coal fires in London to improve air quality. More recently, the health and environmental effects of air pollution were publicized in Henry Obermeyer's 1933 book *Stop That Smoke*! (Harper, New York). In the twentieth century, local or regional regulations concerning emissions became common in order to deal with air quality in many cities. In the 1950s and 1960s, systematic studies were undertaken to identify sources of air pollution. Two important conclusions came out of these investigations: (1) Local sources of pollution had widespread effects outside the region, and (2) a significant portion of air pollution originated from motor vehicles.

The first finding was dealt with by the gradual implementation of regulations at state/provincial and federal government levels. In the United States, the State of California created the California Air Resources Board in 1967, and the federal government established the Environmental Protection Agency in 1970. During the same era, similar government regulatory bodies in Canada, Australia, most of Western Europe, and Japan were also established.

A major focus of such agencies has been the reduction of vehicle emissions by setting standards for new models. Over the years, these standards have become more strict and have required the implementation of new approaches to emission control. Although a number of technological advances have led to a reduction in vehicle emissions over the years, two major developments have played a major role. The first deals with emissions in the engine, and the second deals with emissions in the exhaust gas. In the first case, the *positive crankcase ventilation* (*PCV*) system reduces emissions by directing hydrocarbon-rich fumes from the crankcase into the engine's intake manifold to be burned rather than releasing them to the atmosphere.

Implementation of PCV systems predates the establishment of state and federal regulatory agencies that deal specifically with vehicle emissions. In 1961, pollution concerns prompted the State of California to require that all new passenger vehicles sold in that state must have PCV. By 1964, the implementation of PCV on new passenger vehicles was widespread in the United States.

By the mid-1970s, increasingly strict emission standards (Section 4.3) required a consideration of emissions in the exhaust gas. Exhaust gas emissions are most effectively dealt with using a catalytic converter. This is a device that converts toxic emissions from an internal combustion engine into nontoxic (or at least less toxic) by-products through a stimulated catalytic reaction. Most catalytic converters used on gasoline internal combustion engines are three-way converters, which deal with three types of pollutants in the following ways:

- Reduction of nitrogen oxides to nitrogen and oxygen
- Oxidation of carbon monoxide to carbon dioxide
- Oxidation of hydrocarbons to carbon dioxide and water

Government emission control standards have been effective nationally and even globally and have made substantial reductions in vehicle emissions. Reductions in emissions are discussed in detail in Chapter 4 and are an example of one case where government regulations were effective in dealing with environmental concerns.

Topic for Discussion
Consider the details of the chemical reactions for the three processes that occur in a three-way catalytic converter. What are the health and/or environmental concerns of the chemical compounds that are produced?

Overall, establishing government energy policies that are environmentally sound, technologically feasible, and economically viable for a particular nation or region is a challenge that can provide long-term benefits and lead to a sustainable energy future.

2.5g Integrating New Technology with Existing Infrastructure

Whatever the mix of energy sources that will be adopted for the future, it is clear that changes must be made in how we produce and utilize energy. The energy technologies that have been developed over many years must give way to the new approaches in order to establish a sustainable infrastructure for the future. As outlined in the Preface, this is a formidable task, and this change must be done in such a way as to maintain the effectiveness of our energy systems.

Transportation is an interesting example of the potential challenges to changes in energy production and use, as well as of the need to consider all relevant aspects of reducing our dependence on fossil fuels. If, for example, future transportation systems are based on battery electric vehicles or fuel cell vehicles, then the elimination of the existing fossil fuel vehicle infrastructure and the development of a new transportation infrastructure must be done in a manner that minimizes or precludes disruption. While the way forward may seem straightforward, certain not so obvious issues need to be considered.

It is clear that the existing operational infrastructure of oil wells, refineries, oil pipelines, and gas stations will need to be replaced by electric charging or hydrogen filling stations. Also, the manufacturing infrastructure for the production of fossil fuel vehicles will need to be converted to the manufacture of battery electric vehicles or fuel cell vehicles. However, the necessary changes to the electricity supply system may constitute an even greater task. A summary of energy sources in an industrialized country (the United States) is illustrated in Figure 2.6. At present, roughly one-third of end user energy is for transportation; the remainder is roughly equally divided between industry and residential plus commercial. The energy for transportation comes almost exclusively from petroleum. If fossil fuel vehicles are replaced by electric vehicles (using either batteries or hydrogen as an electricity storage mechanism), then electric generating capacity must be increased to provide for both our current electricity needs and energy for transportation. Even considering the fact that electric vehicles are more efficient than fossil fuel–powered vehicles (Chapter 20), moving from a fossil fuel transportation economy to an electric transportation economy will require a 60–70% increase in electric generating capacity. A long-term goal of reducing fossil fuel use, from the standpoint of either resource availability or environmental impact, will require a careful consideration of resource availability in order to ensure that our energy needs can be fulfilled. Thus any plan to significantly increase the number of electric vehicles with a corresponding reduction in fossil fuel vehicles will call for suitable increases in electric generating capacity to satisfy the increased demand. The full benefits of fossil fuel vehicle reduction will require that additional generating capacity will come from renewable sources.

An important concern related to the integration of alternative energy technologies with our existing electricity supply grid deals with the intermittent nature of most renewable energy sources. This would include, for example, the daily variations in solar energy, the somewhat less predictable variations in wind energy, and the cyclic periodicity of tidal energy. Hydroelectric and geothermal are alternative resources that offer a more predictable supply of energy; however, geothermal availability is probably not sufficient to make any substantial long-term contribution to global energy use. To make the most efficient use of available resources, it is necessary, as discussed in Chapters 17 and 18, to integrate alternative energy technologies with suitable energy storage techniques and energy from base-load fossil fuel, nuclear, or possibly hydroelectric facilities. A long-term goal of eliminating fossil fuels will require intelligent control of energy production and distribution facilities (i.e., smart grid, as discussed in Chapter 17) to ensure a reliable supply of electrical energy.

2.6 Summary

In this chapter, the energy needs of humanity over the years have been reviewed. It was shown that, with the development of technology, came a rapid and significant increase in energy use. The reliance on wood as the major source of energy gave way to coal in the second half of the nineteenth century and then to petroleum in the mid-twentieth century. It has been shown that per-capita energy use in different countries is directly related to economic factors. Climate, population density, and types of industry play an important role in determining energy use. The chapter also summarized the current distribution of energy use, illustrating that fossil fuels account for the vast majority of energy use, followed by nuclear and hydroelectric. Growth mechanisms have been presented, and the analysis of energy resource utilization on the basis of the Hubbert model was presented. This model provides an understanding of the rate at which a resource is utilized in the context of the total quantity of the available resource.

The challenges for the implementation of new energy technologies have been reviewed. These include not only the technological challenges themselves but also those related to a consideration of resource availability, economics, social attitudes, environmental factors, and political policies. Viable energy technologies must be sustainable and economically competitive with traditional alternatives; otherwise, there will be a reluctance to move away from well-established technologies. Public opinion of alternative energy is based on a perception of factors such as cost, environmental impact, and risk and is a major factor in determining the acceptance of traditional and new energy sources. A complete environmental analysis of renewable energy technologies is often difficult and indicates the complexity of fully assessing available energy options. Energy policies are designed to promote sound approaches to future energy production and use, and their implementation is most efficient when dealt with at a variety of different government levels.

Problems

2-1 Consider an island with a population density of 50 people/km^2 (about equal to the present world average). If the annual growth rate is 2%, determine the year in which the population density be equal to 20,000 people/km^2 (approximately the current population density of Macau, the world's most densely populated nation).

2-2 A quantity has a doubling time of 110 years. Estimate the annual percent increase in the quantity.

2-3 The population of a particular country has a doubling time of 45 years. When will the population be three times its present value?

2-4 Assume that the historical growth rate of the human population was constant at 1.6% per year. For a population of 7 billion in 2012, determine the time in the past when the human population was 2.

2-5   What is the current average human population density (i.e., people per square kilometer) on earth?

2-6   The total world population in 2012 was about 7 billion, and Figure 2.11 shows that at that time the actual world population growth rate was about 1% per year. The figure also shows an anticipated roughly linear decrease in growth rate that extrapolates to zero growth in about the year 2080. Assuming an average growth rate of 0.5% between 2012 and 2080, what would the world population be in 2080? How does this compare with estimates discussed in the text for limits to human population?

2-7   The population of a state is 25,600 in the year 1800 and 218,900 in the year 1900. Calculate the expected population in the year 2000 if (a) the growth is linear and (b) the growth is exponential.

2-8   The population of a country as a function of time is shown in the following table. Is the growth exponential?

year	population (millions)
1700	0.501
1720	0.677
1740	0.891
1760	1.202
1780	1.622
1800	2.163
1820	2.884
1840	3.890
1860	5.176
1880	6.761
1900	8.702
1920	10.23
1940	11.74
1960	13.18
1980	14.45
2000	15.49

2-9   Consider a solar photovoltaic system with a total rated output of 10 MW_e and a capacity factor of 29%. If the total installation cost is $35,000,000, calculate the decrease in the cost of electricity per kilowatt-hour if the payback period is 25 years instead of 15 years. Assume a constant interest rate of 5.8%.

2-10   If a quantity has a doubling time of 30 days and its value is 1000 units at noontime today, what will be its value at noontime tomorrow?

2-11   If a facility has a capital recovery factor of 0.12 and the annual interest rate is 5%, what is the payback period?

2-12   (a) If the total amount of a resource available is 10^{10} kg, calculate its lifetime if the rate of use is constant at 10^8 kg/y.

(b) Repeat part (a) if the initial rate of use is 10^8 kg/y and this increases at an annual rate of 5%.

(c) Repeat part (b) for an initial rate of 10^8 kg/y and an increase of 10% per year.

2-13 (a) Assume that human population growth is described as exponential growth and the current population is 7.5 billion. If the current daily increase in population is 250,000, calculate the doubling time.

(b) For the conditions in part (a), what is the current annual growth rate?

2-14 If world energy use increases linearly, use Figure 2.13 to estimate the annual percentage increase for OECD and non-OECD countries from 1990 to 2035 relative to the use in 1990.

2-15 Assume that the combustion of coal is approximated by the burning of carbon. If a coal-fired generating station operates at 40% efficiency, calculate the amount of CO_2 emitted per MWh of electricity produced.

2-16 Repeat problem 2.15 for the combustion of natural gas (methane). How does the use of the HHV affect your answer?

2-17 In the example of energy use on page 36, leading to equation (2.1), consider a family (two people) who use gasoline at the rate indicated but who heat their home electrically in accordance with the efficiencies given in the text above equation (2.1). How would this affect the per person power consumption?

2-18 A person in the United States consumes 2000 Calories of food per day. What fraction of the average per capita energy use, as indicated in Table 2.1, does this represent?

2-19 From the data in Table 2.1 and Figure 2.3, estimate the total energy production per year from hydroelectricity in the United States.

2-20 For an annual growth rate of 1.6%, use the data in Table 2.3 to estimate the year in which the human population will reach 200 per km^2 of land area.

Bibliography

E. Cassedy and P. Grossman. *An Introduction to Energy: Resources, Technology, and Society* (2nd ed.). Cambridge University Press, Cambridge, MA (1999).

R. Hinrichs and M. Kleinbach, *Energy: Its Use and the Environment* (5th ed.). Brooks-Cole, Belmont, CA (2012).

J. A. Kraushaar and R. A. Ristinen. *Energy and Problems of a Technical Society* (2nd ed.). Wiley, New York (1993).

E. L. McFarland, J. L. Hunt, and J. L. Campbell. *Energy, Physics and the Environment* (3rd ed.). Cengage, Stamford, CT (2007).

R. A. Ristinen and J. J. Kraushaar. *Energy and the Environment.* Wiley, Hoboken, NJ (2006).

V. Smil. *Energy at the Crossroads: Global Perspectives and Uncertainties.* MIT Press, Cambridge, MA (2003).

J. Tester, E. Drake, M. Driscoll, M. W. Golay, and W. A. Peters. *Sustainable Energy: Choosing Among Options.* MIT Press, Cambridge, MA. (2006).

R. Wolfson. *Energy, Environment, and Climate* (2nd ed.). Norton, New York (2011).

World Energy Council. *World Energy Resources - 2016*, available online at https://www.worldenergy.org/publications/2016/world-energy-resources-2016/

PART II

Fossil Fuels

For the past 150 years, fossil fuels have formed the largest component of our energy use. During that time, we have used up a substantial fraction of the available fossil fuel resources. These resources are not renewable, and their depletion will ultimately require the implementation of alternative sources of energy. Estimates of the longevity of fossil fuel resources are difficult and depend on a number of factors. The estimates range from a small number of decades for domestic (U.S.) oil to many hundreds of years for coal resources. During the time we have used fossil fuels, we have also made a substantial negative impact on the environment. Some of the damage caused to the earth and its ecosystem may be irreversible.

Clearly, our current use of fossil fuels cannot continue indefinitely. Dwindling supplies will force us to consider other energy options. However, environmental concerns supply the motivation to pursue alternative energy opportunities as a means of reducing the adverse effects of fossil fuel use. The infrastructure of human society is based on the use of substantial quantities of energy. Because most alternative energy sources have relatively low energy densities, developing a sufficient quantity of these resources to meet our needs is an enormous undertaking, as suggested in the Preface. However, to fully appreciate how fossil fuels can be replaced, it is important to understand, in detail, how we currently utilize fossil fuels and the consequences of this use. Chapter 3 provides an overview of our fossil fuel use, and Chapter 4 summarizes its environmental effects.

An offshore oil platform is shown in the photograph. ■

James Jones Jr/Shutterstock.com

Fossil Fuel Resources and Use

Learning Objectives: After reading the material in Chapter 3, you should understand:

- The properties of fossil files and methods for obtaining and processing them.
- The availability of fossil fuels.
- The use of fossil fuels worldwide.
- The application of the Hubbert model to fossil fuel use.

- Enhanced fossil fuel recovery methods.
- The properties and availability of shale oil and tar sands.
- Methods for coal liquefaction and gasification.

3.1 Introduction

Fossil fuels originate from ancient organic matter that has been subject to high temperatures and pressures inside the earth over periods from millions to hundreds of millions of years. Depending on the details of the starting material and the formation conditions, the resulting fossil fuel can be solid (coal), liquid (oil), or gas (natural gas). The age of even the most recent fossil fuel deposits is large compared with the timescale on which we are depleting this resource. Thus it cannot be renewed. Unlike fuels such as wood, which can be used in a manner that is carbon neutral by replacing trees as they are used (Chapter 16), fossil fuel use produces a net release of carbon into the atmosphere. The resulting environmental consequences of fossil fuel use are a serious concern, as discussed in the next chapter. In this chapter, the properties of various fossil fuels are introduced, and the methods by which they are extracted from the earth, processed, and converted into other forms are discussed. Both traditional fossil fuel use, which is currently responsible for about 85% of our energy, as well as possible future methods of enhanced recovery, are presented.

3.2 Oil

As seen in Figures 2.3 and 2.4, oil and its derivatives (e.g., gasoline) are the largest single energy source at present. This is due to the facts that there is an enormous infrastructure for extracting these resources from the earth, processing them, and using them, and that, at present, they remain inexpensive compared with most other sources of energy. One of the major uses of oil and its derivatives is as a fuel for transportation because of their high energy density and the convenience of their liquid (or gaseous) form. Like all fossil fuels, oil was formed as a result of the decomposition of organic plant or animal matter during prehistoric times. Typical oil deposits are about 500 million years old. At that time in the earth's history, life existed primarily in saltwater oceans. and oil deposits are therefore located in regions that were once at the bottom of the seas. Organisms collected on the floors of ancient seas, and these were covered with layers of mud and sand. Over the years, the pressure and temperature resulting from the layers of sediment turned the mud and sand into sedimentary rock and converted the organic material into oil and natural gas. Even the very early stages of the formation of the hydrocarbons associated with fossil fuels take thousands of years. However, the complete process by which petroleum is formed takes many millions of years. During the decomposition process, much of the carbon content of the organic material was lost to the atmosphere in the form of carbon dioxide. Only a very small fraction of the carbon in the original organic matter contributes to the formation of oil. In fact, about 20 tonnes of organic matter is needed to produce one liter of oil. Light hydrocarbon molecules are the constituents of natural gas, and heavier hydrocarbon molecules make up oil. It is common to find these together in the same deposit. As the oil and gas form, they can move through the layers of sedimentary rock and eventually collect together in deposits. A typical deposit might consist of oil and gas in a layer of very porous rock that is trapped between two layers of dense impermeable rock. Figure 3.1 shows a schematic of a typical deposit. As a result of the continental drift that occurs on a timescale of many millions of years, some of the petroleum deposits have remained under the oceans while others have ended up underneath land.

The use of oil as a fuel became established in the mid-1800s, although it did not become a major component of our energy production until after 1900. This trend is seen for the United States in Figure 2.1, and the rapid increase in oil use during the past century is largely a result of the development of the automobile. Titusville, Pennsylvania, is often cited as the location of the world's first oil well in 1859

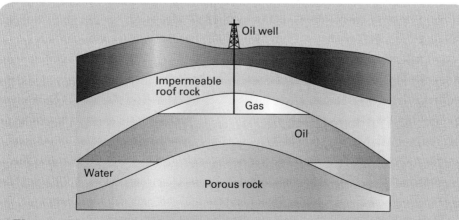

Figure 3.1: Geology of a typical oil deposit where the oil and gas are trapped in porous rock between layers of dense rock. The gas rises to the top, and the oil floats on any water in the deposit.

Example 3.1

Assuming that organic matter has an energy content of about 15 MJ/kg (Chapter 16), estimate the fraction of energy from organic matter that is available in a crude oil deposit.

Solution

From the preceding discussion, 1 L of crude oil is produced from the decomposition of 20 t of organic matter. The original organic matter will have an energy content of

$$(15 \text{ MJ/kg}) \times (1000 \text{ kg/t}) \times (20 \text{ t}) = 3 \times 10^5 \text{ MJ}.$$

The resulting liter of crude oil will have an energy content of 38.5 MJ (from Appendix IV), for a ratio of

$$\frac{38.5 \text{ MJ}}{3 \times 10^5 \text{ MJ}} = 1.3 \times 10^{-4} \text{ (or } 0.13\%).$$

(Figure 3.2). However, it has been claimed that oil production began in Petrolia, Ontario, in 1858 (Figure 3.3). Today several types of commercial oil wells are used for different types of oil deposits. Some examples of land-based oil wells are shown

Figure 3.2: Oil well drilled in Titusville, Pennsylvania, in 1859 by Edwin Drake.

Reproduction, copyrighted in 1890, of a retouched photograph showing Edwin L. Drake, to the right, and the Drake Well in the background, in Titusville, Pennsylvania, where the first commercial well was drilled in 1859 to find oil.

Figure 3.3: Oil well dug in Petrolia, Ontario, in 1858.

Courtesy of Martin Dillon

in Figures 3.4 and 3.5. Facilities for offshore drilling either float or are attached to the seafloor. Drilling platforms for deep water (Figure 3.6) float on the surface and are attached to the seafloor with anchors. Jackup rigs are used in shallower water, typically up to 120 m deep, for drilling or other marine services. Moveable legs are lowered to the seafloor and raise the vessel above the sea surface. Figure 3.7 shows the *Seajacks Kraken,* a jackup rig that is used as a support platform for oil and

Figure 3.4: Small pump jack oil well. This is sometimes called a horsehead pump.

Zeljko Radojko/Shutterstock.com

Figure 3.5: Oil drilling rig in Wyoming, U.S.

Jim Parkin/Shutterstock.com

Figure 3.6: Off-shore oil platform.

Brian McDonald/Shutterstock.com

Figure 3.7: The *Seajacks Kraken* jackup rig in Halifax, Nova Scotia, Canada. The overall length with the Helideck extension (seen to the right in the figure) is about 80 m.

Richard A. Dunlap

natural gas drilling operations. The operation of an oil well (for a pump jack well) is illustrated in Figure 3.8. A downhole pump, located at the bottom of the well in the oil bearing region, is connected to the aboveground pumping mechanism by means of the sucker rod. Oil is drawn into the downhole tube through perforations in the casing and is forced up the tube through the pump valves.

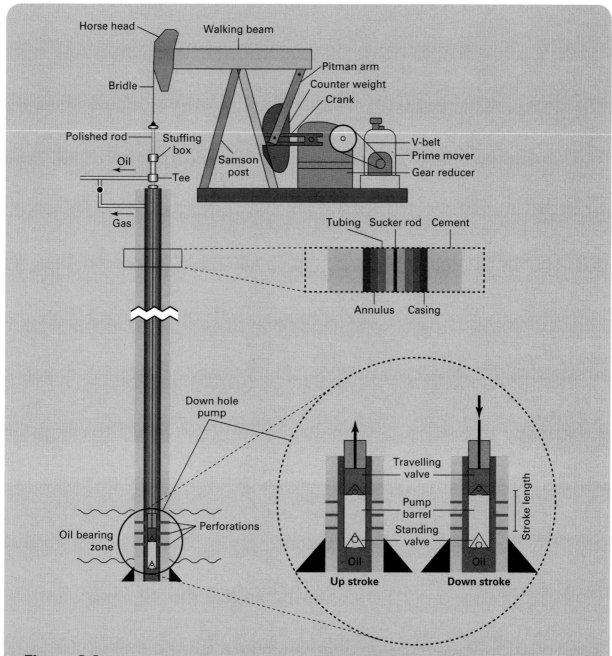

Figure 3.8: Diagram of a pump jack oil well showing the downhole mechanism and its operation.

3.3 Refining

Crude oil, as it is extracted from the earth, is a mixture of a large number of different compounds, mostly hydrocarbons. The hydrocarbons cover a very large range of molecular masses. Table 3.1 lists some common components of crude oil and their properties.

Carbon may bond with hydrogen in many different ways. One important series of hydrocarbons is the alkane series, which has the chemical formula C_nH_{2n+2}. Properties of the alkane series are given in Table 3.2. Some imporant members of this series are methane (the major component of natural gas), ethane (the basis for ethanol, C_2H_5OH), and octane (an important component of gasoline).

As is obvious from Tables 3.1 and 3.2, the larger the number of carbon atoms per molecule, the higher the boiling point of the hydrocarbon will be. This property forms

Table 3.1: Properties of typical hydrocarbons extracted from oil during the refining process.

name	number of carbon atoms per molecule	state at room temperature	boiling temperature (°C)	uses
natural gas	1–5	gas	−165 to 25	gaseous fuel
petroleum ether	5–7	liquid	25 to 90	industrial solvent
gasoline	5–12	liquid	25 to 200	automobile fuel
kerosene	12–16	liquid	175 to 275	stove and jet fuel
fuel oil	15–18	liquid	< 375	diesel and home heating
lubricating oil	16–20	liquid	> 350	lubrication
grease	>17	semisolid	—	lubrication
paraffin	>19	solid	—	candles
tar	large	solid	—	roofing and paving

Table 3.2: Alkane series of hydrocarbons with the formula C_nH_{2n+2}. The heat of combustion is the HHV (Chapter 1).

n	formula	name	boiling temperature (°C)	molecular mass (g/mol)	heat of combustion MJ/kg	heat of combustion eV per molecule
1	CH_4	methane	−164	16	55.5	9.20
2	C_2H_6	ethane	−89	30	51.9	16.1
3	C_3H_8	propane	−42	44	50.3	23.0
4	C_4H_{10}	butane	0	58	49.5	29.8
5	C_5H_{12}	pentane	36	72	48.7	36.3
6	C_6H_{14}	hexane	69	86	48.1	42.9
7	C_7H_{16}	heptane	98	100	48.1	49.9
8	C_8H_{18}	octane	125	114	46.8	55.3

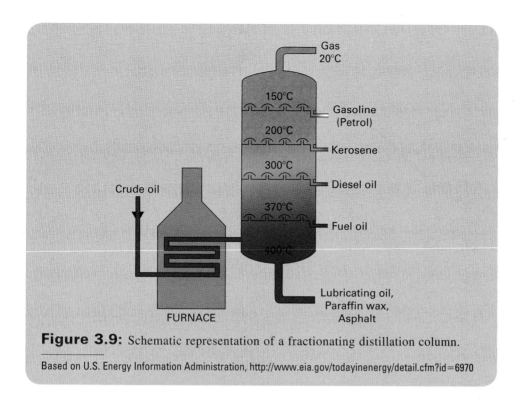

Figure 3.9: Schematic representation of a fractionating distillation column.

Based on U.S. Energy Information Administration, http://www.eia.gov/todayinenergy/detail.cfm?id=6970

the basis of the refining process. Crude oil is heated in a furnace to about 400°C. It then travels up a fractionating column, as seen in Figure 3.9, where its temperature is progressively decreased. At 400°C most of the hydrocarbons are vaporized, and any very heavy hydrocarbons or impurities fall to the bottom of the fractionating column. As the vapor travels up the column, it experiences decreasing temperature, and progressively lighter hydrocarbons condense out and can be extracted. In this way, the crude oil is separated into components of different molecular mass ranges. A photograph of a typical oil refinery showing the fractionating columns appears in Figure 3.10 on the next page.

The proportions of the components produced in the distillation process are determined by the proportions of the different compounds present in the crude oil. Unfortunately, this ratio of these components is generally not compatible with our relative need for them. For example, gasoline and diesel fuel are in much higher demand than, say, paraffin. To make optimal use of our petroleum resources, it is necessary to convert some of the less needed refinery products to more needed forms. In most cases, this involves breaking down heavy hydrocarbons (like tar and paraffin) to make lighter hydrocarbons (like octane) in a process called cracking. Of the several variations on this process, most involve heating the hydrocarbon to a high temperature to break down the chemical bonds. Some common processes are *steam cracking*, where the heavy hydrocarbon is mixed with steam and heated; *hydrocracking*, where the hydrocarbon is exposed to hydrogen gas; and *catalytic cracking*, where the hydrocarbon is exposed to a catalyst like alumina, silica, or a *zeolite*. The proportions of the various by-products of the cracking process depend on the details of the process used and the temperatures involved.

The reverse process, which involves the sticking together light hydrocarbons into heavier ones, is referred to as polymerization. Ethane, propane, and/or butane

Figure 3.10: Fractionating columns at Dartmouth Refinery, Dartmouth, Nova Scotia.

Richard A. Dunlap

may be combined to produce octane or the similar-weight hydrocarbons that comprise gasoline. Light hydrocarbons are heated and react with a catalyst in order to yield the desired reactions. The *Fischer-Tropsch process* (see Section 3.10) is commonly used for this purpose and is referred to as a gas-to-liquid (GTL) process.

Example 3.2

Gasoline is largely made of octane, which has a density of 703 kg/m^3. Estimate the total energy content of a 70-L tank of gasoline (typical of a midsize automobile).

Solution

The 70-L tank holds gasoline with a mass of

$$(70 \text{ L}) \times (0.001 \text{ m}^3/\text{L}) \times (703 \text{ kg/m}^3) = 49.2 \text{ kg}.$$

From the heat of combustion for octane in Table 3.2, this amount of gasoline will have a total chemical energy content of

$$(49.2 \text{ kg}) \times (46.8 \text{ MJ/kg}) = 2.3 \text{ GJ}.$$

Gasoline typically contains about 500 different hydrocarbons that have between 5 and 12 carbon atoms per molecule. Its average density is about 720 kg/m^3. A substantial fraction of these hydrocarbon components of gasoline have molecular masses that are similar to octane. When gasoline is sold, it is commonly labeled with an octane rating. This number does not relate directly to the amount of octane in the fuel. It is rather a measure of the efficiency of the combustion of the gasoline in an internal combustion engine relative to the efficiency of combustion of a standard mixture of octane and other hydrocarbons. It is interesting to note that gasoline sold in, for example, New York might have an octane rating of, say, 89, while if the same gasoline is sold in Denver, it might have an octane rating of 87. The variance results from differences in combustion efficiency that are related to differences in air density (i.e., related to altitude).

Oil and its products must be transported, first from the oil fields to the refineries and subsequently from the refineries to distribution centers. Over land, liquid petroleum products are most commonly transported large distances by pipelines (Figure 3.11). Over sea, liquid petroleum products are transported by ship. There are two basic categories of tankers used for this purpose, crude carriers and product carriers. The first type of vessel carries crude oil from the oil fields to the refinery, and the second category of vessel carries refined liquid petroleum products to distribution centers. The ultra-large crude carriers (ULCC) are some of the largest ships ever constructed (see Figure 3.12) on the next page. These very large vessels were mostly constructed in the 1970s and newer vessels are

Figure 3.11: Section of an oil pipeline in Alaska. Common oil pipeline diameters are 760 mm and 914 mm.

Figure 3.12: The *Knock Nevis* (originally named the *Seawise Giant*) is an ultra-large crude carrier (ULCC) built in 1979 and scrapped in 2010. At a total length of 458 m, this was the longest ship ever constructed. Oil capacity was about 4.2×10^6 bbl or 6.6×10^8 L.

Geof Kirby/Alamy Stock Photo

typically somewhat smaller. Product carriers are typically around 250 m in length. Both oil pipelines and oil tankers pose potential environmental concerns (Energy Extra 4.1).

3.4 Natural Gas

Natural gas is a mixture of light hydrocarbons (typically about 85% methane and 15% ethane) that is gaseous at room temperature (STP). Natural gas may be extracted from deposits in the earth where it is associated with oil, as in Figure 3.1, or from deposits where it occurs on its own. In either case, it is formed in a manner similar to the that of oil from the decomposition of ancient organic matter. The formation of natural gas from organic matter is slightly less efficient than the formation of oil, and only about 0.085% of the carbon in the original organic matter becomes a component of natural gas. Natural gas may also be obtained as a by-product from the distillation of crude oil, as seen in Figure 3.9, where it is extracted as the lightest hydrocarbon component of the oil. The use of natural gas has increased over the years and in the United States is now about on par with the use of coal (Figure 2.1) in terms of net energy production. Natural gas typically contains fewer impurities than heavier hydrocarbons and burns more efficiently, thus creating less pollution. The ideal combustion of methane is given as

$$CH_4 + 2O_2 \rightarrow CO_2 + 2H_2O + 55.5 \text{ MJ/kg.} \tag{3.1}$$

The by-products are nontoxic and have negligible adverse effects other than the role of CO_2 as a greenhouse gas in climate change, as discussed in Chapter 4. Although other possibly undesirable chemicals (e.g., NO_x) can be formed (Chapter 4), the combustion of natural gas is inevitably cleaner than the combustion of gasoline or oil.

Example 3.3

Calculate the mass of CO_2 produced per MJ of energy from the combustion of methane.

Solution

From Table 3.2 and equation (3.1), the combustion of 1 kg of CH_4 produces 55.5 MJ/kg of energy, so 1 MJ is produced by the combustion of 0.018 kg of methane. The molecular mass of methane is 16 g/mol, so that 18 g of methane corresponds to 1.125 moles. Equation (3.1) shows that the combustion of 1 mole of CH_4 produces 1 mole of CO_2. Thus 1.125 moles of CH_4 yields 1.125 moles of CO_2. Because CO_2 has a molecular mass of 44 g/mol, 1.125 moles of CO_2 will have a mass of $(1.125 \text{ mol}) \times (44 \text{ g/mol}) \times (10^{-3} \text{ kg/g}) = 0.049 \text{ kg}$.

Natural gas is most commonly obtained as a by-product of oil wells. In some cases, it occurs primarily on its own and in deposits that are sufficiently well pressurized that pumping equipment is not required. In such cases, the wellhead provides a means of managing the flow of gas from the well in a safe manner and a suitable connection into pipelines to transport the gas. The above-ground portion of such a well is sometimes called a *Christmas Tree* (Figure 3.13).

Figure 3.13: Natural gas wellheads.

rCarner/Shutterstock.com

Figure 3.14: LNG tanker *Arctic Princess*. Built in 2006, this vessel is 288 m in length. The containers for maintaining the pressure and temperature of the natural gas in its liquid state are clearly seen.

Oleksandr Kalinichenko/Shutterstock.com

In its gaseous state, natural gas may be transported easily overland through pipes. Transportation across oceans is somewhat more difficult because, at standard temperature and pressure, natural gas occupies large volumes. Natural gas can be liquefied to produce liquid (or liquefied natural gas, LNG) by lowering the temperature to below about −165°C. In this state, it occupies only about 1/600th of the volume as in its gaseous state and can be transported across oceans in ships designed for this purpose. An example of an LNG transport ship is shown in Figure 3.14. Liquid natural gas should not be confused with *natural gas liquids*, which are heavier hydrocarbons (liquid at room temperature) that are mixed with natural gas from some deposits and which are extracted during the distillation process.

3.5 Coal

Coal is formed, much as oil and natural gas are, over extended periods of elevated pressure and temperature. In the case of coal, however, the organic material originates from terrestrial plant matter. The earliest extensive plant growth on land occurred about 350 million years ago, and these early forests are the origin of the oldest coal deposits. Coal formation is, relatively speaking, an efficient process in which typically about 0.8% of the original carbon in the plant matter ends up as coal.

Coal exists in several varieties depending on age and formation conditions. The oldest and also the hardest coal is anthracite, and this has the highest carbon content.

Table 3.3: Types of coal and their properties. These are typical values; the actual values can be quite variable, and the ranges overlap. Most of the noncarbon content of coal is in the form of volatile compounds, with a significant fraction being water.

type	carbon (%)	moisture content (%)	energy content (MJ/kg)
anthracite	90	10	35
bituminous	55	20	31
sub-bituminous	45	30	23
lignite	25	45	14

Typically, younger coals are softer and have lower carbon content. Examples of the major categories (or ranks) of coal are given in Table 3.3. These are often further divided into subcategories. Bituminous and sub-bituminous coals are the most common (comprising about 71% of coal in the United States), followed by lignite (28%) and anthracite (1%).

Coal may be found at various depths and in deposits of different geometries. It may, as a result of continental movements, exist under the oceans. In this case, it is not an economically viable resource because recovery methods are quite different from those used for a liquid or gaseous resource, like oil or natural gas, that exist beneath the ocean. About half the coal that occurs on land is also not economically recoverable because it is either too deep or occurs in very thin veins.

Example 3.4

Using the data from Figure 2.6 and the fact that coal-fired generating stations in the United States produced 1.22×10^{12} kWh$_e$, determine the average efficiency of generating electricity from coal in 2015.

Solution

According to the figure, 14.3 quads of primary energy from coal were used for electricity generation. This may be converted into joules to give

$$(14.3 \text{ quad}) \times (1.055 \times 10^{18} \text{ J/quad}) = 1.51 \times 10^{19} \text{ J}.$$

The electricity generated from coal-fired generating stations may be converted into joules as

$$(1.22 \times 10^{12} \text{ kWh}_e) \times (3.6 \times 10^6 \text{ J/kWh}_e) = 4.39 \times 10^{18} \text{ J}.$$

The ratio of these two energies gives the average efficiency of coal-fired electricity generation as

$$(4.39 \times 10^{18} \text{ J})/(1.51 \times 10^{19} \text{ J}) = 0.29 \text{ or } 29\%,$$

in reasonable agreement with estimates based on Carnot efficiency.

In the United States, coal was second to wood as a source of energy up until about 1880, when it became the most important energy resource. In the 1940s, the use of oil

Table 3.4: Current use of coal in the United States.	
use	**% of total**
electricity generation	70
coke production	17
export	10
other (residential heating, industrial processes, etc.)	3

(including gasoline) surpassed the use of coal. At present, coal remains about even with natural gas as the second most common source of energy in the United States. In North America, most of the coal mined is used to generate electricity. A rough breakdown of coal use in recent years in the United States is given in Table 3.4.

It can be seen from the table that the second most common use of coal in the United States is the production of coke. Coke is made by heating coal to about 1200°C in the absence of air. Volatile materials are driven off, and more or less pure carbon is left behind. This carbon, or coke, is used primarily in the smelting of steel. Iron ores, which are comprised primarily of iron oxides, are heated to high temperature with the coke. By the reaction

$$\text{Fe}_a\text{O}_b + b\text{C} \rightarrow a\text{Fe} + b\text{CO}, \tag{3.2}$$

the iron oxide is reduced, releasing carbon monoxide and leaving pure iron behind.

Example 3.5

Using the information in Figure 2.6 for 2009 and the information in Table 3.4, estimate the total number of tonnes of coal produced annually in the United States (assume the coal is all bituminous).

Solution

From Figure 2.6, the contribution of coal to the annual electricity generation in the United States is about 18.3 Q $\times$ (1.055 Q/EJ) = 19.3 EJ. Note that this is the primary energy content of the coal burned, not the electric output. Using the energy content of bituminous coal given in Appendix V, 31 MJ/kg = 31×10^9 MJ/t. The coal contribution to electricity generation requires

$$\frac{19.3 \times 10^{18} \text{ MJ}}{31 \times 10^9 \text{ MJ/t}} = 6.21 \times 10^8 \text{ t.}$$

Since this corresponds to 70% of the U.S. coal production (Table 3.4), the total coal production is

$$\frac{6.2 \times 10^8 \text{ t}}{0.7} = 8.9 \times 10^8 \text{ t.}$$

This is in reasonable agreement with the actual value of 1.06×10^9 t, as given in Section 3.6.

As a solid, coal is readily transported over land by train or over sea by ships. A typical coal train might consist of about 100 hopper cars, each with a capacity of about 100 t of coal, for a total capacity of about 10^4 t. This is approximately the amount of coal that is burned in one day by a 1.5 GW$_e$ coal-fired generating station.

3.6 Overview of Fossil Fuel Resources

It is fairly straightforward to determine how much of each of these fossil fuel resources is being used at present. It is also fairly easy to determine how much has been used in the past. It is not so easy to determine how much more is available in the earth for future use. It is even more difficult (assuming we knew how much is remaining) to estimate how long these remaining resources will last. The past and present use for each conventional fossil fuel resource is now discussed, along with estimates of the remaining resources.

3.6a Oil

Oil is a primary energy source, and the energy available per year from the world oil production of 4.46×10^9 Mt or about 3.4×10^{10} bbl (bbl=barrels) is 210 EJ. Table 3.5 summarizes oil production for different countries and regions of the world. North America uses much more oil than it produces. This is a direct result of the fact that the United States imports a significant fraction of its oil. Canada, by comparison, produces very nearly the

Table 3.5: Oil production (in Mt per year) for 2015 (or 2014, as noted*) of major oil-producing countries (> 10 Mt/y). Oil production is often measured in Mt (10^6 t). It is a more consistent means of comparison than volume (in bbl or L), because the density of oil from different regions varies. [1 t in North America corresponds to about 7.15 bbl or 1137 L].

country	Mt per year	country	Mt per year
Saudi Arabia	568.5	Oman	46.6
Russia	540.7	United Kingdom	45.3
United States	519.9*	Azerbaijan	41.7
Canada	215.5	India	41.2
China	214.6	Indonesia	40
Iraq	197	Egypt	35.6
United Arab Emirates	175.5	Malaysia	31.9
Iran	169.2*	Australia	31.6
Venezuela	157.8*	Argentina	29.7
Kuwait	149.1	Ecuador	29.1
Brazil	131.8	Libya	20.2
Mexico	127.6	Thailand	17.2
Nigeria	113	Vietnam	15.5*
Angola	88.7	Congo	14.5*
Norway	88	Equatorial Guinea	13.5
Kazakhstan	79.3	Turkmenistan	12.7
Qatar	79.3	Gabon	11.6
Algeria	68.5	Bahrain	10.1
Colombia	53.1		

Based on data from World Energy Council. World Energy Resources-2016, available online at https://www.worldenergy.org/publications/2016/world-energy-resources-2016/

same amount of oil as it uses. Other regions, such as Africa and the Middle East, export most of their oil production. Some insight into the reasons for this behavior can be gained by comparing the details of U.S. oil production with, for example, oil production in Saudi Arabia. In terms of total production, the United States produces somewhat less oil per year than Saudi Arabia (520 Mt vs. 568.5 Mt). However, there are about 600,000 oil wells in the United States, compared with about 1000 oil wells in Saudi Arabia. Thus U.S. oil wells (on average) produce much less oil than Saudi Arabian oil wells.

Example 3.6

Estimate the average rate of oil production (in L/min) from a typical oil well in the United States and a typical oil well in Saudi Arabia.

Solution
In the United States the average production per well per day will be

$$\frac{(519.9 \times 10^6 \text{ t/year}) \times (1137 \text{ L/t})}{600,000 \text{ wells} \times 365 \text{ days/year}} = 2700 \text{ L/day per well}$$

or

$$\frac{2700 \text{ L/day}}{1440 \text{ min/day}} = 1.9 \text{ L/min,}$$

equivalent to a fairly slow-running water faucet.
    In Saudi Arabia, the production is

$$\frac{(568.5 \times 10^6 \text{ t/year}) \times (1137 \text{ L/t})}{1000 \text{ wells} \times 365 \text{ days/year}} = 1.77 \times 10^6 \text{ L/day per well}$$

or

$$\frac{1.77 \times 10^6 \text{ L/day}}{1440 \text{ min/day}} = 1230 \text{ L/min.}$$

The differences illustrated in Example 3.6 between U.S. and Saudi Arabian oil production are related to oil well design; a typical continental U.S. oil well is shown in Figure 3.4, and Middle Eastern oil wells are similar to the design in Figure 3.5. These design differences result from the fact that oil has been utilized much longer in the United States than in the Middle East, and the oil wells are very mature. Basically, this means that oil reserves in the United States are closer to being depleted.

Trends in world oil production can be viewed quantitatively in the context of the Hubbert model. This analysis requires the knowledge of the total amount of oil available. How much oil has been used is obvious, so it is necessary to estimate how much is remaining. This is not necessarily easy to determine, and the amount depends to a large extent on what is included in the estimate. Sources sometimes give amounts which are described as known reserves, known recoverable reserves, estimated reserves, estimated economically recoverable reserves, and so on. What we want to know is how much is available that we could make use of practically.

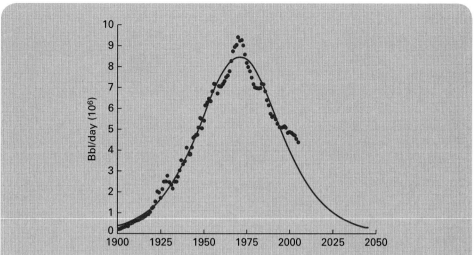

Figure 3.15: Average daily oil production in the continental United States from 1900 to 2010 and the predicted utilization based on the Hubbert model using a total oil resource of 2.1×10^{11} bbl.

Based on The oil Drum, In Defense of the Hubbert Linearization Method, Posted by Sam Foucher on June 24, 2007, http://www.theoildrum.com/node/2689

The average daily oil production in the United States as a function of year since 1900 is shown in Figure 3.15. The figure shows that the Hubbert model is reasonably accurate in predicting U.S. oil use during this period. These data suggest that about 70% of U.S. oil reserves have been consumed up to 2010 and that domestic oil production would be minimal past about 2040.

However, if the data in Figure 3.15 are extended to more recent years, as shown in Figure 3.16, it becomes evident that the oil production in the United States has

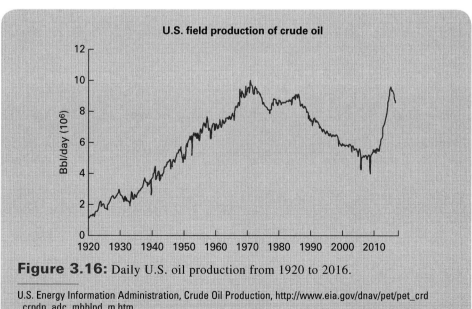

Figure 3.16: Daily U.S. oil production from 1920 to 2016.

U.S. Energy Information Administration, Crude Oil Production, http://www.eia.gov/dnav/pet/pet_crd _crpdn_adc_mbblpd_m.htm

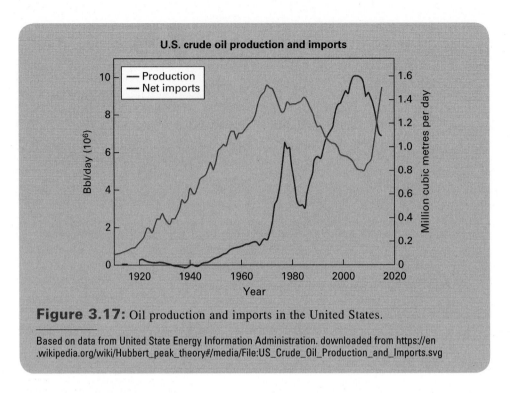

Figure 3.17: Oil production and imports in the United States.

Based on data from United State Energy Information Administration. downloaded from https://en
.wikipedia.org/wiki/Hubbert_peak_theory#/media/File:US_Crude_Oil_Production_and_Imports.svg

increased substantially over what would be expected on the basis of the Hubbert model.
This increased U.S. oil production is also evidenced by the changes in the proportions
of domestic and imported oil (Figure 3.17).

The reasons for the deviations from the Hubbert model (Figure 3.16) can be
understood on the basis of a careful analysis of U.S. production. Figure 3.18 shows

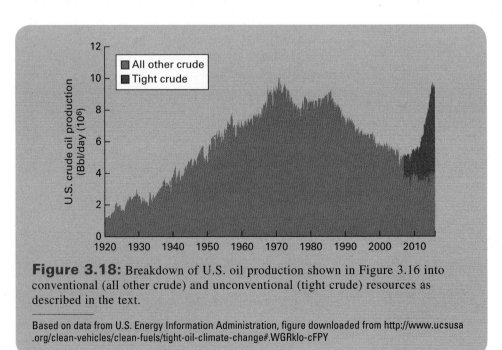

Figure 3.18: Breakdown of U.S. oil production shown in Figure 3.16 into
conventional (all other crude) and unconventional (tight crude) resources as
described in the text.

Based on data from U.S. Energy Information Administration, figure downloaded from http://www.ucsusa
.org/clean-vehicles/clean-fuels/tight-oil-climate-change#.WGRklo-cFPY

a breakdown of the data in Figure 3.16 into conventional and unconventional oil production. In the figure, unconventional oil resources are referred to as *tight oil* (sometimes called light tight oil or LTO). These resources include oil obtained from fracking and other advanced technologies (Energy Extra 3.1 and Section 3.7) and shale oil (Section 3.8), although the term tight oil does not have a precise scientific definition and is not always used consistently. The figure shows, however, that the data for conventional oil resources much more closely follows the shape of the Hubbert curve and that the excess oil production in the United States in recent years is the result of changes in oil recovery technology that have effectively increased the total amount of resource available.

An analysis of world oil production on the basis of the Hubbert model is illustrated in Figure 3.19. Here only conventional oil production is shown; as a result, the total resource remains a constant, and the data show reasonable agreement with the Hubbert model. The data indicate that at present about 40% of the worldwide conventional oil resources have been utilized and that oil production will be minimal sometime after 2100. It can be seen, however, that there are anomalies in the data, particularly around the 1970s. These anomalies can be understood by looking at a semilog plot of these data (Figure 3.20) on the next page. The data show two distinct regions of exponential growth. The decrease in growth that occurred in the 1970s was a result of growing concerns over limited fossil fuel resources and their environmental effects. It was at this time that the first serious energy conservation efforts began (Section 17.8).

The discussion above concerning the utilization of oil resources in the United States and worldwide emphasizes several difficulties concerning the extrapolation of data on the basis of the Hubbert model, in particular the estimation of the longevity of resources. The actual situation is quite complex. On the one hand, the Hubbert model may underestimate the lifetime of a particular resource because of the discovery of additional previously unknown resources or the increase in available resources resulting from the development of new technologies that make additional resources economically

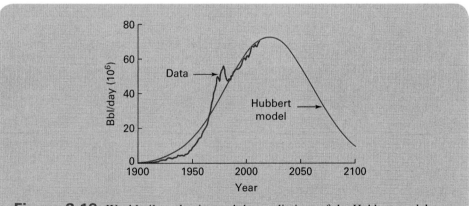

Figure 3.19: World oil production and the predictions of the Hubbert model based on a total resource of 2.5×10^{12} bbl. Data include only conventional oil resources.

Based on data from U.S. Energy Information Administration

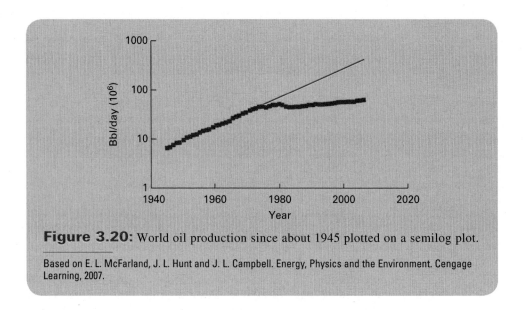

Figure 3.20: World oil production since about 1945 plotted on a semilog plot.

Based on E. L. McFarland, J. L. Hunt and J. L. Campbell. Energy, Physics and the Environment. Cengage Learning, 2007.

viable. In these cases, as Figure 3.16 suggests, the resource utilization curve may be double peaked. Resource availability may also be extended as a result of decreased use due to conservation efforts. On the other hand, the Hubbert model may overestimate the lifetime as a result of increased use due to increasing population and increasing industrialization in developing countries (see Figure 2.18). While graphs such as those in Figures 3.15 and 3.18 provide insight into the longevity of fossil fuel resources, it is important to realize that the actual future use of these resources depends on a number of

Example 3.7

Estimate the annual growth rate in world oil production prior to 1980 and since 1980.

Solution

From the graph in Figure 3.20, we can see that changes in oil production are characterized by two distinct regions, before and after about 1980. Each region is well described by a straight line on a semilog plot, meaning that there are two regions of exponential growth with different growth rates.

Extrapolating the straight line behavior in the two regions between 1940 and 2020, we find that before about 1980 the line has a slope of 0.0328×10^6 bbl/(day·year) and after about 1980 the line has a slope of 0.0063×10^6 bbl/(day·year).

From Chapter 2 it was shown that exponential growth is described in a semilog plot with a slope of a and a growth rate in percent per year is given by

$$R = 100 \cdot [\exp(a) - 1].$$

Substituting the values for a from above gives

$$R = 7.56 \text{ % per year before 1980.}$$

$$R = 1.42 \text{ % per year after 1980.}$$

factors. Ultimately, it may be the desire to move to other alternative energy resources, rather than fossil fuel resources themselves, that determines our future energy use.

3.6b Natural Gas

Natural gas has become attractive in recent years due to the smaller amount of pollution it produces compared to that from the combustion of oil or coal. The advantage of natural gas (methane) by comparison with coal (mostly carbon) in terms of CO_2 emission is clear from a comparison of the energy produced relative to carbon emission shown in equations (1.13) and (1.15). Table 3.6 shows the annual production of the major natural gas-producing nations. The table also gives the estimated proven reserves (as of 2015) and the longevity of the resource at a constant production rate. The total world natural gas production in 2015 was 3.54×10^{12} m^3 and the total world proven reserves are 187×10^{12} m^3. The ratio of reserves to annual production worldwide is about 53 years. This gives a rough estimate of the lifetime of natural gas resources.

The use of natural gas may be considered on the basis of the Hubbert model in Figure 3.21, which shows the natural gas production in the continental United States. Up until the late 1980s the data agree well with a single-peaked Hubbert utilization model. However, increased production since that time has resulted from improved extraction technologies. The estimated lifetime of natural gas reserves in the United States in Table 3.6, about 14 years, suggests a decrease in production within the next few years resulting in a double-peaked curve in Figure 3.21, analogous to expectations for oil production shown in Figure 3.17.

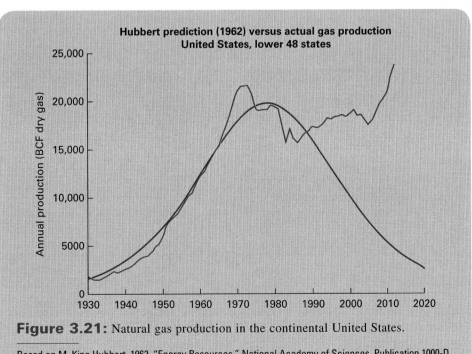

Figure 3.21: Natural gas production in the continental United States.

Based on M. King Hubbert, 1962, "Energy Resources," National Academy of Sciences, Publication 1000-D, p. 81–83

Table 3.6: Annual production for 2015 (or 2014*) of major natural gas-producing countries ($> 10^{10}$ m^3 per year).

country	annual production (10^9 m^3)	proven reserves (10^9 m^3)	reserves/production (years)
United States	767.3	10,440.5	13.6
Russia	573.3	32,271.0	56.3
Iran	192.5	34,020.0	176.8
Qatar	181.4	24,528.1	135.2
Canada	163.5	1987.1	12.2
China	138.0	3841.3	27.8
Norway	117.2	N/A	N/A
Saudi Arabia	106.4	8325.2	78.2
Indonesia	75.0	2839.0	37.8
Turkmenistan	72.4	17,479.0	241.4
Malaysia*	68.2	1169.3	17.1
Australia	67.1	3471.4	51.8
Uzbekistan	57.7	1085.9	18.8
United Arab Emirates	55.8	6091.0	109.2
Mexico	53.2	324.2	6.1
Nigeria	50.1	5111.0	102.1
Egypt	45.6	1846.3	40.5
Netherlands	43.0	674.7	15.7
Pakistan	41.9	542.6	12.9
Thailand	39.8	219.5	5.5
United Kingdom	39.7	206.0	5.2
Trinidad and Tobago	39.6	325.7	8.2
Argentina	36.5	332.2	9.1
Oman	34.9	688.1	19.7
Venezuela	32.4	5617.2	173.2
India	29.2	1488.5	50.9
Bangladesh	26.8	232.2	8.7
Brazil	22.9	423.5	18.5
Bolivia	20.9	281.0	13.5
Myanmar	19.6	528	16.9
Azerbaijan	18.2	1148.3	63.2
Ukraine	17.4	604.1	34.7
Bahrain	15.5	172.1	11.1
Kuwait	15.0	1784.0	119.1
Libya	12.8	1504.9	118.0
Brunei	12.7	276.0	21.7
Peru	12.5	414.2	33.1
Kazakhstan	12.4	936.0	75.7
Colombia	11.0	134.7	12.2
Vietnam	10.7	617.1	57.9
Romania	10.3	110.0	10.7

Based on World Energy Council. World Energy Resources-2016, available online at https://www.worldenergy.org/publications/2016/world-energy-resources-2016/.

3.6c **Coal**

Total world coal production is about 7.7×10^9 t per year, corresponding to about a quarter of world primary energy. The annual production of major coal producing countries is summarized in Table 3.7. Because the expected lifetime for coal is much longer than for oil and natural gas, the extrapolation of coal use into the future is much less certain. Figure 3.22 shows the world coal production from around 1900 until 2014. The application of the Hubbert model to the use of coal (Figure 3.22) shows that about

Table 3.7: Major coal-producing countries (> 100 Mt per year) in 2015. This amounts to about 93% of world coal production of 7700 Mt in 2015.	
country	**coal (Mt)**
China	3747
United States	813
India	677
Australia	485
Indonesia	392
Russia	373
South Africa	252
Germany	184
Poland	136
Kazakhstan	106

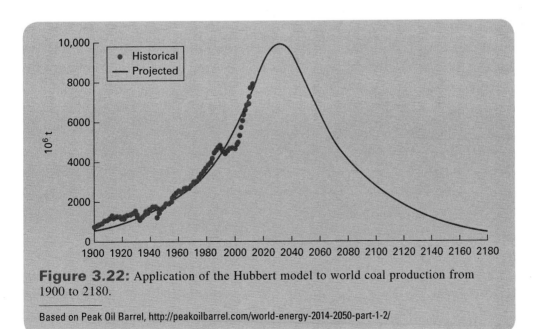

Figure 3.22: Application of the Hubbert model to world coal production from 1900 to 2180.

Based on Peak Oil Barrel, http://peakoilbarrel.com/world-energy-2014-2050-part-1-2/

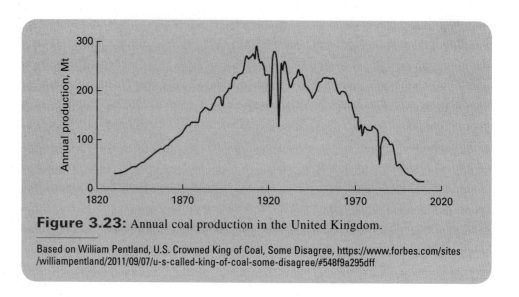

Figure 3.23: Annual coal production in the United Kingdom.

Based on William Pentland, U.S. Crowned King of Coal, Some Disagree, https://www.forbes.com/sites/williampentland/2011/09/07/u-s-called-king-of-coal-some-disagree/#548f9a295dff

a third of coal resources have been used and that the use of coal will peak sometime in the early 2030s. Extrapolation of the Hubbert curve is less certain than it is for oil or natural gas, as these are much more mature resources, but suggests that coal resources will be minimal by the mid-2100s.

An interesting example of the Hubbert model is illustrated by the production of coal in the United Kingdom (Figure 3.23). Here coal production began earlier than in other locations, and the resource was of limited extent. The figure shows a curve characteristic of the shape of the Hubbert curve and indicates that coal production, as predicted by the model, is very close to ending.

Example 3.8

Show that the statement in the chapter that annual coal production accounts for about one quarter of world energy use is valid.

Solution

From Table 3.7 the total annual world coal production is about 7.7×10^{12} kg. The exact primary energy content of this coal will depend on the relative proportions of the different grades. Using the following values for the United States from the text of 71% bituminous and sub-bituminous (at an average energy content of 24 MJ/kg), 28% lignite (at an energy content of 14 MJ/kg), and 1% anthracite (at energy content of 35 MJ/kg) gives an average energy content of about 23 MJ/kg. Thus the total energy content of the annual world coal production will be

$$(7.7 \times 10^{12} \text{ kg}) \times (23 \times 10^6 \text{ J/kg}) = 1.77 \times 10^{20} \text{ J}.$$

From the Preface, the total world energy consumption is about 6.6×10^{20} J per year. Thus, coal accounts for

$$100 \times (1.77 \times 10^{20} \text{ J})/(6.6 \times 10^{20} \text{ J}) = 27\% \text{ of primary energy use.}$$

One method of increasing the productivity of a natural gas or an oil well is referred to as *fracking*, which is short for "hydraulic fracturing." By this method, introduced in the oil and gas industry in the late 1940s, rock around a well is fractured to allow natural gas or oil to flow into the well. The rock is fractured by injecting a fluid at high pressure into the well. Wells often extend vertically downward from the surface, and, once they enter a layer where natural gas or oil resources are present, they turn horizontally to access as much of the deposit as possible. In this horizontal region of the well, fracking is an effective way to increase well productivity. Once the fracture has been created, a *proppant* is injected into the well. The proppant is a material, such as sand, that fills the crack but is porous. This material allows oil or gas to flow into the well but supports the

fracture to prevent it from closing under the pressure of the rocks above. The fracking process is illustrated in the diagram in this box. The fracking technique is in common use in the natural gas industry, and it has been estimated that as many as 90% of the natural gas wells in the United States employ this method.

The liquid most commonly used for fracking is water, but it includes a number of additives to alter its properties. About 750 different chemical additives have been used in the industry, many of which are toxic or carcinogenic. A partial list of additive classes is shown in the table. The proppant itself is commonly sand, which is chosen for particular properties such as grain size and composition to optimize its effectiveness. It may also include a radioactive tracer to allow for observation of its distribution.

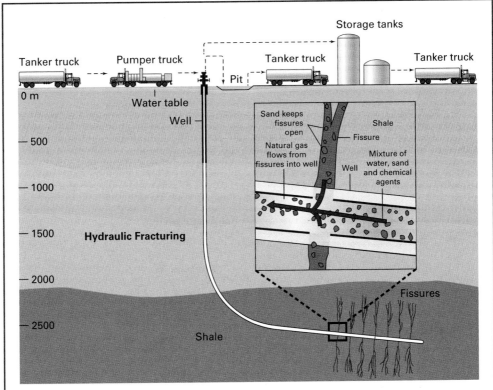

The fracking process

Based on http://www.acus.org/content/fracking-diagram, http://schema-root.org/technology/wells/fracking, http://www.biggerpieforum.org/How-does-fracking-work, http://www.greenphillyblog.com/green-news-politics/fracking-developments-how-you-can-prevent-philadelphias-water-supply-from-contamination/, http://deargreenplanet.blogspot.ca/2012/09/fracking-should-be-shaleved-1.html; http://www.azomining.com/Article.aspx?ArticleID=18

Continued on page 98

Energy Extra 3.1 continued

A partial list of fracking fluid additives		
additive	purpose	example material
acid	improves entry of fluid into rock	hydrocloric acid
surfactant	reduces fluid surface tension	propanol
corrosion inhibitor	reduces corrosion of well equipment	methanol
biocide	reduces bacterial growth	glutaraldehyde
scale inhibitor	inhibits precipitation of minerals from the fluid	ethylene glycol

adapted from information at http://en.wikipedia.org/wiki/Hydrolic_fracturing

Because of these materials, there are health and safety concerns over fracking. The fracking fluid may be returned to surface, or it may be left underground. If it is returned to the surface, it must be dealt with properly because of potential health effects. If it is left underground, it could spread and enter the drinking water supply. Also, methane (natural gas) can propagate through fractures and enter the water supply. Buildings have exploded as a result of water that contained dissolved methane (possibly from a fracked well).

A 2004 study by the U.S. Environmental Protection Agency (EPA) concluded that there was no evidence of health risks due to fracking. As a result, the 2005 Energy Policy Act states that oil companies do not have to make known their (proprietary) formulas for fracking fluids. However, this policy causes problems if toxic chemicals are detected in drinking water near drilling sites because it is difficult to know whether the toxins come from the well or from another source. Although proposals have been recently made in the U.S. Congress, no real regulatory guidelines are in place for fracking. However, due to health concerns, New York has banned fracking. Fracking is used in some (but not all) provinces in Canada, and the public concern is that its use may become more widespread in that country.

Although fracking is an effective way of increasing the productivity of wells, it remains very controversial because the health risks are unclear. Improved regulation and environmental studies are needed to alleviate public concerns over this practice.

Richard A. Dunlap

Public expression of objections to fracking.

Topic for Discussion

The long-term availability of adequate freshwater supplies in the world is an important concern. Fracking uses large quantities of water; in fact, fracking a single natural gas well can utilize 20 million liters of water (or even more). To put this in perspective, make a rough estimate of the length of time that this amount of water could provide the freshwater supply for a typical single-family residence.

3.7 Enhanced Oil Recovery

It is clear from the discussion in the previous chapter that fossil fuel resources are limited and that oil and natural gas are more limited than coal. In this section, several topics related to alternative approaches to fossil fuel use are discussed: (1) methods for extracting a larger fraction of oil from a reservoir, (2) other sources of oil, and (3) methods of converting coal into liquid or gaseous hydrocarbons so that they may be used more conveniently for transportation needs.

Simple, or primary, oil extraction involves allowing the oil to flow from the reservoir under its own natural pressure. Typically only 15–20% of the oil in a deposit can be obtained by this method because there is insufficient pressure to force the remaining oil out of the ground.

Secondary recovery makes use of a second well drilled near the primary well into which water or gas is injected to pressurize the deposit and to force the oil out of the primary well. This method can typically recover an additional 25% or so of the oil in the reservoir and is in common use for wells in the United States. The fraction of the oil in the reservoir that can be extracted depends on the rate at which the oil is extracted. Optimizing the fraction of oil that can be recovered is not generally compatible with an extraction rate that best meets supply needs. This fraction also depends on the nature of the deposit and the nature of the oil. Although all crude oil is fairly viscous, deposits containing lighter varieties of oil allow for more efficient removal. The main difficulty in extracting all of the oil from a deposit is its viscosity and the surface tension between the oil and the porous rock where it resides.

Tertiary recovery involves methods that can effectively deal with these problems. Surfactants, which are chemicals that allow oil and water to mix (like detergents), can be injected into the well along with water, and this enables the oil to get "unstuck" from the pores in the rock. Another technique is to inject steam into a secondary well. This pressurizes the oil, forces it toward the primary well, and heats it, thereby reducing its viscosity and allowing it to flow more easily. Fireflooding is a technique where some of the oil in the deposit is actually burned *in situ*. This heats the remaining oil and forces it out of the well. This approach will be discussed further in the next section. These tertiary techniques tend to be expensive, either financially (as in the case of surfactants) or in terms of energy (as in the case of steam injection). In the latter case, the energy cost is about one-third of the energy value of the oil extracted. The future viability of these techniques is difficult to assess.

3.8 Oil Shale

Oil shale is a sedimentary rock that formed from relatively recent (50 million years ago) hydrocarbon deposits. The hydrocarbons originated from decaying aquatic life in ancient lakes and are in the form of *kerogens*. These hydrocarbons have chain-like molecules that are more complex and intertwined than those found in petroleum. They also tend to contain more impurities, such as sulfur. Heating the kerogens breaks down the chains and yields petroleum-like molecules. It was recognized in the nineteenth century that oil could be extracted from oil shale, and it is now used either by burning the shale directly to thermally generate heat and electricity or to process the shale to produce liquid hydrocarbons for fuel use. Oil shale occurs in a number of locations worldwide and is most commonly mined (similarly to coal) by surface mining, open

Table 3.8: Estimated major shale oil resources by country.	
country	**resources (10^9 bbl)**
United States	6000
China	330
Russia	270
Israel	250
Jordan	100
Democratic Republic of the Congo	100

Based on data from World Energy Council. World Energy Resources-2016, available online at https://www.worldenergy.org/publications/2016/world-energy-resources-2016/

pit mining, or strip mining after which the oil shale may be burned in thermal generating plants or processed to extract the shale oil. Table 3.8 gives recent estimates for major shale oil resources. These numbers may be compared to the total estimated world recoverable oil resources, 2.5×10^{12} bbl. Clearly oil shale is an important resource.

Figure 3.24 shows oil shale production for countries that have significantly utilized this resource. The amount of oil that can be extracted from oil shale depends on the specific deposit, and good-quality oil shale deposit can produce upward of 0.6 bbl (100 L) per tonne of shale.

Example 3.9

Compare the energy content of 1 t of good-quality oil shale with the energy content of 1 t of bituminous coal.

Solution

From Appendix IV, the energy content of 1 L of crude oil is 38.5 MJ. Therefore, 1 t of good-quality oil shale which contains 100 L/t will yield

$$(38.5 \text{ MJ/L}) \times (100 \text{ L}) = 3.85 \times 10^3 \text{ MJ}.$$

One tonne of bituminous coal has energy content as given in Appendix V of 3.1×10^4 MJ. Thus, good-quality oil shale has only about 12% of the energy content of coal.

Table 3.8 shows that the United States has the vast majority of oil shale resources worldwide and commercial interest in developing these resources has grown in recent years. As Figure 3.24 illustrates, more than half of the global oil shale production is from Estonia. In fact, Estonia obtains about 90% of its energy from oil shale and is the only country that derives a major component of its energy from this source. The experiences in this country serve as a good case study for the possible implications of potential larger scale shale oil production elsewhere.

To assess the possibilities for shale oil use as a liquid fuel, it is important to consider the methods by which oil is extracted from oil shale. The shale is heated to about 500°C to extract the kerogens and break down the hydrocarbon chains. The resulting mix of hydrocarbons can be refined in much the same way that crude oil is refined. This

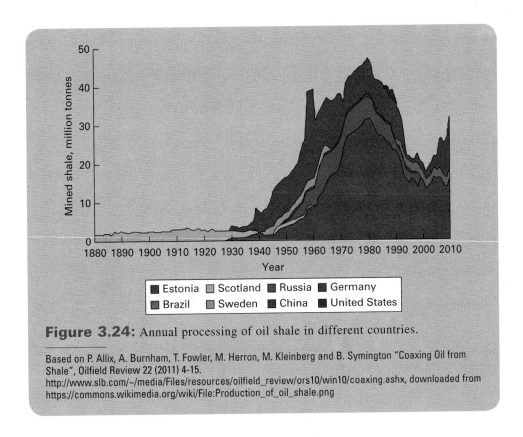

Figure 3.24: Annual processing of oil shale in different countries.

Based on P. Allix, A. Burnham, T. Fowler, M. Herron, M. Kleinberg and B. Symington "Coaxing Oil from Shale", Oilfield Review 22 (2011) 4-15.
http://www.slb.com/~/media/Files/resources/oilfield_review/ors10/win10/coaxing.ashx, downloaded from
https://commons.wikimedia.org/wiki/File:Production_of_oil_shale.png

process requires considerable quantities of water. In fact, using current technology, the production of 1 L of shale oil requires about 3 L of water. As well, the production of any useful quantities of oil requires the processing of considerable amounts of shale. It is also interesting to note that during the process of crushing the shale and extracting the oil, the volume of the shale actually increases by about 35%. The requirement for large quantities of water and the need to dispose of significant waste material are factors that affect the viability of shale oil utilization. An evaluation of the desirability of shale oil production must include (at least) the following points:

- The availability of an adequate water supply.
- The ability to effectively dispose of spent shale.
- The possibility of adverse environmental consequences of mining the shale.
- The cost, both financial and in terms of the energy used to produce shale oil.

In the United States, most of the oil shale deposits occur near the boundaries between Utah, Wyoming, and Colorado. These locations are illustrated on the map in Figure 3.25 on the next page. This region has relatively little rainfall, although there is an important farming industry. Water is provided for agricultural activities by the White River and the Green River. The quantity of water needed for any significant shale oil production could have severe implications for agriculture, and it is neither practical nor economical to transport the quantities of water that are needed from other locations.

The problem of spent shale disposal and restoration of the land after mining must also be considered. Because oil shale deposits are typically near the surface, the environmental consequences can be significant. These consequences can include the exposure of previously buried materials that may contaminate the water supply,

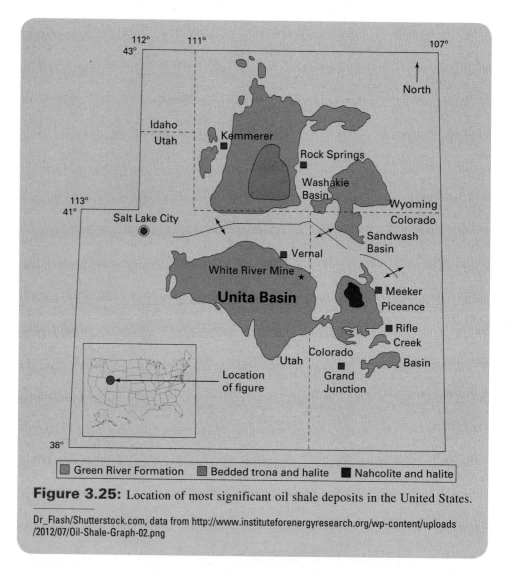

Figure 3.25: Location of most significant oil shale deposits in the United States.

Dr_Flash/Shutterstock.com, data from http://www.instituteforenergyresearch.org/wp-content/uploads/2012/07/Oil-Shale-Graph-02.png

increased erosion, and the distribution of particulates in the air during processing. Mining and the extraction of oil from shale produces potentially harmful emissions, including carbon dioxide. In this respect, it is perhaps significant that in 2002 it was estimated that 97% of the air pollution in Estonia originated from the power industry. By comparison, in 1998, electricity generation in Canada resulted in 20% of the country's SO_2 emissions and 10% of the NO_x emissions. Overall total greenhouse gas emissions from oil shale mining and processing are greater than they are for an equivalent amount of traditional fossil fuel. Because current U.S. government policies prohibit government agencies from purchasing oil that is produced by methods that release more greenhouse gases than conventional petroleum, shale oil initiatives must deal with this difficulty.

From a purely financial standpoint, the viability of shale oil production may be questionable. A study by the RAND Corporation concluded that, as of 2005, the cost of producing shale oil was in the range of $70–95 per bbl. At that time, the world market price for crude oil was about $40 per bbl. Fluctuations in oil prices (often increases), along with technological improvements and long-term amortization of infrastructure costs, may at some point make shale oil economically competitive.

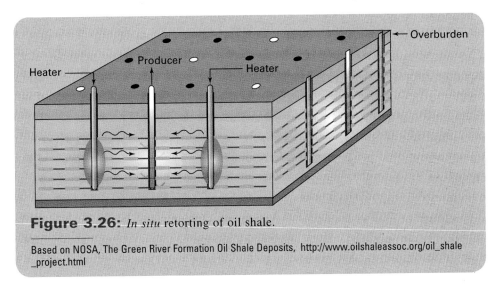

Figure 3.26: *In situ* retorting of oil shale.

Based on NOSA, The Green River Formation Oil Shale Deposits, http://www.oilshaleassoc.org/oil_shale
_project.html

An alternative method, *in situ retorting,* of extracting oil from oil shale that is deep within the earth has been considered. This process is similar to fireflooding. In this case, as illustrated in Figure 3.26, a well is drilled into the oil shale deposit, and a second well is drilled nearby. Air and natural gas are injected into the second well and burned. The heat propagates outward, heating the oil shale and driving the hydrocarbons out through the first well. Although this is a less efficient method of extracting the oil from the oil shale, it avoids the problems and costs associated with mining and transporting large quantities of rock. It is unclear if this approach will provide a suitable means of utilizing oil from this resource.

The shale oil industry in the United States has sometimes been compared with the early stages of the tar sands project in Canada. Although the extraction of oil from tar sands (described in the next section) has met with some degree of economic and environmental success, the long-term usefulness of shale oil remains to be seen.

3.9 Extra-Heavy Oil and Tar Sands

Oil that is extracted from oil wells using conventional recovery methods may vary in its viscosity, ranging from light to fairly heavy. Extra-heavy oil also exists and is more difficult to extract from the earth. This extra-heavy oil is sometimes categorized in two ways: extra-heavy oil and natural bitumen (or tar sands), the latter deposits being a mixture of extra-heavy oil and sand. In both cases, the possibility of using these resources lies in our ability to extract the oil in an economical and environmentally conscious way.

The estimated world resources in extra-heavy oil and natural bitumen are summarized in Tables 3.9 and 3.10. These are each close to the estimated total oil available from

Table 3.9: Estimated total extra-heavy oil resources.	
location	10⁹ bbl
Venezuela	2112
United Kingdom	12
rest of world	26
total	2150

Based on data from World Energy Council 2010 Survey of Energy Resources.

Table 3.10: Estimated total oil resources from natural bitumen (tar sands).	
location	10⁹ bbl
Canada	2434
Kazakhstan	421
Russia	347
United States	53
rest of world	74
total	2256

Based on data from World Energy Council 2010 Survey of Energy Resources.

traditional sources, although it may be uncertain how much of this oil is actually recoverable. It is clear from the tables that virtually all extra-heavy oil exists in Venezuela, whereas about two-thirds of the oil available from tar sands exists in Canada.

Extra-heavy oil deposits are located in the Orinoco Oil Belt in North Eastern Venezuela, as illustrated in Figure 3.27. Commercial production of extra-heavy oil from this region began in 2001. At the beginning of 2016, Venezuela's oil production

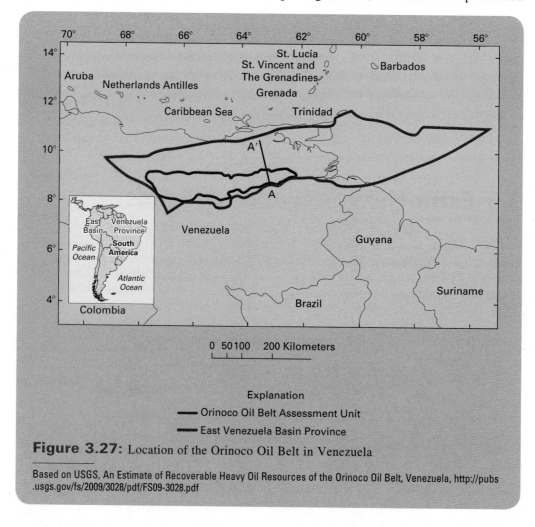

Figure 3.27: Location of the Orinoco Oil Belt in Venezuela

Based on USGS, An Estimate of Recoverable Heavy Oil Resources of the Orinoco Oil Belt, Venezuela, http://pubs.usgs.gov/fs/2009/3028/pdf/FS09-3028.pdf

was about 2.3×10^6 bbl per day but had declined by about 10% by the end of the year due to economic and political reasons.

The existence of tar sands in Canada has been known since the eighteenth century. However, it was not until the middle of the twentieth century that interest in utilizing this resource became serious, and the first commercial production of oil from tar sands began in Alberta in 1967. The Canadian tar sands deposits occur almost exclusively in three locations in Alberta (the Athabasca Oil Sands, the Peace River Oil Sands, and the Cold Lake Oil Sands), as indicated on the map in Figure 3.28.

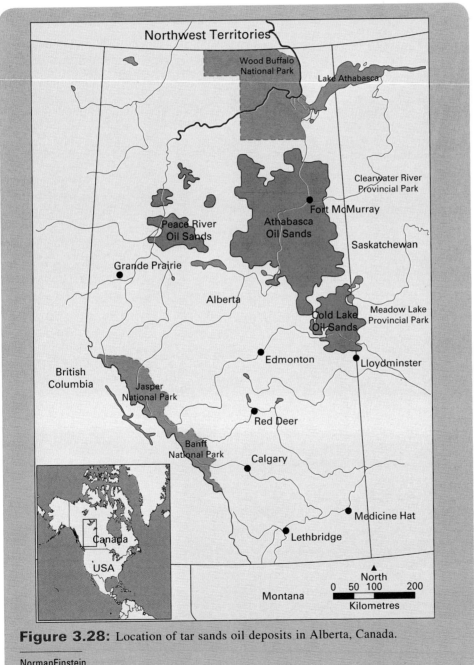

Figure 3.28: Location of tar sands oil deposits in Alberta, Canada.

NormanEinstein

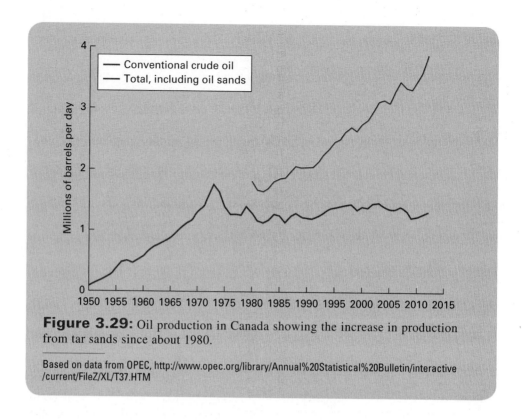

Figure 3.29: Oil production in Canada showing the increase in production from tar sands since about 1980.

Based on data from OPEC, http://www.opec.org/library/Annual%20Statistical%20Bulletin/interactive/current/FileZ/XL/T37.HTM

Production of oil from tar sands in Canada has increased consistently since about 1980, while conventional oil production in that country has remained relatively constant. As illustrated in Figure 3.29, more than half of Canadian oil now comes from tar sands.

3.10 Coal Liquefaction and Gasification

Coal resources are much more plentiful than oil or natural gas resources. Although the use of liquid or gaseous fuels is convenient for transportation applications, the use of a solid fuel is somewhat more problematic. Subject to considerations discussed in the next chapter, coal is a possible means of generating electricity for quite some time. The development of suitable electric vehicle technology could allow coal to be used as a source of energy for transportation. However, the inevitable losses due to intrinsically low conversion efficiencies may be a consideration. Another alternative for the use of coal for transportation purposes involves the conversions of the carbon atoms in the coal into a gaseous or liquid hydrocarbon. These processes are referred to as coal gasification and coal liquefaction, respectively.

The basic process of coal gasification was discovered around 1780. It was commercialized in the early twentieth century and was used to produce gas for distribution to homes and industry for heating and lighting purposes. A similar process by which the carbon in wood could be converted into a combustible gas for use as an automobile fuel was used to deal with gasoline shortages during World War II

Figure 3.30: Automobile manufactured by Adler with wood gas generator attached to the back (Adler Diplomat equipped with Imbert Holzgas Generator, 1941).

Keystone/Stringer/Hulton Archive/Getty Images

(Figure 3.30). Because *coal gas* is of lower energy content and contains more impurities (and hence produces more pollution) than natural gas (which is fairly pure methane), its use was mostly discontinued after natural gas came into common use in the 1940s. In more recent years, coal gasification has not been used extensively because of the potentially low quality of the fuel produced and the comparatively low cost of traditional oil and natural gas that can be used for similar purposes. However, some commercial utilization of coal gas has taken place in South Africa since the 1960s. Current research in the development of coal gasification facilities is aimed at alleviating the problem of dwindling oil and natural gas supplies. A basic description of the coal gasification process follows.

The basic process of coal gasification involves heating the coal in a *gasifier*, which induces the following sequence of processes:

Pyrolysis is the driving off of volatile compounds from the coal, leaving a more carbon-rich material.

Combustion involves the incomplete oxidation of the carbon to produce carbon monoxide by the reaction

$$2C + O_2 \rightarrow 2CO. \tag{3.3}$$

Some carbon dioxide can also be produced, but it is advantageous to minimize this.

Gasification is the process by which the remaining carbon and the carbon monoxide are reacted with steam by the reactions

$$C + H_2O \rightarrow H_2 + CO \tag{3.4}$$

and

$$CO + H_2O \rightarrow CO_2 + H_2. \tag{3.5}$$

Example 3.10

Hydrogen may be produced from carbon by the reaction in equation (3.3) followed by the reaction in equation (3.5).

(a)   What is the mass of hydrogen produced from 1 t of carbon?
(b)   What is the mass of CO_2 emitted?
(c)   What is the volume of water used?

Solution

(a)   From equation (3.3), 2 moles of carbon will yield 2 moles of CO, and from equation (3.5), 2 moles of CO will yield 2 moles of H_2. Thus 1 mole of carbon yields 1 mole of H_2. Using the atomic weights 12 g/mol for carbon and 2 g/mol for H_2, means that 12 g of carbon produces 2 g of H_2. Therefore, 1 t of carbon will yield

$$(1 \text{ t}) \times (2 \text{ g}/12 \text{ g}) = 0.167 \text{ t} = 167 \text{ kg}$$

of hydrogen.

(b)   From equations (3.3) and (3.5), 1 mole of carbon will produce 1 mole of CO_2. Using the atomic weight of 16 g/mol for oxygen gives

$$(1 \text{ t}) \times ((12 + 2 \times 16) \text{ g}/12 \text{ g}) = 3.67 \text{ t}$$

of CO_2.

(c)   Since from equations (3.3) and (3.5) it is seen that 1 mole of carbon requires 1 mole of water, then the amount of water required will be

$$(1 \text{ t}) \times ((2 + 16) \text{ g}/12 \text{ g}) = 1.5 \text{ t or 1500 L of water.}$$

The resulting gas is typically about 50% CO and 50% H_2, which can be burned to produce energy by the reactions

$$2CO + O_2 \rightarrow 2CO_2 \tag{3.6}$$

and

$$2H_2 + O_2 \rightarrow 2H_2O. \tag{3.7}$$

The remaining components of the coal gas, mostly CO_2, add relatively little to the energy production. More complex processes involving catalytic reactions can be used to convert the coal gas into a fuel that is approximately 90% methane.

Coal liquefaction is an alternative approach for producing a transportation fuel from coal. Liquefaction can be by means of either a direct or an indirect process. In the direct processes, coal is ground into a fine powder and then mixed with a solvent to form a slurry containing about one-third to one-half coal. The slurry is then heated in a hydrogen atmosphere under a pressure of about 15 MPa to about 400°C for about an hour. Chemical reactions between the carbon in the coal and the hydrogen atmosphere produce a variety of liquid hydrocarbons. Catalysts can be used to improve the reaction rates. Low-quality coals, such as lignite and sub-bituminous, work best in this process, whereas very hard coals such as anthracite are mostly unreactive. Liquids produced by this process tend to have fairly low molecular masses but can be polymerized to produce fuels similar to gasoline and diesel fuel. Typically, about half the weight of the coal is converted into a liquid fuel.

Indirect liquefaction first produces a gas from the coal by a process involving the reaction of the coal with steam and oxygen at high temperature (as discussed). In the so-called *Fischer-Tropsch process*, coal gas is reacted with a catalyst to produce liquid hydrocarbons that have a wide range of molecular masses.

The direct route is generally more efficient and has been the subject of research efforts in recent years. All coal liquefaction processes release substantial quantities of CO_2, and even if they can be made financially and energetically viable, their desirability must be considered in the context of environmental factors, as discussed in the next chapter.

3.11 Summary

This chapter has reviewed the properties of fossil fuels. Oil has been utilized as a source of energy for more than 150 years and can be refined into various gaseous, liquid, and solid hydrocarbons. Fuel oil (or diesel fuel) and gasoline are the most commonly used oil products.

An analysis based on the Hubbert model suggests that annual oil production worldwide is near its maximum. It seems obvious that oil production rates will begin to decrease in the foreseeable future. This decrease has already been seen in the United States where oil reserves are much more mature than they are in the Middle East. Models suggest that within the next 50 years, traditional oil resources will be nearing the end of their availability.

Natural gas consists primarily of methane. Its use has increased in recent years due to its availability and the fact that it produces less greenhouse gas and pollution per unit energy than do other fossil fuels.

Coal resources are more extensive than oil or natural gas, and an evaluation of their future availability on the basis of the Hubbert model indicates that the longevity of coal is substantially greater than that of other fossil fuels.

Other, less traditional, fossil fuel resources include shale oil and tar sands. Although shale oil resources are extensive, particularly in the United States, the viability of this energy resource is uncertain from both an economic and environmental standpoint. Thus, commercial utilization has been minimal in the past, and the future development of this resource is unclear. Tar sands and extra-heavy oil resources are also very extensive, and their use has been successfully commercialized, particularly in Venezuela (for extra-heavy oil) and in Canada (for tar sands). These resources together may more than double the future availability of oil.

The availability of extensive coal resources, as well as the desirability of liquid and gaseous fuels for transportation use, makes the prospect of coal liquefaction or gasification attractive. The technology for these processes has been known for many years but has seen limited utilization. As long as crude oil prices remain relatively low, the economic viability of coal liquefaction or gasification is questionable.

Problems

3-1 Write down the chemical formulae for the combustion of the alkanes in Table 3.2 with $n = 2$ to $n = 8$. Determine the mass of CO_2 produced per megajoule of energy.

3-2 Locate the current world price of oil (US$/bbl) and the current retail price of regular gasoline in your area (per liter or per gallon, as appropriate). What is the markup in the price per unit energy between crude oil and retail gasoline?

3-3 Locate the current world price of oil (US$/bbl), coal (bituminous coal, per tonne), and natural gas (per 1000 m^3). For each fossil fuel, calculate the cost of energy per gigajoule assuming that the fuel can be converted to usable energy with an efficiency of 100%.

3-4 Assume that all of the United States' annual electricity requirement of 3×10^{12} kWh is produced by coal-fired generating stations operating at a net overall efficiency of 40%.

    (a) How many tonnes of coal are burned per second? Assume the coal is all bituminous.

    (b) Assuming that coal is 100% carbon, how many tonnes of CO_2 will be produced each year?

3-5 On land, coal is transported primarily by train. A typical large coal train may be about 1.5 km long and may consist of 120 cars, each holding 110 tonnes of coal. As each tonne of coal has an equivalent energy content (in terms of the stored chemical energy it contains), a moving coal train represents a flow of (chemical) energy or a power. For a coal train traveling at 100 km/h, calculate the equivalent power in watts.

3-6 How many tonnes of oil shale which produces 120 L/t would be needed to produce the same energy as 1 tonne of bituminous coal?

3-7 Assume that the total energy needs of a person in the United States (Chapter 2) is satisfied by burning coal. If each person is responsible for the mining, transportation, processing, and burning of their own coal, how much coal must each person process, on average, per day?

3-8 If shale oil replaced coal in the United States, how many years would the U.S. resources last at the current rate of use?

3-9 Estimate the total coal produced in the United Kingdom between 1820 and the present.

3-10 For the reaction shown in equation (3.4), calculate the mass of CO and the mass of H_2 produced by the gasification of 1 kg of carbon. What is the energy content of these two reaction products?

3-11 The combustion of pure hydrogen by the reaction

$$2H_2 + O_2 \rightarrow 2H_2O$$

produces 142 MJ per kg of hydrogen. Compare the energy per kg of octane and the total energy from the individual combustion of its carbon and hydrogen atoms.

3-12 (a) Locate, from an appropriate source, the composition for methanol and ethanol. Write down the formulas for the combustion of these two hydrocarbons.
   (b) Calculate the mass of CO_2 produced per kg of methanol and ethanol burned.
   (c) Locate values for the heat of combustion of methanol and ethanol and determine the mass of CO_2 produced per MJ of energy.

3-13 A coal-fired generating station has an efficiency of 38% and produces an average electrical output of 1000 MW_e.

   (a) How much coal does it burn per second? Assume the coal is bituminous.
   (b) How much heat (in joules) is released to the environment per second?
   (c) How many liters of water at 20°C could the heat in part (b) boil (at STP) per second?

3-14 A coal-fired generating station consumes 2×10^6 tonnes of coal per year and produces an average electrical output of 800 MW_e. Calculate the plant's efficiency.

3-15 A large passenger ship weighs 75,000 tonnes and has a cruising speed of 45 km/h. The engines burn diesel fuel (approximately the same energy content as crude oil) with an efficiency of 20%. The drive system has an average efficiency of 55% for converting mechanical energy into propulsion for the ship. Calculate the volume of fuel needed to accelerate the ship from rest to cruising speed.

3-16 A person uses the energy equivalent of 50 bbl of crude oil in one year. What is that person's average power consumption?

3-17 Find the retail cost of gasoline (per unit volume) in your area and find the cost of electricity for residential users in your area.

   (a) Compare the cost of energy (per MJ) from gasoline and from electricity from the public utility.
   (b) The efficiency of a gasoline automobile is about 20% and the efficiency of a battery electric vehicle is about 85%. Discuss the relative transportation costs for gasoline and electric vehicles.

3-18 Liquid natural gas is transported by ship. A typical LNG tanker may have a capacity of 10^5 m^3. If this fuel is used to provide energy for a city with a population of 1 million and each person consumes power at an average rate of 10 kW, how long would the LNG last?

3-19 Using the information in Figure 2.6 and the average composition of coal (lignite, sub-bituminous/bituminous and anthracite) in the United States, calculate the mass of the coal used annually. Use the average energy content for sub-bituminous/bituminous coal.

3-20 A portion of the Keystone Pipeline connecting oil production in Canada and refining in the United States consists of a 0.76-m-diameter pipe that carries 500,000 bbl of crude oil per day over a distance of 1744 km.

(a) Calculate the average velocity of the oil in the pipe.

(b) What is the rate of chemical energy flow through the pipe (in watts) based on the velocity of the oil and its energy content?

Bibliography

G. J. Aubrecht II. *Energy: Physical, Environmental, and Social Impact* (3rd ed.). Pearson Prentice Hall, Upper Saddle River, NJ (2006).

G. Boyle, B. Everett, and J. Ramage (Eds.). *Energy Systems and Sustainability*. Oxford University Press, Oxford (2003).

K. S. Deffeyes. *Beyond Oil: The View from Hubbert's Peak*. Hill and Wang, New York (2005).

K. S. Deffeyes. *Hubbert's Peak: The Impending World Oil Shortage*. Princeton University, Princeton, NJ (2001).

M. E. Eberhart. *Feeding the Fire: The Lost History and Uncertain Future of Mankind's Energy Addiction*. Harmony Books, New York (2007).

J. A. Fay and D. S. Golomb. *Energy and the Environment* (2nd ed.). Oxford University Press, New York (2012).

R. Heinberg. *Power Down: Options and Actions for a Post-Carbon World*. New Society, Gabriola Island, Canada (2004).

R. Hinrichs and M. Kleinbach. *Energy: Its Use and the Environment* (5th ed.). Brooks-Cole, Belmont (2012).

M. King Hubbert. "Energy Resources of the Earth." *Scientific American* **225** (1971). pp. 60–70.

J. A. Kraushaar and R. A. Ristinen. *Energy and Problems of a Technical Society* (2nd ed.). Wiley, New York (1993).

M. Simmons. *Twilight in the Desert: The Coming Saudi Oil Shock and the World Economy*. Wiley, New York: (2005).

V. Smil. *Energy at the Crossroads Global Perspectives and Uncertainties*. MIT Press, Cambridge (2003).

Environmental Consequences of Fossil Fuel Use

Learning Objectives: After reading the material in Chapter 4, you should understand:

- The causes and effects of thermal pollution.
- The causes and types of chemical pollution.
- Principles of the greenhouse effect.

- The reasons for global climate change.
- Methods of carbon sequestration.

4.1 Introduction

Fossil fuel use has numerous possible environmental consequences. A specific example described in the last chapter was the geographic consequences of oil shale mining and its effects on water resources. Similar effects can be related to coal mining and oil drilling. However, this chapter concentrates primarily on three effects of fossil fuel use that are directly related to the production of energy from these fuels: thermal pollution, chemical/particulate matter pollution. Greenhouse gas emissions have, by far, the greatest long-term global effects and are also the most difficult to deal with.

4.2 Thermal Pollution

The production of useful energy from fossil fuels almost always involves combustion. The heat produced by burning fossil fuels is sometimes used directly for residential or commercial heating or for industrial processes. In these cases, the heat eventually finds its way into the environment. The effect of this thermal energy is most noticeable in relation to chemical pollution in urban areas as will be discussed in this chapter. In other cases, the heat produced by the combustion of fossil fuels is, at least partially, converted into mechanical energy in a heat engine. This mechanical energy is most commonly used for transportation (automobiles, trucks, etc.) or for the generation of electricity. In either case, the thermodynamic efficiency of the conversion of heat into mechanical energy is limited by the Carnot efficiency. For an automobile engine, the overall efficiency is fairly low, typically around 17%. For an electric generating station, the efficiency can be as high as 40%. This latter efficiency is similar to that obtained by nuclear power plants, as discussed in the next few chapters, because these also use heat engines to convert the thermal energy released by nuclear reactions into mechanical energy. Thus, the basic principles for heat disposal discussed in this section are applicable to nuclear power plants as well. The more efficient the facility is, the greater the fraction of energy that is converted into mechanical energy and the less the fraction of energy that is transferred into the environment as waste heat. Therefore, efficiency factors, such as those discussed in Chapter 1, are important to consider.

The traditional method of disposing of waste heat from a fossil fuel or nuclear generating station is to use a body of water as the cold reservoir (Figure 1.7). In the simple case, *once-through cooling* is used; cold water is pumped from the body of water, heated in a heat exchanger, and then transferred back into the cooling reservoir. The use of water as the cold reservoir for a heat engine has several attractive advantages:

- Water has a high specific heat, so it can contain a substantial quantity of thermal energy with a minimal increase in temperature.
- Water has a high thermal conductivity, so the transfer of heat to the water is efficient.
- Large reservoirs (such as the oceans) or somewhat smaller ones (such as lakes or rivers) are readily available in many parts of the world.
- The temperature of the water remains relatively stable (in comparison to the temperature of the air) as a function of time of the year.
- The infrastructure required for the use of water as the cold reservoir for a heat engine is simple, straightforward, and relatively inexpensive.

However, the release of heat into a body of water can have significant undesirable environmental effects. This heat release can change the ecology in rivers and more particularly in lakes or enclosed ocean bays. This excess heat is important not only in terms of the possible increase in the average temperature of the body of water but also in terms of the vertical distribution of heat (i.e., the temperature profile) in the water. In a lake, the surface layer consists of warmer water with a layer of cooler, denser water below. The oxygen content of the cooler water is greater. The details of the effects of introducing waste heat into a body of water are complex, but the ecology of the lake can be affected in three clear ways:

- Changes in oxygen content resulting from temperature changes can affect biological processes in organisms in the lake.
- Changes in temperature can affect chemical reaction rates.
- Changes in the temperature profile can affect the natural seasonal mechanisms that mix the water in the lake.

These factors can have profound effects on the ecology of the region and have been a major factor in promoting the use of the atmosphere as a cold reservoir for electricity generating stations. In fact, in some countries, regulations have required the implementation of air cooling for new power plant construction. Air cooling uses cooling towers (Figure 1.11) to transfer waste heat to the atmosphere. There are two major designs of cooling towers: wet cooling towers and dry cooling towers. *Wet cooling towers* use excess heat to evaporate water. The latent heat of vaporization of the water is responsible for carrying away the excess waste heat. In *dry cooling towers*, hot water (carrying the waste heat) is circulated through pipes that are designed to effectively transfer heat to the atmosphere. This is basically like the design of a radiator for an automobile engine. For both methods, air may be circulated through the tower either by natural convection or by fans.

The use of cooling towers is an effective way of mitigating the effects of waste heat. However, their use is not without some environmental consequences. The changes in temperature and humidity around cooling towers (particularly wet cooling towers) can cause a localized region of increased fog and precipitation. Dry cooling towers are the preferred method of waste heat disposal from an environmental (if not economic) perspective. In all cases, there is a trade-off involving construction costs, operational costs, efficiency, and environmental effects.

4.3 Chemical and Particulate Matter Pollution

Chemical and particulate matter pollution from burning fossil fuels can fall into several categories (depending on the chemical involved):

- Carbon monoxide (CO)
- Nitrogen-oxygen compounds (NO_x)
- Hydrocarbons
- Sulfur dioxide (SO_2)
- Particulate Matter

The formation and significance of each are now discussed.

4.3a Carbon Monoxide

Carbon monoxide is produced by the incomplete oxidation of carbon during combustion. This corresponds to the chemical reaction

$$2C + O_2 \rightarrow 2CO. \tag{4.1}$$

This reaction typically results when insufficient oxygen is present to produce CO_2. In the United States, about 10^{11} kg of CO are released to the atmosphere every year as a result of burning fossil fuels. A rough breakdown of the sources of CO in the United States is shown in Table 4.1.

Carbon monoxide is an important factor in pollution because of its toxic properties, which result from its particular reactivity with hemoglobin. CO is about 200 times as reactive with hemoglobin as O_2. This means that inhaled CO tends to displace O_2 that is dissolved in the blood, leading to adverse health effects.

Table 4.1: Sources of CO released to the atmosphere as a result of fossil fuel use.

source	% CO released
vehicles	60
industrial	10
waste disposal	8
agricultural burning	8
forest fires	8
other	6

Example 4.1

If all the CO produced in the United States from the burning of fossil fuels was converted into CO_2 before its release, what would be the mass of this additional CO_2? Compare this with the approximately 3×10^{12} kg of CO_2 that is released directly in the United States per year.

Solution

It is stated in the text that fossil fuel burning in the United States releases 10^{11} kg CO annually.

If this is converted into CO_2, the mass of the CO_2 is given in terms of the molecular masses of CO (28 g/mol) and CO_2 (44 g/mol) as

$$\frac{10^{11}\text{kg} \times 44 \text{ g/mol}}{28 \text{ g/mol}} = 1.57 \times 10^{11} \text{ kg}.$$

This represents about 5% of the total CO_2 emissions. Compare this with CO_2 produced directly from coal burning in the United States as presented in Example 4.5.

Typically, exposure for an hour or so to different concentrations of CO in air results in the following symptoms:

- 100 ppm: headache/confusion
- 300 ppm: nausea/unconsciousness
- 600 ppm: death

The long-term effects of low-level CO exposure are not clearly known. Because the major source of CO is from vehicles, much progress has been made in recent years in reducing the CO emissions from gasoline engines. CO and some other emissions, as specified by U.S. standards for automobiles, are shown in Table 4.2. The implementation of emission control standards has clearly made a significant reduction

Table 4.2: U.S. emission standards for CO, NO_x, and hydrocarbons in grams per kilometer.

year	CO	NO_x	HC
1960 (precontrol estimate)	52.2	2.5	6.58
1970	21.1	3.1	2.5
1975	9.3	1.9	0.93
1980	4.3	1.2	0.25
1981	2.1	0.62	0.25
1983	2.1	0.62	0.25
1994	2.1	0.25	0.16
2001	2.1	0.12	0.078
2004	1.1	0.04	0.056

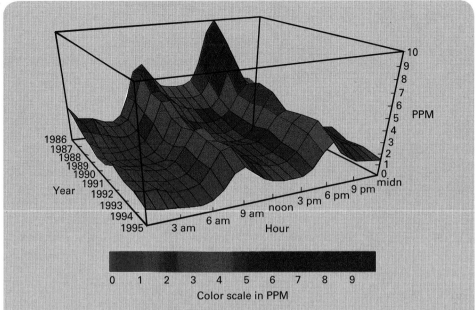

Figure 4.1: Average CO concentration in the air (in ppm per volume) as a function of the time of day in Denver, Colorado for years 1986 through 1995. Data are shown are average weekday values during the winter season (November to February).

U.S. EPA Fig 3-9 on page 3–25 from "Air Quality Criteria for Carbon Monoxide" EPA 600/P-99/001F (2000) At http://www.epa.gov/ncea/pdfs/coaqcd.pdf

in CO emission, but this is to some extent offset by the greater number of vehicles on the road and the greater number of kilometers driven per vehicle per year.

CO levels in urban areas have decreased during the years since emission control standards were in place. There are clear fluctuations during the day that depend on transportation patterns. Figure 4.1 shows average CO concentrations as a function of the time of day in Denver, Colorado. A clear correlation between vehicle activity

Example 4.2

There about 250 million personal motor vehicles in North America. If each vehicle is driven an average of 25,000 km per year, what is the total reduction in CO emissions per year between precontrol vehicles and vehicles that meet the 2004 U.S. emission standard?

Solution

The difference between precontrol and 2004 standard vehicles is

$$52.2 \text{ kg/km} - 1.1 \text{ kg/km} = 51.1 \text{ g (CO) per km.}$$

For 250 million vehicles, each driving 25,000 km, the total annual CO reduction is

$$(0.0511 \text{ kg/km}) \times (2.5 \times 10^8 \text{ vehicles}) \times (25,000 \text{ km/vehicle}) = 3.2 \times 10^{11} \text{ kg.}$$

in the morning and afternoon is seen, and the dispersion of CO into the upper atmosphere is seen on a timescale of a few hours. The graph also illustrates the progress that has been made in reducing vehicular CO emissions.

4.3b Nitrogen Oxides

Nitrogen oxides are formed during the combustion of fossil fuels (or any other materials) when the temperature is sufficiently high (above about 1100°C). This is because nitrogen in the atmosphere becomes oxidized by the reaction

$$N_2 + O_2 \rightarrow 2NO. \tag{4.2}$$

The nitric oxide (NO) formed by this reaction is a colorless gas that is mildly toxic to humans. The NO itself is not the most significant aspect of pollution from nitrogen oxygen compounds. On the timescale of a few hours, the NO reacts with ozone (O_3, which is reasonably plentiful in the atmosphere) to form nitrogen dioxide by the reaction

$$NO + O_3 \rightarrow NO_2 + O_2. \tag{4.3}$$

NO_2 is a highly toxic brown gas. It is responsible for the brownish color of smog that accumulates in urban areas, as Figure 4.2 illustrates. At low concentrations, it is an irritant causing respiratory and eye problems. At higher concentrations, it causes lung, liver, and heart damage. Because NO readily converts to NO_2 in the atmosphere, these species are generally lumped together as NO_x. A rough breakdown of the sources of NO_x is shown in Table 4.3.

The reduction of vehicle emissions of NO_x as a result of U.S. emission standards is indicated in Table 4.2. Daily fluctuations of NO_x levels in an urban area as a result of daily variation in human activity is clearly indicated in Figure 4.3. The

Figure 4.2: Smog covering Shanghai, China.

r.nagy/Shutterstock.com

Table 4.3: Typical NO_x emission sources in the United States.	
source	% NO_x
vehicles	35
coal burning	20
natural gas burning	25
other	20

conversion of NO to NO_2 by equation (4.3) is seen by the increase in NO_2 levels at the expense of the decrease in O_3 levels during the day.

4.3c Hydrocarbons

Hydrocarbons (HC) are released during the burning of fossil fuels as a result of incomplete combustion. This is a respiratory and eye irritant and, in higher concentrations, can cause lung disease. A rough breakdown of the sources of hydrocarbon emissions is given in Table 4.4 on the next page. The reduction in U.S. hydrocarbon emissions from vehicles that has resulted from emission control standards is shown in Table 4.2.

4.3d Sulfur Dioxide

Sulfur dioxide (SO_2) results primarily from the combustion of sulfur-containing fuels and to a lesser extent from industrial processes. A breakdown of SO_2 sources is shown in Table 4.5 on the next page. Coal, as shown, is the major source of SO_2 because all coal contains some concentration of sulfur, and this is released as SO_2 during combustion. Other fossil fuels, such as oil and natural gas, contain smaller concentrations of sulfur impurities, and these impurities can largely be eliminated before the fuel is used.

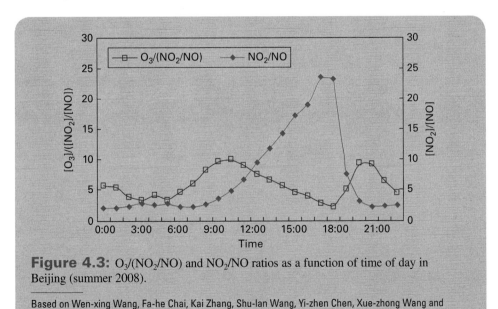

Figure 4.3: $O_3/(NO_2/NO)$ and NO_2/NO ratios as a function of time of day in Beijing (summer 2008).

Based on Wen-xing Wang, Fa-he Chai, Kai Zhang, Shu-lan Wang, Yi-zhen Chen, Xue-zhong Wang and Ya-qin Yang "Study on ambient air quality in Beijing for the summer 2008 Olympic Games" Air Qual Atmos Health 1 (2008) 31–36

Table 4.4: Sources of hydrocarbon pollution.	
source	% HC
vehicles	35
industrial	20
natural gas burning	25
other	20

Table 4.5: Sources of SO_2 pollution.	
source	% SO_2
coal	65
industrial	25
other	10

SO_2 is a respiratory and eye irritant but also has adverse environmental effects for another reason. It reacts with oxygen in the atmosphere by the following reaction:

$$2SO_2 + O_2 \rightarrow 2SO_3. \tag{4.4}$$

This is followed by a reaction with water vapor to produce sulfuric acid:

$$SO_3 + H_2O \rightarrow H_2SO_4 + O_2. \tag{4.5}$$

The formation of sulfuric acid in the atmosphere results in so-called *acid rain*. This has particular adverse effects on buildings, painted surfaces (e.g., automobiles), and other structures (Figure 4.4). The emission of SO_2 from coal generating stations can be reduced by reacting the exhaust gas with CaO or $CaCO_3$, in a device known as a *scrubber*, prior to releasing it to the atmosphere.

Figure 4.4: Long-term effects of acid rain on a statue.

Ryan McGinnis/Alamy Stock Photo

4.3e **Particulate Matter**

Particulate matter consists of dust particles that appear as smoke during the burning of some fossil fuels (or wood). They also result from some industrial processes and natural processes such as forest fires and volcanoes. Particulate matter is relatively unimportant for the combustion of oil, gasoline, and natural gas but is much more significant during the combustion of coal. Particulate matter acts primarily as a respiratory irritant, although, because its chemical composition is uncertain, particles that contain toxic compounds can have more serious health implications. Because most coal that is used for energy production is burned in coal-fired electric generating stations, it is relatively easy to deal with this type of pollution, and most coal-fired stations utilize systems for reducing particulate matter pollution. These systems fall into two major categories; *mechanical filters (baghouses)* and *electrostatic precipitators*. The former are basically filters that remove particles in the exhaust gas. The filters are shaken to collect the particles in a container (Figure 4.5). An electrostatic

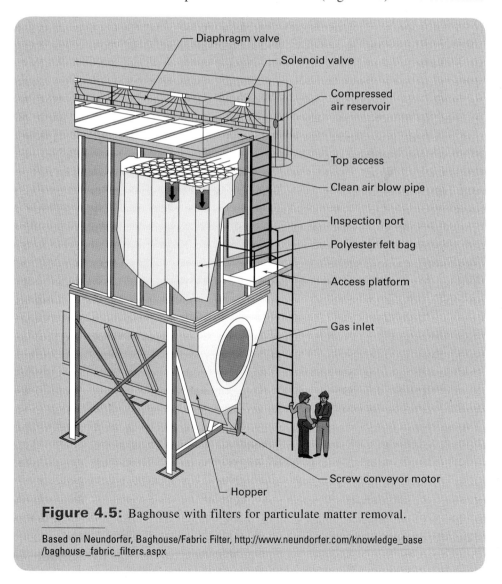

Figure 4.5: Baghouse with filters for particulate matter removal.

Based on Neundorfer, Baghouse/Fabric Filter, http://www.neundorfer.com/knowledge_base
/baghouse_fabric_filters.aspx

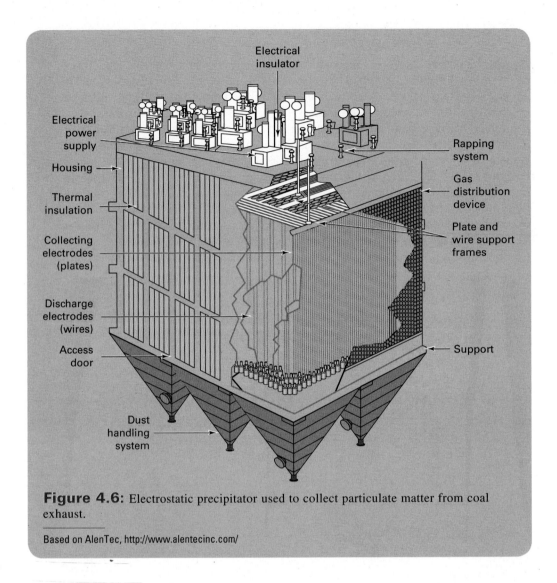

Figure 4.6: Electrostatic precipitator used to collect particulate matter from coal exhaust.

Based on AlenTec, http://www.alentecinc.com/

precipitator (Figure 4.6) uses electrostatically charged plates to collect the particles from the exhaust.

The presence of chemical and particulate matter pollution, particularly in urban areas, is largely the result of fossil fuel burning. The health implications are clear; concentrations of these pollutants in the atmosphere are often near or even exceed government health standards. However, the most serious health problems can occur when pollution combines with certain adverse meteorological conditions. These weather conditions have to do with the temperature as a function of altitude in the atmosphere and greatly influence the dispersion of pollution into the upper atmosphere. The normal conditions present in the atmosphere that are responsible for the dispersion of pollutants, as illustrated in Figures 4.1 and 4.3, will be discussed first.

The temperature as a function of height in the atmosphere is shown in Figure 4.7. Up to a height of about 10 km, which is the region of interest, the temperature shows a relatively linear decrease with increasing altitude. The rate at which the temperature

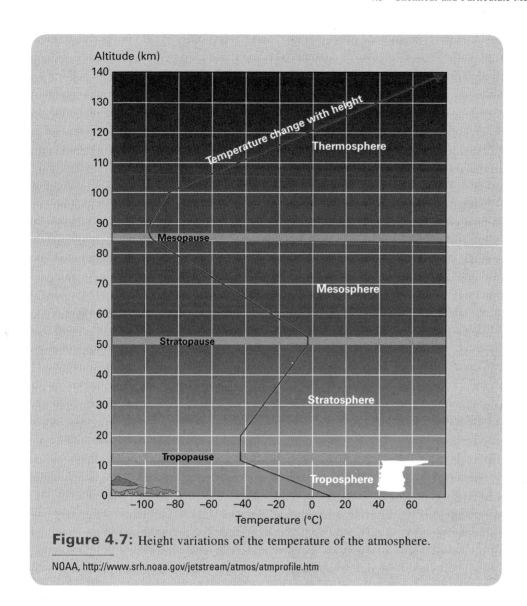

Figure 4.7: Height variations of the temperature of the atmosphere.

NOAA, http://www.srh.noaa.gov/jetstream/atmos/atmprofile.htm

decreases as a function of height is known as the *adiabatic lapse rate* (ALR), and its value depends on several factors, including the relative humidity of the air. Typically the ALR is of the order of about 10°C/km or 0.01°C/m.

To describe the effects of the ALR on air movement, let us consider a parcel of air with a volume, V, at the earth's surface. The pressure, P, and temperature, T, are determined by its equilibrium condition with the surrounding air. If we lift this parcel of air, then, as it rises, it will cool at the ALR in order to remain in equilibrium. It will also expand due to the decreasing atmospheric pressure. Under certain meteorological conditions, the actual temperature of the atmosphere may decrease either more quickly or more slowly than the ALR (Figure 4.8 on the next page). If the actual temperature decreases faster than the ALR, then a parcel of air that is raised from the surface and cools at the ALR will be warmer than the surrounding air. This warmer air will tend to rise and will be replaced by cooler air from above. This promotes mixing and is an unstable condition. If, on the other hand, the actual atmospheric temperature decreases

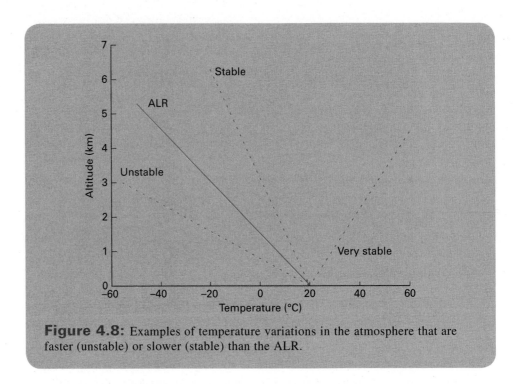

Figure 4.8: Examples of temperature variations in the atmosphere that are faster (unstable) or slower (stable) than the ALR.

more slowly than the ALR, then a parcel of air that is raised above the surface of the earth will be cooler than its surroundings and will tend to sink back to the surface. This inhibits mixing and is a stable condition.

This analysis applies to the warm air released from the exhaust of, say, an automobile or a coal-fired generating station. If the atmospheric conditions are unstable, the warm polluted air rises, and the pollution disperses in the upper atmosphere. If the atmospheric conditions are stable, the warm polluted air remains near ground level. In an extreme case, the temperature near the earth may actually increase as a function of height (up to some altitude). This is a very stable condition and is referred to as a *temperature inversion*. This situation is not uncommon because, during the normal daily warming and cooling cycle, the ground cools faster than the air in the evening, giving rise to a cooler region near the earth's surface. This temperature inversion normally lasts for a few hours and is a contributing factor in trapping pollution near the earth's surface in urban areas. This situation is the reason that exhaust from factories and power plants (which is warm, polluted air) is released from chimneys (smokestacks) above ground level so that it is above any possible near-ground temperature inversions. The world's tallest chimney (Figure 4.9) is taller than the Empire State Building in New York City.

In rare cases, meteorological conditions cause a temperature inversion that may last for several days, trapping pollutants near ground level. This can have serious health effects. A well-known situation of this type occurred in London in 1952, where a temperature inversion lasted for an extended period of time. The correlation between the death rate (above a normal rate of about 250 individuals per day) and the level of pollution is clearly shown in Figure 4.10. The most recent analysis of this occurrence indicates that the excess number of deaths that resulted from the presence of pollution was around 12,000.

Figure 4.9: Chimney at the coal-fueled power generating station in Ekibastuz, Kazakhstan constructed in 1987. At a height of 419.7 m, it is the tallest chimney in the world (as of 2017).

Arpingstone/English Wikipedia

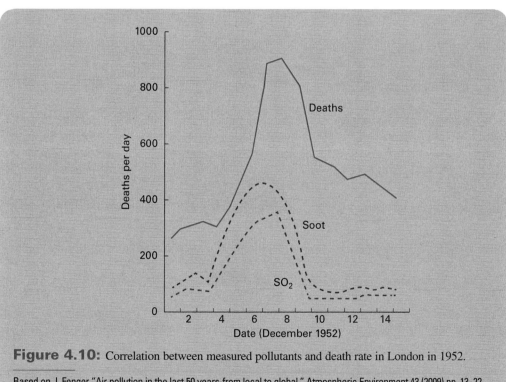

Figure 4.10: Correlation between measured pollutants and death rate in London in 1952.

Based on J. Fenger "Air pollution in the last 50 years-from local to global," Atmospheric Environment 43 (2009) pp. 13–22

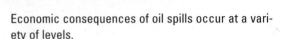

ENERGY EXTRA 4.1
Oil spills

An oil spill is the release of oil (or other liquid petroleum product) into the environment. The terminology is generally used to indicate a spill that occurs as a result of human activity and does not include natural seepage of oil from underground. We generally think of oil spills as those that result from seagoing oil tankers or offshore oil drilling activities. Oil spills, however, also include spills from terrestrial oil wells. Spills are generally accidental, but several notable intentional spills have occurred during wartime. Oil spills have a significant negative impact on the environment. Fish, birds, marine mammals, invertebrates, turtles, and plant life are all at considerable risk directly from the toxic properties of petroleum and indirectly as a result of overall ecological changes.

Economic consequences of oil spills occur at a variety of levels.

The largest accidental oil spills are shown in the table, which does not include intentional spills during the 1991 Persian Gulf War. These intentional spills included damaged oil wells that either gushed oil onto the ground or were set ablaze, as well as oil intentionally discharged from oil terminals or tankers into the ocean. The total oil introduced into the environment from these events has been estimated to be as much as 2 billion barrels (more than 200 times the largest accidental spill). The well-known *Exxon Valdez* disaster in 1989 released between 260,000 and 750,000 bbl of oil and is the largest marine spill in the United States except for the *Deepwater Horizon* accident.

The largest accidental oil spills (>2,000,000 bbl).				
name	type	location/country	year(s)	bbls (approx)
Lakeview Gusher	terrestrial well	California, USA	1910–1911	9,000,000
Deepwater Horizon	drilling platform	Gulf of Mexico, USA	2010	4,500,000
Ixtox I	drilling rig	Gulf of Mexico, Mexico	1979–1980	3,400,000
Atlantic Express/Agean Captain	tankers	Trinidad and Tobago	1979	2,105,000
Fergana Valley	terrestrial well	Uzbekistan	1992	2,090,000

It is sometimes difficult to determine the volume of oil released, particularly for marine accidents. The volume of oil from a tanker is, of course, limited by the volume of the tanker. The *Atlantic Express/Agean Captain* disaster is the most significant tanker disaster and actually involved the collision of two oil tankers, both of which contributed to the spill. The oil released from an offshore well is more difficult to determine but

can be estimated in several ways. One approach is to consider the appearance of the oil on the surface of the ocean. Different thicknesses of oil have different optical properties, and, of course, different thicknesses represent different volumes of oil per unit surface area. The next table summarizes how the appearance of the oil can be used to estimate volume per surface area. It is possible to estimate the surface area covered with oil

Thickness and appearance of oil on the surface of water.		
thickness (nm)	quantity (bbl/km²)	appearance
38	0.233	barely visible
76	0.459	silvery, without color
150	0.944	faint color
300	1.82	bright bands of color
1000	6.10	dull color
2000	12.3	much darker color

Continued on page 127

Energy Extra 4.1 continued

The *Deepwater Horizon* oil spill as viewed from space on May 24, 2010. The extent of the spill is shown in the insert map.

from aerial or satellite imagery (see the photograph of the *Deepwater Horizon* spill as observed from NASA's Terra satellite) and therefore to obtain the total volume of oil. This can result in an underestimate of the amount of oil released because it measures only the oil that is floating on the water's surface.

Topic for Discussion

If an oil tanker spilled 1 million bbl of oil in the ocean and the oil formed a uniform layer 500 nm thick, how large an area would be covered?

4.4 The Greenhouse Effect

Although pollution resulting from the use of fossil fuels is an important problem, the release of greenhouse gases is much more serious and causes more global and long lasting adverse effects. A consideration of the effects of greenhouse gases begins with an analysis of the equilibrium temperature of the earth resulting from the global energy balance. Virtually all the energy that establishes the equilibrium temperature of the earth comes as radiation from the sun. Geothermal energy coming from the interior of the earth accounts for only about 0.1% of the total heat and can be ignored in the present analysis.

The simplest analysis considers a planet without atmosphere and an incident radiation from the sun of S [in watts per square meter (W/m^2)]. Radiative energy from the sun that is incident on the planet is either absorbed by the planet or is reflected back into space. The fraction of the incident radiation that is reflected by a planet is called the *albedo*, *a*, so the fraction that is absorbed is $(1 - a)$. The absorbed energy heats the surface, and this warm surface radiates energy out into space. When the absorbed power (i.e., the incident energy per unit time) and the radiated power are equal, then the temperature of the planet achieves equilibrium. From the viewpoint of the sun, the planet appears as a disk with an area of πR^2, where R is the radius of the planet. Thus the solar power absorbed by the planet is

$$P_{absorbed} = (1 - a)S \, \pi R^2. \tag{4.6}$$

The power radiated into space by the warm surface is given as a function of the surface temperature, T, by the Stefan-Boltzmann law as

$$P_{radiated} = 4\pi R^2 \sigma T^4. \tag{4.7}$$

Here σ is the Stefan-Boltzmann constant (5.67051×10^{-8} W·m^{-2}·K^{-4}), and $4\pi R^2$ is the total surface area of the planet. (The physics of this process will be discussed further in Chapter 8.) Equating (4.6) and (4.7) and solving for temperature gives

$$T = \left[\frac{(1 - a)S}{4\sigma} \right]^{1/4}. \tag{4.8}$$

It is important in this analysis to use temperatures that are measured on an absolute temperature scale (e.g., Kelvin). The solar flux at the outside of the earth's atmosphere is, 1.367 kW/m^2, and the earth's albedo is about 0.3, so equation (4.8) gives the equilibrium temperature of the surface of the earth (without atmosphere) of 254 K (or about –19°C). Clearly this value is too low in terms of our current climate and in terms of the possibility of the evolution of water-based life on earth. What is missing from this approach is the fact that the earth has an atmosphere.

When an atmosphere is present, the problem becomes somewhat more complex. Incident radiation from the sun may be reflected, absorbed, or transmitted by the

---

Example 4.3

The mean orbital diameter of Mars is 1.523 times that of the earth. Calculate the mean surface temperature of Mars if its albedo is 0.25 and there are no greenhouse effects.

Solution

As the solar flux at the distance of the earth is 1.367 kW/m^2, then at the distance of Mars it is

$$\frac{1.367 \text{ kW/m}^2}{(1.523)^2} = 0.589 \text{ kW/m}^2.$$

From equation (4.8), the temperature is calculated to be

$$\left[\frac{(1 - 0.25) \times (589 \text{ W} \cdot \text{m}^{-2})}{4 \times 5.67051 \times 10^{-8} \text{ W} \cdot \text{m}^2 \cdot \text{K}^{-4}} \right]^{1/4} = 210 \text{ K}$$

The actual mean surface temperature on Mars is 210 K, indicating that the greenhouse effects are negligible.

atmosphere and/or reflected or absorbed by the surface of the earth. More importantly, radiation emitted by the planet may be reflected, absorbed, or transmitted by the atmosphere. To properly analyze the effects of the atmosphere, it is necessary to consider the wavelength of the radiation involved. The wavelength of the radiation emitted by a body is related to its temperature (see Chapter 8). The radiation that is incident on the earth is produced by the surface of the sun, which is at about 6000 K. This radiation has a relatively short wavelength, and much of it is in the visible part of the spectrum (see Chapters 1 and 8). The radiation emitted from the surface of the earth comes from a body that is at about 300 K. This radiation has a relatively long wavelength and is in the infrared part of the spectrum. The transparency of the atmosphere is a function of the wavelength of the radiation involved. Certain molecular species that exist in the atmosphere do not interact significantly with short-wavelength radiation but readily absorb long-wavelength radiation. Thus, these molecules allow the short-wavelength sunlight to pass through the atmosphere easily and arrive at the surface of the earth but prevent some portion of the long-wavelength radiation emitted by the surface of the earth from escaping back into space. To maintain the equilibrium condition, requiring equations (4.6) and (4.7) to be equal, the temperature, T, in equation (4.7) must be greater. This means that the temperature of the surface of the earth will be higher than that predicted by the simple model that excludes the atmosphere. Thus, heat is trapped by the atmosphere, giving rise to the so-called *greenhouse effect* and resulting in an average temperature for the earth that is increased by more than 30 K.

Certain molecular species in the atmosphere are most effective at absorbing infrared radiation than others, and these are the major contributors to the greenhouse effect (Table 4.6).

Carbon dioxide and methane are both natural components of our atmosphere that are produced by natural biological processes. Also, carbon dioxide results from the complete combustion of any carbon-containing fuel by the reaction

$$C + O_2 \rightarrow CO_2. \tag{4.9}$$

Nitrous oxide (sometimes referred to as laughing gas because of its medical use) is largely the result of the agricultural use of certain fertilizers. Chlorofluorocarbons come exclusively from human activity (industry, etc.) but are not directly related to the production of energy from fossil fuels. The concentrations of the various greenhouse gases determine the equilibrium temperature of the surface of the

Table 4.6: Greenhouse gases in the earth's atmosphere. The relative absorption is normalized to the absorption per molecule for CO_2. Radiative forcing is a measure of the overall effectiveness of a particular gas at altering the energy balance in the atmosphere, that is, its ability to contribute to global warming. Data are shown for the most abundant chlorofluorocarbon (CFC), CCl_2F_2.

molecular species	current concentration in the atmosphere (by volume)	relative infrared absorption per molecule	radiative forcing (W/m^2)
carbon dioxide (CO_2)	405 ppm	1	1.85
methane (CH_4)	1.82 ppm	25	0.51
nitrous oxide (N_2O)	325 ppb	298	0.18
CFC (CCl_2F_2)	530 ppt	10,900	0.17

Example 4.4

Dry air has the approximate composition of 78% nitrogen, 21% oxygen, and 1% argon. Using the total mass of the earth's atmosphere of 5.148×10^{15} t, calculate the total mass of CO_2 that is presently in the atmosphere.

Solution

According to Table 4.6, the total concentration of CO_2 (molecular mass 44 g/mol) in the atmosphere is 405 ppm by volume or 0.0405%. To convert to weight percent we multiply the volume percent by the ratio of the molecular mass of CO_2 (44 g/mol) to the mean molecular mass of air. Using the molecular masses; nitrogen (N_2) 28 g/mol, oxygen (O_2) 32 g/mol, and argon (Ar) 40 g/mol, gives the mean molecular mass of air as

$$0.78 \times (28 \text{ g/mol}) + 0.21 \times (32 \text{ g/mol}) + 0.01 \times (40 \text{ g/mol}) = 28.96 \text{ g/mol}.$$

This gives the weight percent of CO_2 in the atmosphere as

$$(0.0405\%) \times (44 \text{ g/mol})/(28.96 \text{ g/mol}) = 0.0615\% \ CO_2 \text{ by weight}.$$

Thus the total mass of CO_2 in the atmosphere is

$$(0.000615) \times (5.148 \times 10^{15} \text{ t}) = 3.17 \times 10^{12} \text{ t}.$$

earth. Because the production of energy by burning fossil fuels produces CO_2, it is important to consider the relationship (if any) between human activities and the earth's temperature.

Example 4.5

According to Figure 2.6, 16.6 EJ of energy from coal combustion are used per year in the United States. What is the mass of CO_2 released? What fraction of the total CO_2 emissions does this represent (see Table 4.8)?

Solution

From equation (2.10), the energy release from the combustion of carbon is

$$C + O_2 \rightarrow CO_2 + 32.8 \text{ MJ/kg}.$$

Thus 16.6 EJ of energy represent the combustion of

$$\frac{1.66 \times 101^{13} \text{MJ}}{32.8 \text{ MJ/kg}} = 5.05 \times 10^{11} \text{ kg of carbon}.$$

Since the molecular mass of carbon is 12 and the molecular mass of CO_2 is 44 g/mol, the combustion of this amount of carbon releases

$$\frac{(5.05 \times 10^{11} \text{ kg}) \times (44 \text{ g/mol})}{12 \text{ g/mol}} = 1.85 \times 10^{12} \text{ kg } CO_2.$$

Example 4.5 continued

From Table 4.8, the United States produces 1.477×10^{12} kg of carbon emission per year. Therefore, the value above this represents

$$\frac{5.05 \times 10^{11}\,\text{kg}}{1.477 \times 10^{12}\,\text{kg}} \times 100 = 34\%.$$

4.5 Climate Change

The earth's temperature and the concentration of CO_2 in the earth's atmosphere in the past can be determined from an analysis of air bubbles trapped in the ice in Greenland and in the Antarctic. The results of some of these studies are shown in Figure 4.11. These data indicate a clear correlation between atmospheric CO_2 concentration and the earth's temperature and show the natural cyclic trends over the past half million years or so of the earth's history.

To look at more recent trends in atmospheric CO_2, we can include the results from direct measurements. Figure 4.12 on the next page shows CO_2 levels for the past 1000 years as determined by ice core and direct atmospheric measurements. More detailed direct measurements over the past 50 years or so are shown in Figure 4.13 on the next page. These figures indicate that, in the past century or so, the CO_2 concentration has

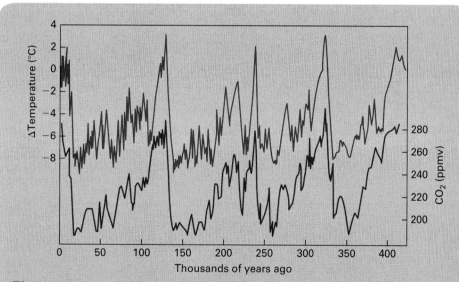

Figure 4.11: Measured temperature fluctuations and CO_2 concentrations from ice core data at Vostok, Antarctica.

Based on data from http://cdiac.ornl.gov/trends/temp/vostok/graphics/tempplot5.gif and http://cdiac.ornl.gov/trends/co2/graphics/vostok.co2.gif

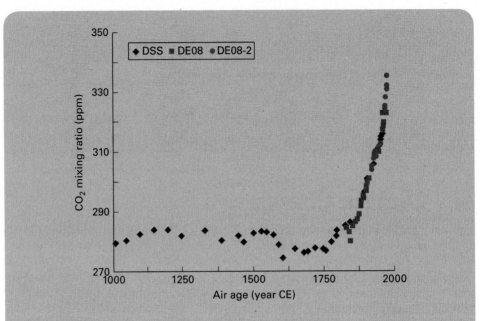

Figure 4.12: Atmospheric CO_2 concentration for the past 1000 years, taken at Law Dome, Antarctica.

Based on http://cdiac.ornl.gov/trends/co2/graphics/lawdome.gif

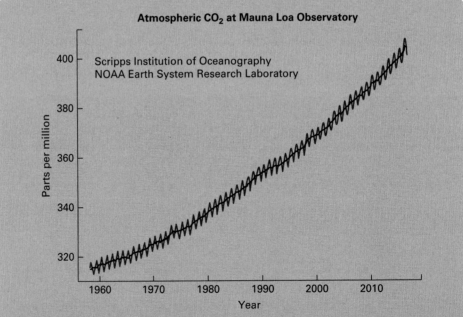

Figure 4.13: Monthly measured atmospheric CO_2 concentration at Mauna Loa Observatory in Hawaii (red line). The oscillations result from the seasonal growth of vegetation (which absorbs CO_2). The black line shows seasonally corrected data.

Earth System Research Laboratory, Trends in Atmospheric Carbon Dioxide, http://www.esrl.noaa.gov/gmd /ccgg/trends/full.html

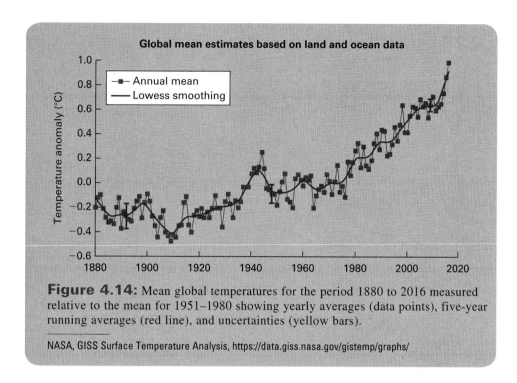

Figure 4.14: Mean global temperatures for the period 1880 to 2016 measured relative to the mean for 1951–1980 showing yearly averages (data points), five-year running averages (red line), and uncertainties (yellow bars).

NASA, GISS Surface Temperature Analysis, https://data.giss.nasa.gov/gistemp/graphs/

risen substantially above even the highest values during the past half million years, and the general continuing upward trend is clear. Compare the recent value of over 400 ppm in Figure 4.13 with the peak values in Figure 4.11 of about 290 ppm.

It is important to try to correlate these changes in CO_2 concentration over the past century or so with changes in the earth's temperature. Figure 4.14 shows the measured mean temperature variations for the past 135 years. Although there is much scatter from year to year, the increasing trend in the data is obvious. This trend is, perhaps, even more clearly indicated by the monthly global temperature averages as illustrated in Figure 4.15 on the next page. These data show the general trend of global warming since the latter part of the 19th century, but also show that, as of the end of 2016, August 2016 was the warmest month ever recorded worldwide and 2016 was the warmest year overall. These trends correlate well with the increase in CO_2 concentration in the atmosphere and is consistent with the general historical temperature–CO_2 trends shown in Figure 4.11.

The gradual increase in average world temperature over the past 135 years is manifested in several ways:

- *A reduction in the size and number of glaciers:* This is evidenced by direct observation; for example, Figure 4.16 on the page 135 shows photographs of a glacier in Glacier National Park (Montana) in 1910 and in 1997. The reduction in size is clear.

- *A reduction in the area and thickness of Arctic sea ice:* There has been a reduction in area of about 9% in the past decade and a reduction of 15–40% in thickness over the past 30 years.

- *An increase in sea level:* This is a result of two factors: the melting of Antarctic ice and the increase in volume of seawater resulting from an

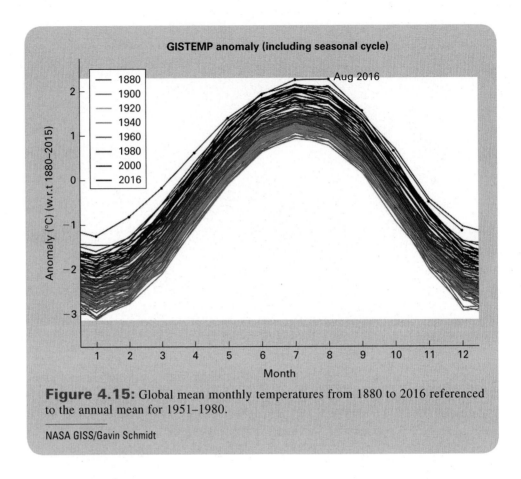

Figure 4.15: Global mean monthly temperatures from 1880 to 2016 referenced to the annual mean for 1951–1980.

NASA GISS/Gavin Schmidt

increase in average sea temperature (related to changes in density). The latter is indicated by direct sea temperature measurements (Figure 4.17 on the page 136).

- *Biological changes:* These include the dates of annual bird migrations and certain stages of plant development, as well as color changes in corals (which are sensitive to temperature).
- *Altered geographical ranges of certain plants and animals.*
- *Thawing of the permafrost in the Arctic.*
- *Weather changes, such as more frequent El Niño events.*

The relationship between average world temperature and CO_2 concentration (and that of other greenhouse gases as well) seems clear. The general scientific consensus is that these changes are linked to human activity, specifically to the release of CO_2 into the atmosphere as a result of fossil fuel burning. Other anthropogenic factors such as deforestation have also been cited as contributing to global warming.

The consequences of global warming can be profound. If a continuation of these trends can be expected, increases in CO_2 levels will yield further increases in average global temperature. Predictions of the severity of these effects depend on a number

Figure 4.16: Reduction in the size of the Grinnell Glacier in Glacier National Park between 1900 (top) and 2008 (bottom).

of parameters and some predictions based on recent climate models are shown in Figure 4.18 on the next page. Figure 4.19 on the page 137 illustrates the correlation between atmospheric CO_2 concentration and global temperature increase as predicted by a study reported by the Intergovernmental Panel on Climate Change (IPCC).

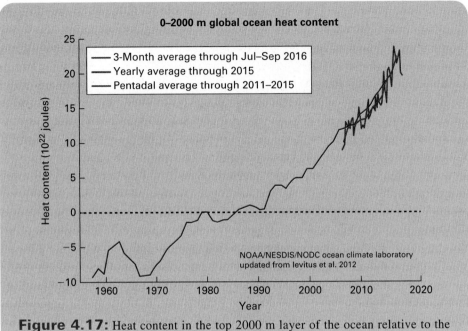

Figure 4.17: Heat content in the top 2000 m layer of the ocean relative to the average for the period 1955–2006.

NOAA, Global Ocean Heat and Salt Content, https://www.nodc.noaa.gov/OC5/3M_HEAT_CONTENT/

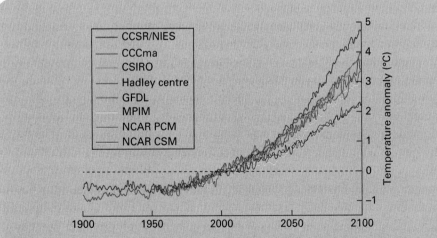

Figure 4.18: Predicted average world temperature increase as a result of CO_2 based on various models. Models are based on the International Panel on Climate Change (IPCC) *Special Report on Emissions Scenarios (SRES)* A2, which assumes a world of independently operating, self-reliant nations; a continuously increasing population; and regionally oriented economic development.

Based on data from http://www.ipcc-data.org

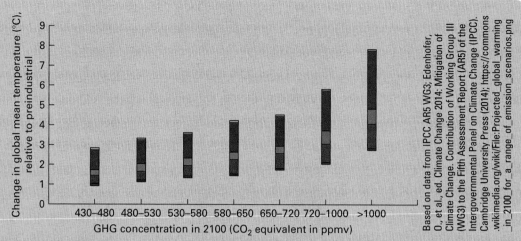

Figure 4.19: Predicted relationship of global temperature increase since preindustrial times and total atmospheric CO_2 concentration in the year 2100. The blue bands represent the mean uncertainty and the orange bands represent the maximum uncertainty of the prediction.

ENERGY EXTRA 4.2
Natural and anthropogenic climate change

It is clear from the data in Figure 4.11 that climatic changes have influenced the earth's temperature throughout its history. Some of these changes are periodic (or quasi-periodic), and the figure clearly shows that a major component of this periodicity consists of cycles in both temperature and carbon dioxide concentration that repeat every 135 million years or so. It is obvious that the reasons for these changes are not anthropogenic.

To better understand the behavior shown in Figure 4.11, it is useful to look at the amplitude of these fluctuations as a function of period, that is, the Fourier spectrum, of these data. This type of analysis is presented in the figure below for data up to about 1970 (when significant short-term changes in the earth's temperature and CO_2 concentration began). Some obvious periodicity in the earth's temperature can be seen in the graph. First, there are variations with a periodicity of one day, due to the earth's

rotation. Second, there are variations with a periodicity of one year due to the earth's orbit around the sun. This annual periodicity is also readily seen in the CO_2 concentrations shown in Figure 4.13.

Two important longer-term periodicities are shown in the graph as blue regions in the spectrum: periodicity of tens of years and periodicity of hundreds of million years. The former can be readily attributed to the observed 11-year and 22-year cycles of solar activity (i.e., the sun spot cycle). The reasons for the latter, very long-term, variations, which are the most obvious in Figure 4.11, are less clear. However, scientists have observed that these long-term fluctuations have approximately half the periodicity of the sun's orbit around the galactic center. It is speculated that, as the solar system passes through the spiral arms of the galaxy, it experiences a higher flux of cosmic rays because of the higher density of cosmic ray sources, such

Continued on page 138

Energy Extra 4.2 continued

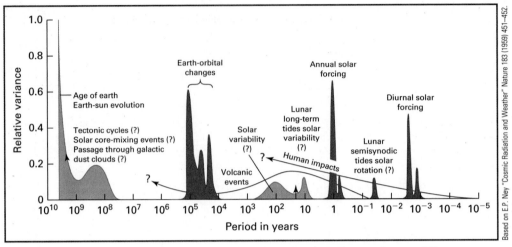

Relative amplitude of climatic changes as a function of period. From reference [2].

as supernovae, in the arms. Edward Ney [2] first suggested a possible mechanism responsible for the relationship between cosmic ray flux and the earth's climate in 1959. Cosmic rays, which are primarily high-energy charged particles such as protons, cause ionization in the atmosphere, and those ions seed clouds by attracting water molecules and condensing them into droplets. The resulting increased cloudiness in the earth's atmosphere reflects incoming sunlight and effectively cools the earth's surface. The historical cosmic ray flux may be determined by the geological observation of certain isotopes that are created in the atmosphere by cosmic rays. The observational correlation between cosmic ray flux and the earth's temperature (as determined by fossil evidence of various organisms) is illustrated in the figure below. While the general features of the model described above explain the observational data, the details of the physics behind these phenomena remain unclear.

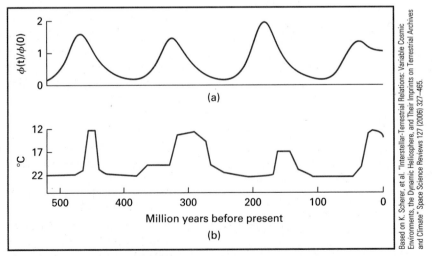

(a) Normalized cosmic ray flux and (b) earth temperature based on fossil evidence as a function of time. Adapted from Scherer et al. [3]

Continued on page 139

Energy Extra 4.2 continued

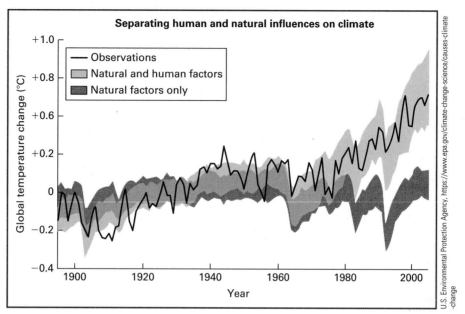

U.S. Environmental Protection Agency, https://www.epa.gov/climate-change-science/causes-climate-change

Observed global temperature changes and a comparison of models including natural and anthropogenic factors.

Solar activity can indirectly affect the earth's climate on a decadal time scale by modulating the cosmic ray flux that reaches the earth's surface. The increased solar magnetic fields associated with times of increased activity deflect the charged particles that constitute cosmic rays and reduce the flux at the earth's surface.

Some have noted that historical temperature changes as seen in Figure 4.11 have preceded corresponding changes in the atmospheric CO_2 concentration. While the question of whether CO_2 concentration drives temperature or temperature drives CO_2 concentration remains, the two quantities are clearly correlated. It is generally theorized that increased temperature leads to increased levels of CO_2, which in turn drive further temperature increases. While it seems that cosmic rays are a likely explanation for past climatological fluctuations that occurred before anthropogenic greenhouse gas emissions, their relevance to the global warming observed during the past century or so remains unclear. Models of global temperature change that include natural and anthropogenic factors, as the figure indicates, show that recent behaviors, particularly within the past forty or fifty years, can only be explained in the context of human activity. The general consensus among climate researchers is that

greenhouse gas emissions from human activities are, by far, the dominant contribution to global temperature trends observed in recent years and that cosmic ray fluctuations play a minor role.

Topic for Discussion

Discuss the significance of the amplitude and time scale of recent global temperature trends and atmospheric CO_2 concentration shown in Figures 4.12 to 4.15 in the context of the features shown in the historic record in Figure 4.11. If cosmic ray flux is responsible for global temperature changes via the mechanisms described above, what trends in average cloud cover might be expected? Is there any evidence in the literature to support this mechanism?

References

1. J.M. Mitchell, "An Overview of Climatic Variability and Its Causal Mechanisms," *Quaternary Research* **6** (1976) 481–493.
2. E.P. Ney, "Cosmic Radiation and Weather," *Nature* 183 (1959) 451–452.
3. K. Scherer, et al., "Interstellar-Terrestrial Relations: Variable Cosmic Environments, the Dynamic Heliosphere, and Their Imprints on Terrestrial Archives and Climate," *Space Science Reviews* 127 (2006) 327–465.

Certainly, these studies show that any significant increase in CO_2 levels beyond the current value of around 400 ppm will lead to a substantial global temperature increase. Predicted CO_2 changes from possible future anthropogenic carbon emission depend on global approaches to energy production and carbon mitigation strategies. However, Figures 4.18 and 4.19 indicate that, unless significant changes in energy use are implemented during the remainder of this century, global temperature increases of 2 to 5°C above preindustrial level can be expected by 2100.

While carbon dioxide is the greenhouse gas that causes the greatest concern in terms of global warming, other greenhouse gases, as indicated in Table 4.6, are also a factor for the earth's climate. Atmospheric methane and nitrous oxide have both shown substantial increases in recent years, as illustrated in Figure 4.20. It is commonly believed that these increases can be linked to anthropogenic causes.

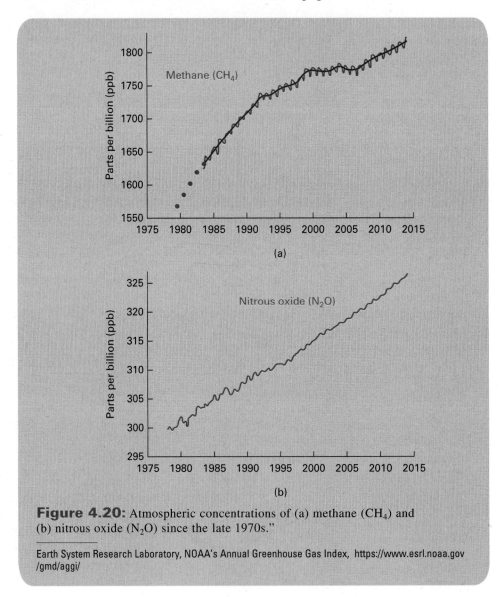

Figure 4.20: Atmospheric concentrations of (a) methane (CH_4) and (b) nitrous oxide (N_2O) since the late 1970s."

Earth System Research Laboratory, NOAA's Annual Greenhouse Gas Index, https://www.esrl.noaa.gov/gmd/aggi/

ENERGY EXTRA 4.3
Life cycle assessment

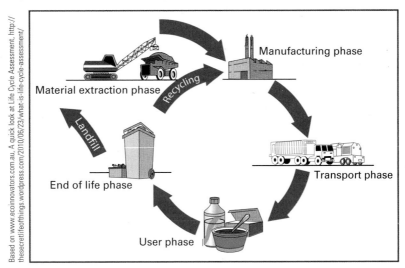

Based on www.ecoinnovators.com.au. A quick look at Life Cycle Assessment, http://thesecretlifeofthings.wordpress.com/2010/06/23/what-is-life-cycle-assessment/

Phases in the life cycle of a manufactured product.

In Chapter 3, it was seen that an energy value is associated with all products that humans use. A more thorough analysis of the environmental impact of a product throughout its life is referred to as *life cycle assessment, life cycle analysis,* or sometimes *cradle-to-grave analysis.* This type of analysis considers all the ways in which the production, use, and disposal of an item influences our environment. As shown in the figure, the life of a product may be divided into several phases. The process begins with the extraction of raw materials from nature. These natural resources are then processed and manufactured into a product that is transported to the user. At the end of the useful life of the product, it is either discarded as waste or recycled (or in a few cases refurbished, reused, etc.).

An assessment of each stage of the life cycle must consider the relevant inputs and outputs. For example, the manufacturing process may be described in terms of material input (e.g., steel and plastic), resource input (e.g., electricity), waste output (e.g., CO_2, waste heat, scrap plastic), and material

output (e.g., an automobile), whereas the use phase may be described by material input (automobile), resource input (gasoline), waste output (CO_2 emissions), and material output (used automobile).

It may seem reasonable to minimize the environmental impact at each stage of the life cycle, but this may not always be the best approach. For example, the production of a more fuel efficient automobile (i.e., lower CO_2 emissions during the use phase) may involve more extensive production processes (i.e., more CO_2 emission in the production phase), and/or less easily recyclable materials at the end of life phase. Thus, for a detailed product assessment, it is important to consider the environmental impact at each phase and how these impacts are interrelated.

The actual life cycle assessment follows these steps:

- *Scope and goals:* Where do we start, and where do we end? For an automobile, do we start with iron ore, or do we start with processed steel? We might also just consider the

Continued on page 142

Energy Extra 4.3 continued

.use phase (i.e., the resources needed to run the car during its lifetime, gas, oil, etc.) and its waste products (the CO_2, NO_x, etc., produced).

- *Inventory analysis:* Can we assess all materials that go into automobile manufacture, or should we concentrate on the major ones (steel, plastic, etc., that make up 98% of the mass)? If the latter, are we ignoring minor components (e.g., electronics) that are energy intensive (per unit mass) to produce?

- *Impact assessment:* What is the environmental impact? For the materials phase of an automobile, we might consider how many kilograms of CO_2 are produced in processing 1 kg of steel. In the use phase, we might consider the kilograms of CO_2 emitted per kilometer driven.

- *Interpretation:* What do these results mean, and can they be used to modify the production and/or use of an item in order to optimize its effectiveness and minimize its adverse impacts?

  Overall, this kind of analysis is not simple or even well-defined, but it is important in understanding how our actions influence the environment around us.

Topic for Discussion

Find information about CO_2 emissions during the production of steel, that is, kilograms of CO_2 emitted per kilogram steel manufactured. Assuming that a 1.5-t family car is made primarily of steel, how much CO_2 is emitted in producing that steel? How does this compare with the CO_2 emissions during one year of average driving?

4.6 Carbon Sequestration

Because the combustion of all carbon based fuels releases CO_2, a reduction in fossil fuel use will result in a reduction in CO_2 emissions. The consequences of this action for changes that have resulted from fossil fuel use are not entirely clear because the reversibility of some changes is not certain. Another approach is to minimize the effects of the carbon released to the environment. Reforestation may have some positive effects, but the sequestration of carbon emissions from fossil fuel burning is essential for CO_2 reduction if we continue the widespread use of these fuels. This is most easily accomplished when dealing with emissions from a relatively small number of major power plants rather than much more numerous portable sources of carbon emissions, such as automobiles.

The magnitude of this problem is emphasized by a simple calculation of the amount of CO_2 produced by fossil fuel combustion. The combustion of coal provides a good example. One kilogram of bituminous coal has a volume of about 7.7×10^{-4} m^3 and contains about 550 g of carbon. If this is completely oxidized, it will produce $(550 \text{ g}) \times (44 \text{ g/mol})/(12 \text{ g/mol}) = 2000$ g of CO_2. At standard temperature and pressure (STP), this will occupy a volume of more than 1 m^3. The total world coal usage of about 5×10^{12} kg per year amounts to an annual emission of about 10^{13} kg of CO_2. This quantity of CO_2 at STP will occupy a cube about 18 km on a side. Four of the major approaches to sequestering or confining this gas are storage:

- Underground.
- In the oceans.
- In solid form.
- Convert it into methanol

Of course, the storage of CO_2 is not ultimately beneficial if the CO_2 escapes, so it is important to consider this possibility for any proposed storage method. The specifics of these methods are now discussed.

Underground Storage CO_2 may be pumped into cavities underground for storage. Possible underground cavities are depleted oil wells and depleted natural gas wells. In fact, at present in the United States about 60×10^6 m^3 of CO_2 per day are pumped underground. Much of this is for the purpose of enhancing oil recovery rather than carbon sequestration as such. Although CO_2 pumped into a depleted natural gas well can be expected to remain there indefinitely (as the well trapped natural gas underground for millions of years), it is uncertain whether CO_2 pumped into a depleted oil well will remain in place. It is possible that over time the CO_2 could diffuse through porous rock and escape. Studies to answer this question are needed.

Ocean Storage CO_2 may be disposed of in the oceans in several ways. It may be pumped into the ocean as gaseous CO_2, it may be liquefied and pumped into the ocean, or it may be released in the form of solid compounds. CO_2 released in the ocean near the surface tends to escape from the surface and return to the atmosphere. CO_2 released at a sufficient depth may be trapped in the lower colder layers of water that mix very little with the warmer surface layers. Two factors need to be considered: the longevity of ocean sequestration and its effects on the ocean environment. Some work has suggested that up to about 20% of CO_2 pumped deep into the oceans may diffuse out to the atmosphere on a timescale of about 300 years. Considerable uncertainty in this behavior still exists. Also, effects on ocean life are uncertain. Some studies suggest that marine organisms tend to avoid regions of increased CO_2 concentration, and other studies have indicated increased organism mortality due to ocean CO_2 sequestration.

The effects of CO_2 on the oceans are, in fact, not only the result of possible intentional ocean sequestration. As the concentration of CO_2 in the atmosphere increases, the equilibrium concentration of dissolved CO_2 in the oceans also increases. This CO_2 reacts with water to form an equilibrium concentration of four chemical species: dissolved carbon dioxide (CO_2), carbonic acid (H_2CO_3), bicarbonate ions (HCO_3^-), and carbonate ions (CO_3^-). These dissolved species decrease the pH of the sea water (which is normally basic, pH > 7) and make it closer to neutral. This process, referred to as ocean acidification, has a number of possible biological consequences. Perhaps the most obvious of these effects is on marine invertebrates' production of shells. Shells are composed of calcium carbonate ($CaCO_3$), which precipitates from the sea water. If the pH of the sea water decreases, the $CaCO_3$ production rate decreases. Precipitated $CaCO_3$ may even dissolve back into the sea water.

Solid Storage CO_2 may be reacted with calcium- or magnesium-containing chemicals to produce fairly stable solids that contain the carbon atoms. The two most likely candidates for this type of storage are calcium oxide and magnesium oxide. These are both naturally occurring minerals that are quite common components of the earth's crust. Table 4.7 on the next page gives some of the relevant properties. Carbon storage occurs as a result of the reaction of these oxides with CO_2 by the processes

$$CaO + CO_2 \rightarrow CaCO_3 \tag{4.10}$$

Table 4.7: Properties of naturally occurring calcium and magnesium oxides as carbon sequestration materials. The *enthalpy of formation* is the heat (or energy) associated with the reaction where the negative sign indicates that the reaction is exothermic.

oxide	% earth's crust	carbonate	enthalpy of formation (kJ/mol)
CaO	4.9	$CaCO_3$	-179
MgO	4.36	$MgCO_3$	-117

and

$$MgO + CO_2 \rightarrow MgCO_3. \tag{4.11}$$

Both processes are exothermic, as indicated in Table 4.7. Therefore, the compounds on the right are stable, and the CO_2 is permanently captured. The major problem with the reactions in equations (4.10) and (4.11) is that they proceed very slowly at room temperature and atmospheric pressure (STP). To capture carbon at a useful rate, it is necessary to increase the rate of the reaction. The successful use of this method for carbon capture requires the development of an industrial-scale process that is efficient, environmentally acceptable, and economical (both from a financial and an energy standpoint).

Example 4.6

A coal-fired generating system operating at a Carnot efficiency of 37% produces 700 MW_e output. What is the mass of MgO needed to sequester the CO_2 produced by the plant each second?

Solution

The total energy production will be

$$(7.00 \times 10^8 \text{ J/s})/(0.37) = 1.80 \times 10^9 \text{ J/s}.$$

For an energy content of 31 MJ/kg of coal, the amount of coal burned per second will be

$$(1.80 \times 10^9 \text{ J/s})/(3.1 \times 10^7 \text{ J/kg}) = 58.1 \text{ kg}.$$

According to the approximate reaction for combustion

$$C + O_2 \rightarrow CO_2,$$

then 1 mole (12 g) of carbon will yield 1 mole (44 g) of CO_2. So 58.1 kg of coal burned per second will produce

$$(58.1 \text{ kg/s}) \times (44 \text{ g/12 g}) = 213 \text{ kg/s of } CO_2.$$

From the reaction in equation (4.11), 1 mole of CO_2 will require 1 mole of MgO (molecular weight, $24 + 16 = 40$), so 213 kg/s of CO_2 will require

$$(213 \text{ kg/s}) \times (40/44) = 194 \text{ kg/s of MgO.}$$

Figure 4.21: George Olah Renewable Methanol Plant in Svartsengi, Iceland

Carbon Recycling International

Methanol Production Carbon dioxide can also be used to produce methanol by reaction with hydrogen (see Chapter 20 for more information on hydrogen) by the reaction

$$CO_2 + 3H_2 \rightarrow CH_3OH + H_2O. \tag{4.12}$$

The hydrogen on the left-hand side of the equation can be produced by the electrolysis of water by the reaction

$$2H_2O + \text{electricity} \rightarrow 2H_2 + O_2. \tag{4.13}$$

Although this process requires the input of energy (i.e., electricity to produce the hydrogen), it has the advantage that it eliminates CO_2 from the atmosphere and produces a useful liquid fuel (see Chapter 16 on biofuels). The company Carbon Recycling International has the world's largest activities in this area and operates the George Olah Renewable Methanol Plant in Svartsengi, Iceland (named after George Olah, the Nobel Prize laureate in chemistry who is a proponent of a methanol economy) shown in Figure 4.21. The plant uses CO_2 gases that are a by-product of geothermal electricity generation (see Chapter 15) and has a capacity of 5 million liters of methanol per year.

Certainly, the viability of carbon sequestration methods must be studied in more detail, and their use to mitigate the effects of continued fossil fuel use has yet to be established.

4.7 International Climate Change Initiatives

Since the 1970s, international efforts have aimed to reduce energy consumption and mitigate the effects of energy production and use on the environment. In more recent years, particularly, emphasis has been placed on reducing carbon emissions with the objective of controlling global warming. In this respect, it is interesting to consider the

Table 4.8: Carbon emissions (in kg of CO_2) for the top 20 CO_2-producing countries in 2015 showing gross domestic product (GDP) and carbon per GDP.

country	CO_2 per year (10^9 kg/y)	GDP per year (10^9 US$/y)	CO_2/GDP (kg/US$)
China	10,358	11,008	0.941
United States	5416	18037	0.300
India	2273	2095	1.085
Russia	1617	1331	1.215
Japan	1236	4123	0.300
Germany	799	3363	0.238
Iran	649	425	1.527
Saudi Arabia	601	646	0.931
South Korea	590	1378	0.428
Canada	557	1551	0.359
Indonesia	535	862	0.621
Brazil	513	1775	0.289
Mexico	473	1144	0.413
South Africa	462	315	1.467
United Kingdom	418	2858	0.146
Australia	400	1339	0.298
Turkey	385	718	0.536
Italy	359	1821	0.197
France	341	2419	0.141
Poland	315	477	0.661
rest of world	7963	—	—
total	**36,262**	—	—

Based on carbon data from http://www.globalcarbonatlas.org/en/CO2-emissions, GDP data from http://databank.worldbank.org/data/download/GDP.pdf

carbon emissions of various countries in the context of their energy use and energy policies. Table 4.8 shows the carbon dioxide emissions of the top 20 carbon-producing nations. We would expect population to be a major factor in determining the carbon emissions of a country. We might also expect that other factors, such as those reflected in the per capita energy consumption discussed in Chapter 2, would influence carbon emissions. These include the degree of industrialization, which is reflected in the per capita gross domestic product (GDP), as well as climate and population density. Taking all these points into consideration, we see some interesting features in this table, particularly when viewed together with the information presented in Figure 2.5. France has the lowest carbon emissions per dollar GDP of any country in the table and scores better than similar neighboring European countries, such as Germany and Italy. Brazil scores quite well compared with other countries with similar climates and similar per capita GDPs, such as Mexico. It may be possible to understand these differences in terms of each country's approach to satisfying their energy needs and their available natural energy resources. Factors that contribute to these features certainly include the facts that France produces nearly all its electricity by nuclear reactors (Chapter 6), and Brazil has a well-developed, successful national commitment to biofuels (Chapter 16).

Clearly the information in Table 4.8 illustrates the diversity of energy use in various parts of the world. It also indicates the complexity of the task of dealing with the continuing effects of human activities on the climate and the environment in general. Global initiatives to deal with climate change began through the United Nations in the early 1990s with the United Nations Framework Convention on Climate Change (UNFCCC), which followed from similar earlier global initiatives to deal with ozone depletion, and continued with agreements made in Kyoto (1997) and Paris (2015), as described below.

Example 4.7

From the data in Table 4.8, calculate the total annual primary energy generated worldwide if carbon emissions are the result of the combustion of pure carbon. Compare with the total world primary energy use.

Solution

From equation (1.13), the combustion of one kg of pure carbon produces 32.8 MJ and

$$(44 \text{ g/mol } CO_2)/(12 \text{ g/mol } C) \times (1 \text{ kg}) = 3.67 \text{ kg of } CO_2.$$

Thus the production of 1 kg of CO_2 yields

$$(32.8 \text{ MJ})/(3.67 \text{ kg } CO_2) = 8.95 \text{ MJ per kg } CO_2.$$

The total world CO_2 emission as given in Table 4.8 (3.6262×10^{13} kg) corresponds to the annual generation of

$$(3.6262 \times 10^{13} \text{ kg}) \times (8.95 \text{ MJ/kg}) = 3.24 \times 10^{14} \text{ MJ} = 3.24 \times 10^{20} \text{ J}.$$

This may be compared to the total world primary energy production from the Preface of 6.0×10^{20} J per year. The difference is due to primary energy derived from other sources (such as hydroelectric and nuclear) and the fact that CO_2 emissions per unit energy are less for hydrocarbons (such as natural gas and petroleum) than for pure carbon (see Chapter 3).

4.7a United Nations Framework Convention on Climate Change (UNFCCC)

The UNFCCC was developed at the 1992 United Nations Conference on Environment and Development in Rio de Janeiro (commonly known as the *Earth Summit*) for the purpose of "stabilizing atmospheric concentrations of greenhouse gases at a level that would prevent dangerous anthropogenic interference with the earth's climate system." The convention was originally signed by 154 nations, and there are currently 197 member states. Although the framework does not provide specific guidelines for greenhouse gas emissions of particular countries, it provides general approaches for future international negotiations concerning climate change. Points agreed to as part of the UNFCCC include the following:

- All parties to the convention should act in such a manner as to protect the climate system.

- Developed countries should take the lead in addressing climate change with the aim of stabilizing greenhouse gas emission.
- Developing countries would implement measures to mitigate climate change to the extent that financial and technical resources were available and that the need for social and economic development would allow.

4.7b Kyoto Protocol

The Kyoto Protocol was the first international agreement developed from the UNFCCC that made specific greenhouse gas emission targets for participating countries. These targets are based on the assumptions that (1) global warming is occurring and (2) this warming is anthropogenic. The protocol was adopted at a meeting of the UNFCCC in Kyoto, Japan in December 1997 and went into effect in February 2005. The first commitment period ran from 2008 to 2012. A second commitment period was negotiated in 2012 in Doha, Qatar and is known as the Doha Amendment. This second period runs from 2013 to 2020. The protocol is based on the common goal of reducing the environmental effects of anthropogenic greenhouse gas emissions and puts the principal obligation for emission reduction on developed countries on the basis that they were largely responsible for the then-current levels of greenhouse gases in the atmosphere. Most industrialized countries have a binding target of 6 to 8% reduction in greenhouse gas emissions relative to a 1990 base year for the first commitment period and a 20% reduction relative to the 1990 base for the second commitment period. Developing countries have no binding targets.

While the parties to the Kyoto Protocol have done well to meet the target commitments specified in the first period of the Kyoto Protocol, it is important to consider emissions from countries without binding targets. The United States has not ratified the agreement. Although it was signed by the Clinton Administration in 1998, it was never considered by the U.S. Senate because of objections that it imposed no binding targets on developing countries. Relative to 1990, U.S. greenhouse gas emissions have increased by about 11%. Two of the largest greenhouse gas–emitting countries, China and India, have no binding targets. Since 1990, China's CO_2 emissions have increased by more than a factor of 4, while India's have increased by more than a factor of 2. Canada was an original signatory to the protocol with a target of 6% greenhouse gas reduction in the first period. However, relative to 1990, Canada's emissions have increased by about 24%. In 2011 Canada withdrew from the agreement before the beginning of the second period.

Several factors affected Canada's decision. First, some countries without binding targets produced much greater greenhouse gas emissions, and their emissions had increased substantially since 1990. These include China and India, because of the terms of the treaty, and the United States, because of its failure to ratify the agreement. Secondly, because of its increased CO_2 emissions relative to 1990, Canada would not be able to meet its target reduction and therefore would be subject to substantial financial penalties in the second period. This situation is due to the fact that the federal government negotiates international agreements, but the provincial governments have jurisdiction over their own natural resources and energy-related matters and, therefore, have principal control over emissions, as discussed in Section 17.2b.

4.7c Paris Accord

While the Kyoto Protocol has made progress in reducing greenhouse gas emissions in many countries, the lack of commitment from some major industrial nations and some major greenhouse gas emitters has been problematic. The Paris Agreement (or Accord de Paris) was adopted in December 2015 and has been an important step forward in bringing global support to this initiative. The Paris Agreement is not an extension of the Kyoto Protocol but an independent instrument aimed at the mitigation of global warming. The agreement has been signed by 195 of the 197 member states of the UNFCCC; Nicaragua and Syria have not signed. As of June 2017, 148 of the signatories had ratified or acceded to the agreement, including China, the United States, India, and Canada, although on 1 June 2017 the United States announced its intentions to withdraw from the agreement. According to the terms of the agreement, the earliest that the U.S. withdrawal could take effect is 4 November 2020. In response, a number of individual states formed the United States Climate Alliance that is committed to upholding the objectives of the Paris Accord on the state level.

The major points of the Paris Agreement are as follows:

- A commitment to limit global temperature to less than 2°C above the preindustrial value (some counties have set a more ambitious goal of less than 1.5° C above the preindustrial level).
- The first global climate agreement that includes both developed and developing nations and has agreement of nearly all nations.
- Financial contributions ($100 billion annually by 2020) provided by developed nations to poorer nations to implement measures limiting greenhouse gas emissions.
- Publication of greenhouse gas targets by all member states every five years, beginning in 2023.
- A carbon-neutral world to be achieved sometime between 2050 and 2100.

The agreement sets greenhouse gas emission standards (based on a percentage of total world greenhouse gas emissions) for all signatories. These standards require ratification by each nation and (for the top 20 emitters as agreed upon at the Paris meeting) are summarized in Table 4.9 on the next page. These values may be compared with actual carbon emissions as shown in Table 4.8.

While the Paris Agreement is the first global initiative in which nearly all countries (including all the major greenhouse gas emitters) made a commitment to take measures to mitigate climate change, the agreement may not be ideal in all respects. Specifically, it has been criticized for the following points:

- The agreement is nonbinding and there is no enforcement mechanism or financial penalties for noncompliance. It is expected that countries will voluntarily control and monitor their greenhouse gas emissions.
- Studies by the United Nations Environmental Programme (UNEP) have predicted that the greenhouse gas emission limits set by the Paris Agreement will give rise to a global temperature increase of about 3°C relative to preindustrial values, resulting in catastrophic climatic changes worldwide. In this context, it is important to consider that current global mean temperatures are already about 1.3°C above preindustrial levels (Figure 4.14).

While global cooperation to mitigate the effects of greenhouse gas emissions on the earth's climate has progressed over the past two and a half decades, it seems likely that more stringent targets and better approaches to limiting emissions will need to be established in the future.

country	maximum % total world greenhouse gas emissions needed for ratification
Table 4.9: Percent total greenhouse gas emissions for the countries with the 20 highest ratification limit as specified in the Paris Agreement. (* = ratified or acceded)	
China *	20.09
United States *	17.89
Russia	7.53
India *	4.10
Japan *	3.79
Germany *	2.56
Brazil *	2.48
Canada *	1.95
South Korea *	1.85
Mexico *	1.70
United Kingdom *	1.55
Indonesia *	1.49
South Africa *	1.46
Australia *	1.46
France *	1.34
Iran	1.30
Turkey	1.24
Italy *	1.18
Poland *	1.06
Ukraine *	1.04
rest of world	22.94
total	**100.00**

4.8 Summary

This chapter reviewed the reasons why energy production and use produce pollution. Any energy generation method that utilizes a heat engine to convert thermal energy to mechanical energy (often with the final objective of producing electricity) generates excess heat that is released into the environment. These methods include nuclear- and fossil fuel–fired generating stations, as well as vehicles operating on internal combustion engines. Although the effects of thermal pollution are typically localized, thermal generating stations that release excess heat into rivers, lakes, or the ocean can have substantial effects on the local ecology. Chemical pollution results from the combustion of fossil fuels, either in the process of generating electricity or for transportation applications. Carbon monoxide results from the incomplete combustion of carbon compounds, and nitrogen oxides are produced when nitrogen in the air is subject to elevated temperatures. Studies have shown that a significant fraction of these pollutants come from vehicle emissions. Most governments have implemented emission control standards to mitigate this problem. Sulfur compounds and particulate matter result from impurities and unburned components of fossil fuels and are most significant during the combustion of coal for electricity generation. Chemical or mechanical methods are in common use to remove these components of exhaust gas from coal-fired stations before it is released to the atmosphere.

Climate change caused by the emission of greenhouse gases is, by far, the most serious consequence of our use of fossil fuels. The greenhouse effect results from the presence of molecules in the atmosphere that selectively absorb radiation at particular wavelengths. Sunlight, which penetrates the atmosphere, is absorbed by the earth and reirradiated at longer wavelengths that are more readily absorbed in the atmosphere, leading to an increase in temperature at the surface of the earth. Carbon dioxide is produced by the combustion of any hydrocarbon and is an effective greenhouse gas. The continued use of fossil fuels will lead to an increase in greenhouse gases in the atmosphere, and this is the major contributor to global warming. Carbon sequestration is a possible means of reducing greenhouse gas emissions. Carbon dioxide from fossil fuel combustion can be stored underground in pressurized caverns or dispersed in the oceans. Perhaps the most attractive method of sequestering carbon is the formation of carbonates by solid state reaction. More work on this technology needs to be done to develop an efficient and economical process.

Problems

4-1 Pluto has no atmosphere and has an average orbital diameter that is 39.2 times that of the earth. The mean surface temperature is 37 K. Estimate the albedo of Pluto.

4-2 Assume that coal has a sulfur content of 3% by weight. If all the sulfur is converted into SO_2 during the combustion process, how much SO_2 is produced per tonne of coal? How much is produced per megajoule of energy produced?

4-3 A typical gasoline-powered vehicle requires about 3.7 MJ of primary energy (i.e., energy content of the gasoline) to travel 1 km. How many kilograms of CO_2 are produced annually by a vehicle that is driven 40,000 km per year?

4-4 Denver has a land area of about 400 km^2. Using Figure 4.1, make the assumption that all CO is uniformly distributed in the atmosphere up to a height of 300 m. Estimate the average total mass of CO in Denver's atmosphere. *Note:* The density of air is 1.204 kg/m^3.

4-5 There are about 250 million personal vehicles in the United States. Assume that each drives 25,000 km per year on average. What is the average reduction in NO_x emissions per square kilometer per year resulting from the implementation of the 2004 U.S. emission standards compared with the situation before any standards were introduced (1960)?

4-6 (a) Calculate the mass of CO_2 produced per kilogram of methane burned.

    (b) At most natural gas wells, excess natural gas is generally burned rather than released to the atmosphere. Methane released during the decomposition of waste at landfill sites is also sometimes burned. Discuss the positive or negative environmental effects of these practices.

4-7 If approximately 0.4% of the carbon emitted from a gasoline internal combustion engine is CO rather than CO_2, how many liters of gasoline would need to be burned to reach a lethal level of CO in a cubic volume 100 m on a side? *Note:* The density of air is 1.204 kg/m^3.

4-8 Suppose 10^{13} kg per year of CO_2 emitted by coal-fired generating stations was to be sequestered by calcium oxide (CaO, density 3350 kg/m^3) according to the reaction

$$CaO + CO_2 \rightarrow CaCO_3.$$

What volume of CaO would be needed?

4-9 (a) The orbit of Venus is at an average distance of 0.723 astronomical units from the sun. Given an albedo of 0.9, calculate the mean surface temperature if Venus has no atmosphere.

    (b) Locate information about the actual surface temperature of Venus and discuss the relevance of your answer to part (a).

4-10 If a gasoline-powered vehicle requires primary energy of 3.7 MJ/km from the combustion of gasoline (mostly octane) and emits 1.7 g/km of CO, what fraction of the carbon in the octane is not fully oxidized?

4-11 The earth's moon has no atmosphere and has an albedo of 0.136. Calculate the total power radiated (in watts) from the moon's surface.

4-12 A coal-fired generating station with a thermodynamic efficiency of 35% produces an average output of 600 MW_e. Assume that coal is pure carbon and that CaO is used to sequester all of the CO_2 that is produced.

    (a) What is the mass of CaO required each year?

    (b) Find the density of CaO and calculate the size of a cubic container required to contain the CaO from part (a).

4-13 A generating station burns methane and produces electricity with an efficiency of 42%. If the facility produces 2.6 TWh of electricity per year, what is the annual level of CO_2 emissions?

4-14 (a) There are approximately 250 million passenger vehicles in the United States. If each is driven 25,000 km per year, calculate the total mass of CO, NO_x, and HC emitted in 1970 if all vehicles just meet the government standard at that time.

    (b) Repeat part (a) for vehicles that just meet the 2001 standard.

(c) Estimate the total reduction in these three pollutants between 1970 and 2001 for vehicles that meet the actual yearly standards compared to the situation where all vehicles remained at the 1970 level.

4-15 Compare the masses and volumes of CaO and MgO required to sequester 1 tonne of CO_2.

4-16 Compare the values of the ALR for the troposphere as calculated from Figures 4.7 and 4.8 with the typical value given in the text.

4-17 Gasoline (approximated by octane) typically contains 0.1% sulfur by weight. If a passenger vehicle requires 3.5 MJ/km of primary energy, calculate the total annual mass of SO_2 produced by 250 million vehicles in the United States, each traveling 25,000 km per year.

4-18 A coal-fired generating station produces an average of 1000 MW_e at an efficiency of 36%. If the coal that is burned contains 2% sulfur by weight, calculate the volume (in liters) of sulfuric acid that is produced daily.

4-19 The total mass of the atmosphere per unit area is equal to the atmospheric pressure at the surface of the earth divided by the gravitational acceleration. This follows from Newton's law $F = ma$. Assume that all carbon in the atmosphere is in the form of CO_2 and that the current concentration (in ppm per volume) is 400 ppm. What is the total mass of carbon in the atmosphere?

4-20 (a) Using the information in Figure 4.11, suggest a quantitative relationship between changes in CO_2 concentration and changes in the average earth temperature.

(b) Use the results of part (a) and the current average temperature of 15°C to estimate the surface temperature of the earth if the atmosphere contained no CO_2.

(c) Compare the results of part (b) with the value given in the text for the equilibrium temperature of the earth without an atmosphere.

Bibliography

R. Alley. *The 2-Mile Time Machine: Ice Cores, Abrupt Climate Change and Our Future.* Princeton University Press, Princeton, NJ (2002).

S. B. Alpert. "Clean coal technology and advanced coal-based power plants." *Annual Review of Energy and the Environment* **16** (1991): 1.

G. J. Aubrecht II. *Energy: Physical, Environmental, and Social Impact* (3rd ed.). Pearson Prentice Hall, Upper Saddle River, NJ (2006).

R. E. Balzhiser and K. E. Yeager. "Coal fired power plants for the future." *Scientific American* **257** (1987): 100.

W. Burroughs. *Climate Change: A Multidisciplinary Approach.* Cambridge University Press, Cambridge, MA (2001).

C. J. Campbell, *The Coming Oil Crisis,* Multi-Science Publishers, London (1988).

W. Collins, R. Colman, J. Haywood, M. R. Manning, and P. Mote. "The physical science behind climate change." *Scientific American* **297** (2007): 64.

B. P. Eliasson, W. F. Riemer, and A. Wokaun (Eds.). *Greenhouse Gas Control Technologies.* Pergamon Press, Amsterdam (1999).

J. A. Fay and D. S. Golomb. *Energy and the Environment.* Oxford University Press, New York (2002).

D. Goodstein. *Out of Gas: The End of the Age of Oil.* W.W. Norton, New York (2004).

J. Hansen. "Defusing the global warming time bomb." *Scientific American* **290** (2004): 68.

R. J. Heinsohn and R. L. Kabel. *Sources and Control of Air Pollution.* Prentice-Hall, Upper Saddle River, NJ (1999).

H. J. Herzog and E. M. Drake. "Carbon dioxide recovery and disposal from large energy systems." *Ann. Rev. Energy. Environ.* **21** (1996): 145.

H. Herzog, B. Eliasson, and O. Kaarstad. "Capturing greenhouse gases." *Scientific American* **282** (2000): 72.

R. Hinrichs and M. Kleinbach. *Energy: Its Use and the Environment* (5th ed.). Brooks-Cole, Belmont, CA (2012).

M. Hoffert, K. Caldeira, G. Benford, et al., Advanced technology paths to global climate stability: Energy for a greenhouse planet." *Science* **298** (2002): 981.

S. Holloway. "An overview of the underground disposal of carbon dioxide." *Energy Conversion and Management* **38** (1997): 193.

J. Houghton. *Global Warming: The Complete Briefing.* Cambridge University Press, Cambridge, MA (2004).

J. A. Kraushaar and R. A. Ristinen. *Energy and Problems of a Technical Society* (2nd ed.). Wiley, New York (1993).

V. A. Mohnen. "The challenge of acid rain." *Scientific American* **259** (1988): 30.

E. Parsons and D. Keith. "Fossil fuels without CO_2 emissions." *Science* **282** (1998): 1053.

K. Schnell and C. Brown. *Air Pollution Control Technology Handbook.* CRC Press, Boca Raton, FL (2002).

V. Smil. *Energy at the Crossroads Global Perspectives and Uncertainties.* MIT Press, Cambridge, MA (2003).

B. Sorensen. *Renewable Energy: Its Physics, Engineering, Environmental Impacts, Economics and Planning* (2nd ed.). Academic Press, London (2002).

K. Wark, C. F. Warner, and W. T. Davis. *Air Pollution, Its Origin and Control.* Addison-Wesley, Reading, MA (1998).

T. Wigley, R. Richels, and J. Edmonds. "Economic and environmental choices in stabilization of atmospheric CO_2 concentrations." *Nature* **379** (1996): 240.

E. J. Wilson and D. Gerard (eds.). *Carbon Capture and Sequestration.* Blackwell, Oxford (2007).

PART III

Nuclear Energy

After fossil fuels, nuclear energy is the largest component of the world's current energy use. In the early days of commercial nuclear fission reactor development, the hope was that nuclear power would alleviate any energy concerns for the future. Part of this conception arose as a result of a 1954 speech to the National Association of Science Writers by the then chairman of the United States Atomic Energy Commission, Lewis L. Strauss. Strauss stated, "Our children will enjoy in their homes electrical energy too cheap to meter," implying that electricity from fission reactors could be distributed without charge or for a flat fee. In an earlier speech, he had said, "[I]ndustry would have electrical power from atomic furnaces in five to fifteen years." This prediction certainly came true, although some have suggested that the term *atomic furnaces* referred to fusion energy, not fission energy. Although nuclear energy has never become "too cheap to meter," it has become economically competitive with many other established energy technologies. The initial growth of the nuclear power industry, however, did not continue much past the mid-1980s. This was particularly true in the United States, where concerns over security, safety, and waste disposal were significant factors forming both public opinion and government policy about nuclear energy. Major nuclear reactor accidents—Three Mile Island, then Chernobyl, and most recently Fukushima—have demonstrated the validity of safety concerns. Also, it became apparent that the simple and easy approach to nuclear reactors would not provide a long-term solution for our energy needs.

An appreciation of the reasons for changing opinions and policies over nuclear energy from fission reactors requires a detailed understanding of their operation, the environmental consequences, the safety risks of their use, and the availability of nuclear resources. This part of the book reviews these concepts. It also provides an overview of possible future approaches to nuclear fission energy that address concerns over safety and the limited longevity of the resource. Finally, the possibilities of fusion energy are presented. We begin with an overview of the basic principles of nuclear physics.

The Monju Fast Breeder Reactor in Japan in the photograph is a sodium-cooled fast breeder reactor with a rated capacity of 280 MW$_e$. Construction of this reactor began in 1983, and it became operational in 1994. However, a sodium leak shortly thereafter caused a fire, and the reactor was shut down. It was restarted briefly in 2010, but, due to a series of incidents (none of which released radioactive material), public opposition, and legal battles, the reactor remained shutdown for most of its life. In fact, it has been estimated that the reactor has generated electricity for only about 1 hour since it was constructed, and its future remains uncertain. ■

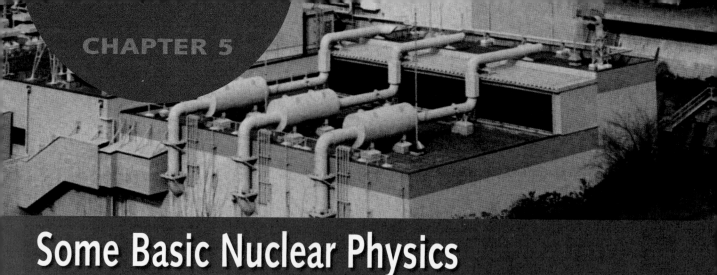

Some Basic Nuclear Physics

Learning Objectives: After reading the material in Chapter 5, you should understand:

- The composition and stability of the nucleus.
- The relationship of the binding energy of the nucleus to its properties.
- The statistics of nuclear decay processes.
- Alpha, beta, and gamma decay processes.
- The reactions between neutrons and nuclei.

5.1 Introduction

The chemical potential energy associated with chemical bonding can be converted into kinetic energy during a chemical reaction such as combustion. The energy release in a chemical reaction corresponds to a few electron volts (eV) per atom or typically a few tens of megajoules per kilogram (MJ/kg) of fuel (see Chapter 1). The nuclear potential energy associated with nuclear bonding can be converted into kinetic energy during nuclear reactions. The energy release in this reaction can be up to about 200 megaelectron volts (MeV) per nucleus, corresponding to about 10^8 MJ/kg of fuel. These numbers emphasize the potential commercial importance of nuclear energy and have motivated the development of both fission reactors (Chapter 6) and the potential development of fusion reactors (Chapter 7).

An understanding of the differences between the interactions that bond electrons to the nucleus of an atom and the interactions that bond the neutrons and protons together inside the nucleus is necessary in order to appreciate the differences in the energy scales between these two types of processes. The basic physics of nuclear interactions leads to an understanding of the properties of the nucleus and to how nuclear processes can be utilized to produce energy in a controlled manner. The present chapter provides an overview of some basic nuclear physics with an emphasis on the properties that are important for the development of commercial power reactors.

5.2 The Structure of the Nucleus

An atom consists of a nucleus and the electrons bonded to it by the Coulombic interaction (or electrostatic interaction for the purposes of the present discussion). The nucleus itself consists of neutral *neutrons* and positively charged *protons* (except the nucleus of the lightest hydrogen isotope, which is just a proton). A nuclear species (or *nuclide*) is a nucleus with a specific number of neutrons, N (the *neutron number*), and a specific number of protons, Z (the *atomic number*). The atomic number determines the identity of the element. The total number of *nucleons, A*, which is a collective name for both neutrons and protons, is

$$A = N + Z \qquad \textbf{(5.1)}$$

and is referred to as the *mass number*. A neutral atom also has Z electrons bound through the Coulombic interaction to the nucleus to cancel the Z positive charges associated with the protons. A nuclide is represented by the terminology $^A_Z E_N$, where A, Z, and N are as previously defined, and E is the name of the element. For example, $^{13}_6 C_7$ would represent the nucleus of a carbon atom with 7 neutrons ($N = 7$) and 6 protons ($Z = 6$), for a total of 13 nucleons ($A = 13$). This terminology is somewhat redundant because the element's name (carbon) specifies the value of Z, and N is related to A by the expression $N = A - Z$. Thus, the terminology ^{13}C provides all the necessary information and is commonly used, although the more complete form is sometimes useful for bookkeeping when dealing with nuclear reactions and decays, as will be explained.

Continuing with the example of carbon, all carbon atoms have nuclei that contain 6 protons. However, not all carbon atoms have nuclei containing 7 neutrons. Some carbon nuclei may have more or fewer neutrons. In fact, carbon nuclei are known that contain between 3 and 10 neutrons (Table 5.1). The family of nuclides with the same number of protons (i.e., the nuclides corresponding to the same element) are referred to as *isotopes* of that element. Thus, the table lists the eight known isotopes of carbon. Some of these isotopes are stable (i.e., they last indefinitely) and occur in nature in the relative proportions given by their natural abundances. The table shows that nearly all carbon atoms occurring in nature have nuclei with 6 neutrons, whereas the remaining have 7 neutrons. Other isotopes of carbon are unstable and spontaneously decay to another nuclide (more about decay processes later). The half-life is the time required for half of a collection of a specific type of nuclei to decay to something else. The shorter the half-life, the more unstable the nucleus is. Having too many or too few neutrons produces an unstable nucleus. Beyond certain limits (i.e., fewer than 3 or more than 10 neutrons for carbon), a nucleus cannot

Table 5.1: Summary of the properties of known carbon nuclei, $Z = 6$.				
N	*A*	mass (u)	natural abundance (%)	half-life
3	9	9.031040087	0	127 ms
4	10	10.01685311	0	19.3 s
5	11	11.01143382	0	20.4 m
6	12	12.00000000	98.9	∞
7	13	13.00335484	1.1	∞
8	14	14.00324199	~0	5730 y
9	15	15.01059926	0	2.45 s
10	16	16.01470124	0	747 ms

form, even for a short period of time. The masses listed in the table are the masses of neutral atoms of carbon of the specified isotopes. These include the masses of the 6 protons, the N neutrons, and the 6 atomic electrons. They also include the mass equivalent of the binding energies (discussed in the next section). Atomic masses are typically specified for different nuclides because they are easier to measure than the masses of the nuclei by themselves. Atomic masses are measured in *atomic mass units* (abbreviated u) where the *u* is defined by setting the mass of the neutral ^{12}C atom to be exactly 12.0 u. In traditional metric units, the atomic mass unit corresponds to $1.6605402 \times 10^{-27}$ kg.

The same general features as in Table 5.1 are also seen for nuclei with other numbers of protons (i.e., different elements). The ability of various combinations of neutrons and protons to form stable nuclei or unstable nuclei or not to form nuclei at all can be summarized by plotting N as a function of Z (Figure 5.1). This graph is referred to as a *Segrè plot*. It is important to note that, for light nuclei, the stable combinations of neutrons and protons occur for the cases where N and Z are more or less equal. This has been seen to be the case for carbon. For heavier nuclei, the Segrè plot shows that stable nuclei occur when the neutron number is greater than the proton number. This is evidenced by the departure of the stability region from the $N = Z$ line in the figure. This feature is an essential factor in the production of nuclear energy.

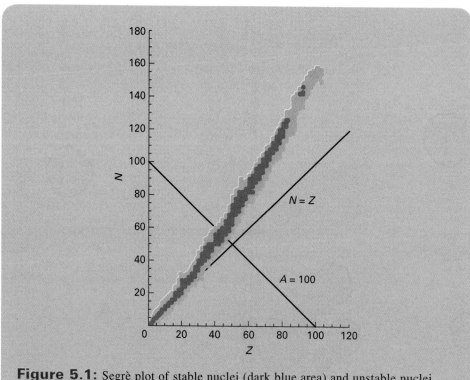

Figure 5.1: Segrè plot of stable nuclei (dark blue area) and unstable nuclei (light blue area).

Based on R.A. Dunlap, An Introduction to the Physics of Nuclei and Particles, Brooks-Cole, (2004).

5.3 Binding Energy

It is important to understand why the particles in a nucleus bond together. There are some parallels between the bonding of nucleons in the nucleus and the bonding of electrons in an atom, where the negatively charged atomic electrons are bonded to the positively charged nucleus by the Coulombic interaction. For the particles in the nucleus, (the neutral neutrons and the positively charged protons), the Coulombic interaction is repulsive (between the protons) and does not influence the neutrons.

Other interactions (besides the electromagnetic interaction) also exist in nature. The gravitational interaction is the one we experience most directly. There are, in fact, four known interactions in nature: (1) the strong (or hadronic) interaction, (2) the electromagnetic interaction, (3) the weak interaction, and (4) gravity (Table 5.2).

Each interaction can be characterized in terms of its relative strength, its range, and the types of objects that it acts on. Gravity is much weaker than any of the other interactions and is of no importance in a description of nuclear physics. It is generally of significance only for large (massive) objects (relative to the nucleus) and at large distances (compared to nuclear dimensions). Interestingly, various interactions act on different types of objects, and we have direct experience of this behavior. For example, gravity acts on masses (independent of whether charges are present). The electromagnetic interaction acts on charges and magnetic moments (which are manifestations of charge) independent of their mass. The gravitational interaction between an object and the earth is manifested as its weight, and the electromagnetic interaction between permanent magnets is manifested as a detectable force between them. This example also demonstrates another property related to interactions: the attractive or repulsive nature of the interaction. Gravity is always an attractive force, whereas the electromagnetic interaction can be either attractive or repulsive.

The positively charged protons that comprise a nucleus experience a repulsive Coulombic interaction, but they (along with the neutrons) also experience a much stronger attractive strong interaction. The strong interaction, which acts only on a certain type of subatomic particle (hadrons), holds the nucleons together in the nucleus. The hadrons are a class of particles that includes neutrons and protons (but not electrons, which are leptons). Examples of subatomic particles are given in Table 5.3 on the next page. As Table 5.2 shows, the strong interaction is very short ranged; that is, it causes a very strong attractive force between the neutrons and protons within the nucleus but drops off to zero very quickly outside the nucleus. The total force holding the nucleus together is the net result of the large, attractive strong component acting between all the nucleons and a much smaller repulsive Coulombic interaction acting between the charged protons. This total interaction represents the binding energy holding the nucleus together. From Einstein's relation,

$$E = mc^2, \tag{5.2}$$

Table 5.2: The four known interactions in nature and some of their properties.			
interaction	relative strength	range (m)	acts on
strong	1	10^{-15}	hadrons
electromagnetic	10^{-2}	long	charges
weak	10^{-5}	10^{-18}	leptons and hadrons
gravitational	10^{-39}	long	masses

particle	class	mass (u)	charge	interactions
neutron	hadron	1.008664904	0	S, E, W
proton	hadron	1.007276470	$+e$	S, E, W
electron	lepton	0.0005485799	$-e$	E, W
positron	lepton	0.0005485799	$+e$	E, W
electron neutrino	lepton	~ 0	0	W
electron antineutrino	lepton	~ 0	0	W

Table 5.3: Some examples of subatomic particles and their classification and properties. [Interactions (gravity excluded): S = strong, E = electromagnetic, W = weak.]

there is an equivalence between mass and energy, so a binding energy, B, is equivalent to a mass m,

$$m = \frac{B}{c^2} \tag{5.3}$$

B is defined to be explicitly positive, and, the larger the value of B, the more strongly the nucleus is bound together. The mass associated with the binding energy decreases the mass of a system of particles. This means that a nucleus comprised of N neutrons and Z protons will have a mass of

$$m_{\text{nucleus}} = Nm_n + Zm_p - \frac{B}{c^2}, \tag{5.4}$$

where m_n and m_p are the masses of the neutron and proton, respectively. The mass of the electrons, Zm_e, and the mass equivalent of the Coulombic binding energy of the electrons, $-b/c^2$, can be added to give the mass of the neutral atom:

$$m_{\text{atom}} = Nm_n + Z(m_p + m_e) - \frac{B}{c^2} - \frac{b}{c^2}. \tag{5.5}$$

A practical application of equation (5.5) is illustrated in Table 5.4. Here the masses of atoms with nuclei containing 49 nucleons are given. These 49 nucleons can be any of the combinations of neutrons and protons, as shown in the table. The table shows that ^{49}Ti has the smallest mass (and hence the greatest binding energy) and is the most stable, that is, it is the only stable nuclide with $A = 49$. As the mass increases for fewer or more neutrons, the mass of nuclei with $A = 49$ increases, and thus the binding energy decreases. The table also shows that this results in increasingly unstable nuclei (as seen by the decreasing half-life). Details of decay processes will be discussed in the next section.

Table 5.4: Atomic masses of nuclides with $A = 49$.

nuclide	N	Z	mass (u)	half-life	decay
^{49}Ca	29	20	48.9556733	8.8 m	β^-
^{49}Sc	28	21	48.95002407	57.5 m	β^-
^{49}Ti	27	22	48.94787079	∞	stable
^{49}V	26	23	48.94851691	330 d	ec
^{49}Cr	25	24	48.95134114	41.9 m	β^+, ec

In nuclear and particle physics, it is generally convenient to use non-SI units in Einstein's relation. Energy in the left-hand side of equation (5.2) is generally defined in MeV, while mass on the right-hand side is expressed in MeV/c^2 or in atomic mass units, u. If mass is given in MeV/c^2, then c^2 is unity. If mass is given in atomic mass units, then c^2 will be in MeV/u as follows. The atomic mass unit (u) is defined as 1/12 of the mass of a neutral ^{12}C atom. In SI units the u is equivalent to $1.6605402 \times 10^{-27}$ kg. Einstein's relation, therefore, gives the energy equivalent of 1 u as

$$E = 1.6605402 \times 10^{-27} \text{ kg} \times (2.997925 \times 10^8 \text{ m/s})^2 = 1.50535 \times 10^{-10} \text{ J} \quad \textbf{(5.6)}$$

or converting to MeV,

$$E = 1.50535 \times 10^{-10} \text{ J} \times 1.602177 \times 10^{-13} \text{ MeV/J} = 931.494 \text{ MeV} \quad \textbf{(5.7)}$$

Thus Einstein's relation may be written as

$$931.494 \text{ MeV} = 1 \text{ u} \times c^2 \quad \textbf{(5.8)}$$

or

$$c^2 = 931.494 \text{ MeV/u}. \quad \textbf{(5.9)}$$

This is a convenient form of units for the speed of light squared (rather than m^2/s^2) that is commonly used in the field of nuclear physics. It will be used in the remainder of the present chapter.

Example 5.1

Calculate the total binding energy associated with a ^{49}Ti atom.

Solution

From equation (5.5), the binding energy may be written as

$$B = (Nm_n + Z(m_p + m_e) - m_{atom})c^2.$$

Using $N = 27$, $Z = 22$, and m_{atom}, as given in Table 5.4, and the masses of the neutron, proton, and electron, as given in Table 5.3, the binding energy is

$$B = (27 \times 1.008664904 \text{ u} + 22 \times (1.007276470 \text{ u} + 0.0005485799 \text{ u})$$

$$- 48.94787079 \text{ u}) \times 931.494 \text{ MeV/u}$$

$$= 426.8 \text{ MeV}.$$

5.4 Nuclear Decays

A collection of unstable nuclei eventually decays into something more stable. If $N(t)$ nuclei of the initial type are present at time t, then the change in the number of nuclei per time interval Δt is defined as

$$\Delta N = -\lambda N(t)\Delta t, \quad \textbf{(5.10)}$$

where the change is negative because the number of nuclei of the initial type is decreasing and λ (the *decay constant*) is a measure of the probability that a given nucleus will decay during the time interval Δt. The greater the value of λ, the more likely the nucleus is to decay, and the faster the decay process will be. The differential form of equation (5.10) can be written as

$$dN = -\lambda N(t)dt. \tag{5.11}$$

This can be integrated to give

$$N(t) = N_0 e^{-\lambda t}, \tag{5.12}$$

where N_0 is an integration constant representing the number of nuclei present at time $t = 0$. The half-life, $t_{1/2}$, is found from equation (5.12) by setting $t = t_{1/2}$ and setting $N(t_{1/2}) = N_0/2$. This gives

$$t_{1/2} = (\ln 2)/\lambda. \tag{5.13}$$

Example 5.2

A nuclide decays with a half-life of 2 days. If 10^8 nuclei of this nuclide are present at the beginning of an experiment, how many remain after 5 days?

Solution

From equation (5.13), the decay constant is given in terms of the half-life as

$$\lambda = \frac{\ln 2}{t_{1/2}},$$

so

$$\lambda = \frac{0.693}{2\ \text{d}} = 0.347\ \text{d}^{-1}$$

Using $N_0 = 10^8$ in equation (5.12) with $t = 5$ d gives

$$N(t = 5\ \text{d}) = 10^8 \times \exp(-0.347\ \text{d}^{-1} \times 5\ \text{d})$$

$$= 10^8 \times \exp(-1.733) = 1.77 \times 10^7 \text{ nuclei.}$$

The rate at which the nuclei decay as a function of time can be found by differentiating equation (5.12) to get

$$\left| \frac{dN(t)}{dt} \right| = \lambda N_0 e^{-\lambda t}. \tag{5.14}$$

Measuring the number of decays per unit time (Figure 5.2) allows for a determination of the half-life, the decay constant, and the number of nuclei present at $t = 0$. The definition of the half-life as the time at which the count rate drops to half of its initial value gives $t_{1/2} = 4.1$ minutes for the example shown in Figure 5.2. From equation (5.13), λ may then be obtained from the measured half-life. Equation (5.14) may be written for $t = 0$ as

$$\left| \frac{dN(0)}{dt} \right| = \lambda N_0. \tag{5.15}$$

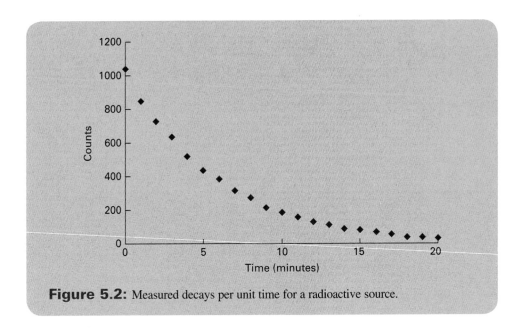

Figure 5.2: Measured decays per unit time for a radioactive source.

This equation allows the value of N_0 to be obtained from a measurement of the number of decays per unit time at $t = 0$.

The details of specific decay processes provide insight into the information given in Table 5.4. It is found that ^{49}Sc decays with a half-life of 57.5 minutes to ^{49}Ti. The ^{49}Sc nucleus has 28 neutrons and 21 protons, and the ^{49}Ti nucleus has 27 neutrons and 22 protons. Thus, it would seem that this decay process corresponds to the conversion of one of the ^{49}Sc neutrons to a proton. This is the *beta decay* (β^- *decay*) process and results in the emission of an electron during the conversion of a neutron to a proton as

$$\mathrm{n} \rightarrow \mathrm{p} + \mathrm{e}^- + \bar{\nu}_\mathrm{e}, \tag{5.16}$$

where the $\bar{\nu}_\mathrm{e}$ is an electron antineutrino (Table 5.3). An early name for an electron was a beta particle, and the observation of electrons as by-products of this decay is the reason for the name of the decay process. This decay is the result of the weak interaction (Table 5.3) because the electron is a lepton and is not subject to the strong interaction. In the case of the β^- decay of ^{49}Sc the process is

$$^{49}\mathrm{Sc} \rightarrow {}^{49}\mathrm{Ti} + \mathrm{e}^- + \bar{\nu}_\mathrm{e}. \tag{5.17}$$

In a decay process, the nuclide on the left-hand side is referred to as the *parent*, and the nucleus on the right-hand side is referred to as the *progeny* (or daughter). Analogously, the conversion of a proton to a neutron, referred to as β^+ *decay*, results in the emission of a positron (or antielectron) and an electron neutrino:

$$\mathrm{p} \rightarrow \mathrm{n} + \mathrm{e}^+ + \nu_\mathrm{e}. \tag{5.18}$$

This process can be seen in the β^+ decay of ^{49}Cr:

$$^{49}\mathrm{Cr} \rightarrow {}^{49}\mathrm{V} + \mathrm{e}^+ + \nu_\mathrm{e}. \tag{5.19}$$

Electron capture (ec, Table 5.4) is equivalent to β^+ decay. The distinction between these processes is not of great relevance to the present discussion and will not be discussed in detail (both processes will be subsequently referred to as β^+ decay). Nuclei that mini-mize the mass (i.e., maximize the binding energy) with respect to Z for a constant value of A are referred to as *beta stable* (or β stable) because it is energetically unfavorable for them to decay by either β^- decay or β^+ decay.

Beta decay can be viewed more quantitatively, and the energy associated with the decay process can be calculated. The difference in the binding energy of the parent and the progeny is converted into kinetic energy according to Einstein's relation

$$E = \Delta mc^2. \tag{5.20}$$

From Table 5.4, the change in mass that occurs during the β^- decay of ^{49}Sc can be converted to energy as

$$E = (48.95002407 \text{ u} - 48.94787079 \text{ u}) \times 931.494 \text{ MeV/u} = 2.01 \text{ MeV}. \tag{5.21}$$

This decay energy is given up as kinetic energy, partly to the progeny nucleus but mostly to the electron and the antineutrino. It is important in such reactions to properly account for all electrons in the problem. It is convenient in the example of β^- decay that no extra electron masses must be included. In other decay processes, such as β^+ decay, the mass of electrons must be considered more carefully.

The series of nuclides with constant A as shown in Table 5.4 is represented in the Segrè plot as a diagonal line with slope of negative 1 (Figure 5.1). The line of β stable nuclei with different N, Z, and A is surrounded on either side by a region of β unstable nuclides. This region of β stable nuclides is sometimes referred to as the beta stability valley.

Another type of decay process that nuclei can undergo is *alpha decay* (α *decay*). The alpha particle is the nucleus of a ^{4}He atom and consists of a bound system of two neutrons and two protons. In α decay, an α particle is given off by a nucleus, thereby reducing N by two and reducing Z by two. A typical example is

$$^{241}\text{Am} \rightarrow {}^{237}\text{Np} + {}^4\text{He}$$

or

$$^{241}\text{Am} \rightarrow {}^{237}\text{Np} + \alpha. \tag{5.22}$$

This process occurs most commonly in heavy β stable nuclei. When calculating the energy of this decay process, it is important to ensure that all electrons are properly accounted for. This is most conveniently done if the atomic masses of the parent and the progeny are used along with the mass of the ^{4}He atom (Example 5.3).

The relationship of β decay and α decay with the data on the Segrè plot is seen by expanding a region of the plot as shown in Figure 5.3. This region is above the α-stability limit, so even nuclei that are β stable are not stable against α decay. Alpha decays are represented by a diagonal line with a length of 2 units and a slope of 1 pointing downward. The example of the decay process given by equation (5.22) is illustrated in the figure. Beta decays are represented by diagonal lines of 1 unit length with a slope of ± 1. For a β^- decay, for example,

$$^{235}\text{Pa} \rightarrow {}^{235}\text{U} + \text{e}^- + \bar{\nu}_\text{e}, \tag{5.23}$$

as shown in the figure, the decay proceeds downward along the line. For a β^+ decay, as an example,

$$^{235}\text{Np} \rightarrow {}^{235}\text{U} + \text{e}^+ + \nu_\text{e}, \tag{5.24}$$

as shown in the figure, the decay proceeds upward along the line.

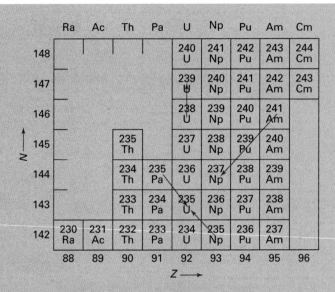

Figure 5.3: Region of the Segrè plot between N = 142 to 148 and Z = 88 to 96, corresponding to A = 230 to 244. An example of α-decay as given in equation (5.22) and examples of β^+ and β^- decays as given in equations (5.23) and (5.24), respectively, are shown. Neutron capture by ^{238}U (discussed in the next section) is also illustrated.

Example 5.3

Calculate the total energy associated with the reaction shown in equation (5.22) if the mass of the ^{4}He atom is 4.00260325 u and the masses of the ^{241}Am and ^{237}Np atoms are 241.0568229 u and 237.0481673 u, respectively.

Solution

Subtracting the masses on the right-hand side of the equation from the mass on the left-hand side of the equation and converting into energy units:

$$E = (241.0568229 \text{ u} - 237.0481673 \text{ u} - 4.00260325 \text{ u})$$

$$\times 931.494 \text{ MeV/u} = 5.64 \text{ MeV}.$$

It is important to note that, if equation (5.22) is written with a ^{4}He atom on the right-hand side, the number of electrons on the two sides of the equation cancel out, and their masses do not have to be included explicitly in the calculation if atomic masses are used.

Nuclei, like atoms, can exist in their ground state or in an excited state. *Gamma decay* (γ-decay) is the emission of a high-energy photon (or gamma ray) from a nucleus when it undergoes a transition from an excited state to a lower energy state. This process often follows β-decay or α-decay because these decay processes often leave the progeny nucleus in an excited state. Normally, excited states of a nucleus undergo gamma decay processes until the nucleus is in its ground state.

5.5 Nuclear Reactions

Decay processes occur spontaneously in unstable nuclei. Generally, one unstable nucleus appears on the left-hand side of equations such as (5.16), and two or more by-products appear on the right-hand side. In a nuclear reaction, a nucleus reacts with a particle (such as an alpha particle or neutron), and two or more by-products result. A simple, generic reaction is

$$a + A \rightarrow B + b. \tag{5.25}$$

This represents a particle (a) incident on a nucleus (A) resulting in the emission of a particle (b) and a resulting nucleus (B). This is often written in shorthand notation as A(a, b)B.

In the simplest case, particles a and b may be the same, and conservation laws require that nuclei A and B also be the same. This is referred to as *scattering*, and the result is that some of the original kinetic energy of particle a may be lost to the nucleus. This may merely impart some kinetic energy to the nucleus, or it can (if sufficient energy is available) leave the nucleus in an excited state (denoted by an asterisk). An example is a neutron incident on a ^{14}N nucleus:

$$n + {}^{14}N \rightarrow {}^{14}N^* + n. \tag{5.26}$$

A more complex case occurs when the incident and the emitted particles are not the same. This means, for conservation reasons, that the nucleus changes identity. Consider, as an example, the case where a neutron is incident on a ^{14}N nucleus. The neutron may be absorbed by the nucleus, and a proton may be emitted. Originally, the ^{14}N nucleus had 7 neutrons and 7 protons. After the reaction, the nucleus is left with 8 neutrons and 6 protons and is now a ^{14}C nucleus. This is the (n, p) reaction given by

$$n + {}^{14}N \rightarrow {}^{14}C + p. \tag{5.27}$$

As with decays, the energy associated with a reaction can be calculated by subtracting the total of the masses on the right-hand side from the total of the masses on the left-hand side and converting the mass difference into energy units. For reactions that involve charged particles—protons, α particles, and the like—it is important, when using atomic masses, to properly account for all electrons. A calculation of the energy for the reaction shown in equation (5.27) gives 0.63 MeV. This means that 0.63 MeV of energy is given up when this reaction occurs and that this excess energy appears as kinetic energy of the proton and ^{14}C nucleus. A reaction that gives up energy is called *exothermic*.

We might also consider the reaction of a proton incident on a ^{13}C nucleus where a neutron is emitted:

$$p + {}^{13}C \rightarrow {}^{13}N + n. \tag{5.28}$$

A calculation of the energy for this reaction gives -3.00 MeV. It is important to note that this energy is negative. This kind of reaction is called an *endothermic* reaction, and energy must be supplied (at least 3.00 MeV of it) to allow the reaction to take place. This energy can come from the kinetic energy of the incident proton if (for example) the proton is supplied by a particle accelerator.

A reaction which will be important for the discussion of nuclear reactors is the (n,γ) reaction, sometimes referred to as *neutron capture*. Here a neutron is absorbed by a nucleus. The nucleus is left in an excited state, and it subsequently decays back to the ground state by γ decay. An important reaction of this type (discussed further in the next chapter) is

$$n + {}^{238}U \rightarrow {}^{239}U + \gamma. \tag{5.29}$$

Example 5.4

Calculate the energy given up in the reaction shown in equation (5.29). Use the atomic masses $m({}^{238}U) = 238.0507826$ u and $m({}^{239}U) = 239.0542878$ u.

Solution

Following equation (5.20), the energy may be written in terms of the difference between the total mass of the left-hand side of the equation and the total mass on the right-hand side. The neutron mass is obtained from Table 5.3:

$$E = [m({}^{238}U) + m_n - m({}^{239}U)]c^2$$

$$= (238.0507826 \text{ u} + 1.008664904 \text{ u} - 239.0542878 \text{ u}) \times 931.494 \text{ MeV/u}$$

$$= 4.81 \text{ MeV}$$

In the case of neutron reactions, such as the (n, γ) reaction shown in equation (5.29), the identity of the element does not change, so the number of electrons on the right- and left-hand sides of the equations is properly accounted for by using atomic masses.

The energy release in the reaction given in equation (5.29) is primarily given up to the γ ray. If the neutron has kinetic energy when it is absorbed by the ${}^{238}U$ nucleus, then this adds to the possible γ ray energy.

5.6 Summary

The present chapter has shown that the nucleus consists of neutral neutrons and positively charged protons. The nucleus is held together by its binding energy, which is the net result of the repulsive Coulombic interaction between the protons and the attractive strong interaction between all nucleons. While light nuclei tend to have approximately equal numbers of neutrons and protons, heavy nuclei require a greater proportion of neutrons to be stable. Nuclei with a ratio of neutrons to protons that is outside of the range of stability, as represented on a Segrè plot, are unstable and decay toward a more stable configuration of nucleons.

This chapter also presented the statistics of decay processes and showed that these are described by exponential behavior with a characteristic lifetime. The various decay processes—alpha decay, beta decay, and gamma decay—by which unstable nuclei decay toward the most stable nucleon configuration have been overviewed in this chapter. Alpha decay corresponds to the release of an alpha particle, or ^{4}He nucleus, from an unstable nucleus. Beta decay corresponds to the conversion of a proton to a neutron (or vice versa), along with the release of an electron (or positron) and an antineutrino (or neutrino). Finally, gamma decay corresponds to the de-excitation of the neutrons and/or protons in a nucleus to a lower energy level with the emission of a photon (gamma ray). The methods by which the energy associated with these reactions can be calculated on the basis of nuclear masses have been described.

Nuclear binding energy can be released through nuclear reactions, and these processes provide a mechanism for harnessing the enormous energy associated with nuclear interactions. This chapter described one of the most important types of nuclear reactions related to nuclear energy production, that is, the reactions between nuclei and neutrons.

Problems

5-1 Tabulate the number of neutrons, protons, and electrons for neutral atoms of the following nuclides: ^{29}Si, ^{44}Sc, ^{49}V, ^{210}Bi, and ^{241}Pu.

5-2 Calculate the total binding energy for a ^{49}Sc atom and a ^{49}V atom. Compare with the results from Example 5.1 for ^{49}Ti, and show that the most stable nucleus has the greatest binding energy.

5-3 A sample of ^{137}Cs (half-life = 30.1 years) decays at a rate of 10^{10} decays per second. How long will it take the decay rate to decrease to 10^7 decays per second?

5-4 Plot the atomic mass for nuclides with $A = 49$ as a function of Z. This is generally referred to as the mass parabola. Show for the ^{49}Ca $\rightarrow$ ^{49}Sc $\rightarrow$ ^{49}Ti decay series that the half-life becomes longer as the mass difference becomes smaller.

5-5 What are the decay products of the β^- decay of the following nuclides: ^{3}H, ^{14}C, ^{64}Cu, and ^{125}Sn?

5-6 Calculate the energy released during the α decay of ^{230}Th. The following atomic masses may be of use: $m(^{230}$Th$) = 230.0331266$ u, $m(^{228}$Th$) = 228.0287313$ u, $m(^{228}$Ra$) = 228.0310641$ u, and $m(^{226}$Ra$) = 226.0254026$ u.

5-7 Nuclide A has a half-life of 4 days, and nuclide B has a half-life of 16 days. At the beginning of an experiment, a sample contains 2^{18} nuclei of nuclide A and 2^{12} nuclei of nuclide B. How long will it take before the numbers of nuclei of the two nuclides are equal?

5-8 ^{64}Cu can decay by both β^- decay and β^+ decay. Draw an appropriate portion of a Segrè plot along the lines of Figure 5.3 showing these two decay processes.

5-9 At time $t = 0$ a sample contains 3.57×10^8 nuclei of a particular nuclide. Exactly one year later the sample contains 3.09×10^8 nuclei of the same species. Calculate the half-life of the decay.

5-10 Using the energy stated in the text for the process in equation (5.28), calculate the atomic mass of ^{13}N.

5-11 The measured decay rates of a radioactive material as a function of time are shown in the table below.

(a) Calculate the half-life of the decay.

(b) Show that a semilogarithmic plot of the decay rate as a function of time is linear.

(c) Derive an analytical form of the linear relationship for part (b).

time (hours)	decay rate (s^{-1})
0	7623
1	7447
2	7275
3	7106
4	6942
5	6781
6	6624

5-12 Calculate the energy associated with the process in equation (5.18)

5-13 Calculate the energy associated with the β^+ decay process in equation (5.19). *Note:* Be careful counting electrons.

5-14 At $t = 0$ there are 4.11×10^{11} nuclei of a particular nuclide in a sample. The decay rate is measured to be $3.21 \times 10^6 \, s^{-1}$. Calculate the half-life.

5-15 The total binding energy is a useful property for comparing the relative stability of nuclei with the same number of nucleons, as in Table 5.4. For nuclei with different numbers of nucleons, as in Table 5.1, it is useful to compare the binding energy divided by the number of nucleons, B/A.

(a) Calculate B/A for ^{11}C, ^{12}C, ^{13}C, and ^{14}C.

(b) Discuss the relevance of the trends in the results for part (a).

5-16 (a) Write the equation for the α-decay of ^{237}Np.

(b) Calculate the energy associated with the α-decay of ^{237}Np.
Note: $m(^{233}Pa) = 233.0402402$ u.

5-17 Using the reaction energy given in the text for equation (5.27) calculate the atomic mass of ^{14}N.

5-18 The average binding energy per nucleon, B/A, for ^{40}Ca is 8.55 MeV. Calculate the mass of a neutral ^{40}Ca atom.

5-19 Assume carbon is 100% ^{12}C.

(a) Calculate the energy required to unbind all the neutrons and all the protons in 1 kg of pure carbon.

(b) Compare the answer to part (a) with the chemical energy obtained by burning 1 kg of carbon.

5-20 A simple method of calculating the age of the earth is based on the natural abundances of ^{235}U and ^{238}U. It is assumed that the quantities of these two isotopes were equal at the time that the earth formed, and measurements show that the natural abundances at present are 0.72% ^{235}U and 99.28% ^{238}U. Using the assumptions above and the half-lives of these two isotopes, 7.1×10^8 for ^{235}U and 4.51×10^9 years for ^{238}U, calculate the age of the earth.

Bibliography

D. Bodansky. *Nuclear Energy: Principles, Practices, Prospects.* American Institute of Physics, New York (1996).

M. G. Bowler. *Nuclear Physics.* Pergamon Press, Oxford (1974).

W. E. Burcham. *Elements of Nuclear Physics.* Longman, London (1979).

W. E. Burcham. *Nuclear Physics: An Introduction* (2nd ed.). Longman, London (1973).

B. L, Cohen. *Concepts of Nuclear Physics.* McGraw-Hill, New York (1971).

W. N. Cottingham and D. A. Greenwood. *An Introduction to Nuclear Physics.* Cambridge University Press, Cambridge (1986).

A. Das and T. Ferbel. *Introduction to Nuclear and Particle Physics.* Wiley, New York (1994).

R. A. Dunlap. *An Introduction to the Physics of Nuclei and Particles.* Brooks-Cole, Belmont, CA (2004).

H. A. Enge. *Introduction to Nuclear Physics.* Addison-Wesley, Reading, MA (1966).

H. Frauenfelder and E. M. Henley. *Sub-Atomic Physics.* Prentice-Hall, Englewood Cliffs (1974).

K. S. Krane. *Introductory Nuclear Physics.* Wiley, New York (1988).

W. E. Meyerhof. *Elements of Nuclear Physics.* McGraw-Hill, New York (1967).

R. L. Murray. *Nuclear Energy: An Introduction to the Concepts, Systems, and Applications of Nuclear Processes.* Pergamon Press, New York (1993).

W. S. C. Williams. *Nuclear and Particle Physics.* Clarendon Press, Oxford (1991).

Energy from Nuclear Fission

Learning Objectives: After reading the material in Chapter 6, you should understand:

- The relationship between nuclear binding energy and the mechanism for extracting nuclear energy by fission.
- Differences between spontaneous and induced fission and the importance of the Coulomb barrier.
- Fission processes in the isotopes of uranium.
- Critical reactions and thermal reactor control.
- The types of thermal fission reactors and their properties.
- The world use of fission energy.
- The availability and production of uranium worldwide.
- Nuclear reactor safety and the reasons and consequences of nuclear accidents.
- Methods for nuclear waste disposal.
- New designs of thermal reactors with improved safety features.
- The principles of operation and advantages of a fast breeder reactor.

6.1 Introduction

One of the first important steps in the development of nuclear energy came in 1932 when James Chadwick discovered the neutron. In 1938, Lise Meitner, Otto Hahn, Fritz Strassman, and Otto Robert Frisch discovered that nuclear reactions involving neutrons could induce fission and liberate substantial amounts of energy. The development of the first nuclear reactor at the University of Chicago by Enrico Fermi in 1942 demonstrated that a controlled fission reaction could be sustained. The development of this reactor, however, was not motivated by the need for sustainable energy but by the desire to produce nuclear weapons. The development of nonmilitary fission reactors for the production of electricity was pursued after World War II. The first fission reactor to generate electricity, the Obninsk Nuclear Power Plant in the USSR, became operational in 1954. It produced about 5 MW_e. The first commercial nuclear power station was opened in 1956 in Sellafield, U.K., and in 1962 the Shippingport Reactor in Pennsylvania became the first operational commercial nuclear reactor in the United States. The idea that nuclear energy could provide inexpensive and virtually unlimited electric power prompted the rapid increase in the number of nuclear power reactor facilities over the next 30 years. Over most of the last 25 years or so, the growth of the nuclear power industry has been much slower and has even been negative in some areas. This chapter describes the physics and engineering of nuclear fission reactors and provides some insight into the reasons for the history of the nuclear power industry, as well as its future.

6.2 The Fission of Uranium

When a decay or an exothermic nuclear reaction occurs, the decrease in the total mass of the constituents involved is converted into energy. This energy is manifested as kinetic energy of the progeny nucleus and emitted particles, and it can, in principle, be converted into useful energy. This decrease in mass is associated with an increase in binding energy [equation (5.3)]. Because these processes conserve the total number of nucleons (i.e., number of neutrons plus number of protons), the average binding energy per nucleon is a good measure of the kinetic energy that can be liberated. In Figure 6.1, the average binding energy per nucleon is plotted as a function of the number of nucleons in the nucleus.

Except for a few anomalous peaks, the figure shows that, as the number of nucleons increases, the average binding energy per nucleon increases sharply for light nuclei, followed by a broad maximum and then a slow decrease for heavy nuclei. This behavior allows substantial amounts of energy to be obtained from certain types of nuclear reactions. Consider, for example, a ^{238}U nucleus. This nucleus has a binding energy of 7.57 MeV per nucleon or a total binding energy for the 238 nucleons of about 1800 MeV. If, for some reason, this nucleus were to split in half, forming two nuclei containing 119 nucleons each, then, according to the figure, each of those nuclei would have an average binding energy per nucleon of 8.50 MeV for a total binding energy of the two smaller nuclei of about 2020 MeV. This difference in binding energies, about 220 MeV, would be given up to the smaller nuclei as kinetic energy. The

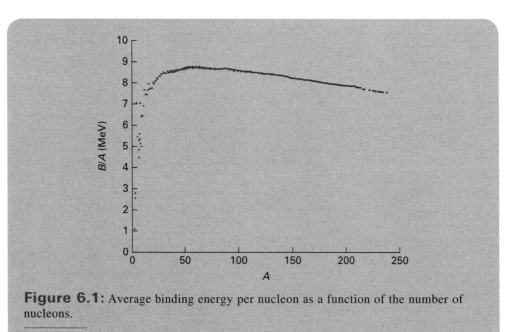

Figure 6.1: Average binding energy per nucleon as a function of the number of nucleons.

Based on R.A. Dunlap, An Introduction to the Physics of Nuclei and Particles, Brooks-Cole, Belmont (2004)

splitting of a heavy nucleus into two smaller fragments is referred to as *fission*. The situation just described is symmetric fission because the two fission fragments have the same number of nucleons. In this simple example, the energy measured in MeV per uranium fission (220 MeV per fission) can be converted to joules per kilogram of uranium to obtain 8.9×10^7 MJ/kg. This can be compared to the burning of carbon, which produces 32.8 MJ/kg. Thus, this simple model shows that nuclear fission produces about 2.7×10^6 times as much energy per kilogram of fuel as combustion. Putting this in different terms, a gram of uranium produces as much energy as almost 3 t of coal. This clearly justifies an interest in nuclear power. Another nuclear reaction—fusion, or the sticking together of light nuclei—deals with the sharp increase on the left-hand part of Figure 6.1 and is discussed in detail in Chapter 7.

It is important to consider why a uranium nucleus might undergo fission. In principle, any process that releases energy (i.e., is exothermic) can occur on its own spontaneously. This can certainly happen for a uranium nucleus because this process is clearly exothermic. However, in a collection of uranium nuclei, it occurs only very rarely. The reason for this can be seen by examining the forces holding the nucleus together. When the nucleons are close together in the nucleus, the attractive strong interaction overcomes the repulsive Coulombic interaction between the charged protons to bind the particles together. This can be represented by viewing the nucleons as sitting in a deep potential well (Figure 6.2). However, to combine separate individual particles (at least protons) to form a nucleus, it is necessary to overcome the Coulombic repulsion so that the particles are close enough together for the strong interaction to take over. The so-called Coulomb barrier responsible for this behavior is shown in the figure. For the particles in a nucleus to split apart (during fission) to form two smaller nuclei, they must get through (or over) the Coulomb barrier. This is fairly difficult for the nucleons in a nucleus, but it does happen occasionally and is referred to as *spontaneous fission*. One way of increasing the likelihood of fission is to provide some additional energy to the nucleons. This can be done by bombarding the nucleus with neutrons.

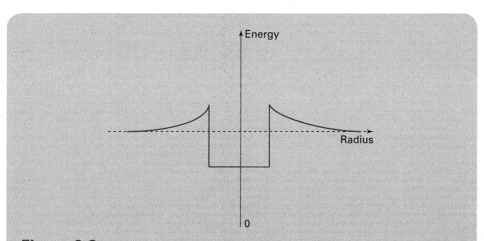

Figure 6.2: Simplified model of the potential energy of a nucleus as a function of distance from the origin. The deep square well results from the attractive strong interaction, and the portion outside the square well results from the repulsive Coulombic interaction for the protons.

To appreciate how this relates to the operation of a nuclear reactor, it is necessary to look in more detail at the properties of uranium. Natural uranium consists of two isotopes; ^{235}U and ^{238}U. These occur with natural abundances of 0.72% and 99.28%, respectively. The greater abundance of ^{238}U compared with ^{235}U is directly a result of its longer half-life, 4.5×10^9 years compared with 7.0×10^8 years. The reaction of a neutron with a ^{238}U nucleus was shown in the last chapter:

$$n + {}^{238}U \rightarrow {}^{239}U + 4.78 \text{ MeV}. \tag{6.1}$$

This provides an additional 4.78 MeV of energy (if the incident neutron has minimal kinetic energy). A similar reaction for a neutron incident on a ^{235}U nucleus is

$$n + {}^{235}U \rightarrow {}^{236}U + 6.54 \text{ MeV}; \tag{6.2}$$

that is, substantially more energy becomes available. An actual calculation shows that the most energetic neutrons and protons in the nucleus are at energies about 6 MeV, below the maximum height of the Coulomb barrier at the edge of the nucleus. The energy provided by an incident neutron (even with very small kinetic energy) to a ^{235}U nucleus is sufficient for the newly formed ^{236}U nucleus to get over the Coulomb barrier. This means that the ^{236}U nucleus is unstable and can undergo fission. This is not the case for ^{238}U unless the incident neutron has sufficient kinetic energy (about 1.4 MeV) to push the nucleus over the Coulomb barrier. Nuclides like ^{235}U that undergo fission when struck by a low-energy neutron are referred to as *fissile*; those that do not are referred to as *nonfissile*. Fission that results from the interaction of the nucleus with an incident neutron is called *induced fission*. Compare the results shown in Example 6.1 for a fissile nuclide with that shown in Example 5.4 for a nonfissile nuclide.

Example 6.1

Use the following known atomic masses of ^{235}U and ^{236}U to obtain the energy release shown in equation (6.2):

$$m({}^{235}U) = 235.0439231 \text{ u}$$

and

$$m({}^{236}U) = 236.0455619 \text{ u}.$$

Solution

The mass of the neutron is given in Appendix II. The energy release in the reaction when the initial kinetic energy on the left-hand side is negligible is given by the energy equivalent of the mass difference between the left-hand side and right-hand side of the equation:

$$E = [m_n + m({}^{235}U) - m({}^{236}U)]c^2$$

$$= [1.008664904 \text{ u} + 235.0439231 \text{ u} - 236.0455619 \text{ u}] \times 931.494 \text{ MeV/u}$$

$$= 6.54 \text{ MeV}.$$

6.3 Nuclear Reactor Design

The fission of uranium just described, particularly that involving ^{235}U, might explain how uranium can be used to obtain energy from nuclear reactions, but a number of important points need to be considered in more detail. Two of these are (1) where do the neutrons come from, and (2) what is the purpose of the ^{238}U? The origin of the neutrons is considered first.

In a piece of ^{235}U, a few nuclei will undergo spontaneous fission; although this is unlikely. However, a 1-kg piece of ^{235}U contains about 2.6×10^{24} uranium nuclei, and a number of these will undergo spontaneous fission, even in a very short time. The simple case of symmetric fission is not very likely. More commonly, one of the fragments is larger than the other. Typically, one fragment with about 137 nucleons and one fragment with about 96 nucleons are produced. All fission processes do not produce the same fission fragments. The distribution of fragment sizes, known as the *fission yield*, is shown in Figure 6.3.

A typical example of an induced fission process in ^{235}U is

$$n + {}^{235}U \rightarrow {}^{236}U \rightarrow {}^{137}I + {}^{96}Y + 3n, \tag{6.3}$$

where the ^{236}U nucleus exists only momentarily before breaking up into the two fragments.

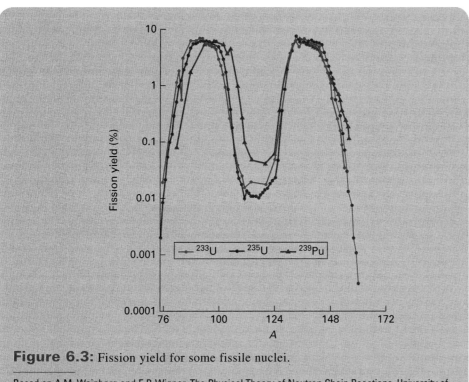

Figure 6.3: Fission yield for some fissile nuclei.

Based on A.M. Weinberg and E.P. Wigner, The Physical Theory of Neutron Chain Reactions, University of Chicago Press, Chicago (1958)

Example 6.2

An induced fission process is

$$n + {}^{235}U \rightarrow {}^{236}U \rightarrow {}^{139}Cs + X + 3n.$$

Determine the identity of the nuclide X.

Solution

We need to balance the number of neutrons and protons on the two sides of the equation. On the left side we have ${}^{236}U$, which has 144 neutrons and 92 protons. On the right side ${}^{139}Cs$ has 84 neutrons and 55 protons. Thus the nuclide X will have

$$N = 144 - 84 - 3 = 57 \text{ neutrons}$$

and

$$Z = 92 - 55 = 37 \text{ protons.}$$

37 protons indicate that the element is Rb and the total number of nucleons will be $57 + 37 = 94$, so the nuclide is $X = {}^{94}Rb$.

Example 6.3

From the atomic mass of ${}^{235}U$ given in Example 6.1, calculate the binding energy per nucleon for this nuclide.

Solution

Each uranium atom has 92 protons and 92 electrons. ${}^{235}U$, therefore, has 143 neutrons, and the total binding energy is

$$B({}^{235}U) = [143m_n + 92(m_p + m_e) - m({}^{235}U)]c^2$$

$$= [143 \times 1.008664904u + 92 \times (1.007276470u + 0.0005485799u)$$
$$- 235.0439231u] \times 931.494 \text{ MeV/u}$$

$$= 1784 \text{ MeV.}$$

Therefore, the binding energy per nucleon is

$$B({}^{235}U)/235 = 7.591 \text{ MeV per nucleon.}$$

The energy calculated in Example 6.4 is slightly less than the simple calculation for symmetric fission. Two features of equation (6.3) are important. First, 3 leftover neutrons are not connected to either fission fragment; second, the fission fragments are not β stable. The reasons for these properties can be readily seen from the Segrè plot in Figure 5.1. Light β stable nuclei have approximately equal numbers of neutrons and protons, whereas heavy β stable nuclei have more neutrons than protons. Thus, when a heavy β stable nucleus breaks into two smaller nuclei, there are more neutrons than are necessary to form two lighter β stable nuclei. Some of these neutrons, typically 2 or 3 (on the average about 2.5 per fission), are released and are

Example 6.4

Calculate the actual fission energy release from the induced fission process shown in equation (6.3), using the following atomic masses:

$$m(^{235}\text{U}) = 235.0439231 \text{ u.}$$

$$m(^{137}\text{I}) = 136.91787084 \text{ u.}$$

$$m(^{96}\text{Y}) = 95.91589779 \text{ u.}$$

Solution

The excess energy is given by the energy equivalence of the difference between masses on the left-hand side of the equation and those on the right-hand side:

$$E = \{[m_n + m(^{235}\text{U})] - [m(^{137}\text{I}) + m(^{96}\text{Y}) + 3m_n]\}c^2$$

$$= [(1.008664904 \text{ u} + 235.0439231 \text{ u}) - (136.91787084 \text{ u} + 95.91589779 \text{ u}$$
$$+ 3 \times 1.008664904 \text{ u})] \times 931.494 \text{ MeV/u}$$

$$= 179.6 \text{ MeV.}$$

referred to as *prompt neutrons* because they appear at the time of the fission (more about this in the next section). The rest of the neutrons are attached to the fission fragments, but there are still too many neutrons compared with the number of protons to produce nuclei that are β stable. These β unstable nuclei undergo β decay (along with some γ decay) until they reach a point where enough neutrons have been converted to protons to make them β stable.

In principle, the extra neutrons produced in equation (6.3) could be incident on other ^{235}U nuclei and induce further fissions, possibly resulting in a chain reaction that could produce substantial quantities of energy. It is important to consider how these neutrons interact with other uranium nuclei and how the chain reaction can be made to continue in a controlled manner. When the neutrons are released by a fission reaction such as equation (6.3), they have kinetic energies of about 2.0 MeV. If these neutrons are released within a piece of uranium, several things can happen to them. Each neutron may undergo one of the following processes:

1. It can be absorbed by a uranium nucleus and induce fission, thereby producing more neutrons.
2. It can be absorbed by a uranium nucleus and undergo the (n,γ) reaction, thereby lowering the excess energy available to the nucleus to a value below the energy of the Coulomb barrier. In this case, the neutron is captured as described in the previous chapter and cannot contribute to further fission.
3. It can exit from the piece of uranium without undergoing reaction 1 or 2, in which case it cannot contribute to further fission.
4. It can interact with other particles in the material and merely lose some of its kinetic energy. In this case, it still has the possibility of interacting as in 1, 2, or 3.

Because, on average, 2.5 fission neutrons are produced for one induced fission event, if one of these neutrons goes on to induce another fission and 1.5 are lost as in processes 2 and 3, then the chain reaction will continue in a controlled way. The reactor is then said to be *critical*. If the piece of uranium is very small, many neutrons will likely escape (process 3). If the uranium is above a certain size (the *critical mass*), then a chain reaction is possible. The critical mass can be calculated by knowing the probability for reaction 1. This is related to the fission cross section, shown as a function of energy for ^{235}U and ^{238}U in Figure 6.4. A spherical piece of pure ^{235}U has a critical mass of about 52 kg, corresponding to a radius of about 8.7 cm. For a mass less than 52 kg, the chain reaction dies out because too many neutrons escape before inducing further fissions. For a mass of ^{235}U greater than 52 kg, the chain reaction gets out of control because more than one neutron from each fission goes on to induce other fissions. In this case, the amount of energy released becomes very large in a very short period of time. It is not possible to maintain the mass of a piece of ^{235}U at just the critical mass so that the reaction is maintained without getting out of control because these processes occur too quickly. The approach to reactor control depends on a detailed understanding of the behavior of ^{238}U and the properties of the fissions fragments. These aspects of a fission reactor are discussed below.

Figure 6.4 shows that the fission cross section for ^{238}U drops off very rapidly below about 2 MeV. This is because ^{238}U is nonfissile, and additional kinetic energy from the neutron is necessary to induce fission. If the neutron loses very much of its original energy, it will not be able to induce fission in ^{238}U. The figure also shows that the greatest chance of inducing fission in uranium occurs when a very low-energy

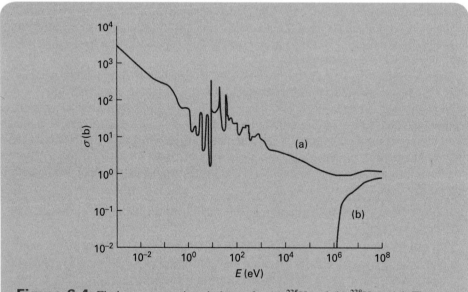

Figure 6.4: Fission cross sections in barns for (a) ^{235}U and (b) ^{238}U nuclei. The cross section is the apparent area of a uranium nucleus as seen by an approaching neutron and is a measure of the probability that a reaction will occur. The cross section is measured in area units of barns, where 1 barn = 10^{-28} m^2.

Based on R.A. Dunlap, An Introduction to the Physics of Nuclei and Particles, Brooks-Cole, Belmont (2004)

Example 6.5

The mean distance traveled by a neutron in a material with a fission cross section σ before inducing a fission event is

$$d = (n\sigma)^{-1}.$$

where n is the number of nuclei per unit volume.

(a) Calculate the mean distance traveled by a fission neutron in a piece of ^{235}U before inducing another fission.
(b) Repeat part (a) for a low-energy neutron (at 10^{-2} eV) entering a piece of ^{235}U after being moderated.

Solution

(a) The fission cross section for neutrons in ^{235}U is shown in Figure 6.4. Fission neutrons have initial energies around 2.5 MeV, so the figure shows that the cross section at that energy will be about

$$1 \text{ barn or } 10^{-28} \text{ m}^2.$$

The number density of ^{235}U nuclei is given in terms of the mass density ρ by

$$n = \rho/m$$

where m is the mass of a ^{235}U nucleus, given by

$$m = (235 \text{ u}) \times (1.66 \times 10^{-27} \text{ kg/u}) = 3.9 \times 10^{-25} \text{ kg}.$$

The mass density of uranium is found on the Web to be $\rho = 1.91 \times 10^4$ kg/m^3, giving a number density of nuclei of

$$n = (1.91 \times 10^4 \text{ kg/m}^3)/(3.9 \times 10^{-25} \text{ kg}) = 4.9 \times 10^{28} \text{ m}^{-3}.$$

Substituting the number density and cross section into the equation given above provides the mean distance traveled as

$$d = [(4.9 \times 10^{28} \text{ m}^{-3}) \times (10^{-28} \text{ m}^2)]^{-1} = 0.20 \text{ m}.$$

This is a good approximation of the diameter of a piece of ^{235}U with the critical mass as given in the text, i.e., 0.174 m.

(b) For a low-energy neutron (10^{-2} eV), Figure 6.4 shows that the fission cross section of ^{235}U is about 10^3 barns $= 10^{-25}$ m^2.

Following the above derivation gives

$$d = [(4.9 \times 10^{28} \text{ m}^{-3}) \times (10^{-25} \text{ m}^2)]^{-1} = 2.0 \times 10^{-4} \text{ m}.$$

This is the reason why fuel rods in thermal neutron reactors do not need to be very large in diameter.

neutron is incident on a ^{235}U nucleus. The probability here is more than 1000 times greater than it is for ^{235}U at high energy or for ^{238}U at any energy. Thus, although in natural uranium, ^{235}U accounts for less than 1% of all uranium nuclei, it is much more likely that induced fission will occur for a low-energy neutron incident on ^{235}U than for a neutron of any energy incident on ^{238}U. So the best approach to producing fission in a predictable way is to slow the neutrons down while preventing them from interacting with any uranium nuclei. Once their energy has been reduced to a very low value, they are allowed to be incident on a collection of uranium nuclei, where they are almost certain to induce fission in a ^{235}U nucleus. This procedure ensures that the neutron will interact with a ^{235}U nucleus and induce fission rather than being lost by undergoing a reaction (as in reaction 2) with either ^{235}U or ^{238}U.

This approach can be implemented by making the pieces of uranium (the fuel elements) in the reactor fairly small (i.e., much less than the critical mass). This means that when a neutron is emitted from a fission event, it quickly exits the fuel element process 3) and travels through a substance (not containing uranium) known as a *moderator* that reduces the energy of the neutron. The moderator lowers the energy of the neutron until the level is comparable to the thermal energy of the atoms in the moderator, around 0.03 eV. The neutrons are then referred to as *thermal neutrons*, and the reactor that operates on this principle is referred to as a *thermal reactor*, or *thermal neutron reactor*. When the neutrons enter another fuel element, they quickly induce fission in ^{235}U nuclei. The flow of neutrons between fuel elements is controlled by control rods made of a material (often cadmium or boron) that easily absorbs neutrons (without undergoing any undesirable nuclear reactions). The general design of a nuclear reactor core is shown in Figure 6.5.

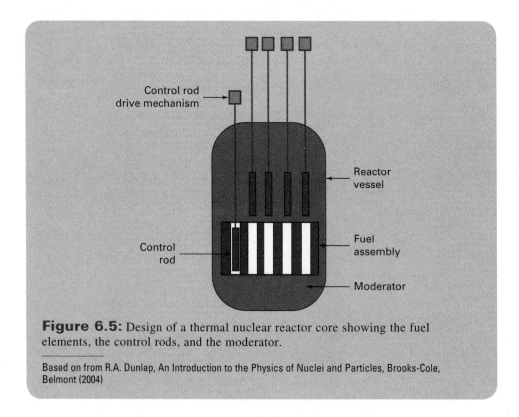

Figure 6.5: Design of a thermal nuclear reactor core showing the fuel elements, the control rods, and the moderator.

Based on from R.A. Dunlap, An Introduction to the Physics of Nuclei and Particles, Brooks-Cole, Belmont (2004)

Each control rod can be moved so that a larger or a smaller portion of each fuel element is exposed to its neighboring fuel elements, thus allowing more or less fission to occur. The neutron density and temperature in the reactor can be monitored, and the control rods can be moved in such a way as to either increase or decrease the neutron flow between fuel elements in order to maintain the chain reaction. This condition is met when exactly 1 fission neutron (on average) goes on to induce an additional fission. How a controlled fission reaction is maintained is discussed in the next section.

6.4 Fission Reactor Control

The basic principles of reactor control require a careful consideration of the timescale of the processes that take place in the reactor. Once a neutron has been absorbed by a nucleus, the actual fission process itself is very fast, on the order of 10^{-8} seconds, and this is the timescale on which a prompt neutron is emitted. The time that a neutron takes to travel through the moderator is much longer, perhaps on the order of 10^{-4} seconds. This is still a very short time compared to how long it takes to physically move a control rod, which is typically done by some mechanical or hydraulic method in response to variations in the operating conditions. If it takes 1 second to move a control rod in response to an increase in neutron flux, the reactor could quickly become unstable. As an example, consider the case when the fission time constant (i.e., the time between neutron emission and subsequent induced fission) is 10^{-4} seconds and the number of fission neutrons inducing fission suddenly changes from 1.00 to 1.01. In the first 10^{-4}-second interval, the number of fission neutrons in the reactor increases by a factor of 0.01 (or 1%); in the next 10^{-4}-second interval, the number of fission neutrons increases to 1.01^2 of its initial value, and so on. After 1 second, the number of neutrons has increased to $1.01^{10,000}$, or about 10^{43} times the initial value. This is much larger than the total number of neutrons associated with all the uranium nuclei in a fuel rod (a kilogram of uranium contains about 3.5×10^{26} neutrons). Thus in this situation all the fuel in the reactor undergoes fission in less than a second, leading to a very undesirable rate of energy release (Section 6.8).

To resolve this difficulty, it is important to look more carefully at the properties of the fission fragments. The fission fragments are inevitably β unstable and have too many neutrons. Normally, these nuclei β decay by converting neutrons to protons until they become stable. Sometimes during this process, one of the neutrons that is part of the nucleus of one of the fragments (and that is not very tightly bound) escapes prior to the occurrence of the β decay process. This is referred to as *delayed neutron emission* and is illustrated for a fission fragment of mass 137 in Figure 6.6 on the next page. It can be seen from the figure that the timescale for delayed neutron emission (in this case) is about 23 seconds. On the average, it is about 10 seconds or so for typical fission fragments and is substantially longer for the timescale for processes involving prompt neutrons. For ^{235}U fission, about 2% of the neutrons produced are delayed neutrons. Reactors are typically designed to be about 99% prompt critical. The additional 1% of neutrons needed to reach the critical condition are delayed neutrons, and these are easily controlled on a timescale compatible with methods for adjusting the control rods.

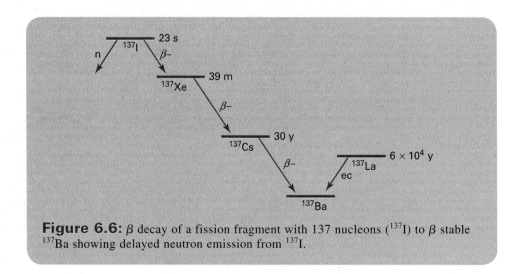

Figure 6.6: β decay of a fission fragment with 137 nucleons (^{137}I) to β stable ^{137}Ba showing delayed neutron emission from ^{137}I.

6.5 Types of Thermal Neutron Reactors

Several types of thermal reactors are in current use worldwide. They differ primarily in the material used as the moderator and the way in which heat is extracted from the reactor. The former point also has some effect on the details of the fuel used. The requirements for a moderator material are as follows:

1. The material must be fairly dense (i.e., liquids and solids are alright, but gases are not).

2. The nuclei in the material must be fairly light. This will optimize the amount of energy lost by a neutron when it interacts with a nucleus in the moderator. (*Note:* This can be understood in terms of simple mechanics. If a neutron collides with a similarly sized object (e.g., a proton), the other object recoils, carrying away some of the neutron's energy. If a neutron collides with a very massive nucleus, it scatters more or less elastically and loses very little of its energy.)

3. The material should merely slow down the neutrons rather than absorb them.

4. The interaction of the neutrons with the moderator should not produce any undesirable or hazardous materials.

5. The material should be as nontoxic as possible, chemically stable, and relatively inexpensive.

Unfortunately, no moderator materials are perfect in all these respects. However, three materials do reasonably well at satisfying these criteria and are in common use in nuclear reactors. These materials are water (H_2O), heavy water (D_2O), and graphite (carbon). A number of different reactor designs are in use worldwide, but most fall into the basic categories of water-moderated reactors (developed in the United States), heavy water–moderated reactors (developed in Canada), and graphite-moderated reactors (developed in the United Kingdom, France, and Russia).

Water-moderated reactors may be boiling water reactors or pressurized water reactors. The former are used as power reactors. The latter may also be used as power reactors or (in smaller versions) for nuclear-powered ships and submarines. The *boiling water reactor* (BWR) is perhaps the simplest nuclear reactor. The uranium (in the form of UO_2) is formed into cylindrical fuel pellets about a centimeter in diameter and a centimeter long. These are packed into metal tubes about 3–4 m long. The tubes are combined into fuel bundles (the fuel elements just referred to) of typically about 100 tubes (Figure 6.7). The fuel bundles are placed in the core, surrounded by the moderator, and separated by control rods. A power reactor contains a few hundred (up to about 800) fuel bundles, and the total mass of uranium in the core is of the order of 100,000 kg. In a BWR, the water (the moderator) is also used to produce steam for running turbines. Because water is not ideal in all respects as a moderator (specifically, it does not ideally satisfy requirement 3), the uranium used in these reactors must have a higher than natural concentration of ^{235}U. Usually, the fuel is enriched to about 3% ^{235}U. A typical design is illustrated in Figure 6.8 on the next page. Reactors of this type typically produce about 1000 MW_e of power. It is important to remember that a nuclear reactor, like a fossil fuel generating station, is a heat engine and that its overall efficiency is limited by the Carnot efficiency, typically around 35–40%, so 1000 MW_e corresponds to the production of about 2.5–3 GW of heat energy by the reactor. Boiling water reactors, such as the Browns Ferry Nuclear Power Plant in Alabama shown in Figure 6.9 on the next page, are in common use in the United States.

Pressurized water reactors (PWRs) are similar to BWRs. The design of the fuel elements is similar except that typically about 200 rods make up the bundle and the core contains about 150–200 bundles (Figure 6.10 on page 187). In a PWR, the water used as the moderator is kept under pressure to prevent it from boiling. The superheated water transfers heat through a heat exchanger to water that is not kept at high pressure and that is allowed to boil and produce steam; this steam in turn operates the turbines (Figure 6.11 on page 187). This increases the complexity of the reactor design (compared to BWRs) but reduces the possibility of radioactive material from the reactor core making its way into the environment. A photograph of a PWR is shown in Figure 6.12 on page 187.

The *heavy water–moderated reactor* was developed in Canada and is known as the CANDU (Canadian Deuterium Uranium) reactor. The design is similar to a PWR; the moderator is held under pressure and does not boil. Rather, it transfers heat through a heat exchanger to water that boils to produce steam to drive the turbines. The fuel bundles are smaller than those in BWRs and PWRs, are about 1.5 m long, and typically contain 37 fuel rods (Figure 6.13 on page 187). This reactor uses the naturally abundant ^{235}U. Traditionally, CANDU reactors have produced about 700 MW_e, although a newer design produces about 1200 MW_e. CANDU reactors, as shown

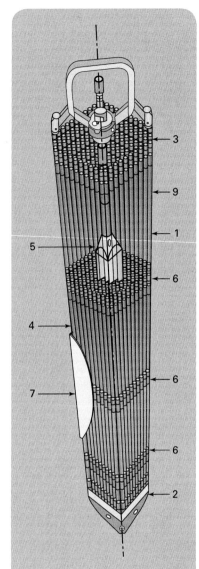

Figure 6.7: BWR fuel assembly. (1) fuel assembly, (2) base plate, (3) head, (4) fuel rod bundle, (5) water channel, (6) spacer grids, (7) fuel channel and (9) full-length fuel rods.

U.S. patent Application, BWR nuclear fuel assembly with non-retained partial length fuel rods, US 20120243652 A1

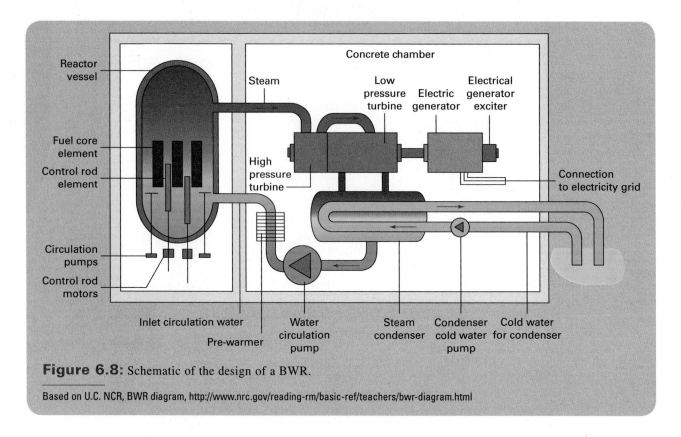

Figure 6.8: Schematic of the design of a BWR.

Based on U.C. NCR, BWR diagram, http://www.nrc.gov/reading-rm/basic-ref/teachers/bwr-diagram.html

Figure 6.9: Browns Ferry Nuclear Power Plant near Decatur, Alabama. The plant consists of three boiling water reactors with a combined generating capacity of 3.3 GW_e. The structure in the lower left of the photograph contains fan-driven cooling towers.

U. S. Nuclear Regulatory Commission

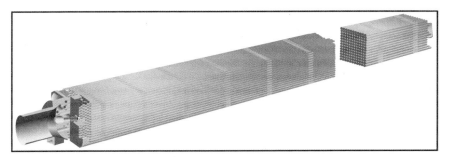

Figure 6.10: Fuel bundle for a PWR.

U.S. Maritime Administration

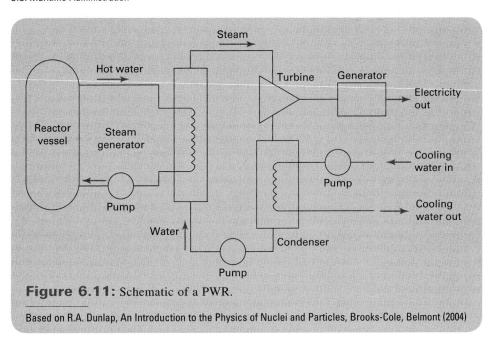

Figure 6.11: Schematic of a PWR.

Based on R.A. Dunlap, An Introduction to the Physics of Nuclei and Particles, Brooks-Cole, Belmont (2004)

Figure 6.12: Seabrook Nuclear Generating Station in New Hampshire U.S., a 1244-MW$_e$ PWR. The reactor core is contained inside the dome-shaped containment building at the right.

Richard A. Dunlap

Figure 6.13: Fuel bundle for a Canadian CANDU reactor.

Atomic Energy of Canada Limited

Figure 6.14: CANDU Nuclear Reactor at Qinshan Nuclear Power Plant at Zhejiang, China

STR/Stringer/AFP/Getty Images

in Figure 6.14, are currently in use in Canada, China, Argentina, Romania, and Pakistan. A similar reactor design based on the CANDU reactor is in use in India.

Graphite-moderated reactors, developed in the United Kingdom and France, use a gaseous coolant to transfer the heat from the reactor to water through a heat exchanger. Both helium and CO_2 have been used as a coolant. A schematic of the design is shown in Figure 6.15. Various designs use either natural or enriched (about 2% ^{235}U) fuel. A gas-cooled graphite-moderated reactor is shown in Figure 6.16.

The water-cooled graphite-moderated reactor has been used extensively in Russia. It is referred to as a RBMK reactor after the Russian name for its design (*Reactor Bolshoi Moschnosti Kanalynyi*). The fuel is ^{235}U enriched to 1.8%, and the fuel elements are very long, typically about 7 m. An illustration of the fuel bundle is shown in Figure 6.17, and a schematic of the reactor design is shown in Figure 6.18 on page 190. An operational RBMK reactor in Russia is illustrated in Figure 6.19 on page 191. The reasons for this design and the details of its operation are discussed later in this chapter with regard to the accident involving a reactor of this type at Chernobyl.

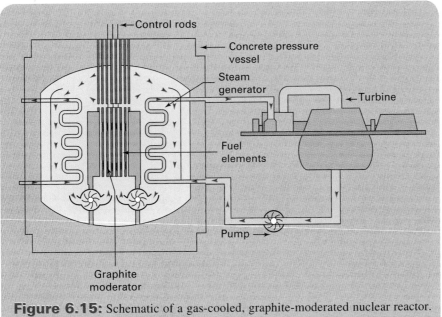

Figure 6.15: Schematic of a gas-cooled, graphite-moderated nuclear reactor.

Based on World Nuclear Association, Nuclear Power Reactors, http://www.world-nuclear.org/info/inf32.html

Figure 6.16: Gas-cooled graphite-moderated nuclear reactor at Torness Nuclear Power Station, Scotland

Jonathan Littlejohn/Alamy Stock Photo

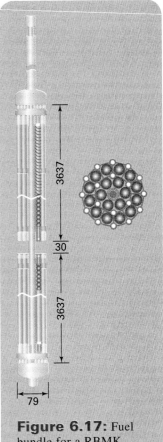

Figure 6.17: Fuel bundle for a RBMK reactor. Dimensions are in mm.

Based on World Nuclear Association, Nuclear Power Reactors, http://www.world-nuclear.org/info/inf32.html

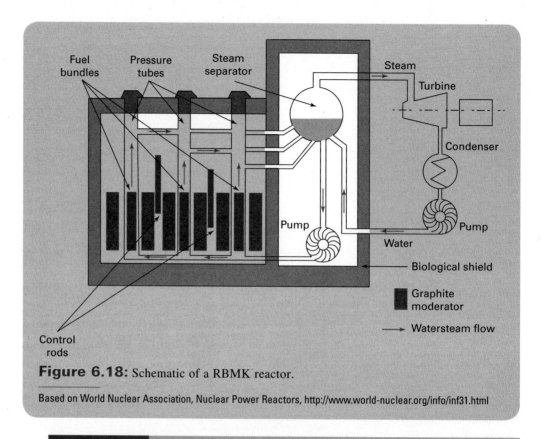

Fuel bundles   Pressure tubes   Steam separator   Steam   Turbine   Condenser   Pump   Pump   Water   Biological shield   Graphite moderator   Watersteam flow   Control rods

Figure 6.18: Schematic of a RBMK reactor.

Based on World Nuclear Association, Nuclear Power Reactors, http://www.world-nuclear.org/info/inf31.html

Example 6.6

If a thermal fission reactor operating at 40% Carnot efficiency produces 1 GW_e output from the fission of ^{235}U in natural uranium, calculate the mass of uranium needed per second to fuel the reactor.

Solution

The induced fission of one ^{235}U nucleus produces about 180 MeV of energy (see Example 6.4). In joules this is

$$(1.8 \times 10^8 \text{ eV}) \times (1.6 \times 10^{-19} \text{ J/eV}) = 2.88 \times 10^{-11} \text{ J}.$$

A 1 GW_e reactor operating at an efficiency of 40% needs to produce 2.5 GW of thermal energy, or 2.5×10^9 J/s. This will require

$$(2.5 \times 10^9 \text{ J/s})/(2.88 \times 10^{-11} \text{ J}) = 8.68 \times 10^{19} \text{ fissions per second.}$$

Since natural uranium is 0.72% ^{235}U then the total number of uranium nuclei needed to fuel the reactor per second will be

$$(8.68 \times 10^{19} \text{ s}^{-1})/(0.0072) = 1.21 \times 10^{22} \text{ s}^{-1}.$$

In moles, this is

$$(1.21 \times 10^{22} \text{ s}^{-1})/(6.02 \times 10^{23} \text{ mol}^{-1}) = 0.02 \text{ mol/s}.$$

Since 1 mole of uranium has a mass of about 238 g, the fuel consumption will be

$$(0.02 \text{ mol/s}) \times (238 \text{ g/mol}) = 4.8 \text{ g/s}.$$

Figure 6.19: The Leningrad Nuclear Power Plant in Sosnovy Bor, Leningrad Oblast, Russia. The facility consists of four 1 GW$_e$ RBMK reactors. Four additional PWRs are scheduled for construction at the site.

6.6 Current Use of Fission Energy

Beginning with the first commercial nuclear power reactors in the late 1950s, there was substantial and consistent growth of the nuclear power industry that spanned the next 30 years, the number of new nuclear power plant construction starts each year in Figure 6.20 on the next page illustrates. This growth was due to the perception that nuclear energy was an almost limitless and reasonably priced source of electricity. This situation changed in the mid-1980s, and development of nuclear energy slowed considerably, although in recent years there has been some renewed growth. These data, however, must be considered in the context of the decommissioning of aging reactors. The total number of operational reactors and the total energy generated may be the best indicators of the utilization of nuclear energy. Figure 6.21 on the next page shows the

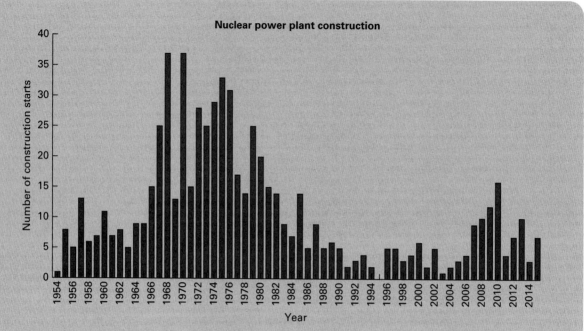

Figure 6.20: Number of nuclear power plant construction starts worldwide as a function of year since 1954.

Based on https://www.oecd-nea.org/press/press-kits/economics.htm and https://www.iaea.org/PRIS/home.aspx

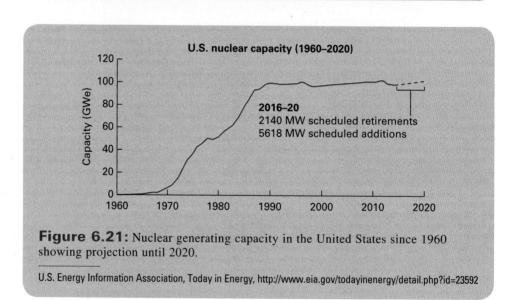

Figure 6.21: Nuclear generating capacity in the United States since 1960 showing projection until 2020.

U.S. Energy Information Association, Today in Energy, http://www.eia.gov/todayinenergy/detail.php?id=23592

total nuclear electricity generating capacity in the United States as a function of year since the beginning of the nuclear power industry, along with predictions for the next few years based on anticipated upgrades and retirements. This figure shows that the use of nuclear energy has been nearly constant in the United States since about 1990.

ENERGY EXTRA 6.1
Watts Bar Nuclear Generating Station

The Watts Bar Nuclear Generating Station in Rhea County in eastern Tennessee is operated by the Tennessee Valley Authority (TVA) and has the most interesting history of any nuclear reactor facility in the United States. It consists of two thermal neutron reactors; Unit 1 was the last nuclear reactor to become functional in the United States in the 20th century, while Unit 2 was the first to become operational in the 21st century. However, the history of these two reactors spans more than 40 years.

Construction of the Watts Bar Nuclear Generating Station began in 1973, at a time before any major commercial reactor accidents and at a time when nuclear power was developing rapidly in the United States and elsewhere. The station consists of two Westinghouse pressurized water reactors with a maximum capacity of about 1.15 GW$_e$. The reactor design is what would now be referred to as a generation II reactor (see Energy Extra 6.3). Although a number of other nuclear power reactors were started in the United States after Watts Bar, the two reactors at this site were the last two to be finished in this country (see bar graph). Watts Bar Unit 1 became operational in 1996, and Watts Bar Unit 2 first produced power in 2016. While proponents of nuclear power may cite Watts Bar 2 as an example of the continued (or revived) interest in nuclear power, critics of nuclear power may cite it as an example of virtually everything that could go wrong with the construction of a power plant.

After about 12 years of construction, work on Units 1 and 2 at Watts Bar Nuclear Generating Station stopped in 1985 as a result of a Nuclear Regulatory Commission (NRC) review that identified a number of deficiencies. At that time Unit 1 was very nearly complete, and Unit 2 was about 80% complete. The deficiencies with Unit 1 were dealt with, and the reactor finally became operational in 1996. Unit 2 was, however, more problematic. By 1985 $1.7 billion had

Watts Bar Nuclear Generating Station

Continued on page 194

Energy Extra 6.1 continued

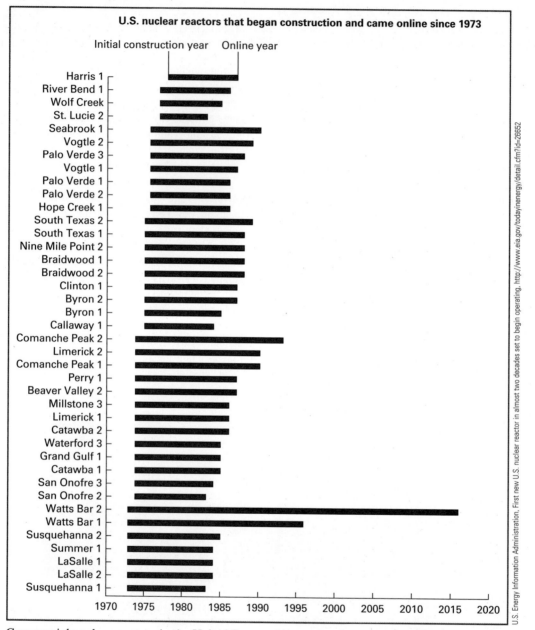

Commercial nuclear reactors in the United States with construction starts in 1973 and later that have been completed.

been spent on the construction of Unit 2, after an initial construction cost estimate of $400 million. The nuclear accidents at Three Mile Island in 1979 and subsequently at Chernobyl in 1986 raised concerns about the safety of nuclear power, and projections of future power needs in the U.S. brought into question the economic viability of completing Watts Bar Unit 2.

In 2007 TVA applied to the NRC for approval to complete Watts Bar Unit 2 with an anticipated operational date of 2013. The completion of the reactor was

Continued on page 195

Energy Extra 6.1 continued

estimated to cost an additional $2.5 billion, and it was estimated that around 250 new jobs would be associated with the reactor. Construction on Watts Bar Unit 2 resumed in October 2007.

As a result of the Fukushima Dai-ichi accident in 2011, the NRC issued more stringent safety regulations for US nuclear reactors. Some of these regulations applied to the construction of Watts Bar Unit 2 (which was basically a generation II reactor design that was 40 years old). As a result of the necessary design modifications (as well as other factors), the completion of Unit 2 took longer than anticipated and cost more than expected. In total, the construction of Watts Bar Unit 2 cost an estimated $6.4 billion and took 43 years. The reactor was completed in 2015 and first reached criticality on May 23, 2016. It was connected to the power grid on June 3, 2016 and became fully operational on October 19, 2016. The reactor was shut down on March 23, 2017 due to the failure of the steam condenser. As of July 2017, it is uncertain when it will be operational again.

Topic for Discussion

Make a simple estimate of the cost per kWh for electricity generated by Watts Bar Unit 2 based on capital infrastructure cost, name plate capacity, and an estimated lifetime of 40 years. The actual cost per kWh will be higher than this simple estimate for a number of reasons. Consider the importance of each of the following factors in determining the net cost per kWh:

- Cost recovery factor
- Fuel cost
- Operating cost
- Capacity factor
- Decommissioning costs

Make estimates of values of parameters when appropriate or locate relevant information on the web.

Figure 6.22 shows the total nuclear electricity generated worldwide since 1971 and indicates that while some growth continued worldwide during the 1990s, the production of nuclear electricity has not changed significantly since then.

The current use of nuclear energy is summarized in Table 6.1. Obviously, nuclear energy is a much more important contribution to the generation of electricity in Europe than in most other parts of the world. The use of nuclear energy is particularly important in France, both in terms of the total generating capacity and as a fraction of total

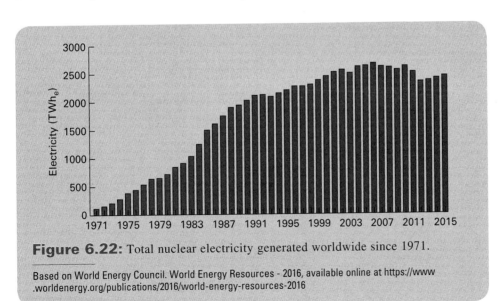

Figure 6.22: Total nuclear electricity generated worldwide since 1971.

Based on World Energy Council. World Energy Resources - 2016, available online at https://www.worldenergy.org/publications/2016/world-energy-resources-2016

Table 6.1: Summary of number and capacity of operational nuclear power reactors and reactors under construction worldwide as of 2015. The total electricity generated and the percentage of total electricity generation is also given (* data for 2013).

country	number operational reactors	operational capacity (GW$_e$)	number reactors under construction	capacity under construction (GW$_e$)	electricity generated (GWh)	% total domestic electricity
Argentina	3	1.6	1	0.025	6519	4.8
Armenia	1	0.4			2571	34.5
Belarus			2	2.2	NA	NA
Belgium	7	5.9			24,825	37.5
Brazil	2	1.9	1	1.2	13,892	2.8
Bulgaria	2	1.9			14,701	31.3
Canada	19	13.5			95,637	16.6
China	31	26.8	24	24.1	161,202	3.0
Czech Republic	6	3.9			25,337	32.5
Finland	4	2.8	1	1.6	22,326	33.7
France	58	63.1	1	1.6	419,022	76.3
Germany	8	10.8			86,810	14.1
Hungary	4	1.9			14,960	52.7
India	21	5.78	6	4.3	34,644	3.5
Iran	1	0.9			3198	1.3
Japan	43	40.3	2	2.7	4346	0.5
Korea (South)	24	21.7	4	5.4	157,199	31.7
Mexico	2	1.4			11,185	6.8
Netherlands	1	0.5			3862	3.7
Pakistan	3	0.7	2	0.6	4333	4.4
Romania	2	1.3			10.710	17.3
Russia	35	25.4	8	6.6	182,807	18.6
Slovakia	4	1.8	2	0.9	14,084	55.9
Slovenia	1	0.7			5372	38.0
South Africa	2	1.9			10,965	4.7
Spain	7	7.1			54,759	20.3
Sweden	10	9.6			54,455	34.3
Switzerland	5	3.3			22,156	33.5
Taiwan	6	5.1	2	2.6	35,143	16.3
Ukraine	15	13.1	2	1.9	82,405	56.5
United Arab Emirates			4	5.4	NA	NA
United Kingdom	15	8.9			63,895	18.9
United States	99	99.2	5	5.6	798,012	19.5
world total	**441**	**382.9**	**67**	**66.4**	**2,441,331**	**11***

Based on data from "*World Energy Council. World Energy Resources–2016,* available online at https://www.worldenergy.org/publications/2016 /world-energy-resources-2016/"

electricity generation. This ambitious approach to non-carbon-producing electricity generation in France may be reflected in the country's carbon emissions (shown in Table 4.8 and discussed in the previous chapter). The table also shows that South Korea has a significant nuclear power program. In the past, Japan had significant nuclear electricity generation, as the number of reactors indicates. However, following the Fukushima accident (see Section 6.8c) operation of nearly all of the reactor facilities in that country has been suspended, resulting in the very low level of energy generation for 2015. Current development in China is likely to give rise to some increase in worldwide nuclear energy production over the next few years.

6.7 Uranium Resources

Like fossil fuels, uranium is a limited resource. At present, as seen in Table 6.1, nuclear energy accounts for about 11% of the world's production of electricity. That corresponds to about 4% of all primary energy consumption. The rate at which uranium is presently being consumed is well-known. An estimate of the longevity of nuclear energy can be made on the basis of this information, predictions for future uranium use, and an estimate of mineral resources.

As with most resources, estimating the amount of available uranium is not necessarily straightforward. Certainly, the more one is willing to pay, the more resource will be available for use. Table 6.2 gives an estimate of available uranium resources for less than US$130 per kg. It is seen that Australia has, by far, the greatest resources.

Table 6.2: Proven resources of uranium as of 2014 recoverable at a cost of < US$130 per kg.

country	uranium resources (t)
Australia	1,174,000
Brazil	155,100
Canada	357,500
China	120,000
Czech Republic	1300
Kazakhstan	285,600
Malawi	8200
Namibia	248,200
Niger	325,000
Romania	3100
Russia	216,500
South Africa	175,300
Ukraine	84,800
United States	207,400
Uzbekistan	59,400
Other	277,500
world total	**3,698,900**

Based on data from "*World Energy Council. World Energy Resources–2016,* available online at https://www.worldenergy.org/publications/2016 /world-energy-resources-2016/"

The remaining resources are available primarily in Canada, Kazakhstan, Russia, the United States, and some African countries. Table 6.3 gives information about the major uranium-producing nations. At present, Kazakhstan leads in uranium production, followed by Canada. These two countries produce nearly 60% of the world's uranium.

Unlike the situation with fossil fuels, where production and use are very nearly the same and differ only by fluctuations in the amount of stored resources, uranium production and use can be substantially different as shown in Figure 6.23. In recent years, demand has exceeded production and the discrepancy has been made up with stored uranium, reprocessed fuel from reactors, and military uranium from decommissioned weapons. Reprocessing is a method by which the fissionable component of spent fuel is extracted for reuse in a reactor. This fuel is largely plutonium, and the reluctance of many governments to reprocess spent fuel but rather to merely store it (Section 6.10) is based on security concerns because plutonium extracted from reprocessed fuel can be used for weapons purposes. This is discussed further in Section 6.12 on fast breeder reactors.

If nuclear reactors continue to supply the same quantity of power as they currently do and production has to meet the reactor requirements, then a simple calculation based on available resources (about 3,700,000 t from Table 6.2) and the current demand (about 70,000 t per year from Figure 6.23) indicates a lifetime for these resources of about 50 years. The use of military uranium, reprocessed fuel, and the development of new technologies for power reactors that would allow for the use of

Table 6.3: Uranium production for 2014.	
country	**uranium production (t)**
Australia	5001
Brazil	231
Canada	9134
China	1500
Czech Republic	193
India	385
Kazakhstan	23,127
Malawi	369
Namibia	3255
Niger	4057
Pakistan	45
Romania	77
Russia	2990
South Africa	573
Ukraine	962
United States	1919
Uzbekistan	2400
Other	36
world total	**56,252**

Based on *World Nuclear Association*, Uranium Markets, http://www.world-nuclear
.org/information-library/nuclear-fuel-cycle/uranium-resources/uranium-markets.aspx

Example 6.7

Assuming a world nuclear generating capacity of about 400 GW$_e$, estimate the lifetime of known uranium resources if all of the fission energy is extracted from the ^{235}U component of the fuel.

Solution

As shown in Example 6.6, a 1 GW$_e$ nuclear reactor uses about 4.8 g of natural uranium per second or

$$(0.0048 \text{ kg/s}) \times (3.15 \times 10^7 \text{ s/y}) = 1.51 \times 10^5 \text{ kg/y for 1 GW}_e.$$

The generation of 400 GW$_e$ will therefore require

$$(400 \text{ GW}_e/\text{y}) \times (1.51 \times 10^5 \text{ kg/GW}_e) = 6.0 \times 10^7 \text{ kg/y}.$$

This is consistent with the current annual uranium requirements shown in Figure 6.23. From Table 6.2 the proven worldwide uranium resources are about 6.3×10^9 kg. So the lifetime of these resources will be

$$(6.3 \times 10^9 \text{ kg})/(6.0 \times 10^7 \text{ kg/y}) = 105 \text{ years}.$$

lower-grade fuel could extend this estimate by a substantial amount. The use of lower-grade ore is discussed further later in this chapter. It should be noted, however, that nuclear energy accounts for only about 4% of world energy use, so even with a greatly increased lifetime, nuclear energy (in its present form) could not satisfy all of our energy requirements for more than a small number of decades.

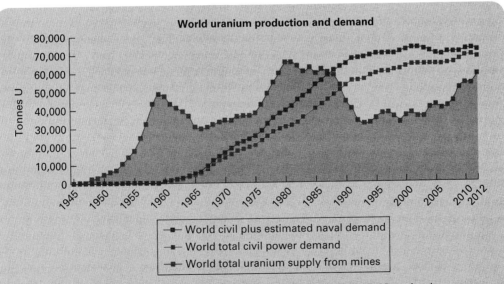

Figure 6.23: World uranium production and demand since 1945. Note that in recent years power reactors have accounted for about 90% of the world's uranium use.

Based on World Nuclear Association, Uranium Markets, http://www.world-nuclear.org/information-library/nuclear-fuel-cycle/uranium-resources/uranium-markets.aspx

6.8 Nuclear Safety

Although nuclear energy production does not release carbon into the environment, as does the burning of fossil fuels, it is not without its environmental concerns. Because electricity is produced from nuclear energy by a heat engine, heat pollution is a local concern around generating stations. The greatest concerns are, however, for the safety of reactors and the disposal of spent fuel, the first of which is discussed in this section. The security of radioactive materials that could have military uses is also a concern, and this is dealt with briefly in ensuing sections.

The safety of nuclear power plants, or at least the public perception of their safety, was a major factor in the subsiding growth of nuclear energy that began more than 20 years ago. The possibility and consequences of a nuclear accident need to be considered very carefully, and the three most significant accidents to date—Three Mile Island, Chernobyl, and Fukushima Dai-ichi—serve as useful case studies for this analysis.

6.8a Three Mile Island

Three Mile Island Nuclear Generating Station is in Eastern Pennsylvania and consists of two PWRs (Figure 6.24). Four large cooling towers are readily seen in the figure, and the two reactor containment buildings appear near the center. Construction on the Three Mile Island plant began in 1968. The first reactor, TMI-1, which is the left-most reactor in the photograph, began commercial operation in 1974. The second reactor, TMI-2, which was involved in the accident, began operation in 1978 and is the farthest to the right in the figure.

Figure 6.24: Three Mile Island Nuclear Generating Station prior to the accident of 28 March 1979.

Centers for Disease Control and Prevention's Public Health Image Library

A typical PWR has three water or steam loops (Figure 6.11). The primary loop contains water that is pressurized to prevent it from boiling. It has three functions: to moderate the neutrons, to keep the reactor core from overheating by transporting heat out of the core, and to transfer that heat to the secondary loop. The secondary (water/steam) loop contains water that is allowed to boil and produce steam, which is used to run the turbines. The third, or cooling, loop contains water and is used to transfer excess heat to the environment. In the case of the Three Mile Island reactor, this cooling loop dissipates heat to the atmosphere by means of cooling towers.

The accident at TMI-2 began at 4:00 a.m. on March 28, 1979 (exactly one year, to the minute, from its initial start-up). The accident began as a failure of the main feed water pump in the secondary (non-nuclear) loop. See the schematic of the reactor design in Figure 6.25. This prevented heat from being removed from the primary loop, and the temperature of the water and hence its pressure increased. The pressurized relief valve in the primary loop opened to lower the pressure by releasing steam. Unfortunately, a failure in the system operating this valve caused the valve not to close when the pressure returned to normal, and excess water in the form of steam was released from the valve. This resulted in a significant decrease in cooling water from the primary loop (known as a LOCA, *loss of coolant accident*), and temperature of the core of the reactor increased. Before the reactor was brought under control, part of the core melted, and radioactive material from the fuel elements contaminated the water in the primary loop. Although the containment vessel around the core of the reactor was not breached, some radioactive material (6.3×10^{11} Bq, Bq = 1 decay per second) was released to the environment as contaminated steam through the pressure release valve. Although TMI-1 was not

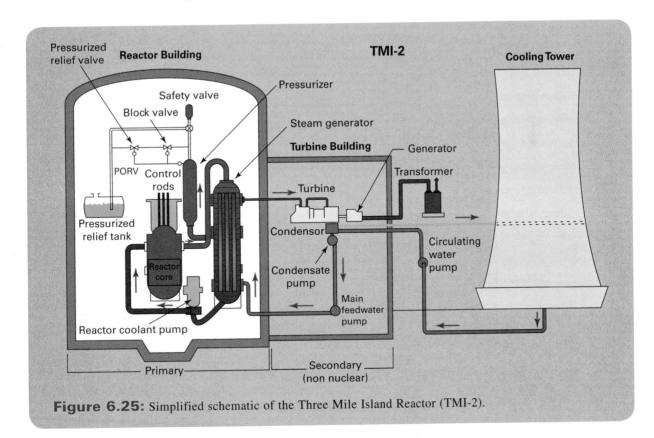

Figure 6.25: Simplified schematic of the Three Mile Island Reactor (TMI-2).

involved in this accident and continues to operate as a commercial power reactor, TMI-2 has been shut down since the accident.

Studies by a number of U.S. government agencies (Nuclear Regulatory Commission; Environmental Protection Agency; Department of Health, Education and Welfare; Department of Energy; and Commonwealth of Pennsylvania) have considered the environmental and health consequences of the Three Mile Island accident. The general conclusion of these studies has been that there were no immediate adverse health effects caused by the accident and that any long-term effects would be negligible. Specifically, it was found that the average radiation exposure to the population of about 2 million in eastern Pennsylvania was about 10^{-5} Sv (Sv = Sievert). By comparison, the radiation exposure from a chest X-ray is about 6×10^{-5} Sv. The maximum exposure for residents in the immediate area of the reactor was 10^{-3} Sv. This value is about the same as the average annual exposure from background radiation in the area, and the additional exposure is about equivalent to spending a year living in a high-altitude city such as Denver. A risk analysis indicates that no additional cancer deaths are expected as a result of the accident.

Three Mile Island is the worst nuclear power reactor accident in the United States and was the result of a series of unfortunate equipment failures that operators did not deal with appropriately. The effect of the Three Mile Island accident on the development of nuclear power in the United States is not entirely clear. A decrease in reactor construction had started in the United States in the early and mid-1970s, and construction of a number of planned reactors had already been cancelled prior to the Three Mile Island accident. However, the Three Mile Island accident was only one of a number of factors that influenced nuclear policy. There were growing concerns over nuclear safety as a result of a number of less serious accidents involving military and commercial reactors, as well as concerns over nuclear security and waste disposal. Other factors were a reassessment of fossil fuel use after the oil crisis in the early 1970s, a shift in the United States toward importation of energy resources rather than domestic production, and changes in federal pollution standards that promoted the use of inexpensive coal resources. Perhaps one of the most significant impacts of the Three Mile Island accident has been a shift in the public perception of nuclear power.

6.8b Chernobyl

A much more serious nuclear power reactor accident occurred at the Chernobyl reactor in 1986. Chernobyl is located in what was then the Soviet Union (and is now the Ukraine near its border with Belarus). The generating station consisted of four 1000-MW$_e$ nuclear reactors of the water–cooled, graphite-moderated (Russian RBMK) type. Two additional reactors were under construction at the site at the time of the accident. The first reactor at Chernobyl became operational in 1977, and the number 4 reactor (which experienced the accident) became operational in 1983. Two major factors contributed to the accident; an intrinsically unsafe reactor design and operator error.

The reactor at Chernobyl was designed to produce both electricity and radioactive ^{239}Pu for nuclear weapons use. To extract the ^{239}Pu from the reactor while continuing to keep the reactor running to produce electricity, certain compromises in the safety of the design were made. Specifically, to remove some of the fuel rods without shutting the reactor down, a large amount of space above the reactor core was needed. To avoid an extensive structure around the reactor core, early RBMK reactors were designed with compromised containment structures. The philosophy of this approach was to deal with safety by minimizing the possibility of an accident without considering the need to contain an accident if one did occur. Another design factor that contributed to the problems at Chernobyl was the way in which electricity needed to run the reactor was supplied.

This electricity was used to run the cooling water pumps and operating the computers controlling the reactor operation. Although the reactor normally operated from power provided through the grid, backup generators were in place to take over in the case of a grid power outage. There were concerns that the delay between a grid power outage and full generator capacity was unacceptable. It was, in fact, the awareness of this situation that prompted a test on the evening of April 25, 1986 to investigate the possibility of using electricity produced onsite by different mechanisms in the event of a power outage. During a routine maintenance shutdown, several safety systems were intentionally disabled in order to conduct the test. The RBMK is inherently unstable at low power levels, and during the test several indications that the reactor was out of control were ignored.

When, in the early morning of April 26, the operators did decide to shut down the reactor for safety reasons, the power output fluctuated, causing pressure to build up in the cooling system and bursting some cooling water pipes. The water and steam that was released reacted with the hot reactor core that contained the graphite moderator and the zirconium tubes containing the fuel rods. This reaction decomposed the water, releasing hydrogen. The hydrogen ignited, causing an explosion that blew off the top of the reactor building. This explosion did not result from the reaction of nuclei, as is the case for a nuclear weapon, but rather from the chemical reaction of hydrogen that had been produced in the reactor. However, a substantial quantity of radioactive materials from the core was released to the environment. The results of this explosion are shown in Figure 6.26. Because graphite (pure carbon) is

Figure 6.26: Chernobyl reactor 4 after the accident of April 26, 1986.

SPUTNIK/Alamy Stock Photo

Figure 6.27: Concrete sarcophagus around Chernobyl reactor number 4.

Time & Life Pictures/Getty Images

flammable, the resulting fire needed to be dealt with. To contain the remains of the reactor core, a concrete enclosure (called the Sarcophagus) was subsequently constructed around reactor 4 (Figure 6.27).

In the immediate aftermath of the Chernobyl accident, 237 people suffered from acute radiation sickness. These were workers and firefighters at the reactor, and 29 of these died from radiation exposure. Two additional deaths resulted directly from the explosion and fire. A study in 2005 identified an additional 25 subsequent cancer deaths that could be directly attributed to the accident.

The longer-term effects of the Chernobyl accident are unclear. A significant amount of radioactive material was released to the environment, about 1.8×10^{18} Bq in total (compare with 6.3×10^{11} Bq for Three Mile Island). A sizable area around the reactor site was evacuated and will remain uninhabitable for many years (Section 6.10 for information about the longevity of radioactive waste). At present, the so-called Exclusion Zone, which was once home to about 120,000 people, encompasses an area about 30 km in radius around the reactor site. Due to weather conditions at the time, the largest portion of radioactive material was transported into what is now Belarus, with smaller amounts into Ukraine and Russia. The distribution of residual radiation as of 1996 is shown on the map in Figure 6.28. The figure shows that contamination has spread over an area several hundred kilometers across and that many areas, even up to more than 100 km from the site, remain uninhabitable. Measurable radioactive material was detected over most of Europe and the eastern part of the Soviet Union. In fact, a large number of people over an extended area have been exposed to an increased level of radiation (Table 6.4). The Chernobyl reactor facility is located on the Pripyat River, which provides water to area reservoir facilities, and elevated radiation levels were recorded in drinking water supplies for up to

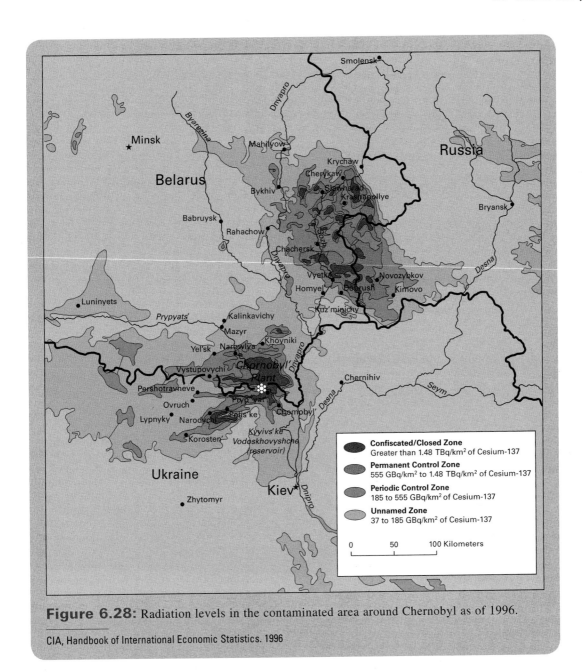

Figure 6.28: Radiation levels in the contaminated area around Chernobyl as of 1996.

CIA, Handbook of International Economic Statistics. 1996

a few months after the accident. The ultimate effects of this exposure on the population of the area are difficult to determine as the relationship between low levels of radiation and increased health risk are uncertain. However, an increased incidence of cancer is unquestionably one of the results of excess radiation exposure. (Risk factors associated with various aspects of energy production are discussed in the next section.) An estimate of the effects of the Chernobyl accident on cancer deaths in Europe and the Soviet Union are summarized in Table 6.4. Although the percentage increase in cancer deaths is quite small (1–2% of all cancer deaths), the actual number of additional deaths is substantial (many thousands).

Table 6.4: Estimated long-term effects of Chernobyl in terms of increased cancer risk in the former Soviet Union and Europe.

region	population affected (millions)	natural cancer deaths	chernobyl cancer deaths	increase in cancer deaths (%)
Soviet Union	279	35,000,000	6500	1.9
Europe	490	88,000,000	10,400	1.2

Based on data from G. J. Aubrecht II, Energy: Physical, Environmental and Social Impact (3rd ed.). Pearson, Prentice Hall, Upper Saddle River, NJ (2006).

6.8c Fukushima Dai-ichi

The most recent serious nuclear accident occurred in March 2011 at the Fukushima Dai-ichi nuclear facility in Japan. This accident followed the 9.0 magnitude Tōhoku earthquake and the tsunami that it caused on March 11, 2011. The details of this accident are still being evaluated, and in many ways this is the most complex nuclear accident to date because it involved six reactors at the facility. The layout of the Fukushima Dai-ichi nuclear reactors prior to the accident is shown in Figure 6.29. It consists of six light water BWRs with a total generating capacity of 4.7 GW_e. The facility is run by the Tokyo Electric Power Company (TEPCO). The facility was protected from tsunamis by the system of seawalls, as shown in the figure. The seawalls were designed to withstand waves up to 5.7 m in height. An analysis of the tsunami that struck the Fukushima facility 51 minutes after the earthquake occurred indicated that the height of the tsunami was 13.1 m. Figure 6.30 shows the impact of the tsunami on the Fukushima nuclear power plant.

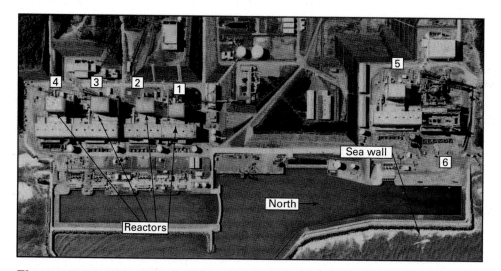

Figure 6.29: Fukushima Dai-ichi nuclear facility in Japan prior to the nuclear accident of 2011. The reactors are the square buildings, with number 6 still under construction. A portion of the seawall structure is seen to the right.

National Land Image Information (Color Aerial Photographs), Ministry of Land, Infrastructure, Transport and Tourism

Figure 6.30: Tsunami approaching Fukushima Dai-ichi nuclear facility near the number 5 reactor on March 11, 2011.

Kyodo News/Contributor/Getty Images

When the earthquake occurred, reactors 1–3 were operational, and reactors 5 and 6 had been shut down for maintenance (i.e., the control rods were fully inserted to eliminate the chain reaction). Reactor 4 had been defueled for replacement of core components, and the fuel rods were placed in storage in a separate pool in the reactor building. Reactors 1–3 shut down automatically at the time of the earthquake, and emergency electrical generators started up to maintain the water pumps and control systems needed to cool the reactors. Fuel in a reactor that has been shut down or even used fuel stored outside the reactor must be cooled because of heat resulting from the radioactive decay of the fission by-products (Figure 6.6). When the tsunami breached the seawall system (Figure 6.30), the power lines connecting the reactor facility to the grid were destroyed, and the generators and control systems were flooded and rendered inoperative.

Following the loss of cooling water, the cores of reactors 1–3 experienced complete meltdowns on a timescale of a few days. Overheating of the zirconium-containing fuel rod tubes released hydrogen gas, as in the Chernobyl incident, according to the reaction

$$Zr + 2H_2O \rightarrow ZrO_2 + 2H_2. \tag{6.4}$$

The hydrogen release led to explosions and fires in these three reactors, causing substantial damage to the reactors and buildings.

Hydrogen buildup in reactor 4 also led to explosions and fires in that building. Current evidence suggests that there is minimal damage to the fuel rods stored in the spent fuel pool in reactor 4. Because no fuel was in the core at the time of the accident, there was no core meltdown, and cooling of the core is not necessary. The damage to reactors 3 and 4 about ten days after the tsunami is shown in Figure 6.31 on the next page.

Reactors 5 and 6 overheated but did not experience a core meltdown. Because building panels were removed to vent the hydrogen, there was no explosion in either

Figure 6.31: Damage to Fukushima Dai-ichi nuclear facility. Reactor 4 is in the upper right and reactor 1 is in the lower center.

DigitalGlobe/Contributor/Getty Images

of these reactors. Cooling water was restored to these two reactors, and they are in safe shutdown mode with no structural damage.

Overall structural damage, radiation leaks, and a generally damaged electrical grid hindered efforts to control the nuclear accident at the Fukushima facility. Complete containment and cleanup of the facility are likely to take many years or even decades. Some estimates have suggested that it might be a century before the fuel can be removed from the facility and properly dealt with.

The environmental impact of Fukushima is certainly substantial, and the release of radioactive material has been significant. This radioactive contamination has had the most substantial effect in Japan, but elevated levels of radioactive material have been observed throughout the world. This material consists primarily of ^{131}I and ^{137}Cs, with half-lives of 8 days and 30 years, respectively. Because of the long half-life of ^{137}Cs, its long-term effects are not known, although some estimates have suggested about a 1% long-term increase in the cancer rate of the population in the accident area. Studies have indicated that the total release of radioactive material to the environment may be in the range of 10–20% of that released in the Chernobyl accident. Two nuclear plant workers were hospitalized for severe (although not life-threatening) radiation exposure, and several dozen workers have been injured in non-radiation-related incidents.

The *International Nuclear and Radiological Event Scale (INES)*, established by the International Atomic Energy Agency (IAEA), is used for the categorization of nuclear events that may have an impact on people or the environment. Chernobyl and Fukushima have both been categorized as Level 7 events (the highest level) and are the only two nuclear accidents to warrant this classification. Although both incidents are categorized as Level 7, the details in terms of their radiation release and long-term effects are quite different. It is generally acknowledged that Fukushima is the most complex nuclear accident because of the number of reactors involved.

ENERGY EXTRA 6.2

Nuclear accidents and the INES classifications

In 1990 the *International Atomic Energy Agency* established the *International Nuclear and Radiological Event Scale (INES)* for the categorization of nuclear events that may have an impact on people or the environment. The scale has eight levels (numbered 0–7), with 7 being the most serious. Levels 4–7 are designated *accidents*, and Levels 1–3 are designated *incidents*. Level 0 is an operating deviation without safety significance. The scale may be viewed as a pyramid, as shown in the figure, where less significant events occur quite frequently, and more significant events occur progressively less often.

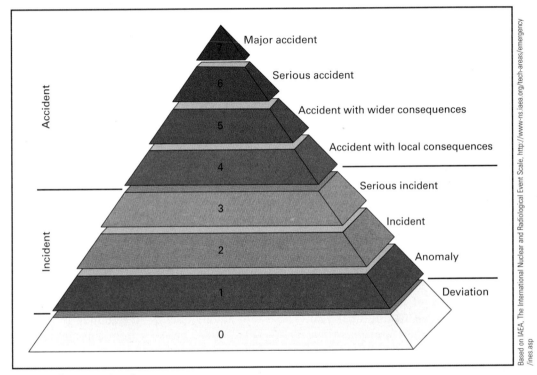

Based on IAEA, The International Nuclear and Radiological Event Scale, http://www-ns.iaea.org/tech-areas/emergency /ines.asp

The INES event triangle.

level	designation	typical characteristics
7	major accident	major release of radioactive material with widespread health and environmental effects
6	serious accident	significant release of radioactive material with impact on people and the environment
5	accident with wider consequences	limited release of radioactive material with some impact on people and the environment
4	accident with local consequences	minor release of radioactive material with possible impact on local food sources
3	serious incident	exposure in excess of ten times the legal annual limits for radiation workers
2	incident	exposure in excess of the legal annual limits for radiation workers
1	anomaly	minor problems with reactor safety components with possible exposure in excess of the legal annual limits for the general public
0	deviation	deviation of reactor operation with no safety significance

Continued on page 210

Energy Extra 6.2 continued

The scale is intended to be logarithmic, similar to the Richter scale used to classify earthquakes. However, unlike an earthquake whose magnitude may be determined by a single quantitative measure of the amplitude of shaking, the evaluation of a nuclear event involves a more subjective analysis based on the study of a number of factors. As a result, a definitive classification of nuclear accidents is possible only well after the incident.

The table lists the INES levels and their characteristics. The evaluation of the severity of an event is based on its effects on people and the environment and on the geographic scope of these effects. The three nuclear accidents at commercial power reactors (Three Mile Island, Chernobyl, and Fukushima) have been the most widely publicized in the media. However, a number of serious (up to Level 6) accidents have occurred at military and research reactors around the world.

Topic for Discussion

Commercial nuclear power has been utilized for more than 50 years. Review the major nuclear accidents that have occurred during that time. Does the Rasmussen report discussed in Section 6.9 provide a reasonable assessment of the risks of nuclear power?

Three Mile Island has been classified as a Level 5 event. There have been no Level 6 events at commercial power reactors, although there has been at least one other Level 5 commercial reactor accident (a fire at a graphite-moderated reactor in the United Kingdom in 1957).

6.9 Risk Assessment

Risk is associated with virtually all human activities. Some have benefits associated with them as well. Assessing risks and benefits is not an easy task, and both of these are difficult to quantify. One way of quantifying risk is by the number of fatalities per year. The use of automobiles is an example. In the United States, there are approximately 42,000 fatalities per year from automobile accidents. The benefits of automobiles include such things as convenience, financial benefits, and perhaps avoidance of fatalities that might otherwise have occurred. However, nearly all North Americans regularly use automobiles and thus, from a personal standpoint, acknowledge that the benefits outweigh the risks. Public perception of risk is a very important factor in deciding which activities are acceptable and which are not. The risk of automobile fatalities is a very well-defined quantity; the year-to-year fluctuations in the number of fatalities are comparatively small. The risk of driving an automobile is (more or less) the same from one year to the next. The public also perceives the risk on a per-event basis. A fatal automobile accident incurs perhaps one or two deaths, maybe a few. There are no catastrophic automobile accidents involving thousands of fatalities. An activity that might result in an accident involving one or two fatalities seems to be acceptable even if it is certain that this will happen tens of thousands of times every year. A single catastrophic event that claims 42,000 lives seems unacceptable even if it occurs only once a year.

The production of energy always involves risk. These risks might be divided into two categories: (1) the risk to workers directly related to the production of energy and

(2) the risk to the general public. The occupational risks involved in the production of electricity from a natural resource can be roughly divided into several categories that include (1) the extraction of the resource from the earth, (2) the processing of the resource, (3) its transportation, and (4) the operation of the generating station.

A comparison of the generation of electricity from coal and from uranium would show some similarities and some differences in the occupational risks. In both cases, there would be a risk to miners from accidents or from exposure to hazardous materials (i.e., coal dust in one case and radioactive materials in the other). In both cases, there would be a risk to workers involved in processing the resource, both from industrial accidents and from exposure to coal dust or radioactive materials. Similarly, in both cases, transportation workers could be involved in accidents. For generating station workers, one might imagine, accidents, fire, explosions, and other risks as factors for the coal-fired station and nuclear accidents for the production of electricity from uranium. It is important to realize, however, in comparing these two cases that one gram of uranium produces as much electricity as 3 t of coal and that this would certainly influence our perspective of, for example, the significance of transportation accidents. Table 6.5 gives the results of one study of the anticipated number of fatalities among energy workers for the production of 1 GWy_e of electricity from coal and uranium. Clearly these estimates suggest that the occupational hazards of producing energy from coal are much greater than those of producing electricity from uranium.

The risk to the general public from different energy production methods is important. The most obvious risk associated with the production of electricity from coal would be air pollution. The long-term effects of possible climate change are difficult to assess and are not considered in this section. Because the effects of air pollution are fairly localized, the influence of pollution from a coal-fired power plant depends considerably on the local population density. The amount of pollution is also dependent on the implementation of processes for cleaning the exhaust. Some estimates of the anticipated number of fatalities per gigawatt-year of electricity produced from coal are shown in Table 6.6. Methods to reduce pollution in the exhaust gas are effective at reducing fatalities, as is the choice of station location.

For nuclear power, three factors should be considered: (1) health effects related to the disposal of radioactive waste, (2) radioactive emissions during normal plant operation, and (3) nuclear accidents. Nuclear waste is discussed in some detail in the next section. Under normal operating conditions, the radioactive material in a nuclear power plant is very well contained, and strict guidelines exist for the amount of radiation that the general public can be exposed to. In fact, studies have

Table 6.5: Anticipated number of fatalities per GWy_e of electricity generated from coal and uranium.

resource	mining	processing	transportation	generating station	total
coal (deep mines)	1.7	0.02	2.3	0.01	4.0
uranium	0.2	0.001	0.01	0.01	0.22

Based on data from Kraushaar and Ristinen, *Energy and Problems of a Technical Society* (2nd ed.) (Wiley, New York, 1993)

Table 6.6: Anticipated fatalities per gigawatt-year of electricity produced by coal under different plant conditions.

location	scrubbers	fatalities
urban	no	74
rural	no	11
urban	yes	28
rural	yes	7

Based on data from H. A. Bethe, "The necessity of fission power," *Sci. Am.* **234(1)** (January 1976) 21–31.

shown that radiation exposure from emissions from a coal-fired power plant may be similar or even greater than that from a nuclear power plant. This is a result of the fact that coal contains uranium and thorium impurities, and some of these are released to the environment with the exhaust.

One way of assessing the risk of different activities is to estimate the resulting average decrease in life expectancy. Table 6.7 shows the results for this type of analysis, suggesting that lifestyle changes can have a much greater effect on the health of the general public than the proliferation of nuclear power.

For most people, the greatest concern related to nuclear power is the possibility of a nuclear accident. Certainly this fear was exacerbated first by the Three Mile Island accident and then by Chernobyl. The Rasmussen report (produced for the U.S. Nuclear Regulatory Commission) considered the safety of the different types of reactors in use in the United States (BWRs and PWRs). This study was made in 1975, before either the Three Mile Island or the Chernobyl accident occurred. It considered the possibility of a nuclear accident in comparison to the possibility of other anthropogenic and natural disasters. Disasters may be categorized by the frequency with which they are anticipated to occur and by the number of resulting fatalities. For example, fires that produce one fatality are frequent, fires that produce 100 fatalities happen every few years, and fires that produce 1000 fatalities are uncommon. Rasmussen's analyses for various anthropogenic disasters and for various natural disasters are summarized in Figures 6.32 and 6.33, respectively.

For the more hazardous events, such as airplane crashes and tornadoes, there are substantial data to compare with the curves in these figures. However, little information

Table 6.7: Decrease in life expectancy due to some risks.

factor	decreased life expectancy
smoking 1 pack per day	7 y
urban rather than rural living	5 y
overweight by 25%	3.6 y
nuclear power plants as of 1970	less than 1 minute

Based on data from Kraushaar and Ristinen, *Energy and Problems of a Technical Society* (2nd ed.) (Wiley, New York, 1993)

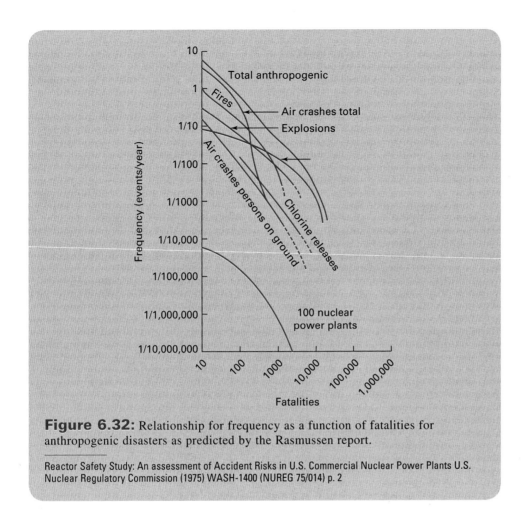

Figure 6.32: Relationship for frequency as a function of fatalities for anthropogenic disasters as predicted by the Rasmussen report.

Reactor Safety Study: An assessment of Accident Risks in U.S. Commercial Nuclear Power Plants U.S. Nuclear Regulatory Commission (1975) WASH-1400 (NUREG 75/014) p. 2

is available for comparison with the predictions for nuclear energy safety. Basically, the only clear accident fatalities are from Chernobyl. The plots in Figure 6.33 would suggest that this event was somewhat anomalous and that further disasters of this magnitude are highly unlikely. There is some controversy over some of the details of the Rasmussen report, but the general consensus is, even nearly 40 years later, that nuclear power is not as unsafe as public perception might make it seem.

The overall risk, to both occupational workers and the general public, of various energy sources has been studied in detail by Inhaber (1979). These results have been adapted to give an overall relative risk factor for the various energy production methods shown in Table 6.8. The analysis presented here is consistent with the comparison of coal and nuclear energy already presented. Perhaps surprisingly, energy sources like solar (photovoltaics) are fairly high on the list. (Solar energy is discussed in detail in Chapters 8 and 9.) The risk is relatively small to the general public but fairly substantial for occupational workers. The basic reason for this feature of solar energy is the very low energy density of solar radiation at the surface of the earth and the fairly low efficiency of photovoltaic cells. Generally speaking, it would require a solar collector with an area of about 40 km^2 to provide the power

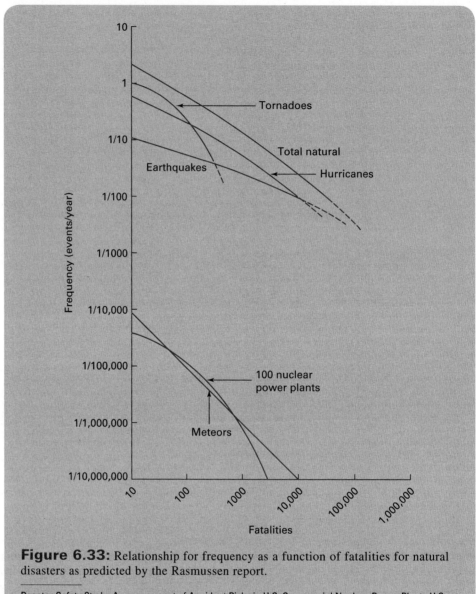

Figure 6.33: Relationship for frequency as a function of fatalities for natural disasters as predicted by the Rasmussen report.

Reactor Safety Study: An assessment of Accident Risks in U.S. Commercial Nuclear Power Plants U.S. Nuclear Regulatory Commission (1975) WASH-1400 (NUREG 75/014) p. 2

produced by an average-sized nuclear power plant. This requires an enormous amount of material to be processed and a large time commitment for construction. The amount of material required and the necessary construction times are compared for different energy sources in Figure 6.34. The figure clearly shows that some of the alternative energy sources to be discussed in later chapters need to be analyzed very carefully.

Table 6.8: Normalized relative total risk (to occupational workers and the general public) of producing electricity by different methods.

electricity source	relative risk
coal	100
oil	67
wind	33
solar (photovoltaic)	23
methanol (biofuel)	10
hydroelectric	1.5
nuclear	0.3
natural gas	0.2

Based on data from H. Inhaber, "Risk with energy from conventional and nonconventional sources," *Science* **203** (1979) 718

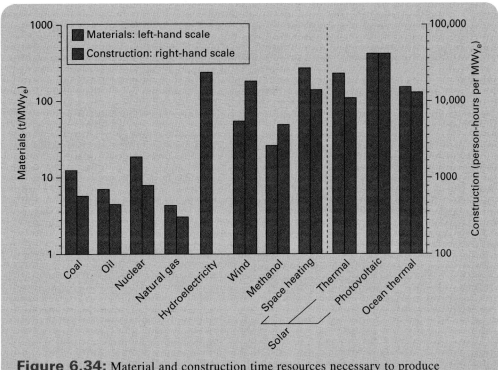

Figure 6.34: Material and construction time resources necessary to produce 1 MWy$_e$ of electricity by different methods. Note that the vertical scale is logarithmic. (Construction time for hydroelectricity is not available.)

Based on H. Inhaber, "Risk with energy from conventional and nonconventional sources," Science 203 (1979) 718

6.10 Waste Disposal

Waste disposal is another serious concern for the use of nuclear energy. Typically, reactors need to be refueled after 6–12 months of operation because the ^{235}U in the core becomes sufficiently depleted and other nuclides are formed that interfere with the normal operation of the reactor. The spent fuel is highly radioactive and must be handled appropriately. As explained previously, some spent fuel is reprocessed and made into new fuel for reactors. At present, the majority is stored in safe locations. Spent fuel contains a mixture of various nuclides that result from the nuclear processes in the reactor. The significance of the different nuclides in the spent fuel depends on their concentration, their decay process, and their half-life. Nuclides with fairly short half-lives are not a concern because the radioactivity decays away rapidly. Very long-lived nuclides are not the biggest concern because they decay very slowly and therefore produce comparatively weak radiation. It is the nuclides with half-lives in the tens to hundreds of years that cause the most concern. These decay rapidly enough that they produce considerable radiation but slowly enough that storing them for many half-lives is a long-term commitment (e.g., hundreds of years).

The spent fuel from a nuclear reactor is a mixture of nuclides of two different types: fission products and actinides. *Fission products* account for about 3% of the mass of the spent fuel (discussed in some detail in the previous chapter) and are exemplified by equation (6.3). The fission products undergo β decay until a stable nuclide is reached. The first few β decays in the chain occur fairly quickly, and it is the last decay or two before β stability is achieved that are the most problematic. Figure 6.6 shows such a sequence of β decays for $A = 137$. As seen there, ^{137}I decays to ^{137}Cs within a few minutes, but the ^{137}Cs β decay to ^{137}Ba has a half-life of 30 years. Table 6.9 gives the properties of the principal, fairly long-lived β unstable nuclides that are found in spent reactor fuel.

Actinides are uranium and related heavy elements that are found in the spent fuel. Approximately 95% of the spent fuel is comprised of ^{238}U from the original fuel, and up

Table 6.9: Some radioactive fission products found in spent nuclear reactor fuel, along with their half-lives and percent yields (Figure 6.3).

nuclide	half-life (y)	fission yield (%)
^{155}Eu	4.76	0.08
^{85}Kr	10.76	0.22
^{90}Sr	28.9	4.5
^{137}Cs	30	6.3
^{151}Sm	90	0.53
^{99}Tc	211×10^3	6.1
^{126}Sn	230×10^3	0.11
^{93}Zr	1.5×10^6	5.5
^{135}Cs	2.3×10^6	6.9
^{107}Pd	6.5×10^6	1.2

Example 6.8

(a) Write the equation for the decay process for ^{137}Cs.

(b) If a sample of ^{137}Cs from fission reactor waste decays at a rate of 5×10^{15} decays per second, calculate the time required for the decay rate to decrease to 1000 decays per second.

Solution

(a) The β^- decay of ^{137}Cs is illustrated in Figure 6.6 and is (by analogy with equation (5.19))

$$^{137}\text{Cs} \rightarrow {}^{137}\text{Ba} + e^- + \bar{\nu}_e.$$

(b) The time dependence of the decay rate is given by equation (5.16) as

$$\left| \frac{dN(t)}{dt} \right| = \lambda N(0)e^{-\lambda t}.$$

where the decay constant is related to the half-life by equation (5.15) as

$$\lambda = (\ln 2)/t_{1/2}.$$

From the half-life of 30 y (9.45×10^8 s) given in Table 6.9, the decay constant is

$$\lambda = (\ln 2)/(9.45 \times 10^8 \text{ s}) = 7.33 \times 10^{-10} \text{ s}^{-1}.$$

The number of ^{137}Cs nuclei at $t = 0$ may be solved from the time dependence of the decay rate as

$$N(0) = \frac{1}{\lambda} \left| \frac{dN(t)}{dt} \right| = (5 \times 10^{15} \text{ s}^{-1})/(7.33 \times 10^{-10} \text{ s}^{-1}) = 6.82 \times 10^{24} \text{ nuclei.}$$

Solving for time as a function of decay rate gives

$$t = -\frac{1}{\lambda} \ln \left[\frac{1}{\lambda N(0)} \left| \frac{dN(t)}{dt} \right| \right].$$

Using a decay rate of 10^3 s^{-1} gives

$$t = -(1/7.33 \times 10^{-10} \text{ s}^{-1}) \times \ln[(10^3 \text{ s}^{-1}/(7.33 \times 10^{-10} \text{ s}^{-1} \times 6.83 \times 10^{24})]$$

$$= 4.0 \times 10^{10} \text{ s,}$$

or about 1266 years.

to about 1% is unused ^{235}U (depending on the degree of enrichment in the original fuel). The remaining 1% or so consists largely of ^{236}U and plutonium, mostly ^{239}Pu. The ^{236}U is produced from ^{235}U by neutron capture, followed by γ decay:

$$\text{n} + {}^{235}\text{U} \rightarrow {}^{236}\text{U} + \gamma. \tag{6.5}$$

The ^{236}U has an α decay half-life of 24 million years and is therefore stable on the timescale of the reactor operation. The ^{239}Pu is produced by neutron capture from ^{238}U by the following reaction:

$$n + {}^{238}U \rightarrow {}^{239}U + \gamma. \tag{6.6}$$

The ^{239}U β decays to ^{239}Np and then to ^{239}Pu on a timescale of a couple of days:

$$^{239}U \rightarrow {}^{239}Np + e^- + \bar{\nu}_e \rightarrow {}^{239}Pu + e^- + \bar{\nu}_e. \tag{6.7}$$

^{239}Pu is β stable and has an α-decay half-life of 24,000 years. The existence of ^{239}Pu in spent reactor fuel causes security concerns for fuel reprocessing because the ^{239}Pu can be used to make weapons. The existence of ^{239}Pu is, however, the basis on which energy can be extracted from ^{238}U, and the method by which the longevity of uranium resources for thermal neutron reactors may be extended. It also forms the basis for the operation of the fast breeder reactor.

Generally speaking, the average life of the activity from fission products is shorter than that of the actinides. So, although the activity comes primarily from fission products at the time the fuel is removed from the reactor, after a sufficiently long time, the actinides are responsible for the majority of the activity (Figure 6.35).

The public perception of the problem of radioactive nuclear waste may not be entirely realistic. Because the production of energy from nuclear reactions requires small amounts of fuel (compared to processes that make use of chemical energy), the actual amount of nuclear reactor waste generated is relatively small in terms of mass or volume. Low-level waste is generated not only from reactors and fuel processing (e.g., uranium refining) but also from medical activities, research, industry and other sources. It is the high-level radioactive waste from reactors that is most problematic. Figure 6.36 shows the total amount of nuclear waste generated by reactors

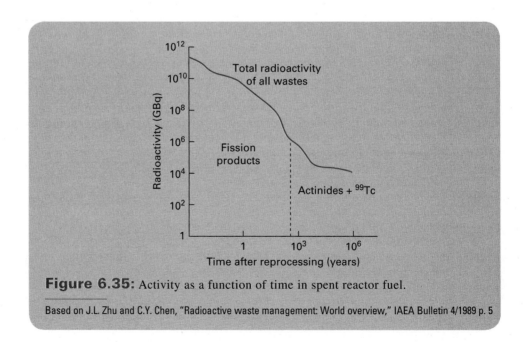

Figure 6.35: Activity as a function of time in spent reactor fuel.

Based on J.L. Zhu and C.Y. Chen, "Radioactive waste management: World overview," IAEA Bulletin 4/1989 p. 5

Example 6.9

Calculate the energy released in the β^- decay of ^{239}U using the following atomic masses

$$m(^{239}\text{U}) = 239.0542878 \text{ u and } m(^{239}\text{Np}) = 239.0529314 \text{ u}.$$

Solution

The decay process is shown in equation (6.7), and the β^- decay energy is given by analogy with equation (5.23) as

$$E = (239.0542878 \text{ u} - 239.0529314 \text{ u}) \times 931.494 \text{ MeV/u} = 1.263 \text{ MeV}.$$

to date and extrapolated far into the future (assuming the continued use of nuclear energy at the current rate and a continuation of current reprocessing policies). The total amount of high-level reactor waste generated to date by the commercial reactor program worldwide (about 300,000 t) amounts to the volume of a cube 40 m on a side; about the volume of coal burned by one typical coal-fired generating station in 4 days. Thus we see that the volume of high-level waste is not very large. It is merely a question of how to safely and securely contain it until it is no longer hazardous. A similar problem exists for hazardous industrial chemicals that are produced in quantities much larger than those of radioactive waste and that, in many cases, are distributed in the environment with relatively little concern for the long-term consequences. A major difference is that an arsenic atom (for example) always remains an arsenic atom and will always be toxic; a radioactive nucleus ultimately decays into something that is not radioactive.

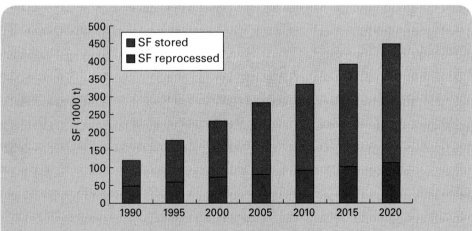

Figure 6.36: Total worldwide mass of nuclear waste produced by commercial power reactors. (SF = spent fuel).

Based on World Energy Council, Survey of Energy Resources, Elsevier (2004) p. 182

Example 6.10

Describe the by-products of the α decay of ^{239}Pu.

Solution

The ^{239}Pu nucleus contains 94 protons and 145 neutrons. Because the α particle (which is the nucleus of the ^{4}He atom) contains 2 protons and 2 neutrons, the conservation of proton and neutron number during α decay requires that the progeny nucleus must have $(94 - 2) = 92$ protons and $(145 - 2) = 143$ neutrons. This is a ^{235}U nucleus; thus the α decay process would be

$$^{239}\text{Pu} \rightarrow {}^{235}\text{U} + \alpha$$

or

$$^{239}\text{Pu} \rightarrow {}^{235}\text{U} + {}^{4}\text{He}.$$

Thus far, there has been little consensus on the best solution for nuclear waste disposal. At present, most radioactive waste produced by commercial power reactors is stored in facilities at the reactor site, although this is not considered to be a final solution to the storage problem. Several approaches can be taken to nuclear waste disposal. The waste can be stored until it has sufficiently low activity to be of relatively little concern or it can be eliminated from our environment. One proposal to eliminate nuclear waste from our environment is to shoot it into space. Another is to use neutron irradiation or some other appropriate means to transmute the radioactive nuclides into stable ones. Both of these are likely to be economically inviable. Figure 6.37 summarizes the possibilities of dealing with nuclear waste. Most likely, continental underground storage will, in the near future at least, be the most practical. One approach that can help to alleviate the storage problem is to separate the relatively short-lived fission products from the generally longer-lived actinides. The fission products will need to

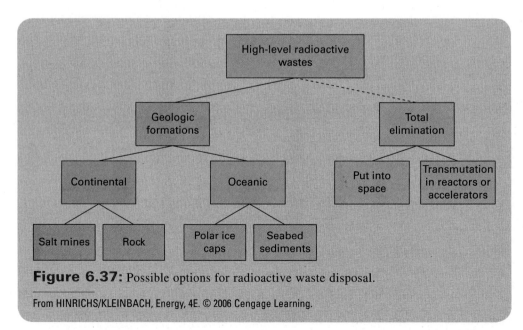

Figure 6.37: Possible options for radioactive waste disposal.

From HINRICHS/KLEINBACH, Energy, 4E. © 2006 Cengage Learning.

be stored for about 300 years before they are nonradioactive enough to be disposed of by simpler means. The actinides will need to be stored (unless they are reprocessed into new fuel) more or less indefinitely. It should be pointed out that, at present in the United States, the quantity of high-level radioactive waste produced by military activities far exceeds that produced by commercial power reactors.

6.11 Advanced Reactor Design

Although there has been little expansion of the nuclear power industry in the past 30–40 years, there has been renewed interest in constructing new reactors in recent years. This interest stems from changes in the public view of viable energy sources for the future and developments that will alleviate much of the concern for public safety. An analysis of the principal nuclear accidents to date indicates that operator error is as major a concern (as at Three Mile Island and Chernobyl) as is faulty design (Chernobyl). However, more recent events have shown that natural disasters (e.g., Fukushima) must also be considered as a possible cause of a nuclear accident. New reactor designs are much more conscious of safety systems to avoid accidents as well as improved systems for containing radioactive materials in the event of an accident.

Also, several more innovative designs for reactors are being developed. Perhaps the most promising of these is the *pebble bed reactor*. This is similar in some ways to the British reactor design because it is a graphite-moderated, gas- (He-) cooled reactor. Rather than being in the form of rods, the uranium fuel elements are in the form of spheres about 0.5 mm in diameter. Each is coated with a temperature-resistant silicon carbide layer (Figure 6.38). About 15,000 of these individually coated fuel spheres are assembled in a graphite matrix to form

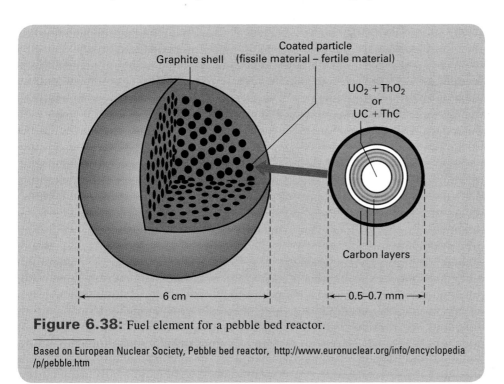

Figure 6.38: Fuel element for a pebble bed reactor.

Based on European Nuclear Society, Pebble bed reactor, http://www.euronuclear.org/info/encyclopedia /p/pebble.htm

a 6-cm-diameter sphere (the fuel pebble). About 300,000 of these fuel pebbles are contained in the core of the reactor (Figure 6.39). The graphite surrounding the uranium, as well as additional graphite spheres not containing uranium, act as the moderator. Helium gas flowing through the reactor core acts as a coolant and is used to transfer heat to drive gas turbines.

The reactor is specifically designed with safety in mind. The high-temperature coating prevents the fuel from melting. Even in the event of a total loss of helium cooling gas, the temperature of the fuel pebbles should not exceed the melting temperature of the carbide coating. These reactors are typically small (150 MW_e or so) and more compact than traditional designs. They may be the basis for thermal reactor designs in the future and a revival of nuclear power.

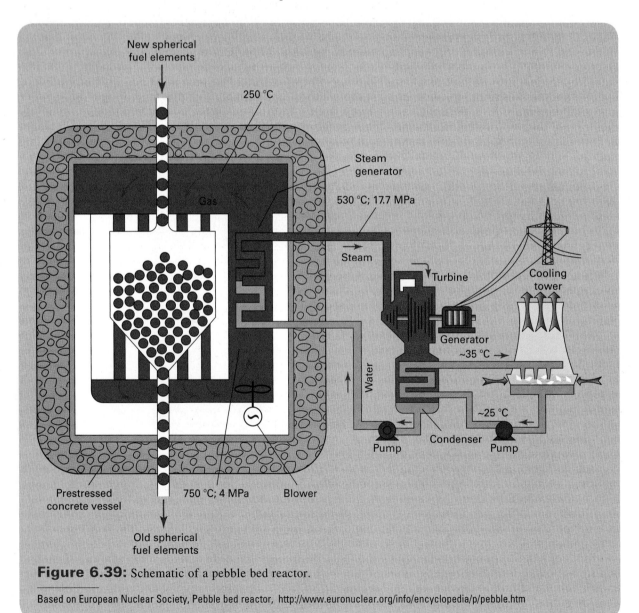

Figure 6.39: Schematic of a pebble bed reactor.

Based on European Nuclear Society, Pebble bed reactor, http://www.euronuclear.org/info/encyclopedia/p/pebble.htm

6.12 Fast Breeder Reactors

In the preceding reactor designs, most of the uranium in a reactor will clearly go unused. This is the ^{238}U. A *fast breeder reactor* (FBR) is designed to make use of much of the potential energy associated with the ^{238}U by converting it into fissile reactor fuel, as in the breeding reactions in equations (6.6) and (6.7). The fast breeder reactor consists of a core containing fissile nuclei, either ^{235}U or ^{239}Pu. The latter material is more practical in the long term because it is produced (bred) by the reactor. The core is surrounded by a breeder blanket of primarily ^{238}U. The core is enriched to about 25% in fissionable nuclei, enabling a controlled fission reaction to be maintained using fast neutrons. The fast neutrons have a higher relative cross section for capture by ^{238}U than do thermal neutrons. Thus some of the excess neutrons can be used to produce new fissionable material in the breeder blanket. To ensure that neutrons in the core remain at high energy, no moderator is used, precluding the use of water (light or heavy) as a coolant because these would both moderate the neutrons. The critical parameter for an FBR is the *breeding ratio*. This is the ratio of the number of fissile nuclei produced from nonfissile ^{238}U in the breeder jacket to the number of fissile nuclei that undergo fission in the core. Ideally, the breeding ratio should be greater than one so that the reactor will produce more fissile fuel than it consumes (in addition to producing heat).

Fast breeder reactors are primarily categorized by the type of coolant they use: gas cooled, liquid metal cooled, and molten salt cooled. Liquid sodium, lead, and a lead-bismuth mixture have been investigated as possible coolants for the liquid metal–cooled variety. All three designs have been considered, and laboratory prototype liquid metal and molten salt reactors have been constructed. However, only the liquid metal fast breeder reactor (LMFBR) using sodium as a coolant has been implemented in a full-scale version. A typical design is shown in Figure 6.40 on the next page. Since the 1960s, approximately 20 functional LMFBRs have been built, and several of these are still operational. Most of these have been fairly small ($\sim 200\ MW_e$) reactors. An LMFBR was constructed in Michigan and became operational in 1963. It ceased operation in 1972, and the U.S. program on fast breeder reactors was subsequently discontinued. Other countries (France, Russia, India, and Japan) have active FBR development programs and, in some cases, operational reactors. Further technological development may be necessary to ensure the reliability and safety of FBRs.

Clearly, the advantage of FBRs is the ability to extract energy from a substantial fraction of the available ^{238}U. Because ^{238}U is 140 times as abundant as ^{235}U, we might expect a similar increase in the longevity of uranium resources if FBRs replace thermal neutron reactors. This means that the world uranium resources would provide energy at the current rate of nuclear energy production for perhaps 5000 years rather than 35, as previously indicated for thermal reactors. The situation is actually much better than this. The requirements for the breeder blanket are less demanding than those for thermal reactor fuel, meaning that lower-quality uranium ore would become viable. Estimates have suggested that at current usage rates our uranium resources could fuel fast breeder reactors for the next 50,000 years. Even with a more active FBR program, resources could be sustained for many thousands of years.

An early experimental fast breeder reactor in the United States is shown in Figure 6.41 on the next page. Currently there are four commercial fast breeder reactors (see Figure 6.42 on page 225) in operation worldwide (two in Russia and two in

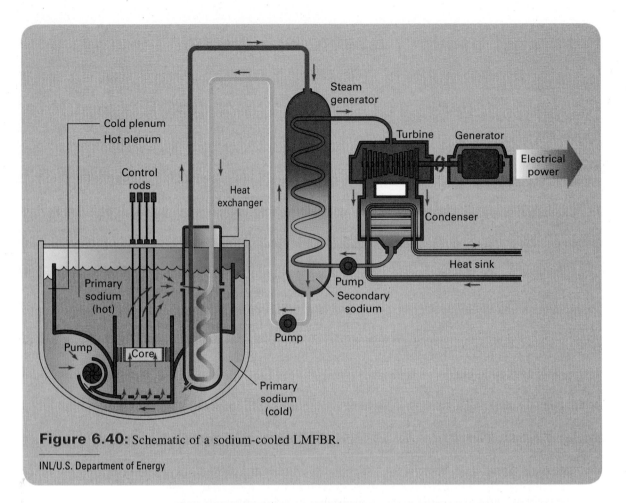

Figure 6.40: Schematic of a sodium-cooled LMFBR.

INL/U.S. Department of Energy

Figure 6.41: Experimental Fast Breeder Reactor (EBR-II) designed, built, and operated by Argonne National Laboratory in Idaho. The reactor ran from 1965 to 1994 and produced 20 MW$_e$.

David R. Frazier Photolibrary, Inc./Alamy Stock Photo

Figure 6.42: The BN-600 fast breeder reactor at the Beloyarsk Nuclear Power Station in Zarechny, Sverdlovsk Oblast, Russia.

Hardscarf, https://commons.wikimedia.org/wiki/File:Beloyarsk_NNP.jpg

India). The fast breeder reactor program in India is part of an overall national energy plan as discussed further in Energy Extra 6.4.

The ability to breed fissile material from nonfissile material opens up another interesting possibility, the use of naturally occurring ^{232}Th as a fuel. ^{232}Th, ^{238}U, and

---

Example 6.11

For an induced fission reaction for ^{233}U of the form

$$n + {}^{233}U \rightarrow {}^{234}U \rightarrow {}^{134}Te + {}^{98}Zr + 2n,$$

calculate the fission energy. Note: $m(^{233}U) = 233.0396282$ u, $m(^{134}Te) = 133.9115405$ u, $m(^{98}Zr) = 97.91274637$ u.

Solution

The fission energy is given in terms of the difference in the mass on the left-hand side of the equation and the right-hand side of the equation,

$$E = [m_n + m(^{233}U) - m(^{134}Te) - m(^{98}Zr) - 2m_n]c^2.$$

Substituting the appropriate values gives

$$E = [1.008664904 \text{ u} + 233.0396282 \text{ u}$$
$$- 133.9115405 \text{ u} - 97.91274637 \text{ u} - 2 \times 1.008664904 \text{ u}] \times 931.494 \text{ MeV/u}$$
$$= 192.5 \text{ MeV}.$$

Since nuclear generated electricity first in appeared in the 1950s, the design of fission reactors has continued to develop. The very earliest reactors are referred to as first-generation, or generation I, reactors, and, aside from a small number of units still operational in the United Kingdom, all of these reactors have been decommissioned. Second-generation (generation II) reactors are presently in the most common use. Generation III reactors are sometimes referred to as advanced reactors, and some have been operational worldwide since the late 1990s. Generation III reactors are evolutionary designs that follow from generation II reactors but incorporate additional design features that include:

- Lowered production costs.
- Increased ease of operation.
- Improved safety and security.
- More rugged design and increased impact resistance.
- Greater utilization of fuel.
- Longer life expectancy.

A number of generation III reactors are currently operational worldwide, and these designs will play the most important role in the development of nuclear power over the next decade or two.

Generation IV reactors, such as the pebble bed reactor, tend to be more radical in their design and strive to further the improved cost and safety aspects of generation III reactors. A number of generation IV designs (like the pebble bed reactor) are in the latter stages of development, and some of them are expected to be online in a few years. A feature that many generation IV reactors have in common is a modular design. Rather than aiming for increased-capacity reactors, many of which are now well in excess of 1-GW$_e$ output, generation IV reactors tend

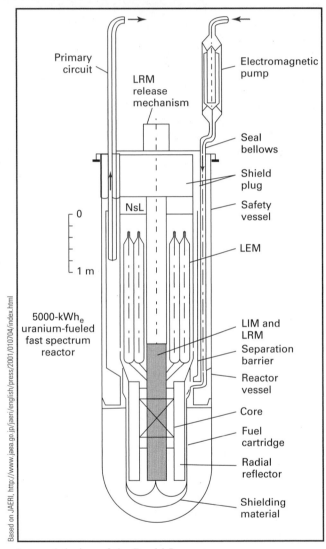

Based on JAERI, http://www.jaea.go.jp/jaeri/english/press/2001/010704/index.html

Internal design of the Rapid-L reactor.

to be in the few-hundred-megawatt range. These individual reactor modules can be combined into larger facilities.

Another interesting concept is the so-called personal nuclear reactor. These designs developed to some extent out of ideas on how a future lunar base might be powered. The requirements for such

Continued on page 227

Energy Extra 6.3 continued

a unit are that it would be easy to operate, self-contained, safe, small, and near zero maintenance and that it would not require refueling for extended periods of time. These same criteria would lead to a small reactor that would be appropriate for powering an apartment or office building or a remote area that could not be connected to the grid. The Tokai Research Establishment, Japan Atomic Energy Research Institute, has promoted the design of the 200-kW$_e$ Rapid-L reactor, shown in the figure. The dimensions on the diagram indicate the small size of the reactor, and its total mass is expected to be around 8000 kg. It is designed to operate continuously for ten years without refueling. Whether such an approach to nuclear power is economical

and practical and whether it gains public support and is consistent with government policies remain to be seen.

Topic for Discussion

A Rapid-L nuclear reactor is expected to be economically viable. It operates with minimal maintenance costs and has a capacity factor of 50%. If the charge for electricity is $0.15 per kWh$_e$ (slightly more than the usual rate for fossil fuel–generated electricity), what is the maximum infrastructure cost if the payback period has to be ten years or less? How does this installation cost, in terms of dollars per W$_e$, compare with typical installation costs for large-scale (i.e., ~1-GW$_e$) nuclear generating facilities.

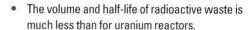

ENERGY EXTRA 6.4
Thorium power reactors

The development of thorium power reactors that operate by producing fissile ^{233}U from the neutron irradiation of naturally occurring nonfissile ^{232}Th, according to equations (6.9) and (6.10), has the potential of extending the usefulness of nuclear power well beyond that which is expected from the use of thermal reactors using ^{235}U or even breeder reactors using ^{239}Pu produced from ^{238}U. The use of thorium as a fuel has a number of potential advantages over the use of ^{235}U in thermal neutron reactors or the use of ^{238}U in breeder reactors. Some of the important advantages include the following:

- Thorium (^{232}Th) is much more abundant in the earth's crust than either ^{235}U or ^{238}U.
- It is much more difficult to produce a nuclear weapon from the fissile material in a thorium reactor (^{233}U) compared with ^{235}U or ^{239}Pu that is present in uranium reactors. (This is primarily due to the difficulty of separating the ^{233}U from the ^{232}U that is present.)

- The volume and half-life of radioactive waste is much less than for uranium reactors.
- ^{232}Th is the only naturally occurring isotope of thorium, meaning that isotopic enrichment, as is necessary for uranium reactors, is not required.
- Some advanced (and safer) reactor designs, such as the pebble bed reactor, are compatible with the use of thorium as a fuel. Reactors that use liquid fuel (which can be rapidly drained from the reactor core in case of an accident) can also use thorium as a fuel.
- Thorium mining (compared to uranium mining) is generally safer and environmentally less damaging because the concentration of thorium in ores such as monazite is greater than the concentration of uranium in uranium ores.

While the design of thorium reactors follows from previous designs of breeder reactors and poses some technical challenges, a careful consideration of the fuel cycle reveals a crucial consideration for the use of thorium as a reactor fuel.

Continued on page 228

Energy Extra 6.4 continued

Doubling time for $^{238}U \rightarrow {}^{239}Pu$ and $^{232}Th \rightarrow {}^{233}U$ fuel cycle for different fuel forms; SDT = simple doubling time, CSDT = compound system doubling time. (Data are adapted from reference [1] and references therein).

cycle	fuel form	SDT (years)	CSDT (years)
$^{238}U \rightarrow {}^{239}Pu$	oxide	25.7	17.8
	metal	12.3	8.5
	carbide	14.7	10.2
$^{232}Th \rightarrow {}^{233}U$	oxide	155.8	108.0
	metal	108.3	75.1
	carbide	101.0	70.0

The conversion of nonfissile ^{232}Th into a fissile fuel that can be used to create a controlled fission chain reaction, as shown in equation (6.9) and (6.10), first requires the irradiation of the ^{232}Th with neutrons to produce ^{233}Th. A suitable reactor fuel results from successive β decays to produce relatively β-stable fissile ^{233}U nuclei. The neutrons required for this process come from previous fission events that, on the average, produce 2 or 3 excess neutrons. Thus, in the ideal case, one fission can go on to breed 2 or 3 new fissile nuclei, leading to a breeding ratio, as described in the text, that is greater than unity. A convenient concept for the understanding of the efficiency of this process is the *doubling time*, that is, the time required for the number of fissile nuclei present in the reactor to increase by a factor of two.

The doubling time depends on the particular nuclides as well as the chemical form of the fuel. Some typical examples of calculated doubling times for different fuel forms for the $^{238}U \rightarrow {}^{239}Pu$ cycle and the $^{232}Th \rightarrow {}^{233}U$ cycle are shown in the table. The simple doubling time is the doubling time for a single reactor. The compound system doubling time is the doubling time for a system of reactors where fissile material from one reactor is used to fuel an additional reactor when sufficient material is accumulated.

It can be readily seen in the table that the doubling times for the $^{232}Th \rightarrow {}^{233}U$ fuel cycle is substantially longer than for the $^{238}U \rightarrow {}^{239}Pu$ fuel cycle. In the best case if new reactors are constructed and fueled when possible, a doubling time of about 70 years has been

calculated for the thorium cycle. In practice, doubling times depend on a number of factors, including the following:

- Plant load factor (i.e., the fraction of time the reactor is actually critical)
- Reactor design and breeding ratio
- Reprocessing and fuel fabrication time
- Reprocessing losses
- New reactor construction delays

It is difficult to accurately predict all of the factors that can influence doubling time. Uncertainties in each of the parameters combine; therefore, there can be substantial error in calculated doubling time values.

The above discussion indicates that the development of a large-scale thorium reactor program requires careful planning and design and needs to be implemented over a significant period of time. Ultimately, a sustainable thorium-based reactor program would use neutrons in equation (6.9) that come from the induced fission of ^{233}U (which is, itself, bred from ^{232}Th). However, as the doubling time analysis indicates, this would be a slow process. Initially the use of neutrons from induced fission of ^{235}U fission or ^{239}Pu bred from ^{238}U would be required. In some countries, e.g., the United States, it has been suggested that a thorium reactor program could be initiated using ^{239}Pu from decommissioned nuclear weapons.

A number of countries have undertaken the construction of experimental thorium reactors in

Continued on page 229

Energy Extra 6.4 continued

the past. In the United States, Oak Ridge National Laboratory ran a thorium reactor research program from 1965 to 1969. A prototype molten-salt reactor using ^{233}U (bred from ^{232}Th) was constructed before the program was discontinued. In Germany the THTR-300 reactor was a 300 MW$_e$ pebble bed reactor using 6-cm diameter fuel particles containing ^{235}U and ^{232}Th. Construction began in the 1970s, and it first went critical in 1983. From 1985 to 1989 it provided power to the grid.

There are presently proposals or development work on thorium-based reactors in a number of countries, including Canada, China, Israel, Japan, Norway, the United Kingdom, and the United States. India has an interesting three-stage plan for thorium reactor development that was first proposed in 1954 by the Indian physicist Homi Jehangir Bhabha. India's large population and current economic development mean that a viable energy policy is essential. Part of India's energy plan is the development of substantial nuclear power, and a major factor in this regard is India's relatively modest uranium resources but extensive thorium resources. The three proposed stages of nuclear power development in India are as follows:

Stage 1 - Thermal neutron reactors: These are conventional thermal neutron reactors fueled by uranium as described in this chapter. India currently has about 6 GW$_e$ in installed capacity, mostly from pressurized heavy water reactors along the lines of the CANDU reactor. Reactors providing an additional 4 GW$_e$ or so of capacity are under construction or planned for the near future. Their completion would bring the total capacity up to about 10 GW$_e$, which meets the Stage 1 goal. For this stage of development, the reactor input is a mixture of ^{235}U and ^{238}U. Neutron capture reactions for these two isotopes are

$$n + {}^{235}U \rightarrow {}^{236}U \rightarrow \text{fission}$$

$$n + {}^{238}U \rightarrow {}^{239}U \overset{\beta}{\rightarrow} {}^{239}Pu$$

leading to the production of fission energy and ^{239}Pu as a by-product.

Stage 2 - Fast breeder reactors: Fast breeder reactors use fissile ^{239}Pu from Stage 1 reactors to produce fission energy and nonfissile ^{232}Th to breed fissile ^{233}U.

The neutron capture reactions leading to fission and breeding are

$$n + {}^{239}Pu \rightarrow {}^{240}Pu \rightarrow \text{fission}$$

$$n + {}^{232}Th \rightarrow {}^{233}Th \overset{\beta}{\rightarrow} {}^{233}U$$

India's first large-scale fast breeder reactor is under construction at Kalpakkam. It was originally scheduled for operation in 2012 but has been delayed, and at present the completion date is uncertain.

Stage 3 - Thorium reactors: In Stage 3 energy is produced by the fission of ^{233}U bred in Stage 2 reactors according to the neutron capture reaction

$$n + {}^{233}U \rightarrow {}^{234}U \rightarrow \text{fission}$$

Additional fissile ^{233}U is bred from ^{232}Th input into the reactor by the breeding reaction

$$n + {}^{232}Th \rightarrow {}^{233}Th \overset{\beta}{\rightarrow} {}^{233}U$$

This process is self-sustaining, and the reactor will require only thorium as an input fuel, thus eliminating the need for external sources of uranium. Due to doubling time and fuel reprocessing requirements, Stage 3 is expected to start 30–40 years after Stage 2 reactors become operational.

International Atomic Energy Agency (IAEA) estimates of thorium reserves

country	thorium reserves (10^3 tonnes)
India	519
Australia	489
United States	400
Turkey	344
Brazil	302
Venezuela	302
other countries	454
world total	2810

While thorium is an attractive alternative source of energy for the future and has many advantages over current nuclear technology that uses

Continued on page 230

Energy Extra 6.4 continued

uranium, the above discussion has shown that this is not an energy solution that can be implemented in the short term, but one that requires careful planning over an extended period of time.

Topic for Discussion

Using an average fission energy of 180 MeV per fission, estimate the maximum total energy production from thorium reserves in each of the countries listed in the table. Make an estimate of the current primary energy use in each of these countries. If thorium is used as the only source of primary energy, what is the life expectancy of the reserves? If world thorium resources are used worldwide, what is their life expectancy?

Reference

1. R. Tongia and V.S. Arunachalam, "India's nuclear breeders: Technology, viability and options" *Current Science* **75** (1998) 549–558.

^{235}U are the only three naturally occurring actinide nuclides. Because of its longer half-life, 1.4×10^{10} years, compared with 4.5×10^9 years for ^{238}U, ^{232}Th is more plentiful on earth than either of the uranium isotopes. It is nonfissile and naturally decays by α decay according to the process

$$^{232}\text{Th} \rightarrow {}^{228}\text{Ra} + \alpha \qquad \text{(6.8)}$$

with an energy release of 4.081 MeV. ^{232}Th may be converted into a fissile nucleus by the following process: Firstly, neutron capture produces ^{233}Th:

$$n + {}^{232}\text{Th} \rightarrow {}^{233}\text{Th} + \gamma. \qquad \text{(6.9)}$$

The ^{233}Th then undergoes two β decays:

$$^{233}\text{Th} \rightarrow {}^{233}\text{Pa} + e^- + \bar{\nu}_e \rightarrow {}^{233}\text{U} + e^- + \bar{\nu}_e, \qquad \text{(6.10)}$$

(Pa = protactinium), with half-lives of 22 minutes and 27 days, respectively. The ^{233}U is β stable and decays naturally by α-decay with a half-life of 1.6×10^5 years. It is therefore stable long enough to serve as a nuclear reactor fuel, and, because it is fissile, can be used to produce energy by induced fission. Thorium is a resource that has attracted interest in recent years and may be a factor in future nuclear reactor development.

6.13 Summary

The chapter dealt with the production of energy from controlled fission reactors. Usable energy can be extracted in the fission of a heavy nucleus because of the dependence of the nuclear binding energy on the nuclear size. The increase in binding energy for decreasing nuclear size for heavy nuclei means that energy is liberated when a heavy nucleus is broken into two lighter nuclei. The spontaneous fission process occurs very slowly because the electromagnetic Coulomb barrier inhibits the process.

However, the fission rate can be increased by inducing fission through neutron reactions. ^{235}U is fissile, meaning that the Coulomb barrier energy is exceeded when a low-energy neutron is absorbed. ^{238}U is not fissile. A controlled chain reaction in the ^{235}U component of the reactor fuel is maintained by regulating the flux of neutrons in the reactor through the use of a moderator and control rods. Reactors operating on these principles are referred to as thermal neutron reactors. The different types of thermal reactors have been discussed, and these are classified on the basis of material used for the moderator.

The chapter also overviewed the world use of fission reactors for the production of electricity. The rapid increase in the use of nuclear energy during the 1960s and 1970s was prompted by the hope that this would provide an inexpensive and plentiful source of energy. Some countries, such as France, developed nuclear power programs that satisfied nearly all of their electricity needs. By the late 1980s, the expansion of the nuclear industry had all but stopped. This change in policy was the result of several factors:

- The realization that electricity from nuclear energy was no less expensive than it was from fossil fuels or hydroelectricity (In fact, these other sources were somewhat less costly.)
- The awareness that uranium supplies were not unlimited and that the longevity of nuclear power from traditional thermal neutron reactors was quite finite
- The concern for the security of nuclear materials, a major factor for policies concerning fast breeder reactors and fuel reprocessing in many countries
- The growing concern for the safety of nuclear reactors (The potential for wide reaching health and environmental consequences of nuclear accidents became apparent after the Three Mile Island and Chernobyl incidents.)
- Concerns over the environmental consequences of nuclear waste disposal

In more recent years, the attitude toward nuclear power has warmed somewhat, even to the point of expansion and/or future planned development of nuclear power capabilities in many parts of the world. This direction is certainly influenced by dwindling fossil fuel supplies and an increased awareness of their environmental effects, the development of new fission reactors that are inherently safer, and a more positive attitude toward fuel reprocessing.

This chapter also discussed the potential risks associated with nuclear reactors. This discussion presented the results of risk assessment studies but also focused on an evaluation of the actual nuclear accidents that occurred at Three Mile Island (1979), Chernobyl (1986), and Fukushima (2011). The recent Fukushima accident in Japan, however, has already had an effect on government policy in some countries. Switzerland and Germany are both phasing out nuclear power activities, and voters in Italy have rejected plans to revive nuclear activities.

The limitations of the availability of uranium resources have been discussed. Thermal reactors make use of only the energy associated with ^{235}U, which comprises less than 1% of naturally occurring uranium. Fast breeder reactors utilize high-energy neutrons to convert nonfissile ^{238}U into fissile ^{239}Pu. By this method, the available energy from natural uranium can be increased by a factor of 100 or more. These reactors are in limited use worldwide, and the possibilities of increasing fast breeder reactor use and of fuel reprocessing have been presented. The possible use of naturally occurring ^{232}Th as a fuel in nuclear reactors was discussed.

Problems

6-1 What is the mass of ^{235}U that, when undergoing induced fission, would produce the same amount of thermal energy as 1 tonne of coal?

6-2 (a) Calculate the mass-energy for one kg of ^{235}U.
(b) Calculate the total nuclear binding energy for 1 kg of ^{235}U.
(c) Calculate the energy from nuclear fission of 1 kg of ^{235}U.

6-3 Using the following masses, show that ^{239}Pu is fissile:

$$m(^{239}\text{Pu}) = 239.0521565 \text{ u.}$$

$$m(^{240}\text{Pu}) = 240.0538075 \text{ u.}$$

Note: This property of ^{239}Pu allows it to be used in a thermal neutron reactor in the same way that naturally occurring ^{235}U is used.

6-4 ^{137}Cs is a common and particularly troublesome fission fragment produced in thermal neutron reactors. The difficulty with ^{137}Cs arises because of its long half-life (30 years) in the β decay chain (Figure 6.6). Calculate the time required for ^{137}Cs in nuclear reactor waste to decay to 10^{-6} of its original activity.

6-5 Calculate the α decay energy of ^{239}Pu [$m(^{239}\text{Pu}) = 239.0521565$ u].

6-6 Low-grade uranium ore contains a fraction of a percent (by weight) of uranium; very high-grade ore (as is found in some deposits in Canada) can contain up to 20% uranium. If energy is extracted from uranium by fission of ^{235}U (0.72% of naturally occurring uranium), then how large a percentage of uranium in ore is necessary to make a tonne of uranium ore equivalent (in terms of energy) to a tonne of coal?

6-7 A possible process for induced fission of ^{235}U is

$$\text{n} + {}^{235}\text{U} \rightarrow {}^{236}\text{U} \rightarrow {}^{138}\text{Xe} + {}^{96}\text{Sr} + 2\text{n}.$$

Using the following atomic masses, calculate the energy released in this process:

$$m(^{138}\text{Xe}) = 137.9139885 \text{ u.}$$

$$m(^{96}\text{Sr}) = 95.92168047 \text{ u.}$$

6-8 Use the information in Figure 6.32 to argue that it is more likely to die in a nuclear accident with 10 fatalities than in one with 1000 fatalities.

6-9 A 800 MW$_e$ nuclear generating station inputs cooling water at a temperature of 5 °C and exhausts it at a temperature of 25 °C. If the station operates at an efficiency of 41%, what is the flow rate of the water in kg/s?

6-10 Assume that the total number of nuclear reactors worldwide has remained relatively constant since the mid-1970s. Is the occurrence of the three major nuclear accidents discussed in this chapter consistent with the predictions of the Rasmussen study shown in Figure 6.32?

6-11 Coal can contain up to about 2000 ppm (by mass) of natural uranium. Compare the chemical energy content of the coal with the available fission energy from the ^{235}U content of the uranium (as used in a thermal reactor) and the total available fission energy from the uranium including ^{238}U (as might be used in a breeder reactor).

6-12 There were 56 total known fatalities from the Chernobyl accident in 1986. On the basis of the Rasmussen report, estimate the anticipated time-scale for a similar accident. Discuss this estimate in the context of the Fukushima Dai-ichi accident.

6-13 A common induced fission process in a thermal nuclear reactor is

$$n + {}^{235}U \rightarrow {}^{236}U \rightarrow {}^{137}Xe + {}^{97}Sr + xn.$$

(a) Determine the number of neutrons, x, on the right-hand side of the equation.

(b) Calculate the fission energy associated with this process. *Note:* $m({}^{137}Xe) = 136.9115629$ u; $m({}^{97}Sr) = 96.926153$ u.

6-14 ^{137}Cs is a typical long-lived waste product from fission reactors. Beginning with the fission process:

$$n + {}^{235}U \rightarrow {}^{236}U \rightarrow {}^{137}I + {}^{95}Y + 4n$$

explain how ^{137}Cs is produced.

6-15 Using the following atomic masses for the isotopes of uranium with $233 \leq A \leq 238$:

$$m({}^{233}U) = 233.0396282 \text{ u},$$

$$m({}^{234}U) = 234.0409456 \text{ u},$$

$$m({}^{235}U) = 235.0439231 \text{ u},$$

$$m({}^{236}U) = 236.0455619 \text{ u},$$

$$m({}^{237}U) = 237.048724 \text{ u},$$

$$m({}^{238}U) = 238.0507826 \text{ u},$$

determine which isotopes are fissile and comment on the trends in your results.

6-16 Calculate the energy of the gamma ray liberated during thermal neutron capture by ^{232}Th. *Note:* $m({}^{232}Th) = 232.0380504$ u; $m({}^{233}Th) = 233.0415769$ u.

6-17 Fissile ^{233}U may be bred from nonfissile ^{232}Th by neutron capture. Write a possible induced fission reaction for ^{233}U (refer to Figure 6.3) and calculate the fission energy. *Note:* atomic masses of the nuclides are available in many nuclear physics texts (e.g., Dunlap, *An Introduction to the Physics of Nuclei and Particles* or Krane, *Introductory Nuclear Physics*); see Bibliography from Chapters 5 and 6 or online at https://www.ncsu.edu/chemistry/msf/pdf /IsotopicMass_NaturalAbundance.pdf

6-18 Consider the fission process in equation (6.3):

$$n + {}^{235}U \rightarrow {}^{137}I + {}^{96}Y + 3n.$$

(a) Calculate the total binding energy for each of the components in the reaction.

(b) Show that the fission energy is the total binding energy on the right-hand side of the equation minus the total binding energy on the left-hand side.

6-19 If ^{232}Th (half-life 1.4×10^{10} years) and ^{238}U (half-life 4.5×10^{9} years) were present in equal amounts when the earth was formed, calculate their relative abundance at present. *Note:* The age of the earth is 4.54 billion years.

6-20 If nuclear energy production and nuclear waste policies remain the same as they are at present, estimate the mass and volume of spent fuel from commercial power reactors in the year 2100.

Bibliography

D. Bodansky. *Nuclear Energy: Principles, Practices, Prospects.* American Institute of Physics, New York (1996).

B. L. Cohen. *The Nuclear Energy Option—An Alternative for the 90s.* Plenum Press, New York (1990).

B. L. Cohen. "The nuclear reactor accident at Chernobyl, USSR." *American Journal of Physics* **55** (1987): 1076.

B. L. Cohen. "Breeder reactors: A renewable energy source." *American Journal of Physics* **51** (1983): 75.

B. L. Cohen. "The disposal of radioactive wastes from fission reactors." *Scientific American* **236** (1977): 21.

R. A. Dunlap. *An Introduction to the Physics of Nuclei and Particles.* Brooks-Cole, Belmont, CA (2004).

M. M. El-Wakil. *Powerplant Technology.* McGraw-Hill, New York (1984).

J. A. Fay and D. S. Golomb. *Energy and the Environment.* Oxford University Press, New York (2002).

S. Glasstone and A. Sesonske. *Nuclear Reactor Engineering: Reactor Design Basics, Volume 1.* Springer, New York (1994).

S. Glasstone and A. Sesonske. *Nuclear Reactor Engineering: Reactor Systems Engineering, Volume 2.* Springer, New York (1994).

D. R. Inglis. *Nuclear Energy—Its Physics and Its Social Challenge.* Addison-Wesley, Reading, MA (1973).

H. Inhaber. "Risk Evaluation." *Science* **203** (1979): 718.

R. A. Knief. *Nuclear Engineering.* Taylor and Francis, Washington, DC (1992).

J. A. Kraushaar and R. A. Ristinen. *Energy and Problems of a Technical Society* (2nd ed.). Wiley, New York (1993).

J. Lamarsh. *Introduction to Nuclear Engineering* (2nd ed.). Addison-Wesley, Reading, MA (1983).

H. W. Lewis. "The safety of fission reactors." *Scientific American* **242** (1980): 53.

J. N. Lillington. *The Future of Nuclear Power.* Elsevier, Amsterdam (2004).

R. L. Murray. *Nuclear Energy: An Introduction to the Concepts, Systems, and Applications of Nuclear Processes.* Pergamon, New York (1993).

R. L. Murray. *Understanding Radioactive Waste.* Battelle Press, Columbus (1989).

R. L. Murray. *Introduction to Nuclear Engineering.* Prentice Hall, Englewood Cliffs, NJ (1961).

A. V. Nero Jr. *A Guidebook to Nuclear Reactors,* University of California Press, Berkeley (1979).

R. G. Steed. *Nuclear Power: In Canada and Beyond.* General Store, Renfrew (2007).

J. Tester, E. Drake, M. Driscoll, M. W. Golay, and W. A. Peters. *Sustainable Energy: Choosing Among Options.* MIT Press, Cambridge, MA (2006).

D. F. Torgerson, K. R. Hedges, and R. B. Duffy. "The evolutionary CANDU reactor—Past, present and future." *Physics in Canada* **60** (2004): 341.

J. Weil. "Pebble bed design returns." *IEEE Spectrum* **38** (2001): 37.

World Energy Council, *World Energy Resources 2016*, available online at https://www.worldenergy.org/publications/2016/world-energy-resources-2016/

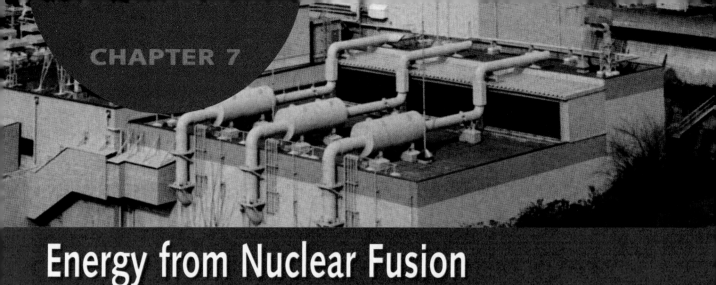

Energy from Nuclear Fusion

Learning Objectives: After reading the material in Chapter 7, you should understand:

- The properties of fusion reactions and the production of fusion energy.
- The design and operation of magnetic confinement reactors.
- The design and operation of inertial confinement reactors.

- The importance of the Lawson criterion.
- The design of a fusion power reactor.
- Progress toward a viable fusion reactor.

7.1 Introduction

In the previous chapter, it was seen that the use of either fast breeder reactors or thermal reactors with effective fuel reprocessing, would provide much of our energy needs for a substantial period of time. The decision to utilize nuclear fission energy must deal with concerns over reactor safety, nuclear waste disposal, and the security of nuclear materials. An alternative approach that makes use of the enormous energy associated with the nuclear force is fusion energy. Figure 6.1 shows that the binding energy per nucleon of a nucleus increases with nuclear size for very light nuclei. Thus, binding together two light nuclei to produce a heavier nucleus (up to about $A = 55$) is an exothermic process and can produce usable energy. Fusion has several significant advantages over fission, such as

- A potentially inexpensive and plentiful supply of fuel.
- Reactions that are inherently easier to control and are therefore much safer.
- Substantially reduced environmental hazards from reactor by-products.

However, at present, fusion power is not technologically feasible because of the fundamental differences between the fission and fusion processes. This chapter reviews the physics of nuclear fusion and overviews the efforts to produce a viable fusion reactor.

AFP/Getty Images

7.2 Fusion Energy

The bombardment of a fissile nucleus with a low-energy neutron provides enough excess energy to put the nucleus in an energy level that is above the Coulomb barrier. This causes the nucleus to undergo induced fission, and the fission fragments are repelled from each other by the repulsive Coulombic interaction between the protons. To fuse two light nuclei, they must be pushed together against the Coulomb force. The nuclei must have enough energy and be in proximity of one another for a long enough time to get through or over the Coulomb barrier. Once the nuclei are close enough together long enough, there is a probability that the strong interaction will take over and fuse the nuclei, thus releasing energy. It is certainly straightforward to accelerate nuclei to these energies, even in very modest particle accelerators, and to collide them with other nuclei to produce fusion reactions. Unfortunately, in such a situation, the energy expenditure for the accelerator is substantially greater than the energy gain from the fusion. Thus, although this is a useful way of learning about fusion reactions in the laboratory, it is not a practical means of obtaining energy. So, to make use of fusion energy, it is necessary to create conditions in the laboratory that are compatible with fusion but that do not require an excessive expenditure of energy. This generally means making a collection of nuclei that is dense enough and hot enough to undergo fusion.

From a practical standpoint, it is desirable to reduce the Coulombic repulsion by using nuclei with as few positively charged protons as possible. Thus, although it might be possible, in principle, to fuse two aluminum nuclei (13 protons each) to form an iron nucleus (26 protons), it is not a productive way to approach this problem.

The simplest fusion process might appear to be the fusion of two protons, or the so-called *p-p process*; that is, the fusion of two hydrogen (^{1}H) nuclei. However, two protons cannot form a bound state, and p-p fusion requires a simultaneous β^+ decay process in which one of the protons is converted to a neutron to give

$$p + p \rightarrow d + e^+ + \nu_e. \tag{7.1}$$

Here d is the deuteron, or the nucleus of a ^{2}H atom, that is, a bound pair consisting of a neutron and a proton. Because β decay is involved (equation 5.18), the weak interaction is, at least partly, responsible for p-p fusion. As a result, it is very difficult to cause this reaction to occur.

Example 7.1

Calculate the energy associated with the fusion process shown in equation (7.1).

Solution

The mass difference between the left- and right-hand sides of the equation is converted into energy by the relationship

$$E = (2m_p - m_d - m_e)c^2.$$

Using values for the masses from Appendix II gives (recall that the positron is the antiparticle of the electron and that their masses are identical)

$$E = (2 \times 1.007276470 \text{ u} - 2.013553214 \text{ u} - 0.000548579903 \text{ u})$$
$$\times 931.494 \text{ MeV/u} = 0.42 \text{ MeV}.$$

Continued on page 238

Example 7.1 continued

Normally the fusion process is followed by the annihilation of the positron with an electron. This annihilation process is

$$e^- + e^+ \rightarrow 2\gamma,$$

where the mass of the photons on the right-hand side of the equation is zero. Thus, the energy released is

$$E = (m_e + m_e)c^2 = (2 \times 0.000548579903 \text{ u}) \times 931.494 \text{ MeV/u} = 1.02 \text{ MeV}.$$

The total energy related to the p-p fusion process is therefore

$$0.42 \text{ MeV} + 1.02 \text{ MeV} = 1.44 \text{ MeV}.$$

Fusion processes involving deuterons (i.e., nuclei of ^{2}H atoms) are of importance, and the simplest of these is *d-p fusion*:

$$d + p \rightarrow {}^3\text{He} + \gamma \qquad (Q = 5.49 \text{ MeV}), \tag{7.2}$$

where, for the purpose of energy calculations, the ^{3}He on the right-hand side refers to the nucleus of a ^{3}He atom. Alternatively, atomic masses may be used for the process ^{2}H + ^{1}H → ^{3}He + γ. In either case, the number of electrons must be conserved in the process because weak interactions are not involved. The energy release, Q, is shown in the equation. The most obvious process involving the fusion of two deuterons is the formation of ^{4}He:

$$d + d \rightarrow {}^4\text{He} + \gamma \qquad (Q = 23.8 \text{ MeV}). \tag{7.3}$$

However, this process is unlikely because the large amount of energy released makes the ^{4}He nucleus unstable and generally results in the release of one of the neutrons or one of the protons. Consequently, there are two more likely modes of *d-d fusion*:

$$d + d \rightarrow {}^3\text{He} + n \qquad (Q = 3.3 \text{ MeV}) \tag{7.4}$$

and

$$d + d \rightarrow {}^3\text{H} + p \qquad (Q = 4.0 \text{ MeV}). \tag{7.5}$$

A final process of importance is the fusion of a deuteron with a triton (i.e., a ^{3}H nucleus). This is *d-t fusion*:

$$d + t \rightarrow {}^4\text{He} + n \qquad (Q = 17.6 \text{ MeV}). \tag{7.6}$$

This process releases a substantial amount of energy and is of particular importance, as will be seen, for the operation of a controlled fusion reactor.

An understanding of the fusion processes that produce energy in the sun is helpful. The sun, like most stars, produces energy primarily by fusing four hydrogen nuclei (^{1}H) together into one helium nucleus (^{4}He):

$$4p \rightarrow {}^4\text{He} + 2e^+ + 2\nu_e, \tag{7.7}$$

where two of the fusing protons must be converted into neutrons by weak β^+ decay processes. The four hydrogen nuclei do not fuse simultaneously to form helium. Instead, the helium forms in a series of steps. The first step of this fusion process is the fusion of two protons, as given by equation (7.1). In principle, two deuterons could then fuse according to equation (7.3) to form a ^{4}He nucleus, although, as explained, this is not likely. A more likely process is *p-d fusion*, as given by equation (7.2), to form ^{3}He. Two ^{3}He nuclei will then most likely fuse according to the reaction

$$^3\text{He} + ^3\text{He} \rightarrow ^4\text{He} + 2\,^1\text{H} + \gamma \qquad (Q = 12.86 \text{ MeV}). \qquad \textbf{(7.8)}$$

This overall process is the most common method of energy production in the sun and is referred to as the *proton-proton cycle* (or *p-p cycle*). Although other processes are going on as well, the p-p cycle produces about 85% of the energy from the sun. The total energy associated with the p-p cycle is $Q = 26.7$ MeV, most of which is eventually converted into solar radiation. All of the steps in this process occur slowly. Equation (7.1) involves the weak interaction; equations (7.2) and (7.8) are limited by the amount of ^{2}H and ^{3}He in the sun and, in the latter case, an increased Coulomb barrier. At the present stage of the sun's evolution, most of the nuclei are still unreacted ^{1}H. However, equation (7.1) is the most limiting factor in the energy production in the sun. This is actually a good situation because it results in an energy output from the sun that is compatible with the requirements for life on earth and ensures that the sun's hydrogen supply is not depleted too quickly.

One might suspect that if the conditions present in the sun could be reproduced in the laboratory, then a functioning fusion reactor would be possible. This is not true. The sun consists of about 10^{57} protons (more or less). Its total energy output is about 3.8×10^{26} W (or J/s). Since one p-p cycle produces about 27 MeV (or 4×10^{-12} J), about $(3.8 \times 10^{26}\text{W})/(4 \times 10^{-12}\text{J}) \approx 10^{38}$ p-p cycles occur every second, corresponding to the fusing of 4×10^{38} protons. As a fraction of the total number of protons in the sun (about 10^{57}), this corresponds to 1 in $(10^{57}/4 \times 10^{38})$, or 1 in 2.5×10^{18}. If the sun produced energy at a constant rate and was able to fuse all of its protons, then it would exist for 2.5×10^{18} seconds, or about 80 billion years. These assumptions are not exactly true, and the sun's life expectancy is somewhat less than this. However, if a fusion reactor filled with ^{1}H nuclei were constructed that approximated the conditions in the sun, then it would take several tens of billions of years to extract all of the available fusion energy from the fuel. This is obviously not practical and is the reason that p-p fusion is not a consideration for fusion energy production on earth.

To properly assess the usefulness of other fusion reactions for the production of energy, it is necessary to examine a bit more of the physics of these processes. It is clearly advantageous to use isotopes of hydrogen rather than helium because of the reduced Coulomb barrier. In general, the greater the number of neutrons, then the greater the strength of the strong interaction that will fuse the nucleons together once the nuclei get past the Coulomb barrier. This is seen for d-d and d-t fusion in Figure 7.1 on the next page. Here the relative probability of fusion (the fusion cross section) is plotted as a function of energy. These reactions correspond to equations (7.4) and (7.5) for d-d fusion and to equation (7.6) for d-t fusion. For d-d fusion, the data in the figure are the sum of both reactions.

The differences between d-d and d-t fusion shown in the figure clearly indicate that achieving d-t fusion in the laboratory should be much easier than achieving d-d fusion. For this reason, laboratory experiments have concentrated largely on d-t fusion, although, as discussed in Section 7.5, the development of laboratory d-d fusion may be considered the ultimate goal of research in this area.

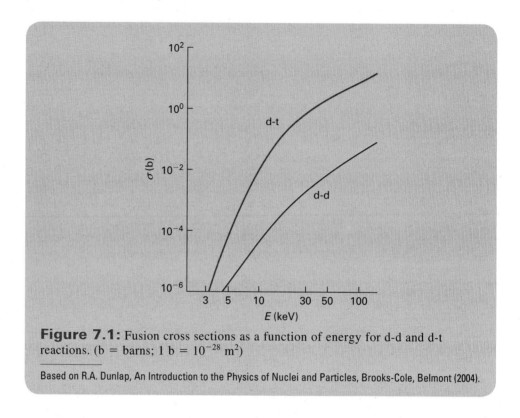

Figure 7.1: Fusion cross sections as a function of energy for d-d and d-t reactions. (b = barns; 1 b = 10^{-28} m^2)

Based on R.A. Dunlap, An Introduction to the Physics of Nuclei and Particles, Brooks-Cole, Belmont (2004).

7.3 Magnetic Confinement Reactors

The basic aim of fusion reactor development is to achieve a situation where the energy input into the reactor that is necessary to maintain fusion conditions is less than the energy that is extracted from the reactor. Thus, there is a net gain of energy from the fusion. The traditional approach to fusion reactors is to achieve the necessary conditions by using a very high temperature. This approach seems to be justified by Figure 7.1, where increasing temperature gives rise to an increase in the kinetic energy of the nuclei and a subsequent increase in the probability of fusion. For the thermal energy of a particle to reach, say, 10 keV, as is typical in the figure, the temperature must be about 100 million degrees (K). The major problem with achieving a useful fusion reaction is to obtain this temperature in order to to get the nuclei close enough together and to keep them in that state until a sustained fusion reaction occurs. At these temperatures, all matter is fully ionized and becomes a plasma; that is, the electrons are stripped from the nuclei, and the negatively charged electrons and positively charged nuclei can move about independently. Although the positive and negative charges in the plasma are not bound together, the plasma as a whole contains the same number of positive and negative charges as the initial neutral atoms and remains electrically neutral.

The obvious difficulty in traditional fusion research is the means by which a plasma at such a high temperature can be contained. All solid materials will melt and vaporize long before this temperature is reached. There are two traditional approaches to containing the plasma and obtaining the necessary fusion conditions: magnetic confinement and inertial confinement.

In a *magnetic confinement reactor*, the charged nature of the electrons and ionized nuclei is utilized. Because the ions and electrons in a plasma are free to move independently, their motion can be controlled by the application of a suitable magnetic field. Magnetic confinement reactors utilize magnetic fields to direct the particles in a plasma in order to prevent them from colliding with the walls of the containment vessel. Two basic geometries are used for these devices: a linear geometry and a toroidal geometry.

The *linear* (or *mirror*) *geometry* uses a plasma column that is pinched or closed off at the ends. The plasma is contained in a cylindrical chamber, and an axial magnetic field is provided by coils around the outside of the chamber. Basically the particles travel in a region of comparatively low field along the length of the cylinder and are reflected from the ends of the cylinder by regions of greater field. In general, progress toward the conditions necessary for a sustained fusion reaction in a mirror confinement device has fallen short of that achieved in other reactor designs. As a result, most current fusion research is directed toward the toroidal reactors and inertial confinement reactors.

In a *toroidal reactor*, the plasma column may be closed in the form of a toroid, in which case, the particles travel along toroidal field lines produced by currents in coils around the toroid [Figure 7.2(a)]. The currents produced by the charged particles in this direction are referred to as poloidal currents. In this geometry, the windings of the coils are closer together on the inside of the torus than on the outside, resulting in a stronger magnetic field near the inside. The consequence of this is that the particles slowly spiral outward, toward the region of weaker field and eventually strike the outer wall of the chamber. To compensate for this effect, an additional (poloidal) field is applied [Figure 7.2(b)]. The net field lines are helical in shape, and the particles avoid interaction with the chamber walls as they follow these curved field lines. Probably the most successful reactor based on this design is the tokamak (an acronym for the Russian name of the device: *to*roidal'naya *ka*mera s *ma*gnitnymi *ku*tushkami, or toroidal chamber with

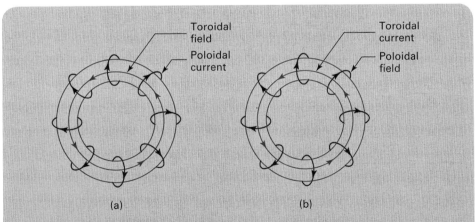

Figure 7.2: Geometry of currents and magnetic field lines in a toroidal reactor: (a) toroidal field produced by poloidal currents; (b) poloidal field produced by a toroidal current.

Based on R.A. Dunlap, An Introduction to the Physics of Nuclei and Particles, Brooks-Cole, Belmont (2004).

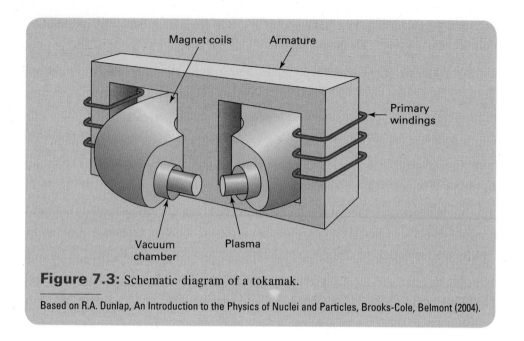

Figure 7.3: Schematic diagram of a tokamak.

Based on R.A. Dunlap, An Introduction to the Physics of Nuclei and Particles, Brooks-Cole, Belmont (2004).

magnetic coils (Figure 7.3). In this device, the toroidal current is actually the current associated with the flow of charged plasma particles around the torus. The photograph in Figure 7.4 shows the interior design of a tokamak. One should not necessarily think of a toroid with the general (donut-like) shape illustrated in Figure 7.3 because many

Figure 7.4: Interior of the Joint European Torus (JET), a tokamak located in Culham, Oxfordshire, England.

EFDA-JET/Science Source

designs have a poloidal diameter that is not much smaller than the toroidal diameter. In fact, many designs do not have a poloidal cross section that is circular. One of the more successful designs has been the spherical tokamak, which resembles a sphere with a circular hole through it (i.e., the so-called cored-apple geometry).

The most recent and significant development in magnetic fusion experiments has been the International Thermonuclear Reactor (ITER), (Figure 7.5). This is a joint project of the European Union, China, India, Japan, Russia, South Korea, and the United States. Canada was originally involved in the project but withdrew. This is a tokamak-type magnetic confinement reactor located in southern France. The project was initiated in 2006, and construction is expected to take until 2019.

These reactors use a combination of deuterium and tritium as fuel, and it is hoped that they will achieve conditions that will initiate a sustained fusion reaction. In this condition, some of the energy produced by the fusion can be used to maintain the conditions of the plasma, and the remainder can be extracted as heat. The design goals of the ITER are to produce 500 MW of heat output for 50 MW of energy input. In an operational power reactor, excess heat is used to generate steam to drive turbine/generators and generate electricity.

Figure 7.5: Cutaway diagram of the ITER reactor.

ITER Organization

7.4 Inertial Confinement Reactors

Inertial confinement refers to the situation where the fusion fuel is confined by inertial forces in the plasma itself. Most experiments that fall into this category are referred to as laser fusion experiments. A pellet of fuel (in most cases, a mixture of deuterium and tritium contained in a millimeter-sized capsule (Figure 7.6) is bombarded by high-energy laser pulses from several directions at once. Often the fuel capsule is contained inside a cylindrical chamber referred to as a hohlraum, which converts the laser radiation into X-rays that are absorbed by the fuel pellet. The fuel pellet heats rapidly to a temperature that is hopefully suitable for fusion to take place. The actual processes that take place in the pellet as it heats are quite complex. Figure 7.7 shows a simplified depiction. In Figure 7.7(a), the laser radiation is absorbed by the fuel pellet, heating it from the outside. In Figure 7.7(b), the heat propagates through the pellet, transforming the outer portions into a plasma. In Figure 7.7(c), this outer plasma atmosphere is driven off as it heats and expands. This process is referred to as *ablation*. In Figure 7.7(d), the remaining core of the pellet is compressed. This results as the outer portion of the pellet expands in an equal and opposite reaction (according to Newton's laws), pushing the inner portion of the pellet inward. These are the inertial forces referred to in the name of the process. Because of the large amount of energy absorbed from the laser beam, the density of the pellet core can be compressed to densities of several thousand times the density of water. A photograph of a laser fusion experiment is shown in Figure 7.8. It is interesting to note that the laser currently being used for this experiment at Lawrence Livermore National Laboratory in California produces an output of 750 TW. That is 50 times the average power consumption worldwide. However, the laser produces energy in pulses that are only 2.4 ns long. One pulse corresponds to 750 TW × 2.4 ns = 1.8 MJ, or about the energy used by a typical automobile to travel a few hundred meters. Thus, although this is not really a lot of energy, it is concentrated into a very small volume during a very short period of time.

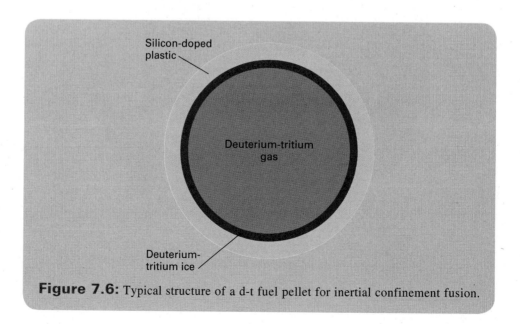

Figure 7.6: Typical structure of a d-t fuel pellet for inertial confinement fusion.

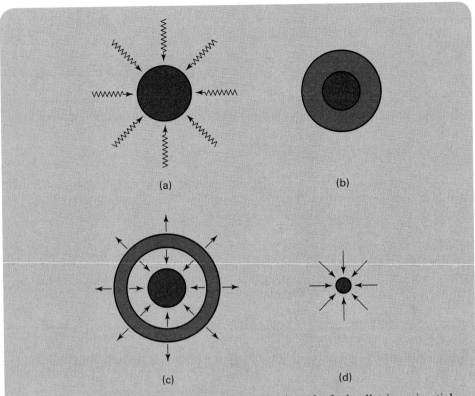

(a)

(b)

(c)

(d)

Figure 7.7: Steps in the heating and compression of a fuel pellet in an inertial confinement fusion reactor. The various stages of fusion are described in the text.

Based on R.A. Dunlap, An Introduction to the Physics of Nuclei and Particles, Brooks-Cole, Belmont (2004).

Figure 7.8: Photograph of the laser fusion (inertial confinement fusion) system at Lawrence Livermore National Laboratory (California). Note workers in lower part of photograph for scale.

Lawrence Livermore National Laboratory

ENERGY EXTRA 7.1
The hohlraum

In many inertial confinement fusion experiments, the millimeter-sized d-t fuel pellet is contained in a centimeter-sized gold-lined hollow cylindrical device called a hohlraum (from the German meaning "hollow area"). The laser enters the hohlraum through a plastic window at one end of the cylinder and is incident on its inner gold walls. The laser rapidly heats the gold to a high temperature and induces transitions of the electron of the gold atoms resulting in the emission of X-rays. The X-rays that are incident on the fuel pellet provide the energy necessary to create a fusion reaction.

The hohlraum provides means of supporting the fuel pellet and a method of cooling it to low temperature. As deuterium and tritium are gases at room temperature (they are both isotopes of hydrogen), they need to be cooled to low temperature ($\sim$ 18 K) in order to remain in the liquid state that is necessary to confine them in a small fuel pellet. The hohlraum provides a convenient means of cooling the fuel to a suitable temperature to ensure that it remains a liquid prior to irradiation. Most importantly, however, is the ability of the hohlraum to transfer energy to the fuel pellet. The X-ray radiation scatters repeatedly from the walls of the hohlraum, which creates a radiative equilibrium around the fuel pellet and ensures that the pellet is heated isotropically. This eliminates the potential problem of laser

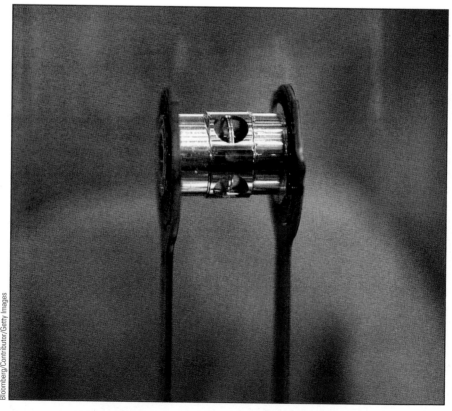

Bloomberg/Contributor/Getty Images

Photograph of a hohlraum used for inertial confinement experiments at the National Ignition Facility at the Lawrence Livermore National Laboratory.

Continued on page 247

Energy Extra 7.1 continued

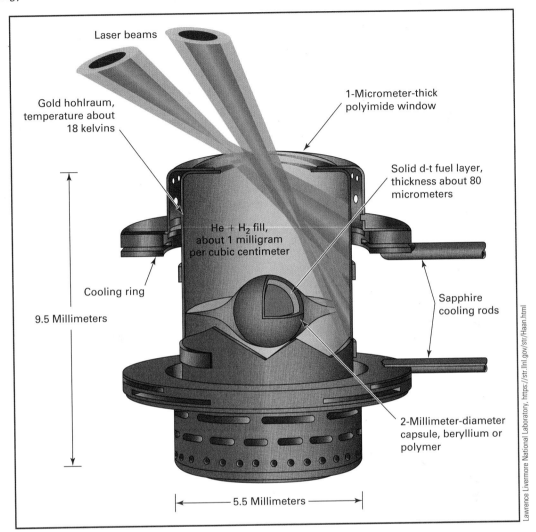

Labels on the diagram:

Laser beams

Gold hohlraum, temperature about 18 kelvins

1-Micrometer-thick polyimide window

Solid d-t fuel layer, thickness about 80 micrometers

He + H₂ fill, about 1 milligram per cubic centimeter

Cooling ring

Sapphire cooling rods

9.5 Millimeters

2-Millimeter-diameter capsule, beryllium or polymer

5.5 Millimeters

Lawrence Livermore National Laboratory, https://str.llnl.gov/str/Haan.html

Diagram of the operation of a hohlraum.

energy being lost by reflection from an isolated fuel pellet and eliminates the need to ensure that the laser beams are precisely focused on the small pellet.

Topic for Discussion

In early 2014 the National Ignition Facility at Lawrence Livermore National Laboratory announced a net gain of energy from d-t fusion. This means that the energy produced by fusion of the deuterium and tritium in the fuel pellet exceeded the X-ray energy incident on the pellet. However, the energy of the X-rays that are incident on the fuel pellet is only about 1% of the energy input into the system. There are losses converting the laser radiation into X-ray radiation, and the lasers are much less than 100% efficient at converting electrical input into laser light. As well, other electrical energy is required at the facility, e.g., cooling pumps and lighting. Finally, as Figure 7.12 shows, the conversion of heat from the fusion process into electricity has thermodynamic limitations. Locate information about anticipated future progress towards laser fusion and consider these predictions in the context of viable electricity production.

7.5 Progress Toward a Fusion Reactor

As explained, it is necessary to achieve conditions for the plasma where the density of nuclei is great enough, the temperature is high enough, and the conditions are maintained long enough to cause sufficient fusion to occur in order to sustain a reaction. These conditions can be quantified in terms of the *Lawson parameter*, $n\tau$, which is the product of the number density of nuclei, n, and the containment time, τ. To sustain a fusion reaction, the temperature must reach at least 200 MK, and the Lawson parameter must satisfy the following relationship (the Lawson criterion):

$$n\tau > 10^{20} \qquad \mathrm{s \cdot m^{-3}}. \tag{7.9}$$

The definition of containment time for inertial confinement fusion is fairly clear but is less obvious for magnetic confinement. For magnetic confinement reactors, the magnetic fields that confine the plasma are pulsed, and the confinement time is the duration of the field pulse. Figure 7.9 shows the relationship of the relevant quantities for different fusion reactor designs.

The development of a fission reactor occurred very rapidly after an understanding of the fundamental physics was achieved. Progress in the development of a viable fusion reactor has been much slower. Work thus far has dealt almost exclusively with the use of deuterium-tritium mixtures for fuel (further comments to follow and in the next section). Experimental laboratory reactors have fulfilled the Lawson criterion, but a situation where there is a net gain in energy from a self-sustaining reaction (sometimes called *ignition*) has not yet been reliably achieved. Progress toward this goal for magnetic confinement reactors and for inertial confinement reactors is illustrated in Figures 7.10 and 7.11, respectively. The light blue area is the region where ignition occurs and net energy is produced. Generally speaking, the progression of data points toward the light blue region represents progress in fusion reactor development as a function of time. For example, in Figure 7.11 the general trend of the points from the lower

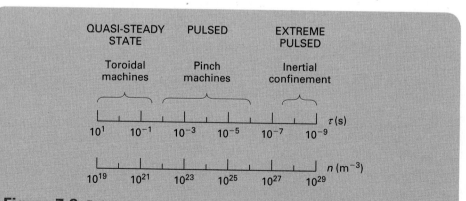

Figure 7.9: Relationship of the quantities in the Lawson parameter for different types of fusion reactors that are necessary to meet the Lawson criterion.

Based on R.A. Dunlap, An Introduction to the Physics of Nuclei and Particles, Brooks-Cole, Belmont (2004).

Example 7.2

For a plasma consisting of deuterons and electrons with a number density of 10^{24} deuterons per m^3, what is the actual mass density compared to air (at STP)?

Solution

For an electrically neutral plasma, there is 1 electron per deuteron, so the total mass per m^3 will be

$$10^{24}\, m^{-3} \times [2.013553214\, u + 0.000548579903\, u] = 2.014 \times 10^{24}\, u/m^3,$$

where the deuteron and electron masses are given in Appendix II. Converting this to kilograms,

$$2.104 \times 10^{24}\, u/m^3 \times 1.6605 \times 10^{-27}\, kg/u = 3.34 \times 10^{-3}\, kg/m^3.$$

The density of air (from the Web) is $1.204\, kg/m^3$. So the plasma is about 0.3% of the density of air.

---

left to the upper right of the figure represents the development of fusion reactors from the late 1960s using milliwatt lasers to the present decade using terawatt lasers. In the case of inertial confinement, the arrow represents a new and hopeful approach to laser fusion. When the fuel pellet (having been irradiated with the laser radiation) is at its

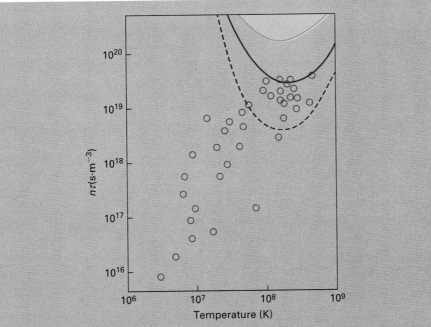

Figure 7.10: Progress toward ignition for magnetic confinement fusion reactors. The broken blue line is the breakeven point if additional energetic particles are injected into the plasma and the red line is breakeven without particle injection.

Based on R.A. Dunlap, An Introduction to the Physics of Nuclei and Particles, Brooks-Cole, Belmont (2004).

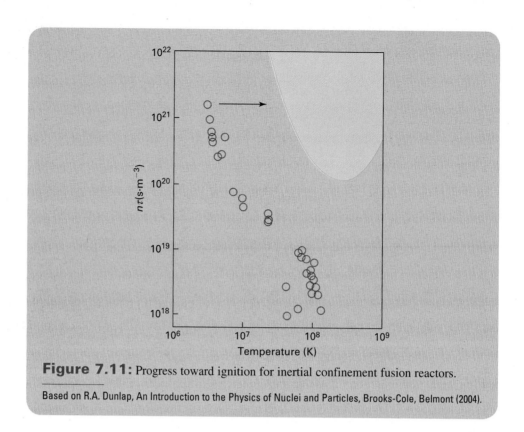

Figure 7.11: Progress toward ignition for inertial confinement fusion reactors.

Based on R.A. Dunlap, An Introduction to the Physics of Nuclei and Particles, Brooks-Cole, Belmont (2004).

maximum density and temperature, a second laser pulse is aimed at the fuel. This drives the center of the pellet to a higher density and temperature, much as the first pulse did, and pushes the conditions closer to ignition.

Much is left to achieve in fusion research before a viable reaction can be obtained. This is presumably, at best, a task that requires effort on a timescale of several decades. Plans for the ITER anticipate that a sustained d-t reaction, producing a ratio of power out to power in of 10 to 1, will be achieved by 2026.

Once a viable reactor has been developed, it can be utilized for the commercial production of electricity. The design of a fusion power reactor is shown in Figure 7.12. This reactor utilizes a deuterium-tritium mixture for fuel and produces energy by means of the d-t fusion reaction. This approach brings up the question of fuel availability. Deuterium (^{2}H) is a naturally occurring isotope of hydrogen and is present in all naturally occurring hydrogen with a natural abundance of 0.0156%, or 1 deuterium atom for every $100/(0.000156) = 6410$ hydrogen atoms. Thus, natural seawater contains deuterium, and this can be readily extracted. If a d-d fusion reactor were feasible, then the energy content of the deuterium in the oceans would fulfill humanity's needs for a period comparable to the life expectancy of the sun. Even taking into account expected increases in energy use and even if only a small fraction of the ocean's deuterium was used, this represents a virtually limitless source of energy.

Because the possibility of d-d fusion in a reactor is uncertain, it is necessary to begin with an assessment of d-t fusion, which requires a consideration of tritium availability. Tritium is not a naturally occurring isotope of hydrogen. It is unstable and decays by β^- decay with a half-life of about 12 years. Thus, any tritium to be used in a fusion reactor must be created artificially.

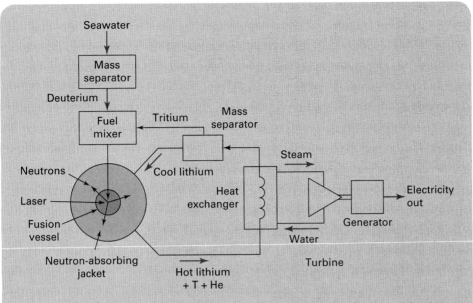

Figure 7.12: Proposed inertial confinement fusion reactor for the production of electricity.

Based on R.A. Dunlap, An Introduction to the Physics of Nuclei and Particles, Brooks-Cole, Belmont (2004).

Example 7.3

The volume of water in the oceans is about 1.3×10^9 km³. Calculate the time that this could serve humanity as an exclusive primary energy source by means of d-d fusion at the current rate of energy use.

Solution

The mass of the water in the oceans (in grams) is

$$m = (1.3 \times 10^9 \text{ km}^3) \times (10^9 \text{ m}^3/\text{km}^3) \times (10^6 \text{ g/m}^3) = 1.3 \times 10^{24} \text{ g(H}_2\text{O)}.$$

The total mass of hydrogen is given in terms of the atomic masses of oxygen and hydrogen as

$$m = \frac{2 \times 1 \text{ g/mol}}{2 \times 1 \text{ g/mol} + 16 \text{ g/mol}} \times 1.3 \times 10^{24} = 1.44 \times 10^{23} \text{ g hydrogen}.$$

The atomic percentage of deuterium is 0.0156%, or a fraction of 0.000312 by weight (the atomic mass of deuterium is twice that of normal hydrogen). This means that the oceans contain

$$1.44 \times 10^{23} \times 0.000312 = 4.49 \times 10^{19} \text{ g deuterium}.$$

Continued on page 252

Example 7.3 continued

This corresponds to

$$(4.49 \times 10^{19} \text{ g}) \times \frac{6.02 \times 10^{23} \text{ mol}^{-1}}{2 \text{ g/mol}} = 1.35 \times 10^{43} \text{ atoms of deuterium.}$$

Because one fusion process requires two deuterium atoms, the number of possible fusions is

$$\frac{1.35 \times 10^{43}}{2} = 6.76 \times 10^{42} \text{ fusions.}$$

Each fusion produces between 3.3 MeV and 4.0 MeV, as given in equations (7.4) and (7.5). Using an approximate value of 3.6 MeV/fusion, the total available energy is

$$(6.76 \times 10^{42} \text{ fusions}) \times (3.6 \text{ MeV/fusion}) \times (1.6 \times 10^{-13} \text{ J/MeV}) = 3.9 \times 10^{30} \text{ J.}$$

Because the current rate of use of primary energy (see the Preface) is 5.7×10^{20} J per year, the energy content of deuterium in the oceans will last (at the current usage rate) for

$$(3.9 \times 10^{30} \text{ J})/(5.7 \times 10^{20} \text{ J/y}) = 6.8 \times 10^{9} \text{ y.}$$

The reactor design shown in Figure 7.12 illustrates that the region where the fusion occurs is surrounded by a jacket containing lithium. This design, along the lines of a liquid metal–cooled fast breeder reactor, transfers the heat produced to water to produce steam, but it also breeds fuel, tritium in this case. Natural lithium consists of about 7% ^{6}Li and 93% ^{7}Li and is a useful material in this respect. Equation (7.6) shows that the d-t fusion reaction produces neutrons. When these neutrons are incident on natural lithium, one of the following reactions can occur:

$$^6\text{Li} + \text{n} \rightarrow \, ^3\text{H} + \, ^4\text{He} \qquad (Q = 4.78 \text{ MeV}) \qquad \textbf{(7.10)}$$

and

$$^7\text{Li} + \text{n} \rightarrow \, ^3\text{H} + \, ^4\text{He} + \text{n} \qquad (Q = -2.47 \text{ MeV}). \qquad \textbf{(7.11)}$$

The first reaction is exothermic and has a large cross section, while the second reaction has a smaller cross section and is endothermic. The first reaction is therefore useful for producing tritium from lithium. The availability of fusion power from the d-t reaction is limited by the availability of ^{6}Li nuclei to breed tritium and is also dependent on the design and efficiency of the lithium blanket.

The lifetime of d-t fusion as a means of supplying our energy needs is unclear. There is considerable uncertainty in the amount of lithium available on earth. At present, the demand for lithium is relatively small, and the price is relatively low. This

Example 7.4

(a) If a 500 MW$_e$ commercial d-d fusion reactor is constructed that produces electricity using a heat engine limited by a Carnot efficiency of 40%, what is the total mass of deuterium utilized per second?

(b) If the deuterium is provided from sea water, what is the volume of water used per second?

Solution

(a) Using the average energy per d-d fusion as the average of the processes given in equations (7.4) and (7.5) as 3.65 MeV or

$$(3.65 \text{ MeV}) \times (1.6 \times 10^{-13} \text{ J/MeV}) = 5.84 \times 10^{-13} \text{ J}.$$

The 500 MW$_e$ will require an actual energy production of

$$(500 \text{ MW}_e)/(0.4) = 1.25 \text{ GWe} = 1.25 \times 10^9 \text{ J/s}.$$

This will require

$$(1.25 \times 10^9 \text{ J/s})/(5.84 \times 10^{-13} \text{ J/fusion}) = 2.14 \times 10^{21} \text{ fusions/s}.$$

As each fusion requires two deuterium nuclei, the reactor will use 4.18×10^{21} deuterium atoms per second. This corresponds to a deuterium mass of

$$(4.18 \times 10^{21} \text{ deuterium/s}) \times (2 \text{ g/mol})/(6.02 \times 10^{23} \text{ deuterium/mol}) = 0.014 \text{ g/s}.$$

(b) Since deuterium constitutes 0.0156% of the hydrogen in sea water, then 4.18×10^{21} deuterium/s corresponds to

$$(4.18 \times 10^{21} \text{ deuterium/s})/(0.000156) = 2.68 \times 10^{25} \text{ hydrogen}$$
atoms per second.

This represents 1.34×10^{25} water molecules with a mass of

$$(1.34 \times 10^{25} \text{ H}_2\text{O/s}) \times (18 \text{ g/mol})/(6.02 \times 10^{23} \text{ H}_2\text{O/mol}) = 400 \text{ g/s or } 0.4 \text{ L/s}.$$

means that there is very little incentive to explore new lithium resources or to develop more efficient methods for its extraction. The best estimate of about 1.7×10^7 t of useful terrestrial lithium suggests that d-t fusion is an energy resource that could supply our total energy needs for between 500 and 1000 years. Improved lithium blanket technology would extend this somewhat, but, as discussed in Chapter 19, other, more immediate uses of lithium may be a competing factor.

A timeline for the development of an operational commercial fusion power reactor is uncertain. Although many estimates put the development of such a facility somewhere around 2050, much basic scientific development is still required before the technological aspects can be predicted with any degree of accuracy.

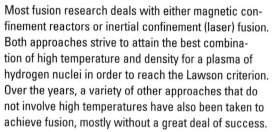

ENERGY EXTRA 7.2
Alternative fusion technologies

Most fusion research deals with either magnetic confinement reactors or inertial confinement (laser) fusion. Both approaches strive to attain the best combination of high temperature and density for a plasma of hydrogen nuclei in order to reach the Lawson criterion. Over the years, a variety of other approaches that do not involve high temperatures have also been taken to achieve fusion, mostly without a great deal of success.

Cold fusion by electrolytic methods was reported in the late 1980s by Martin Fleischmann and Stanley Pons. They claimed to have observed fusion during the electrolysis of heavy water (D_2O) using a palladium (Pd) electrode. These results have never been convincingly confirmed, and their validity is viewed with suspicion by the scientific community. In fact, as a matter of policy, the U.S. Patent Office will not consider applications for any inventions claiming a methodology for cold fusion.

In 2002, sonofusion, sometimes called bubble fusion, was first reported by Rusi Taleyarkhan. In this experiment, high-intensity acoustic waves are transmitted through a deuterated (deuterium-containing) organic fluid. Tiny bubbles in the liquid, caused by a beam of neutrons introduced by the researchers, collapsed under the pressure of the sound waves and caused local heating and high pressure of the deuterium containing fluid. Some estimates have suggested that temperatures of 10 million K can be achieved, and this can result in the claimed d-d fusion reaction. Although other reports of similar results have appeared, reproducibility seems to be, at best, unreliable. Considerable controversy surrounds this work, and, although some work in this area is continuing, much of the scientific community seems to categorize this approach with cold fusion.

A scientifically more interesting approach is muon-catalyzed fusion. The theoretical basis for muon catalysis was presented by Andrei Dmitrievich Sakharov (Nobel Peace Prize, 1975) and coworkers in the late 1940s. Experimental confirmation of this phenomenon has been reported by numerous researchers, including Luis W. Alvarez (Nobel Physics Prize, 1968), who observed muon-catalyzed p-d fusion. A thorough theoretical description of this process was presented by the well-known physicist John David Jackson. There is little doubt about the scientific validity of muon-catalyzed fusion; the question remains, however, as to whether it is of commercial significance for energy production. A basic description of the process follows.

A muon is a subatomic particle with properties similar to an electron except that its mass is about 207 times greater and is unstable with a lifetime of about 2.2×10^{-6} seconds. Muons are created in certain particle reactions and can be created in relatively low energy hadron-hadron collisions (see Chapter 6) in particle accelerators. When a muon enters into a region of matter (e.g., hydrogen), it can displace an electron in an atom and become bound to the nucleus, forming a muonic atom. Because the muon is much heavier than the electron, the muonic atom is much smaller than a normal atom. Two muonic hydrogen atoms can bind together to form a muonic hydrogen molecule in which the hydrogen nuclei are very close together. Because the muonic molecule lasts for a long time (relatively speaking), the nuclei have an increased chance of fusing. This is particularly the case if the hydrogen nuclei are deuterons and/or tritons. If fusion occurs, energy is produced, and the muon is liberated and can catalyze further fusions (until it ultimately decays).

Because it requires considerable energy to produce muons in an accelerator, each muon must catalyze about 300 (d-t) fusions in order to produce a net energy gain. Actually because electricity is produced by a heat engine at about 40% efficiency, approximately 700–800 fusions per muon would be required. In principle, this is quite possible before the muon decays except for the phenomenon of alpha-particle trapping. Because d-t fusion produces alpha particles (^{4}He nuclei), a muon sometimes gets bound to a ^{4}He nucleus. If this happens, the muon cannot escape and catalyzes no further fusions. Experimental investigations of muon catalysis have achieved about 150 fusions per muon. Although these results fall short of the requirements for net energy production, they do show that alternative approaches to difficult problems can sometimes yield interesting and potentially useful

Continued on page 255

Energy Extra 7.2 continued

results. Further progress in this field may lead to net energy production, but it is unknown whether this approach can be made economically viable.

Topic for Discussion

The first (and most important) goal of fusion reactor research is to construct a device that produces more energy than it consumes (i.e., a net energy gain). Once this is accomplished, a commercial fusion power reaction might be possible. However, it is important to consider the economics of such an approach. Discuss the necessary considerations for a successful fusion reactor program. The development and economics of fission power reactors is a good basis for comparison.

7.6 Summary

Fusion energy is an attractive alternative to fission energy because safety, waste disposal, and radioactive material security are not major concerns. The scientific and engineering challenges to the development of fusion power are formidable. This chapter described how the dependence of binding energy on nucleon number for light nuclei allowed for the extraction of energy during a fusion process. From a practical standpoint, the difficulty in fusing two nuclei is the need to overcome the barrier that results from the repulsive Coulombic interaction between the charged protons in the two nuclei. Nuclei must be kept together with a sufficiently high density and energy for a sufficiently long period of time to make the probability of fusion high enough to produce useable energy. Two approaches have been taken to meeting these requirements: magnetic confinement fusion and inertial confinement (or laser) fusion. In a magnetic confinement reactor, ionized gas atoms and their electrons form a plasma, which is confined in a magnetic field. In an inertial confinement reactor, pellets of fusion fuel are heated and compressed to high density by bombardment with intense lasers. The conditions necessary for achieving a fusion reaction are described by the Lawson criterion in terms of the particle density and confinement time.

This chapter discussed the various possible fusion reactions that can produce useable energy. Deuterium-deuterium fusion is the most desirable because it makes use of deuterium nuclei, which are a natural component of all hydrogen. The supply of deuterium in the water of the world's oceans would provide a virtually endless supply of fusion fuel. Unfortunately, deuterium-deuterium fusion is very difficult to achieve, and current experimental investigations of fusion power are concentrating on deuterium-tritium fusion. Deuterium-tritium fusion requires the production of tritium from lithium in order to fuel the reactor. The world's supplies of lithium are limited, and the longevity of d-t fusion as an energy source is likely less than that for uranium fission if fuel is reprocessed or fissile material is bred or if thorium is used as a fission fuel.

Much progress has been made in recent years toward achieving a sustainable fusion reaction, although much work is still needed to make the process viable from an energetic and economic standpoint. A possible design of a fusion power reactor based on a heat engine and generator was presented.

Problems

7-1 If all of the world's energy needs were provided by d-t fusion, how much lithium per year would be utilized to produce tritium? Consider the total energy associated with fusion as being equivalent to the world's current primary energy use.

7-2 The total power output of the sun is 4×10^{26} W. Most of this energy is produced within a radius of 175,000 km of its center. If we were able to reproduce the sun's conditions for p-p fusion in a fusion reactor with an active volume of $10 \times 10 \times 10$ m^3, what would be the reactor's power output? Is p-p fusion a reasonable approach to take for a power reactor?

7-3 Using appropriate masses from Appendix II, show that the energy released from equation (7.7) is 26.7 MeV.

7-4 What is the number density of tritons in a 50% d/50% t plasma that has the same mass density as air at STP (1.204 kg/m^3)?

7-5 Consider a body of water 1 km^2 in area and 100 m deep. The center of mass of the water is raised to a height of 300 m. Compare the gravitational potential energy of this body of water to the fusion energy content of the deuterium it contains.

7-6 The ITER is designed to have a magnetic field pulse duration of up to 480 seconds. What corresponding plasma particle density is needed to achieve the Lawson criterion?

7-7 Tritium-tritium fusion is given by the reaction

$$t + t = {}^4He + 2n.$$

Calculate the energy release for this process. Note the following atomic masses

$$m({}^3H) = 3.016049268 \ u$$

and

$$m({}^4He) = 4.00260325 \ u.$$

Ensure that electrons are properly accounted for on the two sides of the equation.

7-8 What is the kinetic energy of a particle in a plasma at a temperature of 200 MK?

7-9 At the current worldwide rate of primary energy use, determine the potential longevity of d-t fusion as a sole power source for society, based on the availability of lithium. Current estimates of viable lithium resources worldwide are about 1.7×10^{10} kg. Compare this result with the calculation in Example 7.3.

7-10 Verify the energy release given for the process in equation (7.3).

7-11 The sun generates energy through nuclear fusion, which converts mass into energy. If the sun radiates 3.8×10^{26} W of power, calculate the rate at which the sun loses mass (i.e., kg/s).

7-12 Consider a d-t fusion reactor that produces an output of 1000 MW$_e$ at a Carnot efficiency of 35%.

(a) Calculate the mass of fuel (d + t) that is consumed in one second.

(b) Compare your answer to part (a) with the mass of ^{235}U consumed in a fission reactor with the same capacity.

7-13 Write the equation for the β^- decay of tritium and calculate the decay energy. *Note:* See Problem 7.7 for the mass of a tritium atom.

7-14 For 1 m^3 of water, calculate the number of

  (a) H$_2$O molecules.
  (b) HDO molecules.
  (c) D$_2$O molecules.

7-15 The physical radius of a nucleus in units of fm is often approximated by the relation $r = 1.2 \times A^{1/3}$, where A is the total number of nucleons. Compare the actual sizes of the deuteron and triton with the fusion cross sections for d-d and d-t fusion at an energy of 100 keV.

7-16 A 1-m-diameter sphere of d-t plasma in a magnetic confinement fusion reactor just satisfies the Lawson criterion and is at the breakeven point without particle injection. If the sphere of plasma radiates as a black body, calculate the total power radiated from its surface.

7-17 (a) Calculate the total nuclear binding energy of a ^{7}Li nucleus, a triton, and an α-particle.
  (b) Show that the energy associated with the reaction in equation (7.11) can be explained by differences in binding energy.

  Note: The mass of a neutral ^{7}Li atom is $m(^7\text{Li}) = 7.016004049$ u. The mass of a tritium atom is given in Problem 7.7.

7-18 Confirm that the total energy released by the process in equation (7.6) is 17.6 MeV.

7-19 Compare the energy produced per kg of fuel for the combustion of coal, d-d fusion, and d-t fusion.

7-20 The age of the sun has been estimated at 4.6 billion years. If it has radiated energy at the same rate throughout its lifetime, 3.8×10^{26} W, and if its current mass is 1.989×10^{30} kg, what fraction of its original mass has been lost?

Bibliography

U. Columbo and U. Farinelli. "Progress in fusion energy." *Annual Review of Energy* **171** (1992).

R. W. Conn, V. A. Chuyanov, N. Inoue, and D. R. Sweetman. "The International Thermonuclear Experimental Reactor." *Scientific American,* **266** (1992): 103.

R. A. Dunlap. *An Introduction to the Physics of Nuclei and Particles* (Brooks-Cole, Belmont, CA (2004).

H. Furth. "Fusion." *Scientific American* **273** (1995): 174.

G. McCracken and P. Stott. *Fusion: The Energy of the Universe.* Academic Press, New York (2005).

J. Tester, E. Drake, M. Driscoll, M. W. Golay, and W. A. Peters. *Sustainable Energy: Choosing Among Options.* MIT Press, Cambridge (2006).

PART IV

Renewable Energy

There is a clear need to look beyond fossil fuels and nuclear energy for the future. A number of possible viable alternative energy sources are available or are being researched. It is essential to carefully consider the pros and cons of each of these energy technologies. It is not always obvious how much they improve on fossil fuels because of the complexity of fully assessing the environmental impact of infrastructure development and operation and the difficulty of evaluating the economic aspects of particular energy generating methods.

We cannot create energy. Energy production methods are sometimes referred to as energy harvesting technologies because they merely convert primary energy sources from nature (e.g., solar energy, wind, geothermal, etc.) into other forms of energy that we can readily use (e.g., electricity, heat, biofuels, etc.). A major obstacle in the implementation of renewable energy technologies, compared with fossil fuels and nuclear, comes from the low energy density of alternative energy resources. This means in many cases that extensive infrastructure is necessary (e.g., in the case of solar photovoltaics) to match the generating capacity of even modest fossil fuel–fired facilities. Any technology that converts primary energy into usable energy involves a large number of factors that contribute to its environmental and economic desirability. A systematic approach to the assessment of any potential alternative energy technology is to consider how well it fits the CURVE (clean, unlimited, renewable, versatile, and economical). An evaluation of each of these factors provides an objective criterion for determining the viability of certain approaches:

- *Clean*: The environmental impact of an energy source is a crucial factor for its evaluation. A detailed life cycle assessment is beneficial in establishing whether the technology is, in fact, cleaner than fossil fuels.
- *Unlimited*: Of course, no energy source is truly unlimited, but the available quantity of some, like solar energy, is much greater than the total energy use of humankind. The greater the availability, the larger the contribution a resource can make to our future energy needs.
- *Renewable*: Some resources are available in large quantities but do not have an infinite lifetime. This is true, for example, of fossil fuels, which fulfill nearly all our energy needs but will not last indefinitely. They are clearly not renewable. Other resources, such as wave and tidal energy, are renewable but limited in their availability.
- *Versatile*: We use different types of energy for different purposes (e.g., thermal energy for heating buildings or electricity for operating appliances). Of course, one type of energy can be converted into another, but there are always conversion losses, and sometimes the losses can be very large. Although no type of energy can satisfy all our needs, those that have high energy density and are portable fulfill the most demanding applications, such as transportation.
- *Economical*: Any new energy technology must compete with existing technologies, particularly those that are well-established, reliable, and economical. This means that fossil fuels are the first point of reference in determining economics. Technologies like solar photovoltaics face a serious challenge because they do not compete economically.

This part of the text considers the most important approaches to alternative energy and evaluates how well they fit the CURVE and how likely they are to find a place in a future fossil-fuel-free energy economy.

The solar tree in the photograph is located at the Festival Centre, Adelaide, SA, Australia, and similar trees exist in many locations worldwide. Solar trees produce electricity by means of photovoltaic (PV) panels, and this electricity can be stored in a battery system. Because of the limited PV surface area, the solar tree is not intended for large-scale energy production. It is more akin to a sculpture that merges artistic design with modern renewable energy technology and serves to promote alternative energy to the public. ∎

Adrian Lyon/Alamy Stock Photo

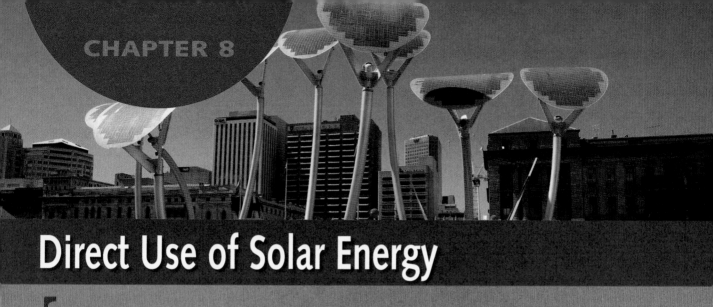

Direct Use of Solar Energy

Learning Objectives: After reading the material in Chapter 8, you should understand:

- The energy content and properties of the sunlight incident on the earth.
- The mechanisms for heat transfer.
- Conductive heat transfer through materials and the use of *R*-values.
- Radiative heat transfer from surfaces.
- The design and operation of a thermal solar collector.

- The energy requirements for residential space heating.
- The description of heating needs on the basis of degree days.
- The storage of thermal energy and the heat capacity of solids.
- The use of passive solar heating techniques as a component of residential heating needs.

8.1 Introduction

Solar radiation is a virtually unlimited source of renewable energy. The simplest technology that makes use of solar energy takes advantage of the heat produced when a material absorbs electromagnetic radiation. This heat may be used directly for space heating or hot water production in residential or office buildings, or it may be stored for use during overcast days or at night. Systems that make use of solar energy for this purpose may be active or passive. *Active systems* use electricity to circulate a fluid, such as water, that absorbs energy from solar radiation and is then used to distribute that energy in the form of heat where it is needed in a building. *Passive systems* merely use the thermal properties of the natural components of a building's structure for the absorption, storage, and distribution of heat in the building.

To understand how solar energy can be utilized, it is first necessary to have knowledge of some of the properties of sunlight, and this chapter begins with a review of the characteristics of solar radiation. It is also necessary to understand the thermal properties of materials to appreciate how solar radiation is absorbed by various materials and converted into thermal energy. This chapter therefore also overviews the thermodynamic properties of the construction materials used in active and passive solar heating systems, as well as those that are appropriate for the storage of thermal energy.

8.2 Properties of Sunlight

The total power output from the sun (as given in Chapter 7) is 3.8×10^{26} W and is emitted isotropically. At the distance of the earth's orbit, this power has a density given by the total power divided by the surface area of a sphere with a radius equal to the average distance between the sun and the earth (1.49×10^{11} m, defined to be one astronomical unit). Minor fluctuations occur during the year because the earth's orbit is an ellipse, not a circle, but these variations are negligible for the present discussion. (Actually, the earth is closer to the sun by about 3% when it is winter in the northern hemisphere.) Thus, the tilt of the earth's axis gives us seasons, not variations in the distance between the earth and the sun. The power density per unit area is therefore given as

$$\frac{P}{A} = \frac{3.8 \times 10^{26}\text{W}}{4\pi \times (1.49 \times 10^{11} \text{ m})^2} = 1367 \text{ W/m}^2. \tag{8.1}$$

This value of the power per unit area is referred to as the *solar constant* and has been measured using sensors on satellites outside the earth's atmosphere. The solar spectrum is very closely approximated by the radiation from a black body at 6000 K. This power spectrum is shown in Figure 8.1. The *irradiance*, as shown in the figure, is the power per unit area per unit wavelength. The total power per unit area at all wavelengths is given by the integrated area under the curve in the figure. The portion of the spectrum in the visible part of the electromagnetic spectrum is indicated. The small anomalies that occur in addition to the smooth curve expected for black body radiation are due to atomic processes that occur in the gas in the outer portions of the sun's atmosphere.

The total solar power incident on the earth is given by the power density times the apparent area of the earth. To the incident solar radiation, the earth appears as a disk with an area πr^2, where r is the earth's radius (6.731×10^6 m), so

$$P_{\text{total}} = (1367 \text{ W/m}^2) \times 3.14 \times (6.371 \times 10^6 \text{m})^2 = 1.73 \times 10^{17} \text{ W}. \tag{8.2}$$

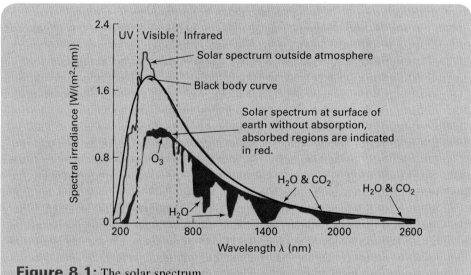

Figure 8.1: The solar spectrum.

Based on G. Li et al. "Recent Progress in Modeling, Simulation, and Optimization of Polymer Solar Cells," Photovoltaics, an IEEE Journal, 2 (2012) 320–340 AND http://www.itacanet.org/the-sun-as-a-source-of -energy/part-2-solar-energy-reaching-the-earths-surface/

Much of the radiation that arrives at the outside of the earth's atmosphere is either reflected back into space or absorbed by the atmosphere. Actually only about half of the total incident sunlight reaches the surface of the earth. The power spectrum of the solar insolation at the earth's surface compared with the spectrum of sunlight outside the earth's atmosphere is shown in Figure 8.1. The characteristic features in the spectrum are caused by absorption at certain wavelengths by various atoms and molecules in the atmosphere. The power per unit area that arrives at the surface of the earth, integrated over all wavelengths, is referred to as the *insolation*. The average insolation at the earth's surface is given by the total power in equation (8.2) times the fraction transmitted through the atmosphere and divided by the total surface area of the earth ($4\pi r^2$):

$$\frac{P}{A} = \frac{0.5 \times 1.73 \times 10^{17}\text{W}}{4\pi \times (6.371 \times 10^6 \text{ m})^2} = 168 \text{ W/m}^2. \tag{8.3}$$

This is the solar insolation on a horizontal surface averaged over the entire surface and averaged over all time (day and night). The average integrated daily insolation is a measure of total incident energy per unit area for an entire day and is expressed as $(168 \text{ W/m}^2) \times (86,400 \text{ s/d}) = 14.5 \text{ MJ/m}^2$ (Section 8.5).

The insolation at any given location on earth at any given time depends on several factors. The factors that can be readily predicted are

- The time of day.
- The day of the year.
- The latitude.

An additional condition that is difficult to predict exactly is the weather. On average, the insolation at a particular location can be determined on the basis of the average typical weather conditions. For the United States, the insolation at different locations averaged over 24 hours per day for the entire year is shown in Figure 8.2. Although this figure shows the general geographical trends for the average insolation, there can be substantial local variations.

The energy of solar radiation can be used in two ways to help satisfy our energy needs: direct use for heating or for the production of electricity. The former use is considered in the present chapter, and the latter is discussed in Chapter 9.

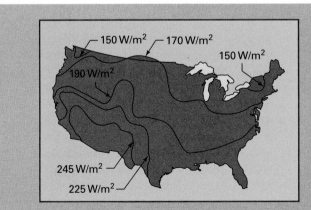

Figure 8.2: Yearly 24-hour averages of insolation on a horizontal surface for the United States.

U.S. Department of Energy

8.3 Heat Transfer

To consider the details of the design of solar collectors, it is necessary to consider how thermal energy is transported. Heat may be transported by conduction, convection, or radiation.

8.3a Conduction

According to the second law of thermodynamics, heat flows from hot to cold, and this basic principle allows us to quantify heat transport by conduction through materials. If a temperature difference exists across a piece of material, then the rate at which energy is transported from the hot side to the cold side (i.e., the power) is:

- Proportional to the magnitude of the temperature difference $(T_h - T_c)$.
- Proportional to the cross-sectional area of the material (A).
- Proportional to the thermal conductivity of the material (k).
- Inversely proportional to the thickness of the material (l).

This relation is expressed mathematically as

$$P = \frac{kA(T_h - T_c)}{l}. \qquad\qquad \textbf{(8.4)}$$

The power is expressed in W, and the thermal conductivity is in $(\text{W} \cdot \text{cm})/(\text{m}^2 \cdot {}^\circ\text{C})$ when the area is expressed in square meters (m^2), the thickness is expressed in centimeters (cm), and the temperature is in degrees Celsius $({}^\circ\text{C})$. For these calculations, it is reasonable to express the temperature in Celsius rather than in degrees measured on an absolute temperature scale (Kelvin) because it is temperature differences and not absolute temperatures that appear in equation (8.4). The thermal conductivities of some common materials (particularly those used in building construction) are given in Table 8.1.

Table 8.1: Thermal conductivities in $(\text{W}\cdot\text{cm})/(\text{m}^2\cdot{}^\circ\text{C})$ for some common materials.	
material	k [$(\text{W}\cdot\text{cm})/(\text{m}^2\cdot{}^\circ\text{C})$]
aluminum	20,100
iron	4600
concrete	170
brick	71
water	60
glass	59
wood (cross grain)	13
sawdust	5.9
cork	4.2
fiberglass insulation	3.8
polystyrene foam	2.84
air	2.3

To quantify the ability of a piece of material to conduct heat (or more precisely, its ability to provide insulation), it is convenient to define the *R-value* of a material, which is its resistance to heat flow. The *R*-value is determined from the thermal conductivity, as given in Table 8.1, and the thickness of the material; it is defined as $R = l/k$, so that the heat flow per unit area, as obtained from equation (8.4), may be rewritten as

$$\frac{P}{A} = \frac{T_h - T_c}{R}.$$

(8.5)

The *R*-value has units of $(m^2 \cdot °C)/W$, sometimes expressed as $(m^2 \cdot K)/W$ or $(m^2 \cdot s \cdot K)/J$. The *R*-value takes into account the thickness of the material, so if a 2-cm-thick piece of material has an *R*-value of, say, $R = 2$, then a 4-cm-thick piece of the same material will have an *R*-value of $R = 4$. (This simple approach deals only with conductive heat losses and is not exact if the radiative losses, to be discussed, are included.) As with resistors in an electric circuit, if several materials are placed one after the other (i.e., as resistors in a series circuit), then the *R*-values are additive:

$$R_{total} = R_1 + R_2 + R_3 + R_4 + \dots.$$

(8.6)

Example 8.1

Calculate the power transferred through a piece of aluminum that is 5 cm by 5 cm by 15 cm long when the temperature on the hot end is 100°C and the temperature on the cold end is 20°C. This is the situation when one end of the bar is in contact with boiling water and the other end is at room temperature.

Solution
From equation (8.4)

$$P = \frac{kA(T_h - T_c)}{l},$$

where the following values are used to calculate the power:

$$k = 20{,}100 \ (W \cdot cm)/(m^2 \cdot °C) \ (\text{from Table 8.1}).$$

$$A = 5 \ cm \times 5 \ cm = 0.0025 \ m^2.$$

$$T_h = 100°C.$$

$$T_c = 20°C.$$

$$l = 15 \ cm.$$

Note that the area of the thermal conductor is given in m^2, whereas the length is given in cm, which, when combined with the value of k, yields power in watts. Why this approach to units is convenient will become apparent when heat transfer through building walls and windows is considered. Thus the power is calculated to be

$$P = 20{,}100 \frac{W \cdot cm}{m^2 \cdot °C} \times (0.0025 \ m^2) \times \frac{100°C - 20°C}{15 \ cm} = 268 \ W.$$

Example 8.2

Compare the power transferred per unit area through 1 cm thick pieces of iron and glass when the temperature is 20 °C on one side and -10 °C on the other.

Solution

From equation (8.4)

$$P/A = k(T_h - T_c)/l$$

where the following values are used for both materials:

$$(T_h - T_c) = (20\,°C - (-10\,°C)) = 30\,°C.$$

$$l = 1\ cm$$

For iron $k = 4600$ [(W·cm)/(m^2·°C)] and for glass $k = 59$ [(W·cm)/(m^2·°C)]. Thus, for iron the power transferred will be

$$P/A = (4600\ [(W·cm)/(m^2·°C)])(30\,°C)/(1\ cm) = 138\ kW/m^2$$

and for glass

$$P/A = (59\ [(W·cm)/(m^2·°C)])(30\,°C)/(1\ cm) = 1.77\ kW/m^2.$$

The iron transfers $(138\ kW/m^2)/(1.77\ kW/m^2) = 78$ times as much power. This ratio can easily be seen from just the ratios of the thermal conductivities $(4600\ [(W·cm)/(m^2·°C)])/(59\ [(W·cm)/(m^2·°C)]) = 78$

Example 8.3

Calculate the effective R-value for a wall comprised of an outer layer of 10 cm of brick, a middle layer of 15 cm of fiberglass insulation, and an inner layer of 2 cm of wood.

Solution

The R-values can be calculated for each of the layers from the values of k in Table 8.1:

$$\text{Brick: } R = \frac{10\ cm}{71(W·cm)/(m^2·°C)} = 0.14\ m^2·°C/W.$$

$$\text{Fiberglass: } R = \frac{15\ cm}{3.8(W·cm)/(m^2·°C)} = 3.95\ m^2·°C/W.$$

$$\text{Wood: } R = \frac{2\ cm}{13(W·cm)/(m^2·°C)} = 0.15\ m^2·°C/W.$$

The total R-value for the wall is $R = (0.14 + 3.95 + 0.15)\ m^2·°C/W = 4.24\ m^2·°C/W$. This can then be used to calculate the energy transferred per unit time per unit area through the wall. Clearly, in this example, the insulating properties of the wall come almost exclusively from the presence of the fiberglass insulation.

Consider a calculation for a piece of window glass. Standard window glass is about 0.3 cm thick, and, using the thermal conductivity for glass given in Table 8.1, one calculates $R = 0.005$. Thus a single pane of window glass provides virtually no insulation on its own. The principal reason that a window reduces conductive heat transfer is that it prevents air flow. A layer of stationary air is created near the surfaces of a pane of glass, and the insulating properties of this nonmoving air are of relevance. When the R-value of this air layer is taken into account, the effective R-value for the pane of glass is about $R = 0.18$. The window also eliminates convective heat transfer from the inside of the building to the outside by preventing air from actually flowing from the inside to the outside. Figure 8.3 illustrates the relationship of

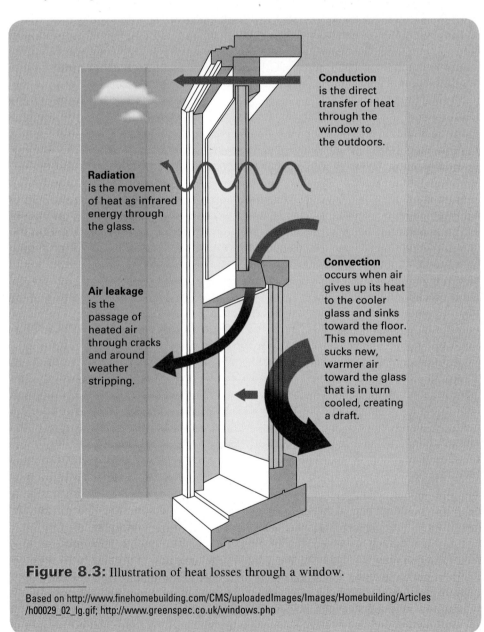

Figure 8.3: Illustration of heat losses through a window.

Based on http://www.finehomebuilding.com/CMS/uploadedImages/Images/Homebuilding/Articles /h00029_02_lg.gif; http://www.greenspec.co.uk/windows.php

different heat loss mechanisms through a window, including conduction (as discussed), convection, and radiation.

8.3b Convection

Convection is an important component in the transfer of heat and is certainly illustrated by the insulating effects of the dead air layer near a pane of glass, as just explained. Convection causes heat transfer by the movement of a fluid. The physics of this process depends on the properties of the fluid, such as density and viscosity. A quantitative description of convection is very complex and is not considered in this text.

8.3c Radiation

All objects that are at a temperature above absolute zero radiate energy. The total power per unit surface area radiated by a black body at a temperature T is given by the Stefan-Boltzmann law as

$$\frac{P}{A} = \sigma T^4, \tag{8.7}$$

where σ is the Stefan-Boltzmann constant 5.67×10^{-8} W·m^{-2}·K^{-4}. Because this expression deals with the absolute temperature of the object (not a temperature difference), it is essential to represent the temperature in Kelvin. This expression shows that the total amount of radiation produced by a black body is a sensitive function of its temperature. In general, objects are not perfect black bodies and therefore do not radiate as much energy as a black body at the same temperature would. The ratio of the energy radiated by a body to that radiated by a black body at the same temperature is called the *emissivity*, ε. Thus equation (8.7) can be written for a real object as

$$\frac{P}{A} = \varepsilon \sigma T^4, \tag{8.8}$$

where ε can have values from 0 to 1 and is a measure of how effective the surface of a particular material is at radiating energy. The emissivity can vary considerably for different materials, and for a particular material it can vary to some extent as a function of temperature. Most common building materials like glass, wood, steel, most painted surfaces, and so on have emissivities of about 0.9 at around room temperature.

In addition to the temperature dependence of the total energy produced by a surface, the wavelength of the radiation produced is also a function of the temperature. The distribution of wavelengths produced by the sun was shown in Figure 8.1. For surfaces at lower temperatures than the sun, the black body spectrum is shown in Figure 8.4 on the next page. Clearly, as the temperature decreases, the total radiated power decreases (as evidenced by the area under the curves in the figure), and the wavelength of the maximum in the spectrum increases. For the sun (a black body at about 6000 K), the peak is in the visible portion of the spectrum. For a surface at, say, 350 K (or about 75°C), the peak is well into the infrared portion of the spectrum.

In addition to emitting radiation, all surfaces absorb radiation that is incident on them. The ability of a surface to absorb energy is its *absorptance, a*, and generally surfaces that are good at emitting radiation are also good at absorbing it. A surface that has a given value of the emissivity, ε, at a certain temperature has an absorptance that is equal to the emissivity, $a = \varepsilon$, for radiation with a wavelength that is characteristic of that temperature.

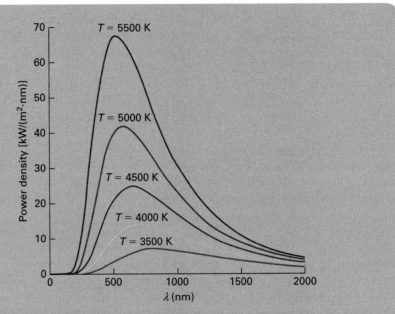

Figure 8.4: Wavelength dependence of black body radiation for surfaces at different temperatures. The vertical axis gives the power radiated per unit area per unit wavelength.

Based on Encyclopaedia Britannica, Wien's law, http://www.britannica.com/EBchecked/topic/643338 /Wiens-law

Example 8.4

Calculate the power emitted by 1 m^2 of the surface of the sun.

Solution

From equation (8.8),

$$\frac{P}{A} = \varepsilon \sigma T^4 \qquad \text{or} \qquad P = A\varepsilon\sigma T^4,$$

where the following values are used:

$A = 1 \text{ m}^2.$

$\varepsilon = 1$ (because the sun behaves as a black body).

$\sigma = 5.67051 \times 10^{-8} \text{ W·m}^{-2}\text{·K}^{-4}$ (from Appendix II).

$T = 6000 \text{ K}$ (from the text).

This gives

$$P = (1 \text{ m}^2) \times 1 \times (5.67051 \times 10^{-8} \text{ W·m}^{-2}\text{·K}^{-4}) \times (6000 \text{ K})^4 = 73.5 \text{ MW}.$$

ENERGY EXTRA 8.1
Optical coatings

Passive solar heating relies on the optical properties of windows. It is generally desirable that windows are as transparent as possible in the visible portion of the electromagnetic spectrum. The natural high transmission in the visible and low transmission in the infrared portions of the spectrum exhibited by window glass give rise to passive solar heating inside a building. However, engineering materials with specific optical properties allow for substantially more control of radiative heat flow through windows at the various wavelengths. The transmittance as a function of wavelength is shown in the figure to the right for plain glass and for glass that has been coated with two different materials, a polymer-based coating and a metal-dielectric coating. Over the visible portion of the spectrum (about 350–650 nm), both coated materials are fairly transparent, transmitting about 70% (compared to 90% for plain glass) of the incident radiation. Although glass only begins to become nontransparent well into the infrared (above 2700 nm), the coated glass becomes nontransparent in the very near infrared, thus blocking a much larger portion of long wavelength re-emitted radiation (see Chapter 4 on the greenhouse effect).

The reflectance of the coated and uncoated glass is shown in the figure to the right. Here the behaviors of the two coated materials are seen to be very different. The polymer-coated glass is fairly similar to glass, while the metal-dielectric–coated glass becomes very reflective in the infrared. These materials may be utilized to optimize passive solar heating in the winter.

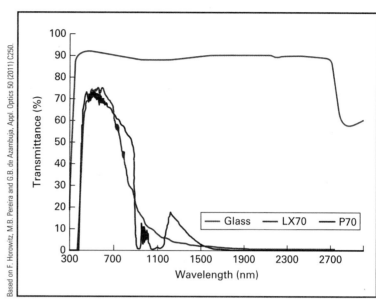

Based on F. Horowitz, M.B. Pereira and G.B. de Azambuja, Appl. Optics 50 (2011) C250.

The transmittance of glass and glass coated with a polymer coating (P70) and a metal-dielectric layer (LX70).

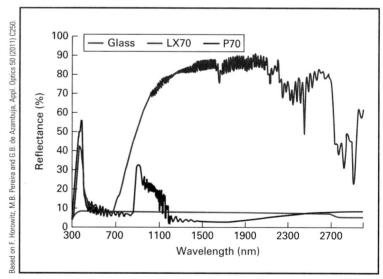

Based on F. Horowitz, M.B. Pereira and G.B. de Azambuja, Appl. Optics 50 (2011) C250.

Reflectance of glass and glass coated with a polymer coating (P70) and a metal-dielectric layer (LX70).

Continued on page 270

As illustrated in the figure, the transmittance of the coated glass allows light to enter but minimizes long-wavelength radiation heat loss back through it.

Topic for Discussion
Solar blinds can make an important addition to high-efficiency windows. These are like venetian blinds with one side coated with a reflective material and the other side coated with an absorptive material. The blades can be oriented with the reflective side out and the absorptive side in, or vice versa. Discuss how these might be used during the day and night and during different seasons to optimize heating efficiency in a home.

8.4 Solar Collector Design

Systems for utilizing solar radiation for heating purposes may be either active or passive. An *active system* is one in which heat produced from sunlight is transported and stored by circulating an appropriate fluid (typically water or air) using pumps or blowers. A *passive system* merely relies upon building design to utilize sunlight to heat the air or other material in the building. Active solar heating systems use either traditional flat plate collectors (Figure 8.5) or evacuated tube collectors. The operation of a flat plate collector is described in the present section. Evacuated tube collectors are discussed in Energy Extra 8.2. The basic design of the flat plate collector is shown in Figure 8.6 on the page 274. The collector is basically an insulated box with a window on the front side to allow sunlight to enter. The sunlight is absorbed by a collector plate, and the heat is transferred to water flowing through pipes attached to the plate. The properties of the collector plate are very important for the operation of the collector. Ideally, the plate should absorb as much of the solar radiation as possible and lose as little energy as possible through reirradiation. The relationship between absorptance and emissivity, as previously explained, needs to be considered carefully in the analysis of the flat plate collector. The radiation that is absorbed in the form of sunlight is produced by the sun and has a spectrum that is basically similar to 6000 K black body radiation (Figures 8.1 and 8.4) with

Figure 8.5: Flat plate solar collector installation.

Richard A. Dunlap

ENERGY EXTRA 8.2
Evacuated tube solar collectors

An alternative design to the traditional flat plate solar collector that has come into use in recent years is the evacuated tube solar collector. The evacuated tube solar collector consists of a series of evacuated glass tubes, each of which contains an absorber. Typically the absorber consists of a metal plate (generally painted black to absorb light) thermally connected to a tube that is closed on both ends, as shown in the figure. The tube contains a quantity of working fluid (a mixture of propylene glycol and water is commonly used) that undergoes a phase transition to the vapor state when heated. The heated vapor rises to the top of the tube, where it transfers heat to water through a heat exchanger (called a manifold). The working fluid then condenses and falls back to the bottom of the tube. The evacuated glass tube serves to reduce heat loss. The water that is heated in the manifold can be used for heating or domestic hot water.

Evacuated tube solar collector.

Close-up of evacuated tube solar collector.

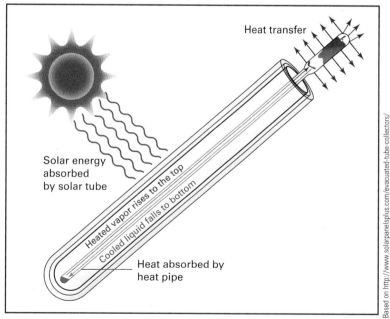

Internal structure of a typical evacuated tube solar collector.

Continued on page 272

Energy Extra 8.2 continued

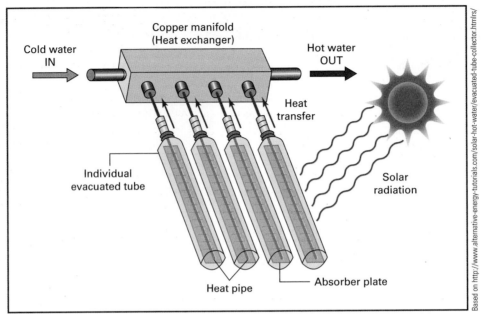

Based on http://www.alternative-energy-tutorials.com/solar-hot-water/evacuated-tube-collector.htmlrs/

Connection of evacuated tube solar collectors to heat exchanger.

In principle, evacuated tube solar collectors can be used in the same applications as flat plate collectors, and each has advantages in certain situations. Evacuated tube collectors have the disadvantage that only about 70% of the total collector area absorbs solar radiation because of the space around the absorber in the glass tube and the spaces between the tubes. Evaporated tube collectors are a newer technology, and over time their long term reliability may be determined. They also tend to be more expensive per square meter, typically about twice the cost of a flat plate system of the same absorber area. They do have the advantage that their more modular design improves serviceability and extendibility.

A major consideration for the comparison of these two different technologies is their relative efficiencies. The graph in the figure shows the outputs of two similarly-sized collectors as a function of the difference between the manifold water temperature (T_m) and the ambient temperature (T_a). In all cases the efficiency decreases with an increasing temperature difference. This is merely a result of the greater heat loss to the environment with the greater temperature difference. However, this decrease is not as significant for the evacuated tube collector as it is for the flat plate collector. This is a direct result of the more effective insulation provided by the evacuated space in the tube. The net result of these features is that there is a crossover point where the evacuated tube collector becomes more efficient for larger temperature differences. The crossover point shifts to a higher temperature difference when there is greater insolation. While the graph shows the general features of the relationship between the efficiency of the evacuated tube collector and the flat plate collector, the details of this behavior, i.e., actual efficiencies and crossover points, depend on specific details of the collector designs.

Continued on page 273

Energy Extra 8.2 continued

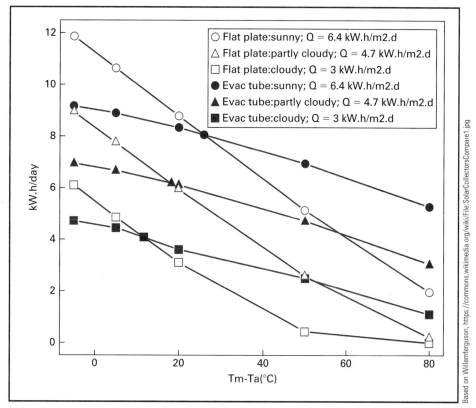

Based on Willemferguson, https://commons.wikimedia.org/wiki/File:SolarCollectorsCompare1.jpg

Comparison of power output from flat plate and evacuated tube collectors for different insolation conditions and different temperature differences. The flat plate collector had an absorber area of 2.8 m², and the evacuated tube collector had an absorber area of 3.1 m².

Topic for Discussion

Consider the economic viability of evacuated tube collectors and flat plate collectors for domestic hot water production for a manifold temperature of 50°C. Provide a quantitative comparison for different climatic conditions, e.g., Montreal in the summer or winter and New Orleans in the summer or winter.

a peak wavelength of about 500 nm. The energy that is reirradiated by the collector plate is characteristic of the temperature of the plate (which might be at around 350 K). This reirradiated radiation is in the infrared portion of the spectrum with a peak at about 8000 nm. The ideal design for the collector plate would be to have a large absorptance at 500 nm and a small emissivity at 8000 nm, so that the collector absorbs substantial energy from sunlight but loses very little of it by reirradiation (Figure 8.7 on the next page). Specially designed collector surface coatings have a high absorptance for visible radiation and low emissivity for infrared radiation. Flat black paint is a simple alternative that works well.

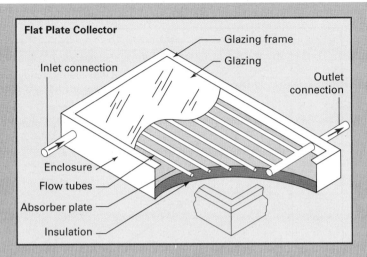

Figure 8.6: Design of a flat plate solar collector.

U.S. Department of Energy

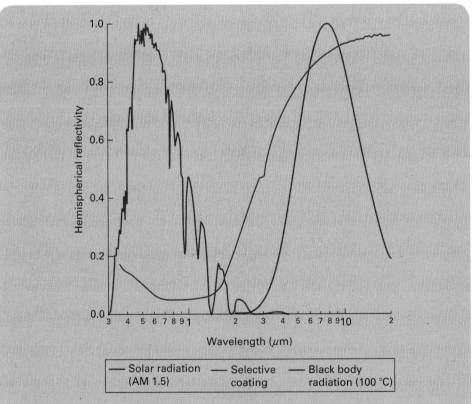

Figure 8.7: Measured reflectance of a suitable solar collector coating material relative to the solar spectrum (AM 1.5) and the black body curve at 100°C (373 K).

Based on P. Oelhafen and A. Schüller, "Nanostructured materials for solar energy conversion," Solar Energy 79 (2005) 110–121.

The glass plate on the front of the collector is also an important factor in the collector operation. Glass has a high transmittance for visible radiation but a low transmittance for infrared radiation. Thus, much of the energy that is reirradiated by the warm collector surface is not transmitted through the glass cover and is trapped inside the collector box. This is the same principle as the greenhouse effect discussed in Chapter 4 and is the reason that the interior of an automobile gets hot on a sunny day.

The efficiency of the collector is the ratio of the heat absorbed by the water to the energy incident on the collector. The loss of energy by reirradiation from the hot collector surface is one factor that limits the efficiency of the collector. Heat loss by conduction through the walls of the box is another contributing factor. Equation (8.8) for radiative heat transfer and equation (8.4) for conductive heat transfer both show that, as the temperature of the collector increases, so does heat loss. At some point, the losses equals the energy incident on the collector, and the efficiency goes to zero. Thus, from an efficiency standpoint, it is advantageous to operate the collector at lower temperatures, and the maximum efficiency occurs when the collector is at the same temperature as its ambient surroundings.

On the other hand, the amount of energy that is actually transported to a home heating system by the water heated in the collector increases with increasing temperature (Section 8.5). From this viewpoint, it would be advantageous to operate the collector at as high a temperature as possible. However, there are practical considerations for the operating temperature based on the properties of the fluid used (e.g., water). If the temperature of the water is high, then the maximum amount of thermal energy is transported through the system. However, when the temperature of the water is high, then the heat losses between the collector and the heating system inside the house are greater. The best compromise between these conflicting criteria is generally to operate the collector at an efficiency of about 50%. This results in a collector plate temperature that is somewhat less than the boiling point of the water.

8.5 Residential Heating Needs

Active solar systems are commonly used for residential space heating and for supplemental water heating. In this section, some requirements for the design of a space heating system are considered. Because solar radiation is present only during the daytime hours, it is necessary to be able to store heat energy for use during the night. Water circulating through the collector is heated and stored in a tank, which also acts as a heat exchanger to provide heat and hot water to the house.

For the design of such a system, it is important to know how much solar energy a particular collector system can produce and how much heat a particular residence needs. The average insolation on a horizontal surface at different locations in the United States shown in Figure 8.2 is not the most useful information for this analysis. First, insolation is less in the winter (because of the angle of the sun), but this is the time of the year when heating needs are the greatest. Second, a horizontal surface is not the ideal orientation for maximizing solar energy collection. Table 8.2 shows the total daily solar energy per square meter for different collector orientations. The perpendicular collector includes a system for moving the collector (during the daylight hours) so that it always remains perpendicular to the incident solar radiation. This provides the maximum amount of energy but adds

Table 8.2: Daily integrated insolation at a typical ~40°N latitude location (Allentown, Pennsylvania, latitude 40.65°N). Monthly averages over a 30-year period from 1961 to 1990 are given for flat plate collectors at various angles. *Vertical* means south facing, and *tilted* means south facing and tilted at the optimal angle. Tracking is a two-axis tracking collector that follows the sun during the day.

month	horizontal MJ/m^2	vertical MJ/m^2	tilted MJ/m^2	tracking MJ/m^2
Jan.	6.8	11.2	11.2	13.7
Feb.	9.7	13.0	14.0	16.9
Mar.	13.3	12.2	16.2	20.2
Apr.	16.9	10.8	18.0	23.0
May	19.4	9.4	18.7	25.2
Jun.	21.6	9.0	19.4	27.0
Jul.	21.2	9.4	19.4	27.0
Aug.	18.7	10.4	19.1	25.2
Sep.	15.1	11.9	17.6	21.6
Oct.	11.2	12.2	15.1	18.4
Nov.	7.2	10.1	10.8	13.0
Dec.	5.8	9.4	9.4	11.2

Based on data from National Solar Radiation Data Base of National Renewable Energy Laboratory

considerably to the cost, complexity, and presumably, reliability of the system. An alternative is to place the collector facing south (in the northern hemisphere) and at an angle from the vertical so that deviations from perpendicular incidence are minimized. A first approximation for the optimal tilt from the vertical is 90° minus the latitude. This approach can be improved by seasonal adjustments where the tilt (from the vertical) is reduced in the winter months and increased in the summer months. Another option is to position the collector vertically and facing south. The final option is to place the collector horizontally. The table shows the daily insolation for each of these cases.

The relative insolation for the different orientations is compared with the ideal perpendicular case for January 21 (typically the coldest time of the year in the northern hemisphere) in Table 8.3. In the interest of economics and simplicity, it is preferable to have a stationary (fixed) collector. The table shows that, at the coldest time of the year, the 60° south-facing collector provides close to 90% of the ideal energy

Table 8.3: Relative insolation compared to a perpendicular collector for different orientations on January 21 at 40°N latitude.

orientation	% of perpendicular
60° south	89
vertical south	79
horizontal	43

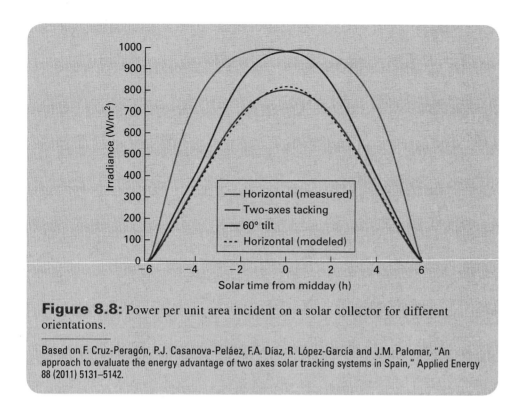

Figure 8.8: Power per unit area incident on a solar collector for different orientations.

Based on F. Cruz-Peragón, P.J. Casanova-Peláez, F.A. Díaz, R. López-García and J.M. Palomar, "An approach to evaluate the energy advantage of two axes solar tracking systems in Spain," Applied Energy 88 (2011) 5131–5142.

and is the best simple option for system design. Clearly the horizontal orientation is not desirable. Figure 8.8 shows the variations throughout the day for different collector orientations.

This discussion, along with an anticipated collector efficiency in the range of about 50%, provides an estimate for the amount of energy that can be collected for a solar collector of a certain size. The amount of energy that a house needs to fulfill its heating needs depends on several factors: the size of the house, the details of its construction, and the local climate. The climate conditions are expressed as *degree days*. One degree day, sometimes expressed as *heating degree days*, is one day during which the average outside temperature is one degree less than the desired inside temperature. In warmer climates, *cooling degree days* are used as a measure of the energy requirements for air conditioning. In this discussion, only heating degree days are considered.

Degree days, expressed for temperatures in Celsius, that is, *degree days (°C)*, are determined on the basis of an ideal inside temperature is 18.3°C. Although this may be less than the target temperature in many buildings, contributions from internal heat sources, as will be discussed, can compensate for this difference. The numbers of degree days for individual days during the year are added together to give the total number of degree days for the year, and this sum can be used as an indication of the annual heating requirements. As an example, consider a cold winter day when the average outside temperature is −5 °C. This day would contribute [18.3°C − (−5°C)] = 23.3 °C degree days towards the annual total.

The average number of degree days per year for various locations in the United States and Canada are shown in Figures 8.9 and 8.10, respectively. More accurate values for some selected cities are given in Table 8.4.

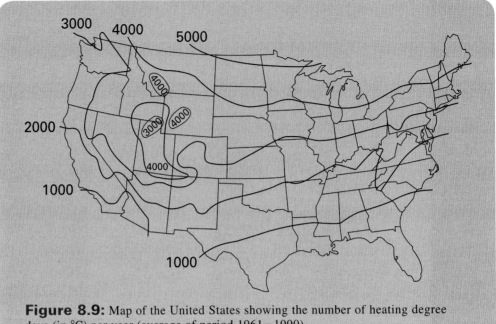

Figure 8.9: Map of the United States showing the number of heating degree days (in °C) per year (average of period 1961−1990).

National Oceanic and Atmospheric Administration (NOAA)

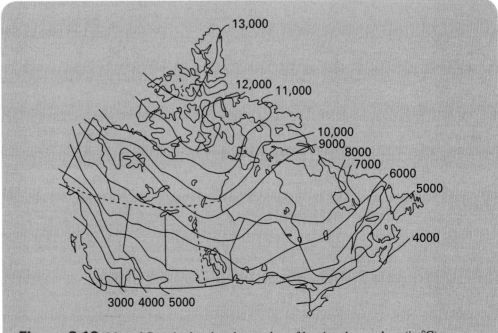

Figure 8.10: Map of Canada showing the number of heating degree days (in °C) per year.

From Hunt, Energy, Physics and the Environment, 3E. © 2007 Cengage Learning.

Table 8.4: Average number of heating degree days (in °C) per year for selected U.S. and Canadian cities. (averages for 1971–2000).

city	state/province	heating degree days per year (°C)
Honolulu	HI	0
Miami	FL	86
Phoenix	AZ	578
San Diego	CA	602
New Orleans	LA	787
San Francisco	CA	1452
Albuquerque	NM	2378
Portland	OR	2426
Baltimore	MD	2574
New York	NY	2635
St. Louis	MO	2643
Philadelphia	PA	2644
Seattle	WA	2665
Vancouver	BC	2926
Boston	MA	3128
Denver	CO	3404
Detroit	MI	3583
Chicago	IL	3607
Toronto	ON	4066
Portland	ME	4069
Halifax	NS	4367
Minneapolis	MN	4379
Montreal	QC	4575
St. John's	NL	4881
Grand Forks	ND	5281
Regina	SK	5661
Edmonton	AB	5708
Winnipeg	MB	5777
Anchorage	AK	5817
Whitehorse	YT	6811
Mt. Washington	NH	7658
Yellowknife	NT	8256
Barrow	AK	11,052
Resolute	NU	12,526

It is now necessary to relate the heat requirements for a building of a particular design to the climate as represented by the number of degree days. We could undertake a detailed calculation of heat loss through the walls, windows, roof, foundation, and so on for the building based on an analysis of the *R*-values of these various building components. However, in 1972, the United States Federal Housing Authority provided an estimate of 67 kJ/m^3 of living space per degree day (°C) for residential heating. Although construction techniques and materials have improved since 1972, this value is useful in acting as a rough guideline for heating needs: The annual heating requirements for a house are given as the volume of the house times the heat requirement per unit volume per degree day times the number of degree days per year for the location of the house.

To design a reasonable solar heating system, it is convenient to look at the daily requirements for the house in Philadelphia in Example 8.5. A winter (February) day, when the average daily temperature is, say, 0 °C, represents 18.3 degree days (°C). This will require

$$700 \text{ m}^3 \times 67 \text{ kJ/m}^3 \times 18.3 = 858 \text{ MJ} \qquad (8.9)$$

of heat on that day. This may be an overestimate of the requirements for at least four reasons:

1. Passive solar heating (Section 8.6) has not been considered.
2. There is heat production within a house as a result of other activities: heat generated by people, heat from cooking, heat generated by appliances and electronics, and so on. Some typical numbers are given in Table 8.5 and can account for something of the order of 10% of the heating needs.
3. Solar heating may work in conjunction with another heating method (e.g., oil or electric) and may not be expected to fulfill all of the heating requirements (discussed further in Chapter 17).
4. Well designed contemporary buildings typically require less heat than the 1972 U.S. guideline.

For our example, it might be reasonable to aim for about 60% of the value in equation (8.9), or about 500 MJ, for a solar heating system that would contribute in a substantial way to home heating needs. From Table 8.2 (appropriate for Philadelphia), we see that a total daily insolation on the order of 14.0 MJ/m^2 can be expected for a 60° south collector. Taking into account a typical collector efficiency of 50% allows for a simple calculation of the area of the collector needed to satisfy these requirements:

$$\frac{500 \text{ MJ}}{14.0 \text{ MJ/m}^2} = 71 \text{ m}^2 \qquad (8.10)$$

Table 8.5: Typical daily contributions to home heating from internal sources.

source	heat per day MJ
lights	11
cooking	7
appliances	33
electronics	5
people	12
other	5
total	73

Example 8.5

(a) Determine the annual heat requirements (in Btu) for a residence in Philadelphia with 250 m^2 of heated living space and 2.8-m ceilings.
(b) Estimate the annual heating costs for electric heat (at 100%) if electricity costs $0.11 per kWh.
(c) Estimate the annual heating costs for oil heat (at 85% efficiency) if oil costs $1.10 per liter.

Solution

(a) From Table 8.4, Philadelphia has 2644 heating degree days per year (°C), so the total heat requirement for a house with a volume of 250 m^2 × 2.8 m = 700 m^3 at 67 kJ/m^3 per degree day (°C) is

$$67 \text{ kJ/m}^3 \times 700 \text{ m}^3 \times 2644 = 1.24 \times 10^{11} \text{ J}.$$

(b) The heating requirement can be converted from J to kWh (Appendix III) as

$$(1.24 \times 10^{11} \text{ J})/(3.6 \times 10^6 \text{ J/kWh}) = 34{,}445 \text{ kWh}.$$

At $0.11 per kWh, this will cost $3789.

(c) From Appendix V, the energy content of 1 L of crude oil (about the same as home heating oil) is 3.85 × 10^7 J. At 85% efficiency, this will yield about 3.27 × 10^7 J/L. Meeting the heating requirement of 1.24 × 10^{11} J will require

$$(1.24 \times 10^{11} \text{ J})/(3.27 \times 10^7 \text{ J/L}) = 3792 \text{ L of oil}.$$

At $1.10 per L, the cost will be about 3792 L × $1.10/L = $4171.

A collector 7 m × 10 m would fulfill these requirements and would approximately cover the roof area of the house in this example.

Example 8.6

(a) Determine the annual heat requirements (in J) for a residence in Winnipeg with 230 m^2 of heated living space and 2.5-m ceilings.
(b) Estimate the annual heating costs for electric heat (at 100%) if electricity costs $0.11 per kWh.
(c) Estimate the annual heating costs for oil heat (at 90% efficiency) if oil costs $0.90 per liter.

Solution

(a) From Table 8.4, Winnipeg has 5777 heating degree days per year (°C), so the total heat requirement for a house with a volume of 230 m^2 × 2.5 m = 575 m^3 at 67 kJ/m^3 is

$$67{,}000 \text{ J/m}^3 \times 575 \text{ m}^3 \times 5777 = 2.23 \times 10^{11} \text{ J}.$$

Continued on page 282

Example 8.6 continued

(b) The heating requirement can be converted from J to kilowatt-hours (Appendix III):

$$\frac{2.23 \times 10^{11} \text{ J}}{3.6 \times 10^6 \text{ J/kWh}} = 6.18 \times 10^4 \text{ kWh}.$$

At $0.11 per kilowatt-hour, this will cost $6800.

(c) From Appendix IV, the energy content of 1 L of crude oil (about the same as home heating oil) is 3.85×10^7 J/L. At 90% efficiency, this will yield about 3.47×10^7 J/L. Meeting the heating requirement of 2.23×10^{11} J will require

$$\frac{2.23 \times 10^{11} \text{ J}}{3.47 \times 10^7 \text{ J/L}} = 6427 \text{ L of oil}.$$

At $0.90 per liter, this will cost $5784.

8.6 Heat Storage

The preceding discussion provided a basis for a usable design of a residential solar heating system. However, it must be realized that the sun does not shine 24 hours a day and that the insolation value used in the previous section is a daily average. In fact, part of the problem in utilizing solar energy for heating is that in the winter, when the most heat is needed, the duration of sunshine is the least. So the energy that is collected during the day must be stored for use during the night. One may also take the approach that energy collected during sunny days can be stored to supplement heating needs during cloudy days. Heat contained in water that has been heated in a solar collector (Figure 8.6) may be stored by keeping the water in an insulated tank or transferring that heat to another material for storage. (The design of such a system is discussed further in Chapter 17; see Figure 17.16). The amount of heat that can be stored in a material is a function of the material's temperature, its mass, *m*, and its thermal properties. If a material is cooled from an initial temperature T_i to a final temperature T_f (assuming no phase transitions are involved), then the heat given up is expressed as

$$Q = Cm(T_i - T_f), \tag{8.11}$$

where C is the *specific heat* of the material. It is clear from equation (8.11) that the specific heat is a measure of the thermal energy content of a material per unit mass per unit temperature change. The *volumetric heat capacity* is defined as the heat contained in a material per unit volume per unit temperature and is obtained from the specific heat by multiplying C times the density. Table 8.6 gives the specific heat and volumetric heat capacity of some common materials. Equation (8.11) may also be written as

$$\Delta T = \frac{Q}{Cm} \tag{8.12}$$

and gives the temperature increase of a material when a quantity of heat is applied to it. From equation (8.12), it is clear that a material with a large specific heat experiences a

Table 8.6: Specific heats and volumetric heat capacities for some common materials.

material	specific heat J/(kg·°C)	density kg/m³	volumetric heat capacity kJ/(m³·°C)
water	4186	1000	4186
aluminum	895	2700	2416
iron	460	7855	3613
wood (pine)	2800	500	1400
stone (solid)	879	2560	2250
stone (loose)	879	~1500	~1300
brick	920	1800	1656
concrete	653	2300	1502
glass	837	2720	2277
sand	816	1600	1306

relatively small increase in temperature when heat is applied and that a material with a small specific heat experiences a much larger increase in temperature. Conversely, materials with large specific heats can release significant amounts of heat without undergoing large temperature changes.

Among the common materials listed in Table 8.6, water clearly has the greatest capacity for storing thermal energy, both per unit weight and per unit volume. In addition, it is convenient because it is nontoxic, relatively nonreactive, and inexpensive, and as a liquid, it provides a suitable means of transporting heat. For a space heating system that uses solar energy, it would be appropriate to store an amount of heat that is comparable to what is used daily during the winter. The following analysis provides an idea of the requirements for a useful heat storage system.

To design a system to store the 1.09 GJ calculated in Example 8.7 in a tank of water, it is necessary to know the maximum temperature of the water, T_i, and the temperature of the water after the 1.09 GJ has been extracted, T_f. The mass of water can then be obtained by rewriting equation (8.12) as

$$m = \frac{Q}{c\Delta T}. \qquad \textbf{(8.13)}$$

Example 8.7

Calculate the amount of energy needed to heat the home in Winnipeg from Example 8.6 on a winter day when the average daily temperature is $-10°C$.

Solution

An average temperature of $-10°C$ over a period of one day corresponds to $[18.3°C - (-10°C)] \times 1$ day = 28.3 degree days (°C). For a house with a volume of 575 m³ requiring 67 kJ/m³ per degree day, the total energy required will be $67{,}000 \ J/m^3 \times 575 \ m^3 \times 28.3 = 1.09 \times 10^9 \ J$.

Example 8.8

Calculate the mass and volume of water needed to store 1.09 GJ of thermal energy if the maximum and minimum water temperatures are 65°C and 30°C, respectively.

Solution

Equation 8.13 gives

$$m = \frac{Q}{c\Delta T},$$

where $Q = 1.09 \times 10^9$ J and $\Delta T = (65°C - 30°C) = 35°C$. Table 8.6 gives $c = 4186$ kJ/(m³·°C) for water, and, since water has a density of 10^3 kg/m³, this is the same as 4186 J/(kg·°C). The mass of water is calculated to be

$$m = \frac{1.09 \times 10^9 \text{ J}}{(4186 \text{ J/(kg} \cdot °C)) \times (35°C)} = 7.4 \times 10^3 \text{ kg}.$$

This mass of water represents a volume of 7.4 m³. This could be contained in a cubic tank about 1.9 m on a side.

Minimizing the size of the tank means maximizing T_i and minimizing T_f. The maximum temperature is limited by the temperature of the water exiting the solar collector (in order to transfer heat from the solar collector water to the storage tank water). The minimum temperature is limited by the temperature of the house (in order to transfer heat from the heating system to the room). Although an absolute maximum temperature might be 90°C and an absolute minimum temperature would be about 18°C, a reasonable range of operating temperatures that provides good efficiency for heat transfer might be between $T_i = 65°C$ and $T_f = 30°C$.

8.7 Passive Solar Heating

All buildings are heated passively by solar radiation to some extent, but the design and placement of the building can maximize these benefits. In the design of a passively heated building, the steps taken to optimize passive solar heating must not ultimately require more heat. Thus, heat gain must be balanced against heat loss. A typical home designed to optimize passive solar heating incorporates large south-facing windows (Figure 8.11). Figure 8.12 shows how building design can be used to maximize passive solar heating during the winter and minimize excess heating during the summer. A massive stone floor absorbs heat during the day (especially if it is dark colored) and reirradiates the heat during the night to distribute the effects of passive heating over a 24-hour period.

Because passive heating results from energy entering through a window and being trapped as a result of the greenhouse effect, it is necessary to consider the heat gained in comparison with the heat lost through the window. It is easiest to consider the heat lost in the context of conductive heat transfer through the window. A standard double-pane window has an R-value of about 0.35, and a wall without a window that

Figure 8.11: Passive solar home showing large south-facing windows.

© iStockphoto.com/EricVega

contains 15 cm of insulation will have an *R*-value of about 3.5. Thus, it is important to ensure that, on average, the energy gain through windows from passive solar heating will outweigh the energy loss resulting from the low *R*-value of the windows. The effects of the *R*-values of windows will be discussed further in Chapter 17.

The efficiency of solar energy collection under the conditions described in Example 8.9 is quite good. This calculation is based on clear sky conditions. The reduction of passive solar heating caused by clouds depends on local climate conditions. For example, if cloud cover reduces the insolation to 50% of its clear sky value, then the

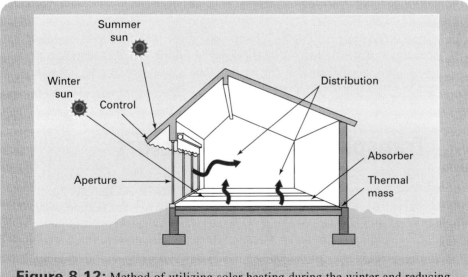

Figure 8.12: Method of utilizing solar heating during the winter and reducing solar heating during the summer.

Based on U.S. Department of Energy, http://www.energysavers.gov/your_home/designing_remodeling /index.cfm/mytopic=10270

Example 8.9

Calculate the net energy gain (per day in January) from passive solar radiation incident on a double-pane window in a house at 40°N latitude. Consider a 1.5 m × 2.5 m south-facing (vertical) window and an average daily temperature of −5°C with an inside temperature of 20°C. What is the efficiency of this heating method?

Solution

From Table 8.2, the average January daily insolation on a south-facing vertical surface is 11.2 MJ/m². For a (1.5×2.5)m² = 3.75 m² window, the total incident energy is (11.2 MJ/m²) × 3.75 m² = 42 MJ per day. The energy loss is obtained from equation (8.5) by multiplying the power per unit area by the area and the time; that is,

$$E = \frac{(T_h - T_c)At}{R}.$$

For a standard double-pane window, the R-value is 0.35 (s·m²·°C)/J. Using area in square meters, time in seconds (86,400 s/d), and the temperature difference in degrees °C yields energy in J. Thus

$$E = \frac{(20°C - (-5°C)) \times (3.75 \text{ m}^2) \times (86,400 \text{s/d})}{0.35 \text{ (s} \cdot \text{m}^2 \cdot °C)/J} = 23.1 \text{ MJ per day.}$$

The net energy gain from passive solar heating is 42.0 MJ − 23.1 MJ = 18.9 MJ per day. This is compared to the total incident energy to give an efficiency of 100 × (18.9 MJ)/(42.0 MJ) = 45 %.

incident energy is 0.5 × 42 MJ = 21 MJ per day, resulting in a net energy loss through the window. This calculation also emphasizes the significance of a high R-value and the effects of lower outside temperatures.

8.8 Transpired Solar Collectors

A transpired solar collector, sometimes called a "solar wall" after the trade name SolarWall© of Conserval Engineering Inc. (Toronto, Ontario and Buffalo, New York), which markets this device, is a means of improving passive solar heating without the complexity and maintenance costs of flat plate or evacuated tube systems. Large industrial or office buildings (rather than small residential buildings) with a large (mostly window-free) south-facing wall, in a cool, sunny region with a central ventilation/heating system are ideal for the application of this technology.

The transpired solar collector consists of a dark perforated metal facade with about a 15-cm-air-space between the collector and the building wall. Figure 8.13 shows

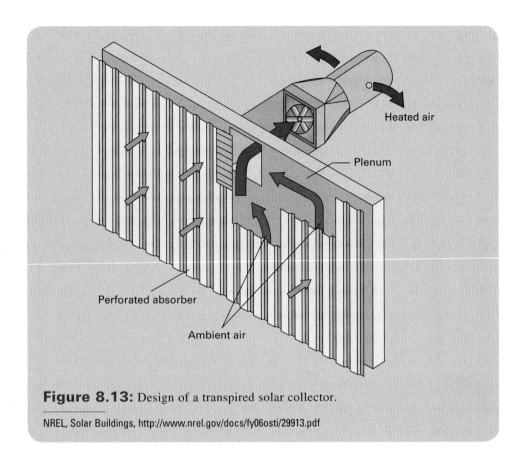

Figure 8.13: Design of a transpired solar collector.

NREL, Solar Buildings, http://www.nrel.gov/docs/fy06osti/29913.pdf

the basic design of a transpired solar collector. Solar radiation heats the dark absorber, and as air is pulled into the building for ventilation and heating, it is heated as it passes through the perforated solar wall. The fan shown in the figure, which draws air through the collector, may be a component of the building's ventilation system or may be an additional fan associated with the collector. Air entering the building can be heated by as much as approximately 22°C compared to the ambient outside air temperature, and the collector can provide up to 2.7 GJ/m^2 per year of additional heat to the building. Preheating ventilation air reduces the heating requirements for the building.

Figure 8.14 on the next page shows the operation of the transpired solar collector in the winter and summer. In the winter, air preheated by the collector is drawn into the building, thereby reducing the load on the heating system. In the summer, ventilation air is drawn directly into the air conditioning system without being heated by the transpired collector.

Transpired solar collectors may be retrofitted to existing buildings that meet the requirements for this technology. They are readily incorporated into the design of many new buildings, and their installation cost, in this case, is partially offset by the reduced cost of a finished surface for the facade. A typical transpired solar collector installation that was incorporated into the original building design is illustrated in Figure 8.15 on the page 289. A transpired solar collector that was retrofitted on the roof surface of a sports center is shown in Figure 8.16 on the page 289.

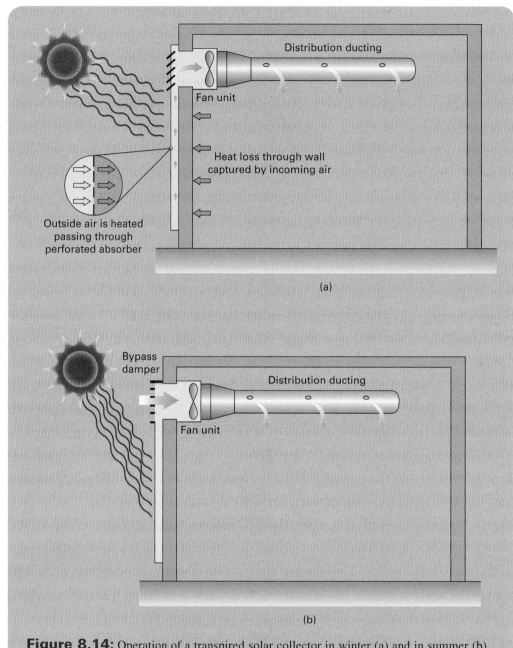

Figure 8.14: Operation of a transpired solar collector in winter (a) and in summer (b).

U.S. Department of Energy, Transpired Collectors (Solar Preheaters for Outdoor Ventilation Air), http://www1
.eere.energy.gov/femp/pdfs/fta_trans_coll.pdf

Figure 8.15: Transpired solar collector on the south- and west-facing walls of the Mona Campbell Building at Dalhousie University, Halifax, Nova Scotia, Canada. This building has LEED Gold certification (see Energy Extra 17.1).

Richard A. Dunlap

Figure 8.16: Transpired solar collector on the Sportsplex, Dartmouth, Nova Scotia. The 300 m^2 collector is the dark section of the roof near the center of the building.

Richard A. Dunlap

8.9 Summary

The chapter has reviewed the properties of sunlight. Based on the total power radiated by the sun, the insolation at the surface of the earth, averaged over time and all locations, was found to be 168 W/m^2.

To describe how solar energy can be used for heating purposes, the chapter reviewed the three heat transfer mechanisms: conduction, convection, and radiation. The power transferred by conduction per unit area of a material is given as $P/A = \Delta T/R$, where ΔT is the temperature difference, and R is the R-value, defined as the thickness of the material divided by the thermal conductivity. R-values for insulating materials allow for the convenient calculation of heat transfer through walls, windows, and other building components. Radiative heat transfer is described by the Stefan-Boltzmann law and is proportional to the fourth power of the absolute temperature of an object.

A flat plate solar collector is a box in which solar energy is transferred to a working fluid for distribution to a building for space heating or hot water needs. The efficiency of the solar collector can be evaluated on the basis of the radiative and conductive heat mechanisms that transfer thermal energy into and out of the collector.

The chapter evaluated the usefulness of solar heating on the basis of heating requirements for a building. The energy needed for heating purposes is a function of the size of the building and the quality of its construction, as well as the local climate. The effects of climate are quantified using degree days, where one degree day is one day during which the outside temperature is one degree below a standard inside temperature. Annual heating needs are determined by the number of degree days integrated over the year. As illustrated in the chapter, the utilization of the roof area of a single-family home for solar collectors is sufficient to make a reasonable contribution to space heating needs.

The periodic nature of sunlight, on both a daily and a seasonal scale, as well as the variability of heating requirements, again on a daily and seasonal scale, means that the efficient utilization of solar energy requires some method of thermal energy storage. The chapter discussed the effectiveness of different materials for heat storage in terms of their heat capacity and the possible mechanisms for transferring heat into and out of the storage system.

The chapter described the effectiveness of passive solar heating in terms of the trade-off between energy gain and energy loss through a window. Passive solar heating features can be incorporated into new building designs, often at little expense by utilizing specific design characteristics, building location, and/or orientation.

Problems

8-1 Calculate the power radiated by a woodstove of dimensions 65 cm high by 55 cm deep by 85 cm wide with a surface temperature of 120°C. Assume that heat is radiated from all surfaces of the stove and that the stove has an emissivity of 1. Note that woodstoves are painted black because black surfaces have high absorptance, and objects with high absorptance also have high emissivity.

8-2 Locate information about the current cost of home heating oil or natural gas in your area (whichever is in common use) and the cost of residential electricity. Assuming an efficiency of 85% for an oil furnace and 100% for electric heat, calculate the relative cost of electric heat compared with oil heat or natural gas if both heating systems require the same net energy to heat a house.

8-3 Consider a vertical south-facing window in a house at 40°N latitude. For an interior temperature of 20°C, make a plot of the minimum R-value as a function of outside temperature from $-30°C$ to $10°C$ for which the passive solar heating exceeds the heat loss through the window.

8-4 Compare the total solar energy received at the surface of the earth in one year to the total annual global energy requirements.

8-5 Compare the R-values of:
  (a) Two pieces of 3-mm-thick glass in thermal contact.
  (b) Two pieces of 3-mm-thick glass with a 1-cm-air-space between them.

8-6 Approximate a house as a cube with an edge length of 7 m. The house loses heat from the four walls and the roof (but not the floor). The average R-value for the walls and roof is $R = 1.2$ (this takes into account walls, windows, doors, etc.). Calculate the heat loss in MJ/m^3 per degree day (°C), and compare this to the estimated residential heating needs discussed in this chapter.

8-7 A 300 liter (U.S.) electric hot water heater uses 9000 W of power to heat water. If the heater is filled with water at an initial temperature of 10°C, how long will it take for the water to reach 60°C? Assume there are no heat losses.

8-8 Compare the masses and volumes of water, concrete, sand, and wood needed to store 1 GJ of heat if the operating temperatures are $T_c = 30°C$ and $T_h = 80°C$. In each case, calculate the edge length of a square storage unit with a height of 2.5 m.

8-9 Consider a house that is approximated as a cube that is 8 m on a side and has no windows. The house loses heat from the four walls and the roof. For an inside temperature of 20°C and an outside temperature of $-20°C$, it is desired to keep the total heat loss less than 5 kW. What is the minimum R_{SI}-value that is required for the walls and roof?

8-10 A cylindrical water tank 1.5 m in diameter and 2.5 m high is surrounded on both ends and the sides with $R_{SI} = 3$ insulation. It is desired to keep the water at 80°C, while the ambient temperature is 20°C. What is the minimum wattage of an electric heater that needs to be placed inside the tank?

8-11 The average monthly temperatures in Anchorage, AK are given in the table below. Calculate the heating degree days per year, and compare the results with the information in Table 8.4

month	average temperature (°C)
January	−8.5
February	−6.5
March	−4.0
April	+2.5
May	+8.5
June	+12.5
July	+15.0
August	+13.5
September	+9.0
October	+1.0
November	−5.5
December	−7.5

8-12 (a) Calculate the power transferred through a 10-cm-diameter iron rod that is 1 m long and is held at 0°C on one end and 20°C on the other end.

(b) It is desired to reduce the power transferred through the rod to 4 W by reducing the diameter of the rod. What should the new diameter be?

8-13 A two storey house in Portland, ME with dimensions 8.5 m by 12.0 m with a height of 7.0 m is heated electrically. The cost of electricity is 0.11 per kWh. Estimate the annual heating cost.

8-14 A house is heated with electricity at a cost of $0.10 per kWh. The building has 20 m^2 of double-glazed windows with an R_{SI} value of 0.4. A mechanism is constructed that covers the windows with 10 cm of polystyrene foam at night (i.e., for 12 hours per day) during six months of the year. If the outside temperature is constant at 0°C and the inside temperature is kept at 20°C, calculate the annual savings from this approach.

8-15 A wood stove has a total surface area of 1.5 m^2 and radiates a power of 1.0 kW. Make a plot of its surface temperature in °C as a function of emissivity.

8-16 A window with an R_{SI}-value of 0.18 receives 300 W/m^2 of insolation. If the interior of the building is at 20°C, what is the minimum outside temperature for which there is a benefit from passive solar heating?

8-17 A wall consists of 3.0 cm of exterior wood and 2.0 cm of interior wood separated by a 15 cm layer of polystyrene foam. If the interior is kept at 19°C, calculate the outside temperature for which the heat loss through the wall is 10 W/m^2.

8-18 A homeowner in Allentown, PA installs a 60° south-facing solar collector that (on the average over the year) replaces 1 kW of electric heat (purchased from the public utility at $0.11 per kWh). The homeowner is considering adding a tracking mechanism to maintain perpendicular alignment of the collector with the solar insolation. If the payback period for the upgrade is expected to be five years, what should the homeowner be willing to pay for the tracking device? Comment on this plan.

8-19 A homeowner sells an electrically-heated 280-m² home with 3.2-m-high ceilings in Minneapolis and purchases an identical home in San Francisco. Estimate the expected savings in annual heating costs.

8-20 A home uses a concrete floor to store passive solar energy. If the floor heats to 40°C during the day and cools to 20°C at night, calculate the thickness of concrete needed to store the amount of energy from Example 8.7 if the floor's area is 150 m².

Bibliography

B. Anderson. *Solar Energy: Fundamentals in Building Design.* McGraw-Hill, New York (1977).

B. Anderson and M. Riorden, *The New Solar Home Book* (2nd ed.). Brick House Publishing, Andover, MA (1996).

B. Anderson and M. Wells. *Passive Solar Energy: The Homeowner's Guide to Natural Heating and Cooling* (2nd ed.). Brick House Publishing, Andover, MA (1996).

A. K. Athienitis and M. Santamouris. *Thermal Analysis and Design of Passive Solar Buildings.* James & James, London (2002).

G. Boyle (Ed.). *Renewable Energy.* Oxford University Press, Oxford (2004).

D. Chiras. *The Solar House: Passive Heating and Cooling* Chelsea. Green Publishing, White River Junction, VT (2002).

J. A. Duffie and W. A. Beckman. *Solar Engineering of Thermal Processes* Wiley, Hoboken, NJ (2006).

J. Fanchi. *Energy: Technology and Directions for the Future.* Elsevier, Amsterdam (2004).

F. A. Farret and M. Godoy Simões. *Integration of Alternative Sources of Energy.* Wiley, Hoboken (2006).

D. Y. Goswami, F. Kreith, and J. F. Kreider. *Principles of Solar Engineering* (2nd ed.). Taylor & Francis, New York (2000).

D. Hawkes and W. Forster. *Energy Efficient Building: Architecture, Engineering and Environment.* W. W. Norton, New York (2002).

R. Hinrichs and M. Kleinbach. *Energy: Its Use and the Environment* (5th ed.). Brooks-Cole, Belmont, CA (2012).

R. W. Jones, J. D. Balcomb, C. E. Kosiewicz, et al. *Passive Solar Design Handbook, Volume Three: Passive Solar Design Analysis,* NTIS, U.S. Dept. of Commerce, Washington, DC (1982).

M. Kaltschmitt, W. Streicher, and A. Wiese (Eds.). *Renewable Energy: Technology, Economics and Environment.* Springer, Berlin (2007).

J. A. Kraushaar and R. A. Ristinen. *Energy and Problems of a Technical Society* (2nd ed.). Wiley, New York (1993).

F. Kreith and J. F. Kreider. *Principles of Solar Engineering.* McGraw-Hill, New York (1978).

F. Kreith and R. West. *Handbook of Energy Efficiency.* CRC Press, Boca Raton (1997).

G. Lof. *Active Solar Systems.* MIT Press, Cambridge, MA (1993).

P. J. Lunde. *Solar Thermal Engineering.* Wiley, New York (1980).

E. Mazria. *The Passive Solar Energy Book.* Rodale Press, Emmaus, PA (1979).

B. Norton. *Solar Energy Thermal Technology.* Springer-Verlag, London (1992).

T. Reddy. *The Design and Sizing of Active Solar Thermal Systems.* Clarendon Press, Oxford (1987).

S. Weider. *An Introduction to Solar Energy for Scientists and Engineers.* Kreiger Publishing, Melbourne (1992).

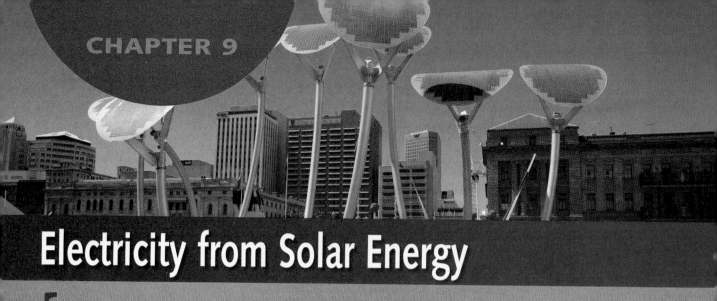

CHAPTER 9

Electricity from Solar Energy

Learning Objectives: After reading the material in Chapter 9, you should understand:

- The generation of electricity from solar energy through heat engines.
- The basic physics of semiconducting materials and the properties of n-type and p-type materials.
- The construction of a semiconducting junction device.
- The production of electricity through the interaction of photons on semiconducting junctions.
- The sensitivity and efficiency of photovoltaic devices.
- The application of photovoltaic devices for electricity generation worldwide.
- The availability of solar energy and economic viability of its utilization.

9.1 Introduction

The extent of the solar energy resource was illustrated in the previous chapter. However, the applications presented thus far, which utilize thermal energy extracted from solar radiation, are most suitable for residential space heating. The conversion of solar energy into electricity allows for the wide-scale distribution of energy on the electric grid, in addition to local residential use. Two different approaches can be taken to the conversion of solar energy into electricity: (1) the conversion of solar radiation into heat, followed by the conversion of heat into mechanical energy by means of a heat engine, and finally the generation of electricity by means of a generator; (2) the direct conversion of radiation into electrical energy by means of a photovoltaic device. These two approaches are discussed in this chapter.

9.2 Solar Electric Generation

A relatively straightforward (at least conceptually) method of producing electricity from solar energy is to use the heat produced in a solar collector to generate electricity by means of a heat engine. This is the most easily accomplished when water is heated above its boiling point to produce steam that can then be used to drive a steam turbine. Focusing collectors are necessary for this purpose because flat plate collectors are not suitable for achieving the required temperatures. Several approaches to large-scale focusing collector designs have been taken in the past. The most notable are:

- Parabolic troughs.
- Parabolic dishes.
- Central receivers.

These three devices are now reviewed briefly.

9.2a Parabolic Troughs

Parabolic trough collectors heat a fluid that is flowing through a pipe located at the focus of a parabolic trough (Figure 9.1). The parabolic troughs rotate to track the sun to ensure that the radiation is properly focused on the fluid-carrying pipes. In this arrangement, the fluid is typically oil, and the heated oil transfers its thermal energy to water through a heat exchanger to produce steam. The overall efficiency of converting solar energy to electrical energy is limited by the Carnot efficiency based on the hot and cold reservoir temperatures. The largest parabolic solar trough project is located in the Mojave Desert in California and has a maximum rated capacity of 354 MW_e. A number of parabolic trough systems have been constructed worldwide, and Table 9.1 lists the major operational facilities. As the table indicates, much of the activity in this field has occurred in the southwestern United States and in Spain.

9.2b Parabolic Dishes

An alternative to the parabolic trough geometry is the parabolic dish geometry (Figure 9.2). Units may be either individual parabolic dishes or arrays of dishes. Instead of a line focus, as is the case for the parabolic trough, the parabolic dish has a point focus. These are tracking collectors and utilize one of two different approaches to converting solar energy into electricity. The Solar Total Energy Project (STEP) in Shenandoah, Georgia heated water to produce steam to generate electricity through

Figure 9.1: Parabolic trough solar collectors at Kramer Junction, California.

RGB Ventures/SuperStock/Alamy Stock Photo

Table 9.1: Major operational parabolic trough solar projects (≥ 150 MW$_e$) as of 2016.

name	location	maximum capacity (MW$_e$)
Solar Energy Generating Station (SEGS)	Mojave Desert, California	354
Mojave Solar Project	Harper Dry Lake, California	280
Solana Generating Station	Gila Bend, Arizona	280
Genesis Solar Energy Project	Blythe, California	250
Solaben Solar Power Station	Logrosán, Spain	200
NOOR 1	Ouarzazate, Morocco	170
Andasol 1	Andalusia, Spain	150
Extresol	Torre de Miguel Sesmero, Spain	150
Solnova Solar Power Station	Sanlúcar la Mayor, Spain	150

Figure 9.2: Array of parabolic dish solar collectors at Maricopa Solar Project in Peoria, Arizona. The facility consists of a total of 60 parabolic dishes.

Brian Green/Alamy Stock Photo

conventional generators. This facility is no longer functional. Another approach is to directly convert the energy content of hot gas produced by the absorption of solar radiation into mechanical energy using a Stirling engine. Figures 9.2 and 9.3 show the Stirling-engine-based parabolic dish array at the Maricopa Solar Project in Peoria, Arizona. The solar-to-electric efficiency for this system is around 26% and is limited primarily by the Carnot efficiency.

Example 9.1

Calculate the average power per unit area that is incident on a 5-cm-diameter focal area for a 10-m-diameter parabolic dish as illustrated in Figure 9.3. Assume a midday insolation of 650 W/m^2 on a sunny day.

Continued on page 297

Example 9.1 continued

Solution

A 10-m-diameter dish has an area of

$$\frac{\pi (10 \text{ m})^2}{4} = 78.5 \text{ m}^2,$$

giving a total insolation on the dish of

$$(78.5 \text{ m}) \times (650 \text{ W/m}^2) = 51 \text{ kW}.$$

If this is focused on an area of

$$\frac{\pi (0.05 \text{ m})^2}{4} = 1.96 \times 10^{-3} \text{ m}^2,$$

then the power per unit area is

$$\frac{51 \text{ kW}}{1.96 \times 10^{-3} \text{m}^2} = 26 \text{ MW/m}^2.$$

Figure 9.3: Parabolic dish solar collectors at Maricopa Solar Project in Peoria, Arizona showing the Stirling engine used to convert heat to mechanical energy.

9.2c Central Receivers

Another possibility for the conversion of solar energy to electricity by means of a heat engine is by means of a central receiver. This is analogous to a single large parabolic dish except that the parabolic dish is replaced with a planar array of computer-controlled mirrors, each of which tracks the sun and reflects the sunlight onto a single point. The individual mirrors are referred to as *heliostats*, and the central receiver is contained in the so-called *power tower*. An early example of this type of design is Solar One in California, which operated between 1982 and 1986. This was later redesigned to become Solar Two (Figures 9.4 and 9.5). Solar Two operated from 1995 to 1999 and achieved its rated output of 10 MW$_e$. A working fluid, circulated through the focal point at the top of the tower, is used to produce steam to either drive a turbine/generator to generate electricity directly or as a mechanism to store thermal energy for later use (Figure 9.6). Solar One used oil for heat storage, and Solar Two used a molten salt.

More recent examples of central receiver systems include the PS10 and PS20 facilities in Spain (Figure 9.7 on the page 300). PS10, rated at 11 MW$_e$, became operational in 2007, and PS20, rated at 20 MW$_e$, became operational in 2009. Both facilities are currently operational.

The largest central receiver solar power plant at present is the Ivanpah Solar Power Facility located at the base of Clark Mountain in the Mojave Desert in California, which became operational in 2014. This facility consists of three central tower receivers (Figure 9.8 on the page 300) and 173,500 heliostats with a total surface area of 2.44 km^2. The total capacity of the three units is 377 MW$_e$.

Figure 9.4: Solar Two in Daggett, California. The scale can be determined from the central tower that is about 90 m high.

Aerial Archives/Alamy Stock Photo

Figure 9.5: One of Solar Two's heliostats. The Solar Power Tower is reflected in the mirror.

DOE Photo/Alamy Stock Photo

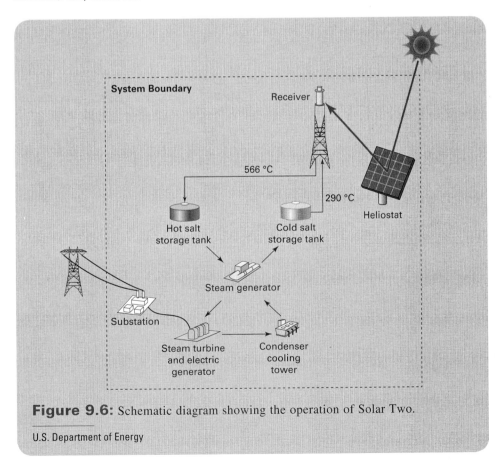

Figure 9.6: Schematic diagram showing the operation of Solar Two.

U.S. Department of Energy

Figure 9.7: The PS10 rated at 11 MW$_e$ (foreground) and PS20 rated at 20 MW$_e$ (background) solar generating stations at Sanlúcar la Mayor near Seville in Andalusia, Spain.

Figure 9.8: Three collector towers at the Ivanpah Solar Electric Generating Station in the Mojave Desert in California.

9.2d Solar Ponds

A solar pond is a saltwater pond, which may be natural or artificial, that provides a means of storing thermal energy from the absorption of solar radiation. The salt forms a natural vertical gradient in the pond, where denser high-salinity water sinks to the bottom while less dense low-salinity water floats on the top. The total salt content is adjusted so that the salt solution near the bottom of the pond is saturated.

When sunlight is incident on the pond, it passes through the water, is absorbed by the bottom, and is converted into thermal energy. This thermal energy heats the lower layer of the water. Normally this heated region of water at the bottom of the pond would rise (since the warmer water has a lower density) and create convection currents that would mix the more saline water with the less saline water above it. However, because the bottom region of the pond is a saturated salt solution, there is no salinity gradient there, and this inhibits convective mixing with the layers above. This feature traps the heat in the lower portion of the pond and creates a thermal difference between the bottom and top layers of the pond, where the bottom can typically reach temperatures of up to 90°C while the top may be at 30°C. While the thermal energy contained in the bottom regions of the pond can be used as a direct source of heat, the temperature difference between the bottom and top of the pond can be used to run a heat engine and generate electricity. Although the hot water does not boil to produce steam to operate a turbine directly, heat exchangers (Figure 9.9) can transfer the thermal energy of the hot water to

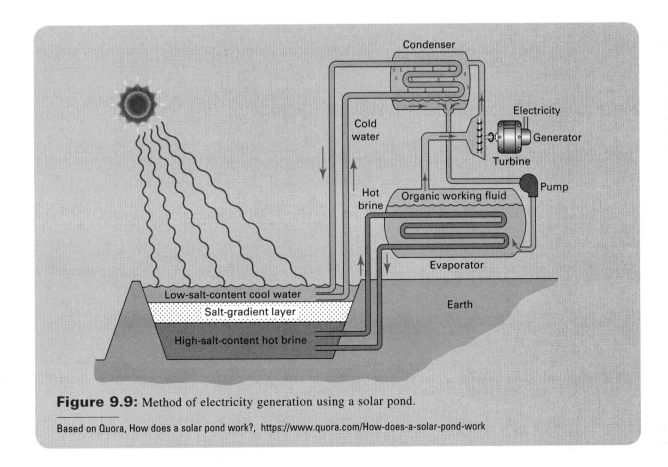

Figure 9.9: Method of electricity generation using a solar pond.

Based on Quora, How does a solar pond work?, https://www.quora.com/How-does-a-solar-pond-work

a working fluid with an appropriate phase diagram that can be used to drive a turbine/generator. The cold surface water serves as the cold reservoir of the heat engine, as the figure shows.

A number of experimental solar ponds have been constructed worldwide. The largest solar pond thus far was operated in Beit HaArava, Israel in the 1980s. It had an area of about 0.2 km^2 and produced an electrical output of 5 MW$_e$. An ongoing experimental solar pond (Figure 9.10) has operated in El Paso, Texas since 1987. It has an area of 3,200 m^2 and has been used as an experimental facility to improve solar pond technology.

One advantage of this approach to generating electricity from solar radiation is that the large thermal mass of the water in the pond evens out fluctuations between sunny and cloudy days and between day and night. On the other hand, the relatively small temperature difference between the hot and cold reservoirs (compared to that for most thermal generation schemes) means that the thermodynamic efficiency is low. Solar ponds are of particular interest in developing countries because the simple technology has relatively low infrastructure and operating costs.

Other approaches to the use of salinity gradients for the production of electricity are discussed in Chapter 14.

Figure 9.10: El Paso solar pond.

Huanmin Lu, Andrew H. P. Swift, Herbert D. Hein and John C. Walton "Advancements in Salinity Gradient Solar Pond Technology Based on Sixteen Years of Operational Experience" J. Sol. Energy Eng 126 (2004) 759–767 doi:10.1115/1.1667977. Courtesy Herbert D. Hein, Jr.

ENERGY EXTRA 9.1
Solar updraft towers

The solar updraft tower is an interesting approach to harnessing energy from the sun. The concept was first proposed in 1903 by Isidoro Cabanyes, a colonel in the Spanish Army. The general design is illustrated in the diagram. A large greenhouse is created by suspending a transparent sheet a few meters above the ground, as the illustration shows. The sun heats the air beneath the canopy, which is drawn through a set of turbines and up a tall chimney or tower. The air is drawn up the chimney because it has lower density (since it is warmer) than the air higher up in the atmosphere (see Figure 4.6). This phenomenon, called the chimney effect or stack effect, has been known for many years. This effect can be quantified, approximately, by the pressure difference, ΔP, driving the air through the turbines as

$$\Delta P = Cah\left[\frac{1}{T_{out}} - \frac{1}{T_{in}}\right],$$

where a is the atmospheric pressure, h is the height of the chimney, T_{out} is the temperature outside of the canopy, and T_{in} is the temperature inside the canopy. The constant a in SI units has the value 0.0342 K/m. This expression emphasizes the need for a tall chimney and an effective greenhouse effect to optimize the temperature difference between the inside and outside.

The solar updraft tower concept was demonstrated very successfully by a prototype device constructed in Spain in 1982. It consisted of a solar tower 195 m in height and 10 m in diameter. The greenhouse covered an area of 0.046 km². While the tower was designed as a three-year test project, it actually operated until 1989, when it collapsed due to rusting support wires. It produced a maximum electrical output of about 50 Kw, corresponding to about 1 W/m² of collector. Compared to an average insolation worldwide of 168 W/m², the experimental solar updraft tower achieved an efficiency of less than 1%.

Low efficiencies are, in general, a drawback of solar updraft towers. Simulations have predicted efficiencies of, at most, a few percent, much less than efficiencies of 15% or more for photovoltaics

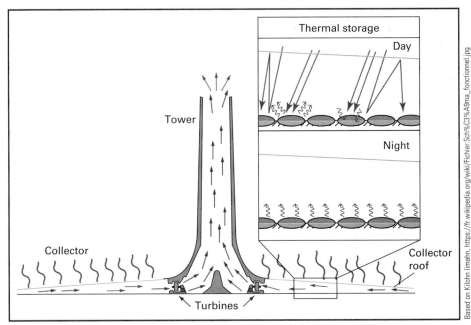

The operational concept of the solar updraft tower.

Based on Kilohn limahn, https://fr.wikipedia.org/wiki/Fichier:Sch%C3%A9ma_fonctionnel.jpg

Continued on page 304

Energy Extra 9.1 continued

Experimental solar updraft tower in Manzanares, Spain.

(see Sections 9.3 to 9.5). As well, infrastructure costs are high, and larger devices are expected to be beneficial from an economic standpoint. On the other hand, updraft towers utilize a relatively simple technology and are expected to have low operational costs.

A potential problem with solar electricity is the requirement for energy storage because of the

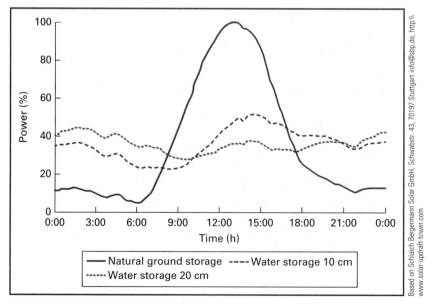

Schlaich Bergermann Solar GmbH, Schwabstr. 43, 70197 Stuttgart.

Continued on page 305

Energy Extra 9.1 continued

intermittent nature of the resource. For the solar updraft tower, the use of thermal storage can even out electricity generation, as the diagram illustrates. Containers of water cover the ground underneath the canopy. During the day solar energy heats both the air under the canopy, thereby generating electricity, and the water. At night the warm water transfers heat to the cooler atmosphere, providing a temperature differential to draw air through the turbines. This approach evens out the fluctuations in insolation during the day.

Topic for Discussion

A recent study has considered the possibility of constructing a 200 MW$_e$ solar updraft tower in Western Australia. The chimney would be 1 km high, and the greenhouse would have a diameter of 10 km. The estimated capital cost would be about $1.67 billion. Discuss the efficiency of the device based on its area, the average insolation, and the rated output. If operating costs are minimal, discuss the payback period, including the cost recovery factor, such that the price of electricity per kWh would be competitive with other renewable technologies.

9.3 Photovoltaic Devices

An alternative approach to the production of electricity from sunlight using a heat engine is *photovoltaics*. An attractive aspect of photovoltaics is the ability to implement this technology on a wide range of scales from milliwatt devices suitable for running watches and pocket calculators (Figure 9.11) to multimegawatt installations (Figure 9.12 on the next page). The operation of a photovoltaic device results from how light interacts with the electrons in a semiconducting material. The description of this behavior begins with an overview of some basic semiconductor physics.

Electrons that are associated with an atom may be described in terms of energy levels. For the hydrogen atom, the electron can occupy quantized levels with energies that are given by the expression

$$E = -\frac{me^4}{8\varepsilon_0^2 h^2 n^2},$$
(9.1)

Figure 9.11: Photovoltaic device (black area above the model number) used to power a pocket calculator. The 1.5-cm^2 cell produces about 10 mW$_e$ in bright sunlight.

Richard A. Dunlap

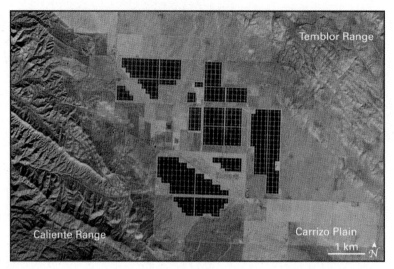

Figure 9.12: Image from space of the Topaz Solar Farm, a 550 MW$_e$ photovoltaic installation covering 25.6 km^2 on the Carrizo Plain in southern California. Note the distance scale on the lower right in the image.

NASA

where e is the charge on the electron, m is the mass of the electron, ε_0 is the permittivity of vacuum, h is Planck's constant, and n is an integer that defines the energy level. Equation (9.1) is sometimes written as $E = -hcR/n^2$, where R is the Rydberg constant ($R = 1.097 \times 10^7$ m^{-1}). Increasing values of n correspond to increasing energy [Figure 9.13(a)]. In an atom that contains many electrons, the interactions between the electrons cause the energy levels to split into sublevels [Figure 9.13(b)]. The sublevels are labeled s, p, d, f, and so forth, as shown in the figure, and correspond to different values of the orbital angular momentum. An s-sublevel can hold 2 electrons, a p-sublevel can hold 6, a d-sublevel can hold 10, an f-sublevel can hold 14, and so

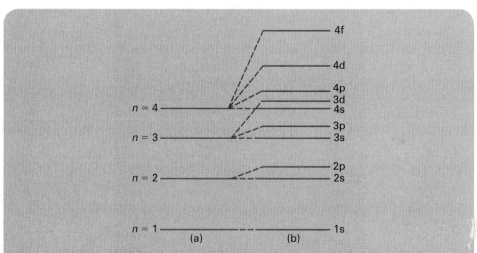

Figure 9.13: Energy levels in (a) hydrogen atom and (b) a multielectron atom.

Based on R. A. Dunlap, An Introduction to the Physics of Nuclei and Particles, Brooks Cole, 2003.

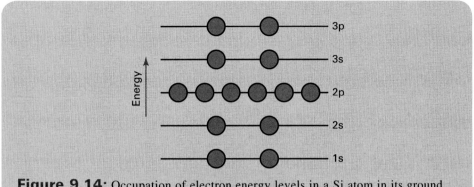

Figure 9.14: Occupation of electron energy levels in a Si atom in its ground state. The circles represent electrons.

on. Silicon (Si), which has 14 electrons, is a good example of a common semiconducting material. The electrons in their lowest energy configuration (ground state) can be expressed as

$$1s^2 2s^2 p^6 3s^2 3p^2, \qquad (9.2)$$

where 1s, 2s, 2p, and so on represent the energy sublevels, and the superscript after the level name gives the number of electrons in that level [Figure 9.14]. The two 3s electrons and two 3p electrons are not as firmly bound to the Si atom as the other electrons and are referred to as *valence electrons.*

If a large number of Si atoms are assembled to form a solid, then the interactions between the electrons associated with one atom and the electrons associated with another atom cause the energy levels to change a little—sometimes a bit lower, sometimes a bit higher. Thus, the energy levels get smeared out into *bands* (Figure 9.15). For a single Si atom, the 1s level can hold 2 electrons. For a piece of

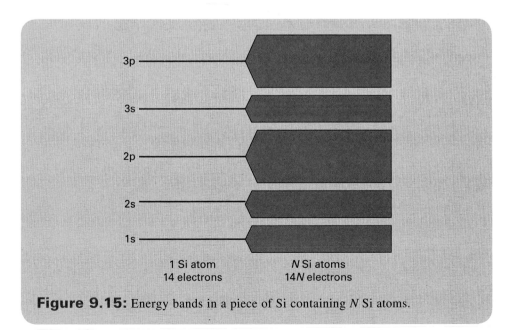

Figure 9.15: Energy bands in a piece of Si containing N Si atoms.

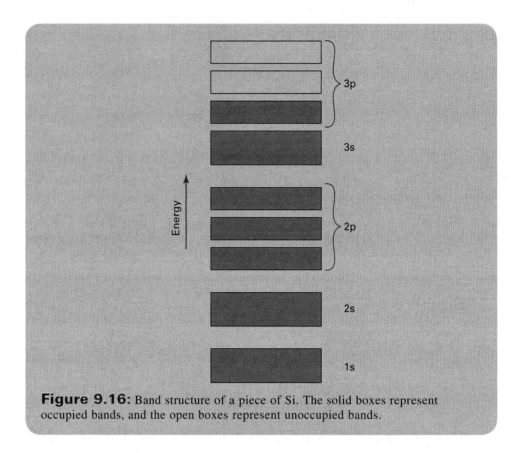

Figure 9.16: Band structure of a piece of Si. The solid boxes represent occupied bands, and the open boxes represent unoccupied bands.

Si containing N Si atoms, the 1s band can hold $2N$ electrons. Similarly, in the piece of Si, the 2s band holds $2N$ electrons, the 2p band holds $6N$ electrons, and so on. It turns out that the interactions in the solid also cause the 2p and 3p bands to split into three sub-bands, each of which can contain $2N$ electrons. Thus, the electrons in a piece of Si fill up the 1s, 2s, 2p, 3s, and the first of the three 3p sub-bands (Figure 9.16). The $2N$ electrons in the 3s band and the $2N$ electrons in the lowest of the 3p sub-bands are the valence electrons in a piece of Si containing N Si atoms. The second 3p sub-band in Si is called the *conduction band* and, as shown in the figure, is empty in the ground state (i.e., contains no electrons). The spaces (in energy) between the bands are regions where no energy levels exist and are forbidden energy regions for the electrons. The forbidden space between the valence band and the conduction band is called the *energy gap* (or *band gap*) and has a width corresponding to an energy E_g. The size of the energy gap depends primarily on the kind of material; for Si, it is 1.1 eV.

The valence electrons contribute to the bonding between the atoms in a piece of solid Si, as represented by the simple two-dimensional picture of the atoms in a Si lattice in Figure 9.17. Each Si atom shares its four valence electrons with its four neighbors.

It is now possible to describe how light interacts with electrons in a solid. As described in Chapter 1, light is quantized in the form of photons. Each photon has an energy related to the frequency of the light (the greater the frequency, the greater the

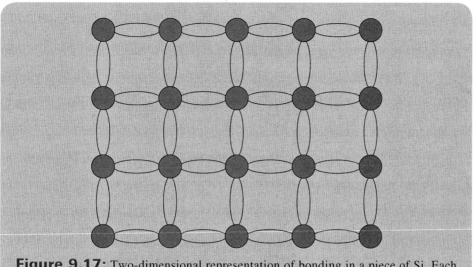

Figure 9.17: Two-dimensional representation of bonding in a piece of Si. Each of the lines between the atoms represents a 3s or 3p electron involved in bonding.

energy per photon [see equation (1.26)]. From equation (1.25), the energy per photon can be related to the wavelength of the light, λ, as

$$E = \frac{hc}{\lambda}.$$ **(9.3)**

In customary units, this is

$$E = \frac{1240 \text{ eV} \cdot \text{nm}}{\lambda},$$ **(9.4)**

where E is in eV, and λ is in nm. If a photon of sufficient energy is incident on a piece of Si, it can impart that energy to an electron in the valence band of the material and cause it to move into an energy level in conduction band (Figure 9.18 on the next page). When the electron moves to the conduction band, it leaves a vacant energy state in the valence band. This vacant state has an effective positive charge and is referred to as a *hole*. To form the electron-hole pair, the photon must have energy greater than the energy gap. In Si, the energy gap of 1.1 eV corresponds [according to equation (9.4)] to a wavelength of 1130 nm. From Figure 8.1, it can be seen that most of the solar spectrum (including the entire visible region) satisfies this condition. A simplified picture of the lattice, illustrating electron-hole pair formation, is shown in Figure 9.19 on the next page. The electron formed in this way is not bound to any particular atom and is free to move about in the material. Similarly, the hole can move about by exchanging places with one of the valence electrons associated with the other Si atoms in the material. If a number of photons are incident on a piece of Si and form a number of electron-hole pairs, then these electrons and holes can move through the material and constitute an electric current. In this way, the energy associated with the photons in the radiation can be converted into electrical energy.

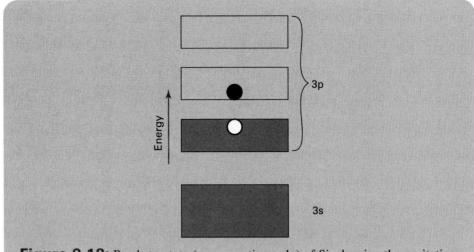

Figure 9.18: Band structure (upper portion only) of Si, showing the excitation of an electron from the valence band to the conduction band caused by the absorption of a photon of sufficient energy.

The preceding description may suggest a means of making a photovoltaic cell, but this approach presents a major problem. As the electron is negatively charged, the hole is effectively positively charged, and they can both move about in the material; these unlike charges attract and at some point are likely to combine and cancel each other out. This is called *recombination* and corresponds to a free electron in the conduction band (Figure 9.18), losing energy and falling across the energy gap to fill a hole in the valence band. This can be seen in the picture of the lattice in Figure 9.19, where an electron can fill a broken bond, thus eliminating both the free electron and the hole.

The design of a functional photovoltaic cell requires the elimination of recombination (as much as possible). Understanding how this may be done requires a

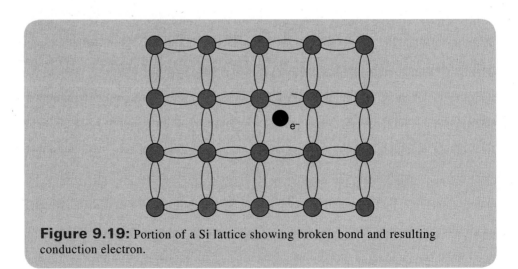

Figure 9.19: Portion of a Si lattice showing broken bond and resulting conduction electron.

Example 9.2

Estimate the range of photon energies that corresponds to the visible portion of the solar spectrum.

Solution

Figure 8.1 shows that the visible portion of the solar spectrum covers wavelengths between about 400 nm and 700 nm. From equation (9.4), these wavelengths correspond to the following energies:

For 400 nm: $\dfrac{1240 \text{ eV} \cdot \text{nm}}{400 \text{ nm}} = 3.1$ eV at the violet end of the visible spectrum.

For 700 nm: $\dfrac{1240 \text{ eV} \cdot \text{nm}}{700 \text{ nm}} = 1.8$ eV at the red end of the visible spectrum.

consideration of the effects of impurities in a semiconducting material. The behavior of an impurity phosphorus atom in a silicon lattice is considered first. Phosphorus has 15 electrons in the configuration $1s^2 2s^2 2p^6 3s^2 3p^3$. When a phosphorus atom replaces a silicon atom, the two 3s electrons from the valence band of the phosphorus atom take the place of the two 3s electrons from the valence band of the silicon atom that was removed. Two of the phosphorus 3p electrons take the place of the two Si 3p electrons, but this leaves one phosphorus 3p electron left over. There is no place for this to go in the silicon valence bands because they are filled up, so it goes into the conduction band [Figure 9.20(a)]. The difference between this picture and Figure 9.18 is that, in the present case, an electron appears in the conduction band

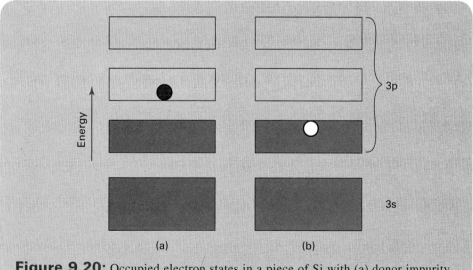

Figure 9.20: Occupied electron states in a piece of Si with (a) donor impurity and (b) acceptor impurity.

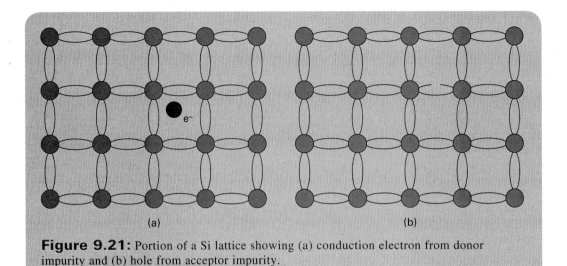

Figure 9.21: Portion of a Si lattice showing (a) conduction electron from donor impurity and (b) hole from acceptor impurity.

without the corresponding creation of a hole in the valence band. This situation is illustrated in Figure 9.21(a), where the phosphorus atom loses one of its localized valence electrons, and that electron is free to move about in the material. This leaves the phosphorus atom (now a positive phosphorus ion) with a missing electron and a corresponding positive charge. It is important to note that, as a whole, this material is electrically neutral because it is made up of atoms that were originally neutral. However, it is different than, say, a piece of pure silicon because a negatively charged electron is free to move about in the material, and a corresponding positively charged ion is fixed in its location in the lattice. The phosphorus atom is referred to as a *donor* because it gives up, or donates an electron to the conduction band, and the material is referred to as an *n-type* semiconducting material because there are free negatively charged electrons in the material.

In contrast, an aluminum impurity in silicon behaves differently. Aluminum has 13 electrons in the configuration $1s^2 2s^2 2p^6 3s^2 3p^1$. Thus there is one electron too few to fill up the 3s band and the first of the 3p valence bands. The lack of an electron results in the appearance of a hole in the 3p valence band without the corresponding appearance of an electron in the conduction band. This situation [Figure 9.20(b)] is in contrast to the electron-hole pair formation shown in Figure 9.18. Figure 9.21(b) shows the resulting formation of a hole associated with a missing bond in the silicon lattice. In general, this mobile hole moves through the material by exchanging places with localized electrons that form bonds between the silicon atoms. The movement of the hole means that the aluminum atom (now a negative aluminum ion) binds up an additional electron and becomes a negative ion. Again, the material is electrically neutral overall, but there will be a positive hole that can move about freely and a fixed negative ion to compensate for it. Aluminum in silicon is called an *acceptor*, and this material is a *p-type* material.

Semiconductors with impurities that behave as just described are referred to as *doped* semiconductors, compared with pure semiconducting materials, which are referred to as *intrinsic* semiconductors. Virtually all semiconducting devices

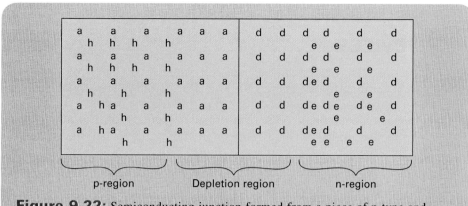

Figure 9.22: Semiconducting junction formed from a piece of p-type and n-type semiconducting material showing acceptor impurities (a), holes (h), donor impurities (d), and electrons (e).

Based on R. A. Dunlap, Experimental Physics. Modern Methods. Oxford University Press Inc, 1988

are constructed from combinations of n-type and p-type materials. The simplest arrangement of n-type and p-type materials is the semiconducting junction (or *diode*) (Figure 9.22), where a n-type material is in electrical contact with an p-type material. The charges in the system are shown in Figure 9.22: the negatively charged acceptor ions, the positively charged donor ions, the negatively charged electrons, and the positively charged hole. The neutral silicon atoms are not shown. Recall that the charged electrons and holes are free to move around, and the charged acceptor and donor ions are fixed in the lattice. The negatively charged electrons in the n-type material tend to move away from the junction because they are repelled by the negatively charged acceptor ions on the other side. Similarly, the positively charged holes in the p-type material tend to move away from the junction because they are repelled by the positively charged donor ions on the other side. Thus, as shown in the figure, a region around the junction (called the *depletion region*) is formed where there are no free electrons or holes.

If a photon of sufficient energy is incident on the depletion region, it can create an electron-hole pair as just described for pure silicon. The electron is repelled from the p-type material and attracted to the n-type material, moving to the n-type material and joining the other electrons in the region on the right of the figure. Similarly, the hole is repelled from the n-type region and attracted to the p-type region, and it joins the other holes on the left of the figure. Thus, the properties of the junction separate the electrons and the holes formed from incident electromagnetic radiation and reduce the probability that they will recombine. This movement of electrons and holes results in a net current flow across the junction, this electric current can be provided to an external device connected to the junction. This is the basis of operation for the *photodiode*, or *photovoltaic cell*. The spatial extent of the photosensitive region can be increased by creating the equivalent of a large depletion region (i.e., charge carrier–free region) by sandwiching an intrinsic (pure) semiconducting layer between the p-type and n-type regions.

9.4 Application of Photovoltaic Devices

Because the operation of the photovoltaic cell requires that the photons that are incident on it have sufficient energy to produce electron-hole pairs, there is, as described, a maximum wavelength of light that produces this effect. The spectral response for some photovoltaic cells is illustrated in Figure 9.23. The cutoff at long wavelengths is due to the fact that the photon energy for longer wavelengths is less than the energy gap and is insufficient to create electron-hole pairs. Figure 8.1 shows that about 23% of the photons received from the sun do not have sufficient energy to be converted to electricity by a Si photovoltaic cell. In addition, much of the electromagnetic energy is converted into heat rather than electricity. The spectral response of a photovoltaic cell can be improved by utilizing a semiconducting material with a smaller energy gap, thus enabling more of the solar spectrum to be effective in producing electricity. This is seen in Figure 9.23 by the longer wavelength corresponding to the cutoff for germanium (E_g = 0.67 eV) compared with Si (E_g = 1.1 eV).

The maximum theoretical efficiency of photovoltaic cells based on different semiconducting materials is summarized in Table 9.2. Typical efficiencies for Si-based cells are around 15–17%, with higher quality cells giving up to about 23%. Techniques such as concentrating the light and splitting the light into different spectral components

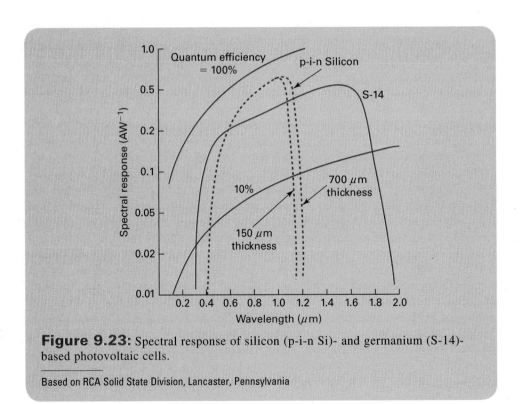

Figure 9.23: Spectral response of silicon (p-i-n Si)- and germanium (S-14)-based photovoltaic cells.

Based on RCA Solid State Division, Lancaster, Pennsylvania

Example 9.3

Calculate the cost per kWh averaged over an operational period of 10 years for a photovoltaic system with an installation cost of $3.00 per W if the system operates with a capacity factor of 8% (this is the net result of the photovoltaic efficiency and the fraction of time that is daylight). (Do not include the cost recovery factor.)

Solution

Ten years correspond to $(10 \text{ y}) \times (365 \text{ d/y}) \times (24 \text{ h/d}) = 87{,}600 \text{ h}$. Over this time period, one watt of photovoltaic installation, operating at an 8% capacity factor, produces a total of

$$87{,}600 \text{ h} \times 0.08 \text{ W/W} = 7008 \text{ Wh/W} = 7.0 \text{ kWh/W}.$$

At a total cost of $3.00 per W, this corresponds to

$$\frac{\$3.00/\text{W}}{7.0 \text{ kWh/W}} = \$0.43 \text{ per kWh}.$$

that are incident on cells with specific spectral responses have yielded efficiencies in excess of 40%.

Photovoltaic cells have been implemented on a variety of size scales. These may be considered in two different categories: off-grid and on-grid. These designations refer to systems that are not connected to the power utility grid and those that are, respectively.

Table 9.2: Energy gaps and maximum theoretical efficiency for solar radiation of some semiconducting materials for photovoltaic cell construction. Values are for room temperature. Some elemental components of these materials are toxic, such as arsenic and cadmium, and the world's resources of indium and gallium may be insufficient for large-scale use (Chapter 2).

material	E_g (eV)	maximum theoretical efficiency (%)
CdS	2.42	18
AlSb	1.6	27
CdTe	1.49	27
GaAs	1.43	26
InP	1.35	26
Si	1.11	25
Ge	0.67	13

ENERGY EXTRA 9.2
Triple-junction photovoltaic cells

Most commercial photovoltaic cells have efficiencies for converting energy from solar radiation into electrical energy of 12–18%. A number of factors affect efficiency, but one of the most important is the spectral sensitivity of the semiconducting material used in the cell (Figure 9.23). The principle semiconductor property that governs the spectral response is the energy gap. Numerous semiconducting materials are known, and the energy gap (and hence the spectral response) of a material can be tuned by adjusting the composition of the material. The problem is that, if the energy gap is tuned to give the material a high efficiency for photons in the red end of the electromagnetic spectrum, then the efficiency at the blue end of the spectrum is sacrificed, and vice versa.

One approach to improving the efficiency of a photovoltaic cell over a broad spectral range is to make a sandwich (a multijunction) of materials that are sensitive over different portions of the spectrum. A triple junction consisting of one semiconducting material that is the most sensitive to the red photons, another that is the most sensitive to green photons, and a third that is the most sensitive to blue photons has been very successful in creating photovoltaic cells with overall high efficiencies to the solar spectrum.

An example of the triple-junction design is shown in the top right figure. In this example, the top layer is made of amorphous silicon (the designation "i a-Si" in the figure refers to a layer of intrinsic amorphous silicon sandwiched between layers of p-type and n-type silicon). The i a-Si layer has an energy gap of about 1.8 eV and is most sensitive to photons in the blue region of the solar spectrum. The middle layer is intrinsic amorphous silicon-germanium (i a-SiGe), with about 80% Si and 20% Ge (again sandwiched between p-type and n-type regions). This material has an energy gap of about 1.6 eV and is the most sensitive to photons in the green portion of the spectrum. The bottom layer is also i a-SiGe with about 60% Si and 40% Ge. This

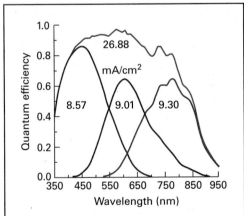

Geometry of triple-junction photovoltaic cell.

composition reduces the energy gap to about 1.4 eV and makes the material the most sensitive to red photons. The ordering of the layers is important because the top layer utilizes the highest-energy

Spectral responses of the three layers of a triple-junction and the total response.

S. Guha, "High-Efficiency Triple-Junction Amorphous Silicon Alloy Photovoltaic Technology" National Renewable Energy Laboratory, Report NREL/SR-520-26648 (1999)

Continued on page 317

Energy Extra 9.2 continued

photons and allows the lower-energy photons to propagate through to the layers beneath it, where they are utilized the most efficiently.

The bottom right figure on the previous page shows the spectral responses of the three layers; blue to red photons correspond to the horizontal axis from left to right in the figure. The total spectral response is illustrated by the total curve in the figure and shows a high efficiency over a wide range of photon energies (or wavelengths).

This approach has been successful in producing photovoltaic cells with overall efficiencies in the range of 40% (more than twice the efficiency of typical commercial silicon-based cells). Unfortunately, the cost of producing these cells,

at least at the present, makes them economically viable only for specific applications. These applications include space vehicles, where weight is a more important consideration than cost, and devices where radiation is focused onto a small area and the solar cell size is small compared to the total collector area.

Topic for Discussion

If triple-junction photovoltaic (PV) devices could be produced at the same cost as simple Si-based devices, how would the cost of PV-generated electricity compare to wind, nuclear, and fossil fuels?

Off-grid photovoltaic cells typically fall into one of the following categories:

- Small units used to recharge batteries in portable electronic devices (Figure 9.11).
- Medium-sized units used for camping, emergency battery charging (Figure 9.24), portable road signs (Figure 9.25 on the next page), or power sources in remote areas (Figure 9.26 on the next page).
- Larger installations used for residential electric power (Figure 9.27 on the next page).

Individual cells are fairly small (typically less than about 10 cm on a side), but they can be combined in series (to increase voltage) and/or in parallel (to increase current). Individual cells can be rectangular (Figure 9.24) or circular (Figure 9.25).

The Solar Settlement in Freiberg, Germany (Figure 9.27) is a sustainable model community designed to produce net zero carbon emissions. Many of the homes, as seen in the figure, incorporate photovoltaic arrays to provide electricity.

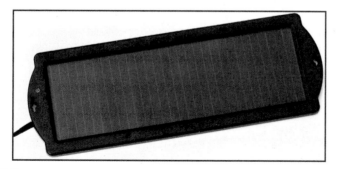

Figure 9.24: Portable 1.8-W_e solar panel. The dimensions of the active area are approximately 9 cm × 30 cm.

Richard A. Dunlap

Figure 9.25: Portable road sign at construction site that utilizes a photovoltaic array (about 1 m × 2.5 m) to charge batteries.

Richard A. Dunlap

Figure 9.26: Photovoltaic array for providing power at a remote radio transmitter station.

Richard A. Dunlap

Figure 9.27: Photovoltaic solar panels on homes in the ecological Vauban district in Freiburg, Baden-Wuerttemberg, Germany.

imageBROKER/Alamy Stock Photo

ENERGY EXTRA 9.3
Dye-sensitized solar cells

Clearly, the major problem with electricity generated by photovoltaic devices is the cost per kWh. One approach to dealing with this difficulty is to produce photovoltaic cells with higher efficiency. This, of course, has the advantage that the power generated per unit land area is greater. However, because more efficient photovoltaic cells are more expensive, this is not an obviously viable trade-off between productivity and price. The opposite approach is to design cells that are as inexpensive as possible without being so concerned with efficiency, thus taking a different approach to photovoltaic materials and photocell design. Dye-sensitized solar cells (DSSCs) are a type of organic photovoltaic cell that has attracted considerable interest in recent years. These cells utilize a very different method of cell construction and manufacture that may lead to innovative and practical devices.

A traditional photovoltaic cell consists of a sandwich of an n-type material and a p-type material between two conducting electrodes. When a photon of sufficient energy (i.e., greater than the energy gap) enters into the region near the semiconducting junction, an electron-hole pair can be formed, resulting in a current through an external circuit. The geometry is somewhat limited because the electron and hole must reach the electrode before recombining. However, the photosensitive region can be made larger and, in principle, more economical, by using the heterojunction design. In a heterojunction device, the semiconducting region consists of a mixture of different materials as shown in the following figure. In a DSSC, the heterojunction is based on a piece of nonporous TiO_2. TiO_2 has a relatively poor efficiency for creating electron-hole pairs in response to photons because of its large energy gap. However, the pores in TiO_2 are filled with a very photosensitive organic dye. The top electrode is often made of a halide-doped tin oxide that is both optically transparent and conducting. This forms the

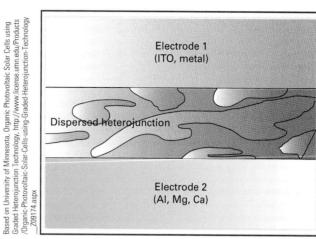

Schematic of dispersed heterojunction photovoltaic cell.

electrical connection to the cell and allows light to enter into the photosensitive region. If the size of the dispersed grains in the heterojunction is comparable to the charge carrier diffusion length, then the charges formed in response to a photon have a good chance of reaching a semiconducting interface before recombination takes place.

The earliest DSSCs produced in the mid-1990s were primarily sensitive only to the high-energy (UV) part of the solar spectrum. In recent years, materials (dyes) that are sensitive to the entire optical spectrum have been developed. Because these materials are effective at absorbing all optical wavelengths,

Structure of triscarboxy-ruthenium terpyridine (black dye).

Continued on page 320

Based on University of Minnesota, Organic Photovoltaic Solar Cells using Graded Heterojunction Technology, http://www.license.umn.edu/Products /Organic-Photovoltaic-Solar-Cells-using-Graded-Heterojunction-Technology _Z09174.aspx

Energy Extra 9.3 continued

they appear black and are thus referred to as *black dyes*. The most common black dyes are based on organic ruthenium-containing complexes as shown in the figure on the previous page at bottom right.

DSSCs have been produced with efficiencies in the range of 11%. This is less than the efficiency for traditional Si-based semiconducting photovoltaic cells, which can be close to 20% efficient, and the element ruthenium is expensive. However, the hope is that new, economically viable materials (that do not include ruthenium) may be developed and that this approach to photovoltaic cell design may yield devices that are competitive in terms of cost per kW/m^2.

Topic for Discussion

The problem of the cost and availability of materials is prevalent when dealing with the development of new technologies. The use of ruthenium in DSSCs is only one example of a great need to develop new materials that have competitive properties but that are less expensive and more readily available. Find information about the market value of ruthenium and the quantity produced each year. An excellent website dealing with elemental resources that has useful information and links is http://minerals.usgs.gov /minerals/pubs/commodity/. Discuss the importance of considering how science, technology, and economics must all play a role in designing our future.

Experimental automobiles (Figure 9.28) and even airplanes (Figure 9.29) that are powered by photovoltaic cells have been constructed. Although these vehicles are interesting engineering challenges, the power density in sunlight is insufficient to make them practical.

On-grid photovoltaic facilities exist in a number of countries and provide electric power for distribution through the power grid. A typical system for integrating a photoelectric array with the power grid is illustrated in Figure 9.30. The thermal generator (i.e., coal or nuclear generating facility) provides the primary AC power to the grid. DC power from the photovoltaic array is stored in a battery system. DC power from the photovoltaic array and from the battery storage system is converted to AC by the inverter. Voltage is matched to the grid power from the thermal generator by a step-up transformer. The synchronizing breaker connects the power from the photovoltaic array to the grid when the voltages are matched in frequency, amplitude, and phase.

Typically, large photovoltaic facilities have rated capacities of hundreds of MW_e (see Figure 9.12). At present the world's largest photovoltaic facility is the 850 MW_e

Figure 9.28: Photovoltaic-powered automobile.

Stefano Paltera, the NASC's official photographer

Figure 9.29: Photovoltaic-powered airplane.

NASA photo by Nick Galante/PMRF

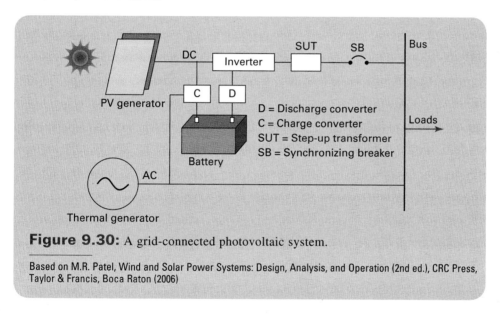

Figure 9.30: A grid-connected photovoltaic system.

Based on M.R. Patel, Wind and Solar Power Systems: Design, Analysis, and Operation (2nd ed.), CRC Press, Taylor & Francis, Boca Raton (2006)

Longyangxia Dam Solar Park in Gonghe, Qinghai, China. This facility was constructed in two phases, with an initial 320 MW_e installation completed in 2013 and an additional 530 MW_e installation completed in 2015. The largest photovoltaic generating station in the United States is presently the Solar Star Power Plant in California, completed in 2015 with a rated capacity of 579 MW_e (see Figure 9.31 on the next page).

The cost of photovoltaics has decreased considerably in recent years, as Figure 9.32 on the page 323 shows. These values are for the cost per watt of photovoltaic cell, and it should be noted that the cost for a complete installation would also include a suitable electrical storage method (i.e., batteries) and appropriate electronics (e.g., inverters) as discussed later in the present chapter.

Example 9.4

Using the average (cloud-free) insolation on earth (Chapter 8), calculate the area of a horizontal photovoltaic array with an efficiency of 20% that would be needed to satisfy the residential electric needs of a city of 250,000 people with an average of 2.6 people per household.

Solution

This city would consist of $250,000/2.6 = 9.6 \times 10^4$ homes. Based on the discussion in Chapter 2, we estimate the average power requirement per home as 1370 W. The total requirement for the city would be

$$(9.6 \times 10^4) \times (1370 \text{ W}) = 1.32 \times 10^8 \text{ W}.$$

The average cloud-free insolation averaged over the year and at all locations on earth is (from Chapter 8) 168 W/m^2. Thus the city's electrical needs would require (at a 20% efficiency)

$$\frac{1.32 \times 10^8 \text{ W}}{0.2 \times 168 \text{ W/m}^2} = 3.9 \times 10^6 \text{ m}^2,$$

or a square array about 2 km on a side.

The cost of photovoltaics may be viewed as primarily a capital cost with little or no operating costs compared with, for example, coal power, which requires a continuous supply of fuel. However, the capital costs of a photovoltaic system must be amortized over its expected lifetime to determine the actual cost per unit energy (Chapter 2). The life expectancy of a photovoltaic cell is probably about 20 years, although the life expectancy for the batteries in the storage system might be less. Table 9.3 summarizes the

Figure 9.31: Solar photovoltaic panels at the 579 MW$_e$ Solar Star Power Plant in Los Angeles County, California. In contrast to most other large photovoltaic power stations, Solar Star uses single-axis tracking collectors rather than fixed collectors.

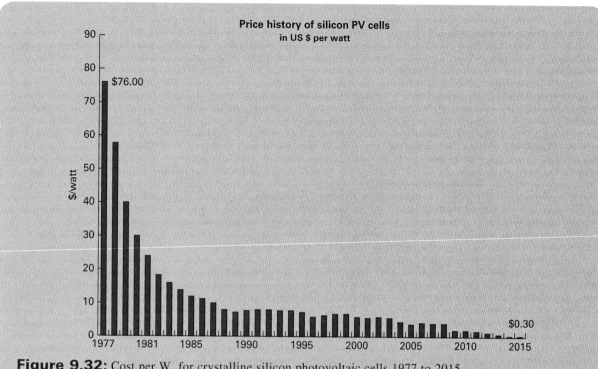

Figure 9.32: Cost per W_e for crystalline silicon photovoltaic cells 1977 to 2015.

Based on data: 1977–2013: Bloomberg, New Energy Finance, (archived) 2014: based on average sales price of $0.36/watt on 6 June 2014 from EnergyTrend.com 2015: based on average sales price of $0.30/W on 29 April 2015 from EnergyTrend.com

Table 9.3: Projected cost per MWh generated for utility scale generating stations entering service in 2022. Government subsidies are not included in the estimates.	
technology	**US$/MWh**
solar-thermal	235.9
wind-offshore	158.1
advanced coal	139.5
natural gas combustion turbine	110.8
nuclear	102.8
solar-photovoltaic	84.7
hydroelectric	67.8
wind-onshore	64.5
natural gas-combined cycle	58.1
geothermal	45.0

Data adapted from U.S. Energy Information Administration, http://www.eia.gov/outlooks/aeo/pdf/electricity_generation.pdf

estimated cost per unit energy for photovoltaics compared with other energy production methods. These estimates from the United States Energy Information Administration represent the estimated cost per MWh generated by stations entering service in the United States in 2022. The unusually high value for electricity produced from coal (compared to the low value for currently operating power stations) results from the inclusion of 30% carbon sequestration. The table indicates that photovoltaics are expected to compete favorably with other energy technologies in the relatively near future.

9.5 Global Use of Photovoltaics

Installed photovoltaic capacity has increased dramatically worldwide in recent years, as Figure 9.33 illustrates. This increase resulted from increasing concern over diminishing fossil fuel resources and their environmental impact, coupled with the decreasing cost of photovoltaic cells (Table 9.3). On the basis of current trends, growth of photovoltaic capacity is expected to continue. Table 9.4 summarizes the total installed capacity of the countries with more than 2 GW_e photovoltaic capacity. China and Japan have seen the greatest growth in absolute terms in recent years. However, Germany has maintained the greatest per-capita capacity in recent years.

The longevity of sunlight as an energy source is, for all practical purposes, infinite, but it is necessary to consider its availability. From equation (8.2), the total insolation on the outside of the atmosphere of the earth is 1.8×10^{17} W. On average, about half of this is transmitted through the atmosphere, giving a total insolation at the surface of 9×10^{16} W. Considering a modest photovoltaic efficiency of 15%, this gives the potential for 1.3×10^{16} W_e from photovoltaic generation worldwide. In the Preface, it was seen that the total primary energy use worldwide is 6.0×10^{20} J per year for an average power

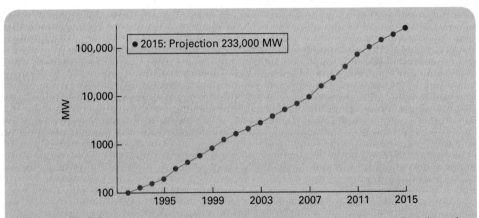

Figure 9.33: Cumulative installed world photovoltaic capacity as a function of year since 1992. Note the logarithmic vertical scale.

Based on data from: 1992-1995: Wikipedia article growth of photovoltaics–collected figures of 16 main markets, including Australia, Canada, Japan, Korea, Mexico, Western European countries, and the United States. 1996–1999: BP-Statistical Review of world energy, 2014 (XLS-spread-sheet). 2000-2013: SPE (former EPIA) Global Market Outlook for Photovoltaics (Report, June 2014, archived). 2014: Snapshot of Global PV 1992–2014 (archive) and IHS estimates/projections for worldwide deployment in 2014 and 2015 (March 2015). 2015 (projection only). Averaged from several different forecasting groups, companies and organizations. Data listed in article Growth of photovoltaics

Table 9.4: Total photovoltaic capacity and capacity per capita as of 2015 for countries with greater than 2 GW$_e$ capacity.

country	total capacity (MW$_e$)	population (10^6)	capacity per capita (W$_e$)
China	43,530	1376.0	31.6
Germany	39,700	80.7	491.9
Japan	34,410	126.6	271.8
United States	25,620	321.8	79.6
Italy	18,920	59.8	316.4
United Kingdom	8780	64.7	135.7
France	6580	64.4	102.2
Spain	5400	46.1	117.1
Australia	5070	24.0	211.3
India	5050	1311.1	3.9
South Korea	3430	49.8	68.9
Belgium	3250	11.3	287.6
Greece	2613	11.0	237.5
Canada	2500	35.9	69.6
Czech Republic	2083	10.8	192.9

consumption of $(6.0 \times 10^{20} \text{ J/y})/(3.156 \times 10^7 \text{ s/y}) = 1.9 \times 10^{13}$ W. Thus, the utilization of only about 0.15% of the available solar energy would fulfill all of our energy needs. Certainly, the use of photovoltaic arrays would be impractical or uneconomical at some locations on earth. Terrestrial locations would probably be more practical than most of the oceans, while northern or southern latitudes would provide very low efficiency. The fraction of land area necessary for solar energy to provide all energy needs can be considered on a country-by-country basis. Population density and per-capita energy use, as well as latitude and climate conditions, would be the relevant factors. A comparison between the United States and Canada can be considered as an example, and a summary of this analysis is given in Table 9.5. The insolation for the United States shown in Figure 8.2 has a typical value of about 200 W/m^2. For Canada, it is assumed that solar collectors would be located in fairly southern areas where the average insolation is in the range of about 150 W/m^2. The table shows that a relatively small amount of the land area of the United States and Canada would need to be dedicated to the production of electricity by solar photovoltaics in order

Table 9.5: Factors determining the fraction of land area needed to provide all primary energy needs by the use of photovoltaics in Canada and the United States. Photovoltaic efficiency of 15% has been assumed.

country	total annual energy use (J)	average power consumption (W)	typical insolation (W/m^2)	land area needed (km^2)	total land area (km^2)	% land area needed
United States	1.0×10^{20}	3.2×10^{12}	200	1.1×10^5	9.63×10^6	1.1
Canada	1.3×10^{19}	4.1×10^{11}	150	1.8×10^4	9.98×10^6	0.18

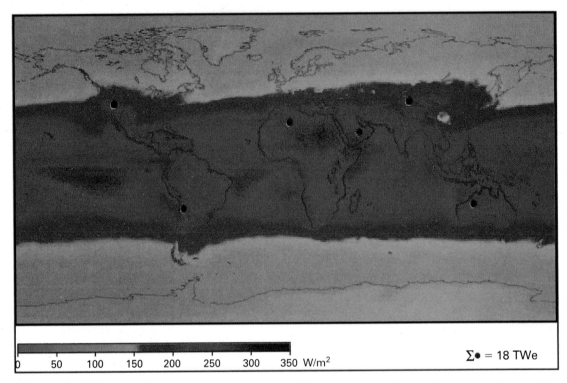

Figure 9.34: Land area needed to supply the world's energy needs from solar. Average insolation is illustrated.

Matthais Loster, http://www.ez2c.de/ml

to satisfy all primary energy requirements. For the United States, a square about 300 km on a side would be required; that is an area about the size of Ohio. For Canada, a square about 135 km on a side (about one third the area of Nova Scotia) would be required. A number of smaller facilities with an equivalent total area would likely be more practical than a single large facility. About six 300 km × 300 km facilities located worldwide in relatively sunny, low-population-density areas as shown in Figure 9.34 (or a comparable number of smaller facilities), would satisfy all of the world's energy needs.

Solar energy is the only single-energy resource that has the capability to provide enough energy to fulfill all of our needs and that is indefinitely renewable. The technology for doing this is within our means but would require that we adapt our energy infrastructure for the exclusive use of electricity as an energy source. The continued availability of (relatively) inexpensive fossil fuels does not provide the financial incentive for this development.

9.6 Summary

Solar radiation is the only source of nonfossil fuel energy that is plentiful enough to fulfill all of society's energy requirements, both for the present and for the foreseeable future. Although the direct use of solar energy for applications such as space heating contribute to energy needs on a residential heating scale, it is the production of electricity from solar radiation, as described in this chapter, for distribution over the grid, that has the potential to satisfy the bulk of our energy requirements in the future. The ways in which solar energy can be used to generate electricity were described. There are basically two approaches: heat engines and photovoltaics. This chapter discussed systems for focusing sunlight onto a small area where it is used to heat a working fluid. The fluid is then used to run a heat engine to produce mechanical energy, which then generates electricity by means of a generator. Experimental facilities use one of three geometries: parabolic troughs, parabolic reflectors, or central receivers. The efficiency of these devices is ultimately limited by the Carnot efficiency.

This chapter provides an introduction to the basic semiconductor physics needed to describe the operation of a photovoltaic device. Both n-type and p-type semiconducting materials are formed by the inclusion of donor or acceptor impurities in a host material. These materials have either an excess electron in the conduction band or an excess hole in the valence band, respectively. A junction between an n-type and a p-type material is sensitive to solar energy and constitutes the basic design of a photovoltaic cell. Photons incident on the junction produce electron-hole pairs that provide a current through an external circuit, thus providing usable electrical energy.

Different semiconducting materials are sensitive to different regions of the electromagnetic spectrum, and this is a major factor in limiting the efficiency of photovoltaic devices. There is a trade-off between efficiency and economy. High-efficiency devices can have efficiencies of about 40% for conversion of solar energy to electrical energy, but they are expensive. More economical devices have efficiencies of about 18%.

Although the photovoltaic installations themselves would seem to be environmentally neutral, the overall environmental concerns of solar electricity are complex. The manufacturing processes needed to produce photovoltaic materials are intensive (per installed watt power capability) because the energy density of sunlight is low, and these processes must be included in an overall evaluation of environmental impact. Some elements used in photovoltaic cells are toxic, and some, such as indium, are of limited worldwide availability. All of these factors contribute to the present high cost of solar electricity, have possible implications for future developments, and must be considered in any overall analysis of the viability of photovoltaics. However, solar photovoltaics are still an attractive alternative to diminishing fossil fuel supplies, and there has been a substantial increase in their use in recent years. Germany has a very ambitious development program in this area and has been the leader in implementing this technology for providing electricity to the grid. The development of photovoltaic cells with good efficiency that can be manufactured economically using low–cost, nontoxic components is an area of active research, and the future availability of such materials would benefit the growth of this renewable energy resource.

Problems

9-1 Consider a 1-m-wide parabolic trough reflector with a 1-cm-inside-diameter pipe filled with oil at its focus. If all of the solar radiation is converted into heat, estimate the time required at midday on a sunny day to raise the temperature of the oil from 20°C to 100°C.

9-2   Assume that all of our energy is obtained from solar by means of photovoltaic devices with an efficiency of 18% using horizontal panels. If all energy is produced in the United States on a state-by-state basis, calculate the percent of land area needed to supply energy in Rhode Island and Idaho. Use an average annual per-capita energy requirement of 350 GJ. Comment on the suitability of this approach.

9-3   A house at 40°N latitude has a roof of dimensions 8 m by 15 m. This area is covered with photovoltaic panels at the optimal fixed angle with an efficiency of 17%. Assuming average sky conditions, what is the annual energy production from this installation? If this electricity were generated by a coal-fired thermal plant operating at 38% efficiency, how much coal would be needed? On the basis of this comparison, what is the annual reduction (in kilograms) of CO_2 that results from this use of photovoltaics?

9-4   If all of the energy for humanity is generated using horizontal solar photovoltaic arrays with an efficiency of 15% and which occupy 1% of the land area of the earth, what is the maximum population density that can be supported?

9-5   Estimate the diameter of a mirror array associated with a solar power tower, as illustrated in Figure 9.6, that would replace a 1-GW_e coal-fired generating station. Be sure to consider the method by which electricity is generated, as illustrated in Figure 9.8. Discuss the practicality of such a design. Consider the relationship between the height of the central tower and the arrangement of mirrors near the outer portions of the array. Each mirror must have a clear view of the central receiver that is not obscured by the mirror in front of it. Does this place a restriction on the size of the array in the context of the tower height?

9-6   What is the maximum wavelength of light (in nanometers) that would produce an output from a photovoltaic device manufactured from a semiconductor with an energy gap of 1.02 eV?

9-7   What is the maximum wavelength of the light that produces voltage in a CdS photovoltaic device? Discuss this in terms of the spectrum of sunlight as presented in Chapter 8.

9-8   Consider a typical family vehicle powered by photovoltaics, similar to the vehicle shown in Figure 9.28, where the top surface is covered with 20% efficient cells. Compare the power available in such a vehicle on a sunny day in comparison with a typical gasoline-powered vehicle. Some typical vehicle characteristics are given in Table 19.5.

9-9   A metal plate is placed on a horizontal surface on a sunny day when the solar insolation is 500 W/m^2. Calculate its equilibrium temperature as a function of its emissivity, and make a plot of these results for values of the emissivity from $\varepsilon = 0.1$ to $\varepsilon = 1.0$.

9-10 If the focal area in Example 9.1 is made of a material with an emissivity of 0.9 and this loses heat only by radiation from the 5-cm-diameter area, calculate its equilibrium temperature.

9-11 Calculate the energy of an electron in the 1s energy level in a hydrogen atom.

9-12 What is the maximum frequency of light that will produce an output from a Ge-based photovoltaic cell?

9-13 The northern part of the United States, as Figure 8.2 shows, has a yearly 24-hour average solar insolation that is typical of the average value worldwide. Using this worldwide average, consider the economics of a photovoltaic installation. If the photovoltaic efficiency is 20% and a payback period of 3 years is expected (exclusive of the cost recovery factor), what is the maximum acceptable installation cost per W? Assume electricity costs $0.10 per kWh.

9-14 Compare the photovoltaic capacity for the United States given in Table 9.4 with the energy use shown in Figure 2.5. Comment on these results.

9-15 Using the energy of electrons in a hydrogen atom as a function of n, as given by equation (9.1), and following along the lines of the graph in Figure 9.13, determine all combinations of values of n for which electronic transitions will produce optical photons.

9-16 Water flows through a 5-cm-diameter pipe at the focus of a 2-m-wide parabolic trough reflector that is 5 m long at a rate of 0.2 L/s. If the insolation is 500 W/m^2 and all of the solar energy is converted into heat, what is the increase in the temperature of the water after it has traversed the length of the reflector?

9-17 A 15-m-diameter parabolic dish utilizes a Stirling engine for electricity generation. On a sunny day the dish receives an insolation that is 47% of the value of the solar constant. Estimate its output in W_e.

9-18 For a photovoltaic efficiency of 20% and an average peak insolation of 350 W/m^2, calculate the additional number of km^2 of solar collectors that would need to be installed in the United States to equal the per-capita photovoltaic capacity of Germany.

9-19 A photovoltaic panel with an efficiency of 19% receives an average insolation equal to the average worldwide average insolation on a horizontal surface. This panel is used with an appropriate storage system (100% efficient) to meet the average household electricity requirements (no electric heat) presented in Chapter 2.

(a) What is the required area of the panel?

(b) What is the annual savings in CO_2 emissions compared to the same amount of electricity supplied by 40% efficient coal-fired generating stations?

9-20 Assume that the portable solar panel shown in Figure 9.25 has an efficiency of 19%.

(a) What is the insolation needed to provide the rated output power?

(b) A camper connects this panel, in sunlight according to part (a), to an immersion heater in a 1-L container of water. How long will it take to increase the temperature of the water from 15°C to 100°C (without boiling)? Assume there is no heat loss.

Bibliography

M. D. Archer and R. Hill. *Clean Energy from Photovoltaics.* Imperial College Press, London (2001).

E. G. Boes and A. Lague. "Photovoltaic concentrator technology." In *Renewable Energy*, T. B. Johansson, H. Kelly, A. K. N. Reddy, and R.H. Williams (Eds.). Island Press, Washington, DC (1993).

G. Boyle (ed.). *Renewable Energy.* Oxford University Press, Oxford (2004).

P. DeLaquil, D. Kearney, M. Geyer, and R. Diver. "Solar-thermal electric technology." In *Renewable Energy*, T. B. Johansson, H. Kelly, A. K. N. Reddy, and R. H. Williams (Eds.). Island Press, Washington, DC (1993).

R. A. Dunlap. *Experimental Physics—Modern Methods.* Oxford University Press, New York (1988).

M. Green. *Solar Cells.* Prentice-Hall, Englewood Cliffs, NJ (1982).

R. Hinrichs and M. Kleinbach. *Energy: Its Use and the Environment* (5th ed.). Brooks-Cole, 2012.

R. J. Komp. *Practical Photovoltaics.* Aatec Publications, Ann Arbor, MI (1995).

J. A. Kraushaar and R. A. Ristinen. *Energy and Problems of a Technical Society* (2nd ed.). Wiley, New York (1993).

T. Markvart. *Solar Electricity* (2nd ed.). Wiley, Hoboken, NJ (2004).

J. A. Merrigan. *Sunlight to Electricity* (2nd ed.). MIT Press, Cambridge, MA (1982).

J. Nelson. *The Physics of the Solar Cell.* Imperial College Press, London (2003).

M. R. Patel. *Wind and Solar Power Systems: Design, Analysis, and Operation* (2nd ed.). Taylor & Francis, Boca Raton, FL (2006).

S. Roberts. *Solar Electricity: A Practical Guide to Designing and Installing Small Photovoltaic Systems.* Prentice-Hall, London (1991).

J. N. Shive. *Physics of Solid State Electronics.* Merrill, Columbus, OH (1966).

F. C. Treble (Ed.). *Generating Electricity from the Sun.* Pergamon Press, Oxford (1991).

F. C. Treble. Solar Electricity: A Lay Guide to the Generation of Electricity by the Direct Conversion of Solar Energy (2nd ed.). The Solar Energy Society, Oxford (1999).

Wind Energy

Learning Objectives: After reading the material in Chapter 10, you should understand:

- The different designs of wind turbines.
- The physics of the energy content of wind.
- The efficiency of different wind turbine designs and the importance of the tip speed ratio.
- The geographic distribution of wind energy.
- The design of wind farms.
- The utilization of wind energy worldwide.
- Advantages of offshore wind farms.

10.1 Introduction

Along with solar energy, wind energy is the only alternative energy source that is prevalent virtually everywhere in the world. It is also one of the first sources of energy to be utilized by humanity. The earliest use of wind energy was in mechanical devices (e.g., for pumping water or grinding grain) and for transportation (e.g., sailing ships). Since the late nineteenth century, however, wind energy has also been utilized to generate electricity, and in recent years it has become a very attractive alternative to fossil fuels. Wind-generated electricity is based on a reasonably cost-effective, well-established technology that has minimal environmental impact. Wind energy will certainly play an important role in the development of alternative energy strategies for the future. This chapter overviews the history, scientific foundations, and recent technological developments of wind energy.

10.2 Wind Turbine Design

Wind is moving air. The air moves because of pressure gradients caused by thermal gradients that result from the heating and cooling of the atmosphere. A typical example is the formation of winds along the shore of the ocean or a lake in the morning. The morning sun heats the land faster than it heats the water. The air above the land is heated more quickly and rises. The rising air results in a pressure gradient that pushes air from above the water onto the land. This is the typical onshore wind that is commonly seen in the morning. In the evening, the opposite happens, resulting in the typical offshore breeze.

Figure 10.1: The *Royal Clipper,* the largest operational sailing ship as of 2014.

Orlica

Like solar energy, wind energy is not constant over time. Although the average wind velocity at a particular location may be known, the variations are somewhat less predictable than the variations in solar insolation. Thus, the full utilization of wind energy must include some type of storage mechanism.

Wind energy has been harvested for literally thousands of years. One of the earliest uses of wind energy was for transportation, as shown in Figure 10.1. A large sailing ship of this type could produce close to 8 MW of power in a strong wind.

The use of wind energy for other purposes certainly dates back hundreds, if not thousands, of years. The use of the traditional so-called *Dutch windmill,* as shown in Figure 10.2, for the grinding of grain was widespread in Europe more than five

Figure 10.2: Traditional four-blade Dutch windmill in the Netherlands.

Vincent

Figure 10.3: Post windmill in Britain.

Paste

centuries ago. Although many designs of traditional windmills suffered in their efficiency because they were in a fixed orientation, some designs incorporated a mechanism for rotating the mill so that it always faced into the wind. Such a design, known as a *post windmill*, is illustrated in Figure 10.3. This type of windmill was mounted on a central post, and the entire building could be rotated manually when the wind direction changed. Later windmills were used extensively for pumping water (for irrigation, etc.). The *American multiblade* design (Figure 10.4 on the next page) was in common use for this purpose in North America during the nineteenth century.

Although there is current interest in the recreational use of wind energy (e.g., sailboats), the principal effort in wind energy development is in the production of electricity. Modern windmills (referred to as *wind turbines*) have been developed for this purpose. One modern design (developed in the 1930s) is the *Darrieus wind turbine*, or *Darrieus rotor* (Figure 10.5 on the next page). These devices have two clear advantages over the more common designs discussed in this chapter: (1) The generator mechanism is at the bottom, making it easier to service, and (2) they do not have to be oriented to face into the wind. They have, however, some serious disadvantages: They are not self-starting, that is, if they stop due to the lack of wind, they need to be manually restarted when the wind returns; they also have a lower efficiency than other designs and are subject to excessive stresses that can result in mechanical failure. Despite numerous experimental devices of this type over the years, they are not as commercially attractive as other designs.

The modern *three-blade wind turbine* (Figure 10.6 on the page 335) is the most common design for electricity production. A *high-speed two-blade wind turbine* is shown in Figure 10.7 on the page 335. Although the two-blade design is intrinsically more efficient, it is more difficult to balance, and, because it runs at a higher speed than

Figure 10.4: American multiblade windmill in Maine.

Richard A. Dunlap

Figure 10.5: Darrieus rotor on Îles de la Madeleine, Québec.

Richard A. Dunlap

Figure 10.6: Two General Electric 1.6-82.5 wind turbines each rated at 1.6 MW. The hub height is 80 m and the rotor diameter is 82.5 m.

Richard A. Dunlap

Figure 10.7: Modern high-speed two-blade wind turbine.

NASA Glenn Research Center

Figure 10.8: Nacelle of a 600 kW Turbowinds T-600 wind turbine.

Richard A. Dunlap

Figure 10.9: Nacelle of a medium-sized wind turbine. This is a 100 kW Northern Power Systems NPS100-21 wind turbine. The hub height is 37 m and the rotor diameter is 21 m.

Richard A. Dunlap

the three–blade, it is more susceptible to failure. For this reason, most commercial wind turbines are of the three-blade design.

Typically in a wind turbine (Figure 10.6), the electric generator, along with associated drive systems and a gearbox to match the rotational speed of the wind turbine to the characteristics of the generator, are contained in a housing (called a *nacelle*) at the top of the wind turbine tower (Figures 10.8 and 10.9). Typically, a *yaw drive* is used to keep the wind turbine aligned into the wind.

10.3 Obtaining Energy from the Wind

The power available from the wind can be calculated by means of an analysis of the energy associated with a moving parcel of air. Figure 10.10 shows a cube of air with an edge length d moving with velocity v. The parcel of air has a mass $d^3\rho$, where ρ is the density of air (1.204 kg/m^3). Thus, the kinetic energy of the moving parcel of air is

$$E = \tfrac{1}{2} mv^2 = \tfrac{1}{2} d^3\rho v^2.$$ **(10.1)**

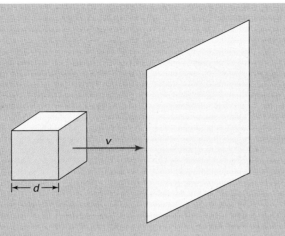

Figure 10.10: Cubic parcel of air with dimensions $d \times d \times d$ traveling with velocity v and passing through a plane parallel to one of its faces.

Richard A. Dunlap

Example 10.1

Calculate the energy (in Joules) of 1 m³ of air moving at 4 m/s. How long could this energy be used to illuminate a 60-W light bulb if it could be converted to electricity with 40% efficiency?

Solution

From equation (10.1),

$$E = \tfrac{1}{2}d^3\rho v^2.$$

Using $d = 1$ m, $v = 4$ m/s, and $\rho = 1.204$ kg/m³ gives

$$E = (1/2) \times (1\ \text{m}^3) \times (1.204\ \text{kg/m}^3) \times (16\ \text{m}^2/\text{s}^2) = 9.6\ \text{J}.$$

Since 1 J = 1 W·s then at 40% efficiency,

$$(0.4) \times (9.6\ \text{W·s})/(60\ \text{W}) = 0.064\ \text{s}.$$

If this parcel of air passes through a plane, as shown in the figure, then the time required for the cube to pass through the plane is $t = d/v$. The power is then written as

$$P = \frac{E}{t} = \tfrac{1}{2}d^2\rho v^3. \tag{10.2}$$

The power per unit area is

$$\frac{P}{A} = \frac{P}{d^2} = \tfrac{1}{2}\rho v^3. \tag{10.3}$$

Using the density of air in kg/m³ and the velocity in m/s, this expression may be written as

$$\frac{P}{A} = (0.602\ \text{kg/m}^3)v^3, \tag{10.4}$$

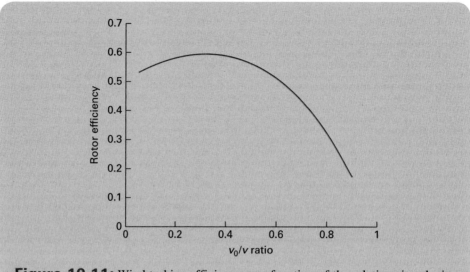

Figure 10.11: Wind turbine efficiency as a function of the relative air velocity after passing through the wind turbine.

where P/A is in W/m². Note that the units on the right-hand side are (kg/m³) $\times$ (m³/s³) = kg/s³. Since J = kg·m²/s² and W = J/s, then W/m² = (kg·m²/s²)/(s·m²) = kg/s³, as expected.

The objective of the wind turbine is to extract as much of the kinetic energy content of the wind as possible for conversion into mechanical energy and then, typically, into electricity. To extract all of the kinetic energy from the parcel of air, it would be necessary to stop the air. This would require a wind turbine design that did not allow the air to pass through the blades to the other side. If the wind turbine blocked the motion of the air entirely, the air would tend to flow around the wind turbine—not the ideal situation. On the other hand, if the wind turbine only very slightly reduced the velocity of the air, then only a small portion of the kinetic energy would be extracted. The best situation is somewhere in between, and a detailed analysis provides the results shown in Figure 10.11, where the efficiency, sometimes called the power coefficient, is plotted as a function of the ratio of the air velocity after passing through the wind turbine, v_0, to the air velocity before the wind turbine, v. This figure shows that the maximum theoretical efficiency that can ever be achieved by a wind turbine, referred to as the *Betz limit*, is 59%.

Example 10.2

What is the velocity of the wind that provides ten times the power provided by wind with a velocity of 5 m/s?

Solution
Equation (10.4) shows that the power in a certain cross-sectional area is proportional to the wind velocity cubed. So, to increase the power by a factor of ten, the velocity has to be increased by a factor of the cube root of 10; that is, the velocity would be

$$(5 \text{ m/s}) \times (10)^{1/3} = 10.8 \text{ m/s}.$$

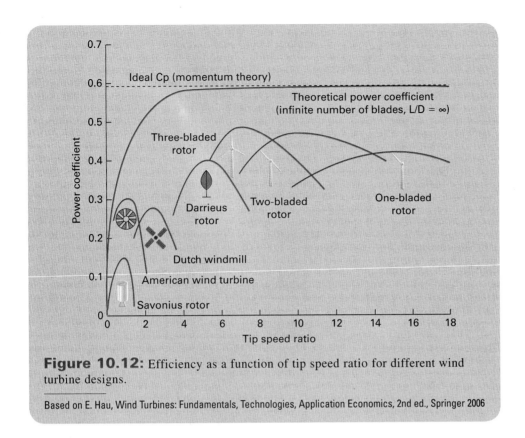

Figure 10.12: Efficiency as a function of tip speed ratio for different wind turbine designs.

Based on E. Hau, Wind Turbines: Fundamentals, Technologies, Application Economics, 2nd ed., Springer 2006

The efficiencies of different types of wind turbines are shown in Figure 10.12. The *tip speed ratio* is the ratio of the speed of the fastest moving tip on the blade or rotor to the actual wind speed. This ratio is determined by the design of the wind turbine and, in particular, by the details of the design of the blades. The best actual efficiencies are for the high-speed two-blade design and are about 45%. The modern three-blade wind turbine is slightly less efficient at around 40–42%. Thus, at best, actual wind turbines achieve about two-thirds to three-quarters of the theoretical maximum efficiency.

On the basis of equation (10.4) and Figure 10.12, the power output of a wind turbine of a given size can be calculated as a function of wind speed. For example, typical three-blade turbine will have an efficiency (η) of about 40% ($\eta = 0.40$). A turbine with a rotor diameter of 20 m in a wind of 5 m/s produces

$$P = (0.602 \text{ kg/m}^3)\eta A v^3$$
$$= (0.602 \text{ kg/m}^3) \times (0.40) \times (3.14) \times (10 \text{ m})^2 \times (5 \text{ m/s})^3 = 9450 \text{ W}. \quad \textbf{(10.5)}$$

This equation shows that the power is related to the cube of the wind velocity and increases substantially with increasing wind speed. For example, the wind turbine just described would produce 255 kW in a 15-m/s wind.

The power of an actual wind turbine differs from this simple analysis in several ways. A typical example of the output of a real wind turbine is shown in Figure 10.13 on the next page. Below a certain speed (the *cut-in speed*), the wind turbine is inefficient and produces little power. From the cut-in speed to the rated speed, the behavior

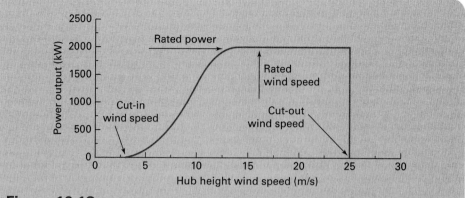

Figure 10.13: Power output for a 2-MW wind turbine as a function of wind speed.

Based on World Energy Council, "2010 Survey of Energy Resources"

Example 10.3

A large Dutch windmill can have blades 30 m in diameter. If such a windmill operates at optimal tip speed ratio in a wind with a velocity of 10 m/s, what is the mechanical power output in kW?

Solution
From equation (10.5), the power is

$$P = (0.602 \text{ kg/m}^3)\eta A v^3.$$

The maximum efficiency of a Dutch windmill, as illustrated in Figure 10.2, is about 27%. The area of the rotor in square meters is

$$A = 3.14 \times (15 \text{ m})^2 = 707 \text{ m}^2.$$

The wind velocity in meters per second is

$$v = (25 \text{ mph}) \times [0.447 \text{ (m/s)/mph}] = 11.2 \text{ m/s}.$$

The power in watts is then

$$P = (0.602 \text{ kg/m}^3) \times (0.27) \times (707 \text{ m}^2) \times (10 \text{ m/s})^3 = 115 \text{ kW}.$$

is fairly well described by the v^3 expression just derived. Above the rated speed, there is a risk of damage to the blades or the generator mechanism as a result of excessive speed. Thus, at this point the speed of the wind turbine is regulated to avoid damage, typically by adjusting the angle (pitch) of the blades to make them intentionally less efficient. At some point (the *cut-out speed*), when the wind speed becomes too great, the wind turbine is shut down by applying a brake to avoid damage.

The amount of power that can be obtained from a wind turbine depends, of course, on the size and design of the wind turbine and on wind conditions. Details of wind resources are discussed in the next section, but some general considerations for the design of wind energy systems are appropriate at this point. The design of the wind turbine should match the wind conditions prevalent at the location. On one hand, the wind turbine should not operate below the cut-in speed too much of the time because it will not be producing power. On the other hand, the wind turbine should not be operating above the rated speed too much of the time because the efficiency will have to be limited to avoid damage. Although each location is associated with an average wind speed, it also has a well-defined distribution of wind speeds. Because the power output of the wind turbine goes up or down with the cube of the wind speed, it is important to know this distribution in order to obtain the maximum total energy production over a period of time. Typical wind speed distributions for different average wind speeds are shown in Figure 10.14. Because the energy is related to the cube of the speed, the peak in the energy distribution occurs at higher speeds than the peak in the speed distribution (Figure 10.15 on the next page). A consequence of this is that the greatest portion of energy produced over a period of time is generally produced in a relatively small fraction of the time when the wind velocity is well above average.

A factor that plays an important role in the power output of a wind turbine is altitude. Up to heights of a few hundred meters, the principal reason for the height dependence of the wind speed typically is the friction between the moving area and surface features of the ground. An empirical relationship for wind speed may be written as

$$v(h) = v(h_0)\frac{\ln\left(\dfrac{h}{\delta}\right)}{\ln\left(\dfrac{h_0}{\delta}\right)},\qquad\text{(10.6)}$$

where h_0 is a reference height and $v(h_0)$ is the wind speed at the reference height. The parameter δ is referred to as the roughness length and characterizes the type

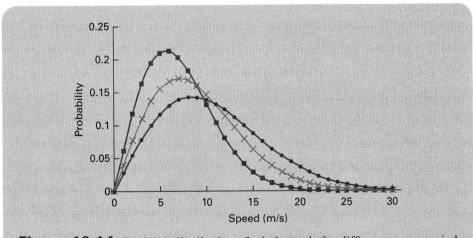

Figure 10.14: Rayleigh distribution of wind speeds for different average wind speed values.

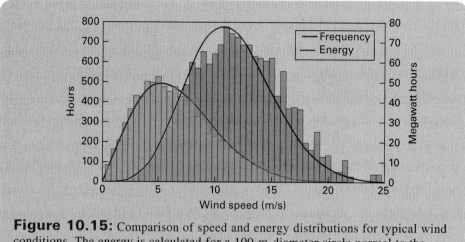

Figure 10.15: Comparison of speed and energy distributions for typical wind conditions. The energy is calculated for a 100-m-diameter circle normal to the wind direction with an efficiency equal to the Betz limit.

Based on Sandia National Laboratories, "New Mexico Wind Resource Assessment," Lee Ranch, http://windpower.sandia.gov/other/LeeRanchData-2002.pdf

of terrain. Table 10.1 gives vales of the roughness length for different types of ground surfaces.

Figure 10.16 illustrates the effects of height and surface roughness on wind speed. The figure shows the importance of positioning the wind turbine as high above the ground as reasonably possible. Thus, locating the turbine at the top of a tall tower is certainly advantageous. However, there is a trade-off between power output and the cost and serviceability of the installation. It is also clear that locating a turbine on the top of a hill rather than at the bottom of a valley is advantageous. The effects of roughness on the height dependence of the wind speed, as illustrated in the figure, show that height considerations are less important for smoother surfaces, especially above water.

For installations that are close to the seacoast, the wind is stronger offshore than on land (Figure 10.17). Placing wind turbines a few kilometers offshore can provide

Table 10.1: Definitions of roughness classes in terms of roughness length and typical terrain characteristics.

roughness class	roughness length, δ (m)	typical terrain
0	0.0002	water
1	0.03	open flat agricultural land
2	0.1	fields with some trees
3	0.4	forests or villages
4	1.6	large cities

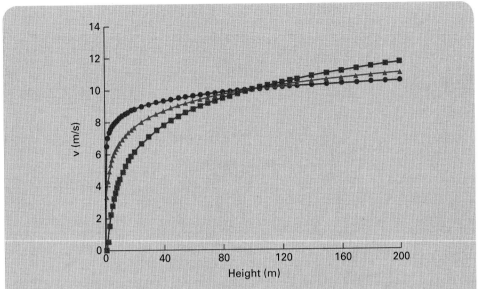

Figure 10.16: Variations in wind speed as a function of altitude for different roughness classes: roughness class 0 (circles), roughness class 2 (triangles), and roughness class 4 (squares). The reference wind speed in equation (10.6) is 10 m/s at a reference height of 100 m.

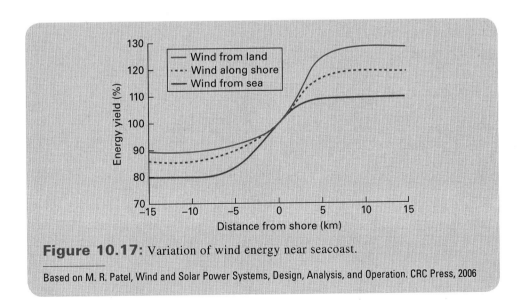

Figure 10.17: Variation of wind energy near seacoast.

Based on M. R. Patel, Wind and Solar Power Systems, Design, Analysis, and Operation. CRC Press, 2006

more energy and also help to minimize the environmental impact of the facility. This is further discussed later in the chapter.

For installations containing multiple wind turbines (generally referred to as *wind farms*), the relative placement of the wind turbines is important. If they are too far apart, then the facility will occupy more land area than necessary. If they are too close together, each wind turbine will not achieve its optimal output. Figure 10.18 on the next page shows the optimal spacing of wind turbines in a wind farm.

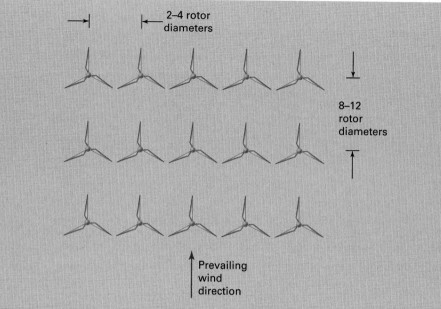

Figure 10.18: Optimal spacing of wind turbines in a wind farm. Distances are shown as a function of the rotor diameter.

Example 10.4

What fraction of the total daily energy produced by a wind turbine is produced during the daylight hours if the wind velocity is 12 m/s during 12 hours of day and 3 m/s during 12 hours of night?

Solution

The energy produced by the turbine over a time t is given by equation (10.5):

$$E = (0.602 \text{ kg/m}^3)\eta A v^3 t.$$

Thus for a specific wind turbine, the energy produced during a specific time interval is

$$E = (\text{constant})v^3 t.$$

So the energy produced during the day is

$$E = (\text{constant})(12 \text{ h})v_{\text{day}}^3,$$

and during the night it is

$$E = (\text{constant})(12 \text{ h})v_{\text{night}}^3,$$

for a total energy production of

$$E = (\text{constant})(12 \text{ h})(v_{\text{day}}^3 + v_{\text{night}}^3).$$

Thus, the ratio of fraction of energy produced during the day is

$$\text{fraction} = \frac{v_{\text{day}}^3}{v_{\text{day}}^3 + v_{\text{night}}^3} = \frac{(12 \text{ m/s})^3}{(12 \text{ m/s})^3 + (3 \text{ m/s})^3} = 0.98.$$

This shows that the majority of the energy is produced when the wind is blowing the strongest.

10.4 Applications of Wind Power

Like solar power, wind power has an essentially infinite longevity. It is important, however, to consider how much power is available. The average wind speed (in m/s) at a height of 80 m for various locations in the United States is shown in Figure 10.19. There are regions of high-wind-power density in the mountainous regions (Rockies and Appalachians), the Western Plains, and offshore on both the Atlantic and Pacific coasts. These are the regions in which the development of wind energy would be the most viable.

Wind power installations may be small residential facilities for generating electricity in individual homes (Figure 10.20 on the next page), or they may be large stand-alone wind turbines (Figure 10.21 on the next page) or wind farms (Figure 10.22 on the page 347) for the production of electricity for the grid. The former may actually be connected to the grid so that excess energy can be sold back to the public utility when production exceeds demand. A typical grid connection for a wind turbine is shown in Figure 10.23 on the page 347. The details of this system follow very much along the lines of the grid-connected photovoltaic system shown in Chapter 9.

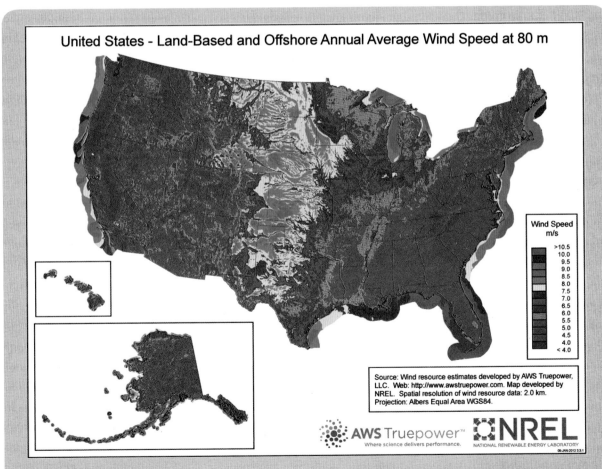

Figure 10.19: Average wind speed at a height of 80 m for onshore and offshore sites in the United States.

NREL

Figure 10.20: Small residential wind turbine.

Richard A. Dunlap

Figure 10.21: An Enercon E-92 WTG 2.3 MW turbine. The hub height is 98 m and the rotor diameter is 92 m.

Richard A. Dunlap

Figure 10.22: Wind farm in Amherst, Nova Scotia. The farm consists of fifteen Suzlon S97 turbines rated at 2.1 MW each.

Richard A. Dunlap

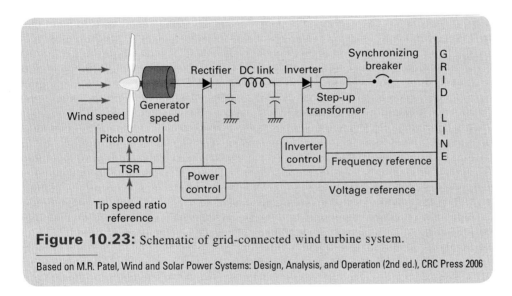

Figure 10.23: Schematic of grid-connected wind turbine system.

Based on M.R. Patel, Wind and Solar Power Systems: Design, Analysis, and Operation (2nd ed.), CRC Press 2006

We can make a rough estimate of the amount of energy available from the wind. As an example, for a typical value of 350 W/m^2 (corresponding to regions where it would be beneficial to develop wind energy), the power produced by a single wind turbine with 50-m-diameter blades operating at 40% efficiency is

$$P = 350 \text{ W/m}^2 \times 3.14 \times (25 \text{ m})^2 \times 0.4 = 275 \text{ kW}. \tag{10.7}$$

Using the mean spacing shown in Figure 10.18 of 3 rotor diameters in one direction and 10 rotor diameters in the other direction gives an area of 150 m × 500 m (or 0.075 km^2) per wind turbine. This means that about 275 kW/0.075 km^2 = 3.6 MW$_e$/km^2 can be generated from wind energy. Note that, to a first approximation and within certain limits, the power produced by a wind turbine is proportional to the square of the rotor diameter. The optimal spacing of wind turbines in a wind farm is proportional to the rotor diameter in both north-south and east-west directions (Figure 10.18). Thus, the number of wind turbines that can be accommodated within a given area of land is inversely proportional to the rotor diameter. The total power is therefore roughly independent of the rotor diameter. However, when other factors, such as the typical altitude dependence of wind velocity are considered, larger wind turbines have clear advantages.

The increase in the utilization of wind energy is most apparent in Denmark, Spain, and Germany, but active development is taking place in many areas throughout the world, including North America, India, and China. These developments have motivated designs for larger and larger wind turbines. A simple analysis of Figure 10.18 might suggest that a wind farm consisting of several large wind turbines would produce the same energy as one consisting of a larger number of small turbines as long as the total area of the rotors is the same. However, large wind turbines are advantageous for two basic reasons. One is that larger turbines are taller, and, as Figure 10.16 suggests, the power density in the wind increases with height (at least up to a point). Another reason is that the infrastructure cost includes not only the cost of the turbine but also the cost of the control systems and grid connections. These costs are minimized if the number of wind turbines is smaller.

The size of the largest common wind turbines as a function of time is illustrated in the next figure:

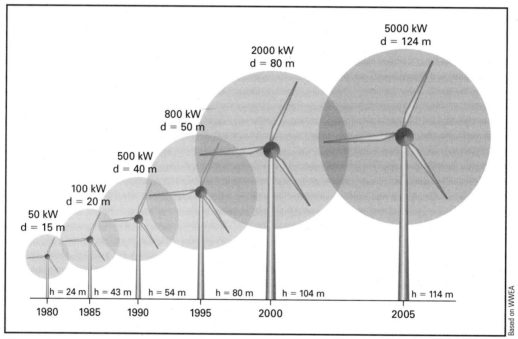

Size and capacity of the largest wind turbines over time (d = diameter, h = height).

Currently, the largest wind turbine is the Enercon E-126 (introduced in 2007) with a rated capacity of 7.58 MW and a total height of 198 m, more than half the height of the Empire State Building. As the capacity of a wind turbine is scaled up, the height increases and the mass of the nacelle increases. This requires the tower to be not only taller but also more massive to support both itself and the weight of the nacelle. The size of the Enercon E-126 is probably the limit of wind turbine size that can be constructed with current designs and materials technology.

One approach to scaling up wind turbine size is to reduce the mass of the generator by using superconducting winding (see Chapters 2 and 18 for more information on superconductors). The American Superconductor Corporation has

Continued on page 349

Energy Extra 10.1 continued

collaborated with the U.S. Department of Energy to investigate the design of a 10-MW superconducting wind turbine. The proposed design would have a mass of about 120 t, in comparison to a nonsuperconducting 10-MW turbine mass of around 300 t. It is possible that this approach could be scaled up to produce a 20-MW wind turbine.

Another approach to higher-capacity wind turbines is to abandon the currently adopted horizontal axis geometry. Vertical axis wind turbines may be an approach that will allow the generator to be in a stationary structure below the rotor and may help to alleviate some of the design limitations of horizontal axis turbines. Some examples of experimental vertical axis turbines are shown in the photograph below.

Topic for Discussion

In addition to structural considerations, the transport and assembly of components can also limit turbine size, particularly for the turbine blades that are manufactured and transported as single units to the turbine site. A standard transport trailer in North America is 14.6 – 16.2 m in length; in the United Kingdom, it is about 13.7 m. What capacity wind turbine could be constructed from components transported on the highway using standard trucking methods? The transport of blades for today's larger wind turbines requires special consideration. You can usually find some interesting videos if you search for "wind turbine blade transport" on YouTube. Discuss the compatibility of the highway system with potentially ideal wind farm locations.

Martin Bond/Science Source

Examples of experimental vertical axis wind turbines.

As an example of land requirements for wind power, consider the region that is most likely the best candidate for large-scale wind production in the United States, the Great Plains. A strip through the central United States that includes North Dakota, South Dakota, Nebraska, Kansas, Oklahoma, and Texas has an area of about 1.7×10^6 km^2. At a power density of 3.6 MW$_e$/km^2, this gives a total power capacity for the Great Plains area of $(1.7 \times 10^6 \text{ km}^2) \times (3.6 \times 10^6 \text{ W}_e/\text{km}^2) = 6.1 \times 10^{12} \text{ W}_e$. A comparison with the estimate of the total power requirement for the United States as obtained from the data in Chapter 2 of about 3.8×10^{12} W indicates that more than half of this portion of the country would need to be covered with wind turbines at their optimum spacing to fulfill all of the country's energy needs. Because many locations would be inappropriate for wind turbines, the possibility of supplying all U.S. energy needs from the wind is very questionable. However, the current cost of wind power (Table 2.6) makes it an attractive alternative that can supplement other energy sources.

Because wind conditions in all locations are somewhat unpredictable and cannot be forecast with any certainty more than a few days in advance, it is difficult to provide a stable base load of power to the grid from wind energy. To ensure an adequate supply of energy at all times, wind energy needs to be combined with other sources. As with photovoltaics, it is necessary to integrate a suitable energy storage system with any large-scale wind facility. Energy storage will be discussed in Chapter 18.

The current status of wind power development in those countries with the greatest installed capacity is summarized in Table 10.2. A few countries have been very serious about developing wind power. In absolute terms, China has installed the greatest wind capacity in recent years. Northern and Western European countries have also been very active in pursuing the development of wind power. In particular, Denmark has the greatest per capita generating capacity, and wind power presently accounts for about 30% of its electricity generation. Offshore wind farms in Denmark (Figure 10.24) are particularly noteworthy. These take advantage of the greater offshore wind energy and help (perhaps) to alleviate some of the environmental concerns.

Wind energy has some (although perhaps not serious) environmental concerns. Like any so-called clean energy source, it is truly clean only if the manufacturing of its equipment and its construction and maintenance all use energy that comes from environmentally friendly sources. The environmental impact of wind energy production itself is fairly minimal. The major concerns that have been expressed are:

1. Noise.
2. Effect on wildlife.
3. Effect on land use.
4. Aesthetics.

All wind turbines make some noise. Small wind turbines are the type most commonly used for residential electricity production, and they make the most noise. This is because of the greater rotational frequencies that are necessary in order to achieve a tip speed ratio compatible with efficient operation. Large wind turbines are typically quieter and produce noise at lower frequency. This can be a nuisance for residents living near these facilities, and the possible long-term health effects of the noise need to be carefully considered.

Birds and bats are at risk from wind turbines, and this has been a public concern for some time. All relevant studies have shown that, even if the use of wind power increases substantially, the risk to birds, and probably to bats, from plate-glass windows, vehicles, and power lines is several orders of magnitude greater than it is from wind turbines.

Table 10.2: Installed wind power as of 2015 for countries with more than 2 GW$_e$ capacity.

country	installed capacity (MW$_e$)	population (10^6)	installed capacity per capita (W$_e$)
China	145,104	1376.0	105.5
United States	72,578	321.8	225.5
Germany	44,947	80.7	557.0
India	25,088	1311.1	19.1
Spain	23,008	46.1	499.1
United Kingdom	13,855	64.7	214.1
Canada	11,200	35.9	312.0
France	10,358	64.4	160.8
Italy	9126	59.8	152.6
Brazil	8715	207.8	41.9
Sweden	6025	9.78	616.1
Poland	5100	38.6	132.1
Denmark	5063	5.67	892.9
Portugal	5034	10.3	488.7
Turkey	4694	78.7	59.6
Australia	4187	24.0	174.5
Netherlands	3431	16.9	203.0
Romania	3244	19.5	166.4
Mexico	3073	127.0	24.2
Japan	3035	126.6	24.0
Ireland	2486	4.69	530.1
Austria	2411	8.54	282.3
Belgium	2229	11.3	197.3
Greece	2152	11.0	195.6

Figure 10.24: Danish offshore wind farm.

Tony Moran/Shutterstock.com

ENERGY EXTRA 10.2
Wind turbine accidents

Table 6.8 shows that the generation of electricity from wind carries a greater risk to humans than other forms of renewable energy, although not as significant as fossil fuels. Like solar energy, wind energy is a low-energy density resource, compared to other resources such as hydroelectric and nuclear. Therefore, considerable manufacturing resources are required per unit installed capacity, as Figure 6.34 illustrates. As the number of wind turbines worldwide increases, so does the number of accidents. This trend is shown in the figure.

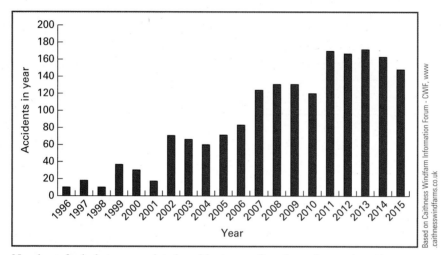

Based on Caithness Windfarm Information Forum - CWIF, www .caithnesswindfarms.co.uk

Number of wind energy–related accidents as a function of year since the mid-1990s.

Wind energy–related accidents fall into several different categories. There are accidents related to the construction of wind turbine components and their installation. Transportation accidents are significant because of the size of the components that need to be moved.

Ali Kemal Akan/Anadolu Agency/Getty Images

Transportation accident involving a wind turbine blade in Afyonkarahisar, Turkey.

Continued on page 353

Energy Extra 10.2 continued

Accidents involving operating wind turbines generally fall into three categories. Most common are the result of failure of the turbine blades. Such failures may result from defective components or inadequate maintenance but are often associated with failure of control mechanisms designed to limit rotational frequency. A major danger associated with blade failure is the ability of the turbine to throw blade fragments considerable distances.

The second most common type of wind turbine accident is fire associated with the nacelle, particularly the generator. Fires of this nature are most commonly the result of lightening strikes but can also be caused by electrical failure inside the nacelle. In forested regions, turbine fires can spread to surrounding areas. A major difficulty associated with wind turbine fires is that the nacelle in most commercial turbines is sufficiently high that firefighting methods are not capable of accessing the fire.

The third type of accident that can affect wind turbines is structural failure of the tower, which may result from structural defects or poor maintenance but may also be the consequence of the failure of other components.

A number of wind energy–related accidents have resulted in human fatalities. Since the 1970s, 164 deaths have occurred as a result of wind energy activities. Of these, 96 were occupational fatalities of

Wind turbine damaged by fire.

workers in the wind energy industry, and the remaining 68 were of members of the general public. The number of fatal accidents per year since 2000 is shown in the figure. The number of fatalities is somewhat larger than the number of fatal accidents, for in some

Damaged wind turbine blades.

Continued on page 354

Energy Extra 10.2 continued

Damaged tower of a wind turbine near Leisnig in Central Saxony, Germany.

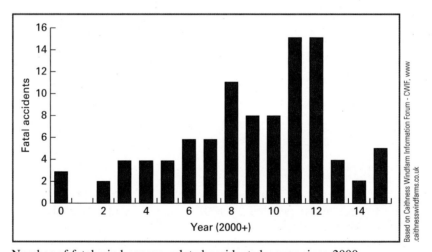

Number of fatal wind energy–related accidents by year since 2000.

cases an accident resulted in multiple deaths. While the number of fatalities per year has increased over the years as a result of the increase in the number of wind turbines, there has been a decrease in the past few years. Further data are needed to determine the statistical significance of this trend.

Topic for Discussion

Based on data for wind energy–related fatalities in this Energy Extra box and information in this chapter about GWy$_e$ wind energy generated, discuss the fatality rate for wind energy compared to coal and nuclear as given in Table 6.5.

Figure 10.25: San Gorgonio Pass wind farm in California. The rows of wind turbines are clearly seen.

Tim Roberts Photography/Shutterstock.com

The deforestation of wilderness areas for the development of wind power is an important concern, but the use of agricultural land can be compatible with wind power. The spacing between rows of wind turbines in a wind farm along the prevaling wind direction can be close to a kilometer (Figure 10.25). An attractive option is the dual use of land in the Midwest for wind power generation and agriculture.

Wind farms certainly change the appearance of the landscape (Figure 10.22), and this may be considered a relevant factor in assessing the effects of different energy production methods. Offshore wind farms influence the aesthetics of the ocean (Figure 10.24) and, of course, must be located so as to minimize possible adverse effects on navigation.

10.5 Summary

Wind energy has attracted much interest as an alternative to fossil fuel energy sources. Wind has a number of important advantages over some other alternatives and relatively few disadvantages. The attractiveness of wind as an environmentally conscientious energy source has resulted in its growth in recent years. Wind resources that are suitable for development are available in nearly all regions of the world. The basic technology for the utilization of wind energy is reliable and well-established and is fairly cost-effective. Wind energy is relatively safe, and overall environmental concerns are minimal.

This chapter began with an overview of the different types of wind turbines. These include vertical axis turbines, such as the Darrieus rotor, as well as the more common horizontal axis turbines.

The chapter presented a basic development of the physics of wind energy and shows that the energy content of moving air is proportional to the cube of the velocity. Wind turbines extract energy from the wind by slowing it down but they cannot extract all the available energy because they cannot stop the air altogether. The maximum amount of energy that can be extracted from moving air is about 59%, as given by the Betz limit. Actual wind turbines more commonly have efficiencies up to about 40%. Efficiency varies with the tip speed ratio, that is, the ratio of the speed of the blade tip to the speed of the wind. Each wind turbine's geometry has an optimal tip speed ratio.

The chapter reviewed wind properties and their relationship to turbine operation. The normal distribution of wind speeds is given by the Rayleigh distribution. Because the energy is proportional to the cube of the velocity, the majority of the energy is available during brief periods of high wind speed. Up to some point, wind speed increases with increasing altitude. This suggests clear advantages to locating turbines atop high towers, although there are trade-offs with structural design and maintenance costs. Wind energy is also typically less over land and increases with distance up to about 5 km over the ocean. Over land, wind energy resources are related to a variety of geographical and climatic features. In the United States, the greatest wind resources are available over the mountainous regions and across the central plains.

The design of wind farms was described in this chapter. To extract the optimum amount of wind energy, turbines should be spaced in accordance with the rotor diameter. The low density of energy in the wind means that fairly large land areas are needed to accommodate wind farms. Offshore wind farms have the advantage of eliminating the need to utilize land resources, as well as taking advantage of the higher energy density of offshore winds.

There have been ambitious efforts in recent years to utilize wind energy. Substantial development has occurred in the European Union, particularly in Denmark, Spain, and Germany. India also has an active wind energy program and has seen significant increases in capacity in recent years.

Problems

10-1   For a 10-m-diameter wind turbine with an efficiency of 45% located on land of roughness class 2, calculate the power produced as a function of hub height from 20 m to 120 m for a wind velocity of 10 m/s at a height of 50 m.

10-2   (a) Consider a wind farm consisting of 10-m-diameter wind turbines with efficiencies of 40% on a grid given by the average spacing shown in Figure 10.18. For a wind velocity of 6 m/s, what is the power output in kW per km^2?

   (b) Compare the result of part (a) with the 24-hour average output of a 1-km^2 photovoltaic array with an efficiency of 18%. Assume average (worldwide) insolation conditions.

10-3   A wind turbine with a 40-m-diameter rotor produces 287-kW output in a 10-m/s wind. What is its efficiency?

10-4   A wind turbine with an efficiency of 42% produces 1-MW output at a wind velocity of 13 m/s. What is the turbine rotor diameter?

10-5   Wind velocities are not constant throughout the day. The daily average power produced by a wind turbine is the power averaged over the wind velocity for the day. Calculate the average power for a turbine with a diameter of 20 m and an efficiency of 37% if, during a 24-hour period, the wind velocity is

 •   2 m/s for 4 hours.
 •   8 m/s for 16 hours.

- 14 m/s for 3 hours.
- 17 m/s for 1 hour.

10-6 A homeowner installs a wind turbine with a rotor diameter of 2 m to supplement electricity from the public utility. The cost of the turbine, the associated electronics, and energy storage system (batteries) is $10,000. If the turbine has an efficiency of 35% and the energy is utilized and or stored at an efficiency of nearly 100%, what is the payback period for the investment? Assume that maintenance costs are minimal, the capital recovery factor is unity, electricity from the public utility costs $0.10 per kWh, and the wind velocity is constant at 12 m/s.

10-7 A Darrieus rotor has an area of 1500 m^2 and operates with the optimal tip speed ratio. What is its power output in a wind with a velocity of 20 m/s?

10-8 A country with good wind resources decides to make a national commitment to the development of wind energy. It is decided that 10% of the land area will be devoted to wind energy with a goal of 1.5 kW$_e$ per-capita wind capacity. This is roughly the electricity used by one person in an industrialized nation (Chapter 2). Two-megawatt wind turbines with rotor diameters of 80 m (Energy Extra 10.1) are placed at the average spacing shown in Figure 10.18. Average output based on wind conditions is 650 kW$_e$ per turbine. Estimate the maximum population density that is consistent with this plan. Locate geographical information about Germany, Spain, India, and Denmark (all of whom have active wind power programs), and discuss the possibility of achieving this goal in each of these countries.

10-9 (a) Two locations have the same average wind velocity. One location has a constant wind velocity of 10 m/s, while the other location has a wind velocity of 5 m/s for 12 hours per day and a wind velocity of 15 m/s for 12 hours per day. Compare the total energy produced per day for identical wind turbines at the two locations.

   (b) Consider part (a) for a location with a wind velocity of 0 m/s for 12 hours per day and 20 m/s for 12 hours per day.

10-10 Wind with a velocity of 12 m/s is incident on a three-blade wind turbine with a rotor diameter of 90 m.

   (a) What is the power of the wind that intercepts the turbine?
   (b) What is the maximum power that can be extracted from the wind?
   (c) What is the angular velocity (ω in s^{-1}) for the rotor when the maximum power is produced?

10-11 Estimate the ratio of wind velocity 10 km offshore to the wind velocity at the shore for

   (a) wind from the land,
   (b) wind along the shore, and
   (c) wind from the sea.

10-12 The *Royal Clipper* (Figure 10.1) has a total sail area of 5200 m^2. Show that the estimate for the total power of a large sailing ship as given in the text is reasonable.

10-13 Consider two identical wind turbines with a hub height of 40 m, one located in flat agricultural land and one located above a forest. What is the relative output power for the two turbines if the wind velocity is 10 m/s at a height of 100 m in both cases?

10-14 (a) Show that for a wind turbine spacing of 3 rotor diameters crosswind and 10 rotor diameters downwind, the power generated per unit land area for a wind farm is independent of rotor diameter and is equal to

$$\frac{P}{A_{\text{land}}} = \frac{\pi \rho \eta}{240} v^3,$$

where ρ is the air density and η is the turbine efficiency.

   (b) Show that when the velocity is measured in m/s, the power per unit area is in kg/s^3.

   (c) Discuss the relevance of hub height on the answer to part (a).

10-15 A three-blade wind turbine with a 100-m-rotor diameter operates with a tip speed ratio of 10.

   (a) What is the efficiency?
   (b) What fraction of the Betz limit is achieved?
   (c) What is the actual power output in kW for a wind speed of 11 m/s?
   (d) For the wind speed in part (c), what is the actual blade-tip velocity?
   (e) For the wind speed in part (c), how long does it take the rotor to make one rotation?

10-16 A high-speed two-blade wind turbine operates with a tip speed ratio of 14. It produces 1.0 MW_e in a 12-m/s wind. What is the rotor diameter?

10-17 Dresden, Germany has a population of 700,000 and a population density of 2200 km^{-2} in the urban area. Assume that the electrical needs of Dresden are an average of 1.0 kW_e per capita and these are satisfied by a wind farm consisting of 100-m-diameter three-rotor wind turbines operating at optimal efficiency and with optimal spacing. If the wind velocity is constant at 7 m/s, what is the area of the wind farm relative to the area of the urban area it serves?

10-18 A country would like to provide all of its residential and industrial electricity needs by wind power at a rate of 2.5 kW_e per capita. It plans to install wind farms covering 10% of the land area consisting of 2 MW wind turbines with a spacing of 1.0 km in the prevailing wind direction and 0.5 km in the orthogonal direction and with an overall capacity factor of 24%.

   (a) Estimate the maximum population density that can be accommodated by this approach?
   (b) In which European countries would this be feasible?

10-19 A three-blade wind turbine with a rotor diameter of 50 m, produces an output of 570 kW in a 10-m/s wind. Calculate the wind velocity after passing through the turbine.

10-20 A Darrieus rotor with a swept area of 1000 m^2 produces 240 kW output in a 10-m/s wind.

(a) What is its tip speed ratio?

(b) If the height of the rotor is 60 m, estimate the time for one rotation. What is its rotational speed (in rotations per second)?

Bibliography

G. Boyle (Ed.). *Renewable Energy*. Oxford University Press, Oxford (2004).

T. Burton, D. Sharpe, N. Jenkins, and E. Bossanyi. *Wind Energy Handbook*. Garrad Hassan and Partners, Bristol (2003).

D. M. Eggleston. "Wind power." In *CRC Handbook of Energy Efficiency*, F. Kreith and R. E. West (Eds.). CRC Press, Boca Raton, FL (1997).

D. M. Eggleston and F. S. Stoddard. *Wind Turbine Engineering Designs*. Van Nostrand Reinhold, New York (1987).

F. R. Eldridge. *Wind Machines* (2nd ed.). Van Nostrand Reinhold, New York (1980).

F. A. Farret and M. Godoy Simões. *Integration of Alternative Sources of Energy*. Wiley, Hoboken, NJ (2006).

L. L. Freris (Ed.). *Wind Energy Conversion Systems*. Prentice-Hall, London (1990).

P. Gipe. *Wind Energy Comes of Age*. Wiley, New York (1995).

P. Gipe. *Wind Power: Renewable Energy for Home, Farm, and Business* (rev. ed.). Chelsea Green Publishing, White River Junction, VT (2004).

E. W. Golding. *Generation of Electricity by Wind Power*. E. & F. N. Spon, London (1955).

S. Heier. *Grid Integration of Wind Energy Conversion Systems*. Wiley, New York (1998).

R. Hinrichs and M. Kleinbach. *Energy: Its Use and the Environment* (5th ed.). Brooks-Cole, Belmont, CA (2012).

G. L. Johnson. *Wind Energy Systems*. Prentice Hall, London (1985).

J. F. Manwell, J. G. McGowan, and A. L. Rogers. *Wind Energy Explained: Theory, Design, and Application*. Wiley, Chichester, U.K. (2002).

J. Park. *The Wind Power*. Cheshire Books, Palo Alto, CA (1981).

M. R. Patel. *Wind and Solar Power Systems: Design, Analysis, and Operation* (2nd ed.). Taylor & Francis, Boca Raton, FL (2006).

P. C. Putnam. *Power from the Wind*. Van Nostrand Reinhold, New York (1948).

J. Reynolds. *Windmills and Watermills*. Hugh Evelyn, London (1970).

R. A. Ristinen and J. J. Kraushaar. *Energy and the Environment*. Wiley, Hoboken (2006).

D. Spera (Ed.) *Wind Turbine Technology: Fundamental Concepts of Wind Turbine Engineering*. ASME Press, New York (1994).

R. Wolfson. *Energy, Environment, and Climate* (2nd ed.). W. W. Norton, New York (2011).

World Energy Council. *World Energy Resources 2016*, available online at https://www.worldenergy.org/publications/2016/world-energy-resources-2016/

A. J. Wortman. *Introduction to Wind Turbine Engineering*. Butterworth, Boston (1983).

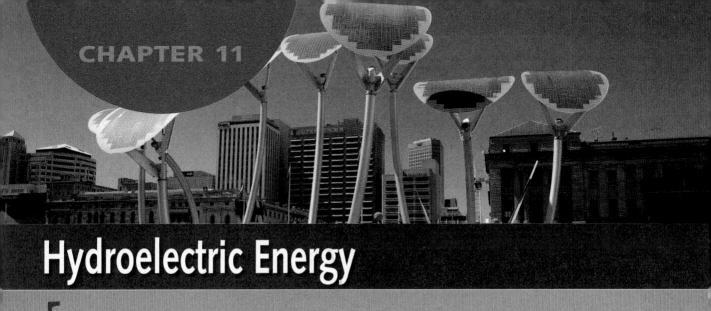

Hydroelectric Energy

Learning Objectives: After reading the material in Chapter 11, you should understand:

- The potential energy associated with water.
- The kinetic energy associated with water.
- The different types of water turbines and their applications.
- The design and properties of high head hydroelectric systems.

- The design and properties of low head and run of the river hydroelectric systems.
- The availability and utilization of hydroelectric energy worldwide.
- The effects of hydroelectric energy on the environment and risks to society.

11.1 Introduction

Hydropower has been used for many centuries. Like wind power, it was first used as a source of mechanical power and was typically used for grinding grain and for sawing wood. Figure 11.1 shows a reconstruction of a waterwheel used in the nineteenth century to obtain mechanical energy. During the past century, hydropower has been primarily used as a source of electricity. Hydroelectric power has been and continues to be the most prevalent method of generating electricity that is generally referred to as renewable and carbon-free. The degree to which this description of hydroelectric power is accurate depends largely how this resource is utilized and includes factors such as geography and climate. The scientific and technological aspects of hydroelectric power are presented in this chapter. The chapter also overviews the extent of hydroelectric resources, their utilization, and the associated risks and environmental impact.

11.2 Energy from Water

The energy associated with water running downhill can be harnessed in two ways: (1) The potential energy of water confined behind a dam can be used to run turbines at the bottom of the dam, or (2) the kinetic energy of flowing water in a river can be used to operate turbines. The first case is generally used for the construction of large hydroelectric generating facilities (Figure 11.2). These typically have capacities of hundreds

Figure 11.1: Waterwheel used to convert the kinetic energy of moving water to mechanical energy for sawing wood (reconstruction of nineteenth-century facility in Nova Scotia, Canada).

Richard A. Dunlap

of megawatts electrical but can be up to 10,000 MW$_e$ or more. They are referred to as *high head* hydroelectric facilities.

Facilities that use the kinetic energy of the water are referred to as run-of-the-river systems, although many are associated with small (i.e., *low head*) dams. They typically have a capacity of a few megawatts electrical (up to about 10 MW$_e$), although small versions can be less than 100 kW$_e$ (Figure 11.3 on the next page). Medium head systems are intermediate between these two extremes, although, in practice, the distinction between these designations is not always well-defined.

Figure 11.2: Three Gorges hydroelectric facility in China. This is the world's largest hydroelectric facility, currently rated at 22,500 MW$_e$. The environmental consequences of the Three Gorges Dam are discussed in Energy Extra 11.2.

PRILL/Shutterstock.com

Figure 11.3: Small run-of-the-river hydroelectric facility on the Musquash River in New Brunswick, Canada, with output of about 7 MW$_e$. The cylindrical tank is a surge tank used for evening out fluctuations in water flow to the turbines.

Richard A. Dunlap

A high head hydroelectric dam blocks a river to create a reservoir. Water is allowed to flow through turbines at the bottom of the dam, and the height of the water in the reservoir above the height of the turbines is called the *head*. The potential energy of water near the surface falls through that distance and is converted into kinetic energy as it flows through the turbines that convert it into electrical energy. The potential energy of the water is

$$E = mgh,$$ (11.1)

where m is the mass of the water, g is the gravitational acceleration (9.8 m/s^2), and h is the head, or height, of the water above the turbine. The power generated is determined by the rate at which energy is generated:

$$P = \frac{E}{t} = \frac{m}{t}gh.$$ (11.2)

If the rate at which water flows through the turbine (i.e., volume per unit time) is φ, expressed in cubic meters per second (m^3/s), then $(m/t) = \rho\varphi$, where ρ is the water density. Thus

$$P = \rho\varphi gh.$$ (11.3)

For equation (11.3), the power is expressed in watts when the density is in kg/m^3, the flow rate is in m^3/s, g is in m/s^2, and the height is in m.

In a run-of-the-river system, water with a mass, m, flowing in a river at a velocity, v, will have kinetic energy

$$E = \frac{1}{2}mv^2.$$ (11.4)

This represents a power-generating capacity of

$$P = \frac{1}{2}\left(\frac{m}{t}\right)v^2,$$ (11.5)

Example 11.1

One cubic meter of water is dropped from a vertical height of 100 m. If the change in potential energy is converted into electricity with an efficiency of 85%, how long would this output fulfill the electrical needs of a typical North American single-family home? (See Chapter 2, and do not include electric heating.)

Solution

One cubic meter of water has a mass of 1000 kg, so the energy given by equation (11.1) is

$$E = (1000 \text{ kg}) \times (9.8 \text{ m/s}^2) \times (100 \text{ m}) = 980,000 \text{ J}.$$

At 85% efficiency, this represents a total of $(980,000 \text{ J}) \times (0.85) = 833,000$ J of electrical energy. From Chapter 2, the average annual residential electric use is approximately 1.2×10^4 kWh $= 4.32 \times 10^{10}$ J. This represents an average power of $(4.32 \times 10^{10} \text{ J/y})/(3.15 \times 10^7 \text{ s/y}) = 1370$ W. Thus, 833,000 J will last for

$$\frac{833,000 \text{ J}}{1370 \text{ J/s}} = 608 \text{ s}.$$

Example 11.2

What is the flow rate in terms of volume per unit time and in terms of mass per unit time that would be required for a 90% efficient hydroelectric facility with a head of 50 m to satisfy (on average) the electrical needs of a typical single-family home?

Solution

Following from Example 11.1, the average power requirement is 1370 W. At 90% efficiency, this represents $(1370 \text{ W})/0.9 = 1522$ W total. From equation (11.3), the flow rate (in m^3/s) can be found to be

$$\varphi = P/(\rho g h).$$

Substituting appropriate values gives

$$\varphi = \frac{1522 \text{ W}}{1000 \text{ kg/m}^3} \times (9.8 \text{ m/s}^2) \times (50 \text{ m}) = 3.11 \times 10^{-3} \text{ m}^3/\text{s}.$$

Multiplying by the density of water gives the flow rate in mass per unit time:

$$(3.11 \times 10^{-3} \text{ m}^3/\text{s}) \times (1000 \text{ kg/m}^3) = 3.11 \text{ kg/s}.$$

or

$$P = \frac{1}{2}\rho\varphi v^2.$$ **(11.6)**

For water flow through an opening with a cross-sectional area A, equation (11.6) reduces to the form given in equation (10.3) for the power per unit area for moving air:

$$\frac{P}{A} = \frac{1}{2}\rho v^3.$$ **(11.7)**

Numerically, equation (11.7) differs substantially from equation (10.3) because the density of water is much greater than the density of air.

Example 11.3

A small river, 40 m wide and 6 m deep, flows at a velocity of 2 m/s. If 20% of the flow of the river is diverted through a run-of-the-river hydroelectric system that generates electricity at an efficiency of 90%, what is the output in watts electrical?

Solution

The cross-sectional area of the river is (40 m) × (6 m) = 240 m². The volumetric flow rate is

$$\varphi = (240\ \text{m}^2) \times (2\ \text{m/s}) = 480\ \text{m}^3/\text{s},$$

and 20% of this flow corresponds to

$$(480\ \text{m}^3/\text{s}) \times (0.2) = 96\ \text{m}^3/\text{s}.$$

From equation (11.6), the power output is

$$P = \frac{1}{2}\rho\varphi v^2 = (0.5) \times (1000\ \text{kg/m}^3) \times (96\ \text{m}^3/\text{s}) \times (2\ \text{m/s})^2 = 192{,}000\ \text{W}.$$

At 90% conversion efficiency to electricity, the output is

$$(192{,}000\ \text{W}) \times (0.9) = 173\ \text{kW}_\text{e}.$$

The following sections describe the details of the design of the turbine and specific design considerations for high head and low head systems.

11.3 Turbine Design

To produce electricity from the energy content of water, a turbine is used to produce rotary mechanical energy, and this, in turn, is used to produce electricity by means of a generator. There are two basic types of turbines for use with water: *reaction turbines* and *impulse turbines*. In the first case, the flowing water is contained in an enclosure around the turbine and experiences a pressure drop as it passes through the turbine. In the second case, a free jet of water is incident on the turbine and does not experience a

Table 11.1: Some important water turbine designs and their operating conditions.			
name	type	suitable head (m)	maximum power (MW$_e$)
Kaplan	reaction	2–40	200
Francis	reaction	10–350	800
Pelton	impulse	50–1500	400
Turgo	impulse	50–250	5

pressure drop. There are numerous designs of both of these types of turbines. The most common are summarized in Table 11.1.

Perhaps the most obvious design for a water turbine is one that resembles a wind turbine. The *Kaplan turbine* (Figure 11.4) is a reaction turbine and is the closest to this design. Figure 11.5 on the next page shows a close-up of the rotating hub and blades of the turbine (i.e., the *runner*). The runner must be enclosed in a cylindrical tube, through which the water flows, in order to prevent the water from being diverted around the blades and thus reducing the turbine's efficiency.

The *Francis turbine* is a reaction turbine and probably the most commonly used type of turbine in the electric power industry (Figure 11.6 on the next page); the details

Figure 11.4: A Kaplan turbine being assembled.

U.S. Army

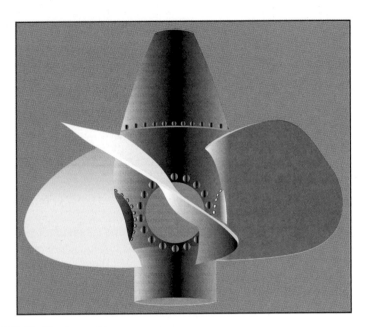

Figure 11.5: Kaplan turbine runner.

M.Fuksa/Shutterstock.com

of the runner are shown in Figure 11.7. In this type of turbine, the water enters radially from the sides and is guided through the blades, causing them to rotate, and then exits axially from the center of the turbine.

Impulse turbines are suitable for high head applications where the water at the base of the dam is at high pressure. The water is allowed to exit through a small opening, forming a free jet of water at high velocity. This high-velocity water jet is incident on the runner of the turbine. An impulse turbine runner has what look like

Figure 11.6: Francis turbine at the Grand Coulee Dam. The inlet scroll is clearly seen in the photograph.

U.S. Bureau of Reclamation photo archives

Figure 11.7: Francis turbine runner.

Esa Hiltula/Alamy Stock Photo

buckets along the outer edge that catch the high-velocity jet of water, deflecting it and imparting momentum to the runner. In the *Pelton turbine*, the buckets are U-shaped [Figure 11.8(a)]. Water is ejected through a nozzle and is incident on the runner [Figure 11.8(b)]. The *Turgo impulse turbine* is basically one-half of a Pelton turbine. The water jet enters into the half-U-shaped bucket at an angle and exits in a different direction thus transferring momentum to the runner (Figure 11.9 on the next page).

A summary of the operating ranges of the various types of turbines is shown in Figure 11.10 on the next page. Although the efficiency of turbines and the associated generators depends on the specific design and operating conditions, efficiencies of

(a)

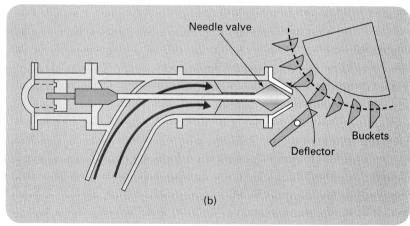

(b)

Figure 11.8: (a) Photograph of a Pelton runner; (b) diagram of a Pelton turbine.

LOOK Die Bildagentur der Fotografen GmbH/Alamy Stock Photo; Based on INFORSE, Hydro Power, http://www.inforse.dk/europe/dieret/Hydro/hydro.html

Figure 11.9: A micro-Turgo runner.

Joseph Hartvigsen

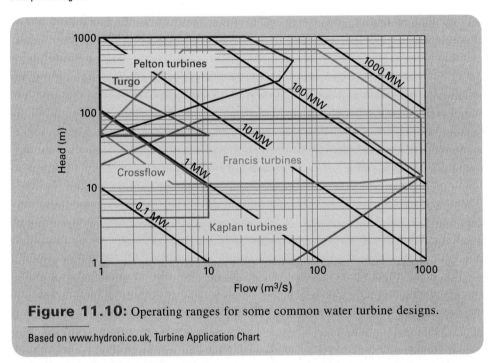

Figure 11.10: Operating ranges for some common water turbine designs.

Based on www.hydroni.co.uk, Turbine Application Chart

85–90% for the conversion to electricity of the kinetic or potential energy associated with water are common for modern systems.

11.4 High Head Systems

For the present discussion, *high head systems* are defined as those that impound water behind a dam for the purpose of producing a reservoir to create a head of water. Most high-capacity hydroelectric facilities are of this type.

Figure 11.11: Grand Coulee Dam in Washington State, United States.

U.S. Bureau of Reclamation photo archives

One of the best known examples of a high head system is Grand Coulee Dam in Washington State (Figure 11.11). The dam is 1592 m in length and 168 m in height. The size of this structure is indicated by the fact that the volume of concrete in the dam is sufficient to construct a road 2.5 m wide and 10 cm thick around the equator of the earth.

Typically, in a high head system there is an underwater water intake in the dam. The water flows through a pipe (the *penstock*) to the turbine near the bottom of the dam (Figure 11.12). The general features of a typical high head hydroelectric dam are

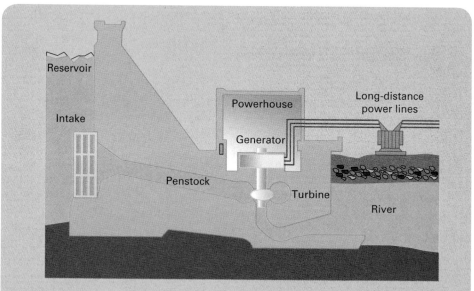

Figure 11.12: Schematic of a typical high head hydroelectric system showing dam, penstock, and turbine.

Based on Texas Comptroller of Public Accounts, Hydropower, http://www.window.state.tx.us/specialrpt /energy/renewable/hydro.php

Figure 11.13: The Libby Dam in Montana.

U.S. Army Corps of Engineers

shown in Figures 11.11 and 11.13. The portion of the dam on the right side of both photographs is the *spillway*, which allows excess water to bypass the turbines. On the left side of the photographs, the structure near the bottom of the dam houses the turbine/generator assemblies. A vertical axis configuration (more on this in the next section) is the most common arrangement for turbines in high head installations, and Francis turbines are in common use in these applications (Figure 11.6).

11.5 Low Head and Run-of-the-River Systems

Low head and *run-of-the-river* hydroelectric facilities do not impound a significant quantity of water. While most of these facilities have small capacities (Figure 11.14), some are much larger (Figure 11.15). As shown in the figure, a dam may be constructed across the entire width of a river in order to make use of the energy of the flowing water without the formation of a significant reservoir. Another approach, sometimes taken, is a diversion system, where part of the flow of a river is diverted into a penstock (Figures 11.16 and 11.17 on the page 372) and allowed to flow downhill through the penstock to the generating system. After passing through the generator turbines, the water is returned to the river or, in some cases, output into a lake or the ocean. Figure 11.10 shows that Kaplan turbines are the most suitable for these low head applications. The turbine and generator arrangement may be on a horizontal axis (Figure 11.18 on the page 373) or on a vertical axis, as is customary for high head applications (Figure 11.19 on the page 373).

Figure 11.14: Milltown Hydroelectric Station is the oldest operational hydroelectric station in North America. It has a capacity of 4 MW$_e$ and has been in operation since 1881. It is situated on the U.S.-Canada border between Maine and New Brunswick, and electricity is distributed to both United States and Canada.

Richard A. Dunlap

Figure 11.15: Chief Joseph Dam in Washington, a low head hydroelectric facility with a capacity of 2620 MW$_e$.

U.S. Army Corps of Engineers

Figure 11.16: Penstocks used to divert part of a river's flow through a generating station.

Erick Margarita Images/Shutterstock.com

Figure 11.17: Penstock providing water to a 1.4 MW$_e$ hydroelectric generating station in Smyrlabjargaárvirkjun, Iceland.

Richard A. Dunlap

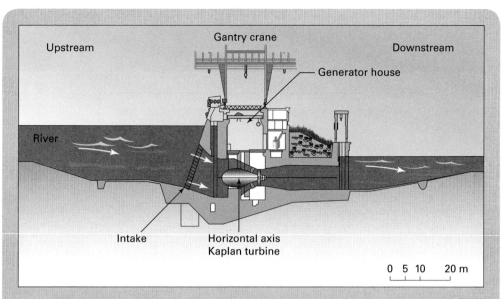

Figure 11.18: Low head or run-of-the-river hydroelectric generating facility utilizing a horizontal axis bulb (Kaplan-type) turbine.

Based on http://de.wikipedia.org/w/index.php?title=Datei:Kraftwerk_Ottensheim-Wilhering_Querschnitt_Krafthaus.jpg&filetimestamp=20051130202844

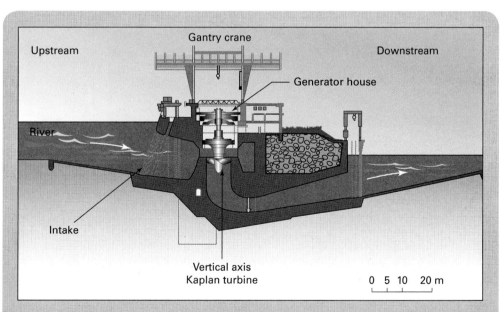

Figure 11.19: Low head or run-of-the-river hydroelectric generating facility utilizing a vertical-axis Kaplan turbine.

Based on http://en.wikipedia.org/wiki/File:Kraftwerk_Wallsee-Mitterkirchen_Querschnitt_Krafthaus.jpg

Panoramic view of Niagara Falls.

Niagara Falls is one of the best known waterfalls in the world. It is also a major source of hydroelectric power. The falls are on the Niagara River on the border between the United States and Canada and have a vertical drop of 51 m. The falls, shown in the photograph, consist of the American Falls on the left of the image, the (small) Bridal Veil Falls just to the right of the American Falls, and the Horseshoe Falls (also known as the Canadian Falls) at the right in the photograph. The falls have been used as a source of mechanical power since 1759 and as a source of hydroelectric power since 1881.

Because Niagara Falls is a popular (and economically important) tourist attraction, its utilization for

The Robert Moses Niagara Falls Generating Station.

Continued on page 375

Energy Extra 11.1 continued

hydroelectric generation is managed in order to minimize its impact on the appearance of the falls. Water from the Niagara River is diverted upstream from the falls and directed through canals or underground tunnels that are up to 10 km or more in length to hydroelectric generating stations and then back to the river downstream from the falls. Eight hydroelectric generating stations are associated with Niagara Falls, with a total generating capacity of about 5 GW. The largest of these is the Robert Moses Niagara Generating Station, shown in the above photograph, which has a maximum generating capacity of about 2.5 GW.

The use of the Niagara Falls hydroelectric resources is governed by a 1950 treaty between the United States and Canada. The United States can utilize a maximum of 920 m³/s of water from the Niagara River for hydroelectric generation; Canada can utilize a maximum of 1600 m³/s of water. The inequality of these amounts compensates for other hydroelectric utilization agreements for resources along the border that favor the United States. These amounts are subject to the condition that at least 50% of the flow of the Niagara River must go over the falls during the daylight hours between April and October (the prime tourist season). At other times, a minimum of 25% of the river's flow must go over the falls. Niagara's hydroelectric stations are an interesting example of the management of a significant energy resource in a way that optimizes its economic benefits.

Topic for Discussion

Estimate the financial cost (per year) of maintaining Niagara Falls as a natural attraction. Specifically, what would be the value of the additional electricity that could be generated by diverting all of the flow of the Niagara River through turbines?

11.6 Utilization of Hydroelectric Power

Of all the sources of energy that are generally considered to be renewable, hydroelectricity is the only one that has seen widespread use. A summary of the major producers of hydroelectric power by country is shown in Table 11.2. The installed capacity is the maximum power output from all hydroelectric facilities in each country. These facilities produce energy, on average, at a rate that is less than their rated maximum capacity. In general, the output is determined by rainfall, and, because of variability in conditions, hydroelectric power is generally used in conjunction with other energy sources

Table 11.2: Capacity and production of hydroelectric power by major producers in 2015.

country	installed capacity (GW$_e$)	actual production (TWh$_e$)	capacity factor	approximate % of domestic electricity production
China	319.4	1126.4	0.40	20
United States	101.8	250.1	0.28	6
Brazil	91.7	382.1	0.48	77
Canada	79.2	375.6	0.54	65
Russia	50.6	160.2	0.36	16
Japan	49.9	91.3	0.21	10
India	47.6	129.0	0.31	13
Norway	30.6	139.0	0.52	98

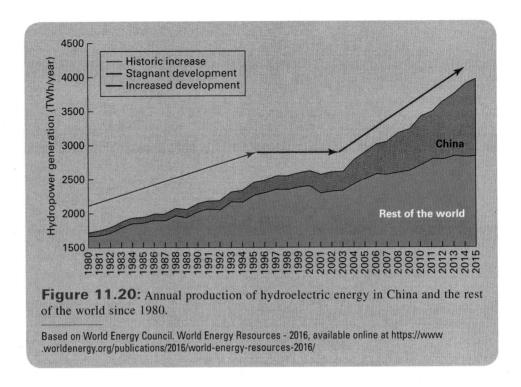

Figure 11.20: Annual production of hydroelectric energy in China and the rest of the world since 1980.

Based on World Energy Council. World Energy Resources - 2016, available online at https://www .worldenergy.org/publications/2016/world-energy-resources-2016/

in order to provide a reliable and consistent energy supply. Energy storage methods (discussed in Chapter 18) are also useful in this respect. The capacity factor is the fraction of maximum capacity that is actually produced on the average. Typical capacity factors for hydroelectric facilities are between 0.3 and 0.6 (or 30–60%). Table 11.2 also gives the percentage of the national electricity use that is produced by domestic hydroelectric power. Certainly, a few countries (e.g., Canada, Brazil, and Norway) rely heavily on hydroelectric power. Canada and Norway have traditionally had substantial hydroelectric facilities, while Brazil has installed considerable capacity in the latter part of the twentieth century. Overall, hydroelectric energy production accounts for about 16% of the world's electricity, or about 6% of the world's total energy.

The trends in hydroelectric energy production as a function of time are interesting to consider. Figure 11.20 shows the total hydroelectricity production worldwide for the past 35 years or so. While there was fairly consistent growth during the 1980s and early 1990s, increase in production was rather slow in the late 1990s and early 2000s. The increase in more recent years as shown in the figure is largely due to the development of hydroelectric resources in developing countries, particularly China. However, much can be understood about the future of hydroelectricity from trends in countries where this industry is much more mature. The trends in Canada are shown in Figure 11.21, and the trends in the United States are shown in Figure 11.22. For the past half century or more, the fraction of electricity coming from hydroelectricity in these two countries has been declining. Although the actual amount of electricity from hydroelectric sources has not changed significantly since about 1970, increased demand, due to both a growing population and changes in per-capita electricity consumption, has been met primarily by increased production from fossil fuel and nuclear sources. The graphs show that in the United States hydroelectricity accounts for about 6% of the electrical supply, whereas in Canada it is currently over 60%. Traditionally,

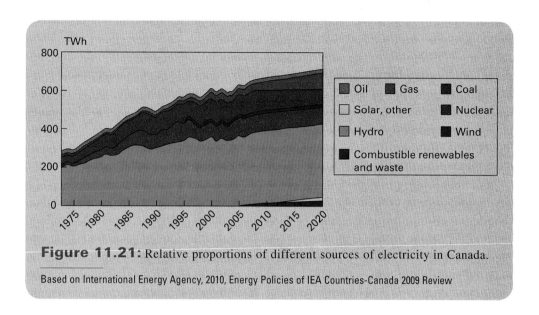

Figure 11.21: Relative proportions of different sources of electricity in Canada.

Based on International Energy Agency, 2010, Energy Policies of IEA Countries-Canada 2009 Review

most of Canada's electricity has come from hydroelectric generation. In fact, in Canada the term *hydro* is synonymous with electricity supplied by a public utility regardless of how it is generated, and the majority of provincial electric companies have names like Québec Hydro and Manitoba Hydro, even though the electricity is produced by a variety of methods.

The reasons for these trends can be readily understood in terms of the availability of hydroelectric power. Hydroelectric power is not an unlimited resource because the locations where it can be developed on a large scale in an economically viable and

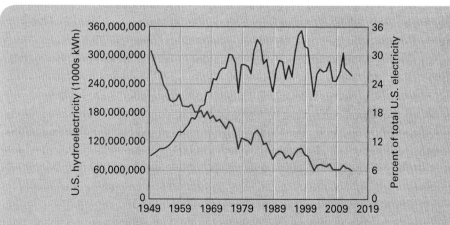

Figure 11.22: Fraction of electricity from hydroelectric generation in the United States as a function of year (yellow line), along with the total hydroelectric generation (blue line).

Based on U.S. Energy Information Administration, Electric Power Monthly, http://www.eia.gov/electricity/monthly/

Example 11.4

(a) From the installed capacity and annual production figures for Norway in Table 11.2, confirm the capacity factor in the table.

(b) Assuming that electricity accounts for about 30% (typical of industrialized countries) of the primary energy use in Norway, confirm that essentially all electricity in this country is generated using hydroelectric facilities. The population of Norway is 5.08 million.

Solution

(a) The capacity factor will be given by the ratio of the annual production to the installed capacity times the number of hours in the year. That is,

$$\text{capacity factor} = (1.39 \times 10^{14}\ \text{Wh}_e/y)/[(3.06 \times 10^{10}\text{W}_e) \times (8760\ \text{h/y})] = 0.52.$$

(b) The annual hydroelectric generation in Norway is expressed in joules as

$$(1.39 \times 10^{14}\ \text{Wh}_e/y) \times (3600\ \text{J/Wh}_e) = 5.00 \times 10^{17}\ \text{J/y}.$$

The per-capita annual hydroelectric generation is

$$(5.00 \times 10^{17})/(5.08 \times 10^{6}) = 9.8 \times 10^{10}\ \text{J/y}.$$

If hydroelectric generation is approximately 85% efficient, then the per-capita hydroelectric energy corresponds to about

$$(9.8 \times 10^{10}\ \text{J/y})/(0.85) = 1.15 \times 10^{11}\ \text{J/y}.$$

From Figure 2.4, the per-capita primary energy use in Norway is 4.3×10^{11} J/y. If electricity accounts for 30% of this energy use, then the per-capita primary energy that is used as electricity will be

$$(0.3) \times (4.3 \times 10^{11}\ \text{J/y}) = 1.29 \times 10^{11}\ \text{J/y},$$

in reasonable agreement with the per-capita hydroelectric energy.

environmentally acceptable way are limited. (See the next section for environmental considerations.) In fact, in the more industrialized regions of the world, such as North America and Europe, the largest portion of the potential for hydroelectric power has already been developed. Figure 11.23 shows the developed and undeveloped hydroelectric potential in different regions of the United States. On average, over half of the hydroelectric potential is already in use. In Canada, the fraction is even greater. Worldwide, the potential for use of hydroelectricity is illustrated in Figure 11.24. The greatest possibility for increased hydroelectric use is in Asia and Africa. Considerable development in Asia, particularly in China, is currently underway or in the planning stages. Overall, an increase in hydroelectric capacity worldwide of 50% would probably be economically viable and would produce energy that would be competitive with energy from other existing sources. An increase by a factor of 2 to 3 would be technically feasible but may not be economically viable at this time. Thus, at some point in the future, it might be reasonable to expect that the contribution of hydroelectricity to world energy needs could increase to perhaps a maximum of 10–15%. Much of the new

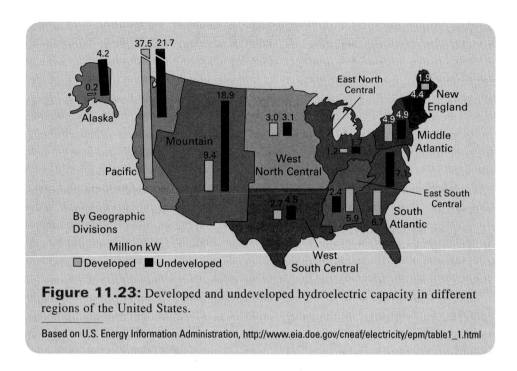

Figure 11.23: Developed and undeveloped hydroelectric capacity in different regions of the United States.

Based on U.S. Energy Information Administration, http://www.eia.doe.gov/cneaf/electricity/epm/table1_1.html

development is likely to be from low head installations due to the cost and environmental concerns of high head facilities, as well as the minimal availability of viable undeveloped high head resources.

It is important, as well, to consider the longevity of hydroelectric resources. Hydroelectricity is promoted as a renewable resource, but it may not be reasonable to interpret this classification in the same way as for solar or wind energy. Silt carried downstream accumulates behind dams and reduces their ability to provide power. At some point, maintenance and operating costs outweigh energy production, and a dam ceases to be economically viable. This is typically a more important factor for

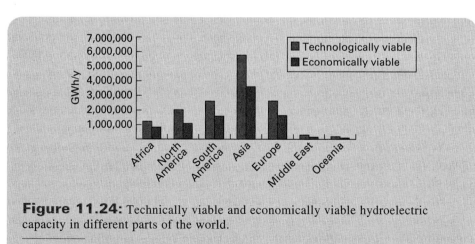

Figure 11.24: Technically viable and economically viable hydroelectric capacity in different parts of the world.

Based on 2010 World Energy Council, Survey of Energy Resources, http://www.worldenergy.org /documents/ser_2010_report_1.pdf

---

Example 11.5

A penstock carries water at a velocity of 4 m/s from a river to a hydroelectric facility with a head of 150 m. If the turbine/generator is 85% efficient and the output is 15 MW$_e$, what is the necessary diameter of the penstock?

Solution

From equation (11.3) the power is given as

$$P = (0.85)\rho\varphi g h.$$

Solving for the flow rate gives

$$\varphi = \frac{P}{(0.85)\rho g h}.$$

Using the values from above gives

$$\varphi = (1.5 \times 10^7 \text{ J/s})/[(0.85) \times (1000 \text{ kg/m}^3) \times (9.8 \text{ m/s}^2) \times (150 \text{ m})] = 12 \text{ m}^3/\text{s}.$$

For a velocity of 4 m/s, a flow rate of 12 m^3/s requires a cross-sectional area of the penstock of

$$(12 \text{ m}^3/\text{s})/(4 \text{ m/s}) = 3 \text{ m}^2,$$

or a diameter of

$$d = (4 \times 4 \text{ m}^2/\pi)^{1/2} = 2.26 \text{ m}.$$

---

large-scale facilities than for small dams and run-of-the-river systems. It is also a sensitive function of the geography of the region. Overall, the longevity of some high head hydroelectric facilities may be limited to somewhere in the range of 50 to 200 years, making them clearly not indefinitely renewable.

11.7 Environmental Consequences of Hydroelectric Energy

Although hydroelectric power is often considered an environmentally friendly source of energy, it is important to consider the details of its impact on the environment. Large-scale high head dams are the most concern, whereas run-of-the-river facilities are not as invasive for many reasons. The construction of a large dam is a major undertaking (Figure 11.25 on the page 382), and the dam can produce a very sizable reservoir upstream. For example, the reservoir associated with the Three Gorges Dam in China extends 600 km upriver. The replacement of land, which in most cases supported vegetation (e.g., trees, etc.), with a reservoir has important implications. Trees sequester carbon and help to reduce greenhouse gases (i.e., CO_2). Vegetation that has been flooded decays and produces greenhouse gases, notably CO_2 and methane. Although other pollutants such as sulfur compounds, NO_x, and particulates are not produced by a hydroelectric facility, studies have shown that in tropical regions the release of greenhouse gases associated with reservoirs can be more than that

ENERGY EXTRA 11.2
Three Gorges carbon payback period

Although alternative energy sources are attractive as replacements for fossil fuels and as a mechanism for reducing the adverse environmental effects, carbon release is always associated with energy production. Large-scale hydroelectric power is an interesting example of the carbon cost associated with renewable energy. Although an analysis of the environmental impact of changes in ecology due to land reallocation is generally not simple, a life cycle analysis of the materials involved in dam construction is reasonably straightforward.

Hydroelectric dams are constructed largely of concrete, which is comprised of three components—cement (CaO), sand, and filler (typically gravel)—in a ratio of about 1:2:3, respectively. While sand and gravel have a relatively small carbon cost, cement contributes carbon to the environment for three reasons: (1) CaO is made by the decomposition of limestone ($CaCO_3$) by heating ($CaCO_3 + heat \rightarrow CaO + CO_2$): (2) furnaces used to heat the limestone use primarily

fossil fuels; and (3) carbon emissions from transportation, etc. The last factor is relatively unimportant, but the first two, together, contribute to an emission of about 0.15 kg CO_2 per kg of concrete. Consider as an example the construction of the Three Gorges Dam in China. This dam used 6.5×10^{10} kg of concrete and emitted $(6.5 \times 10^{10}$ kg$) \times (0.15) \approx 10^{10}$ kg of CO_2 to the atmosphere. The payback period for this carbon release to the environment can be calculated by equating the electric generation from the dam to the generation from a coal-fired plant that would produce the same emissions.

The burning of coal produces 0.11 kg CO_2/MJ. At a Carnot efficiency of about 40% a coal fired station produces about (0.11 kg/MJ)/0.40 = 0.28kg CO_2/MJ$_e$ or about 1 kg CO_2/kWh$_e$. The release of 10^{10} kg of CO_2 would therefore correspond to the generation (by coal) of 10^{10} kWh$_e$.

The following graph shows the actual average monthly power output of the Three Gorges Dam in

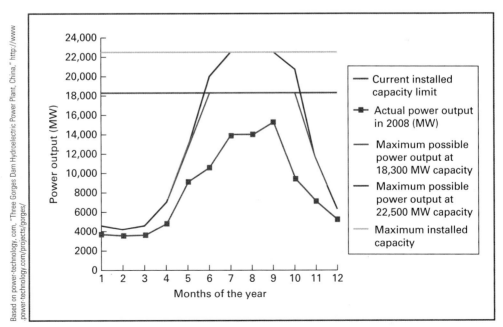

Based on power-technology.com, "Three Gorges Dam Hydroelectric Power Plant, China," http://www.power-technology.com/projects/gorges/

Legend:
- Current installed capacity limit
- Actual power output in 2008 (MW)
- Maximum possible power output at 18,300 MW capacity
- Maximum possible power output at 22,500 MW capacity
- Maximum installed capacity

Monthly average power output of the Three Gorges hydroelectric station in 2008.

Continued on page 382

Energy Extra 11.2 continued

2008. Integrating data over the year yields the total energy generated during 2008: 7.4×10^{10} kWh$_e$. Thus, the carbon payback period for the concrete used in the construction of the Three Gorges Dam is about $[10^{10}$ kWh$_e]/(7.4 \times 10^{10}$ kWh$_e$/y$) = 0.14$ years (about a month and a half). Other material used in dam construction (e.g., steel) would need to be considered in a similar way in a more detailed analysis. This type of analysis assumes that the dam is carbon-neutral once it is constructed.

This analysis shows that the carbon payback period for hydroelectric dam construction can be quite reasonable. It is the long-term environmental effects of large scale hydroelectric generation that need to be carefully assessed.

Topic for Discussion

The Three Gorges Dam contains 4.6×10^8 kg of steel. Estimate the additional carbon payback period for this component of the dam.

produced by a comparably sized fossil-fuel generating plant. In temperate regions, this is not the case, and emissions are typically less than 10% of that from equivalent fossil-fuel generation. Changes in habitat may have adverse effects on wildlife and particularly on wildlife diversity. Fish mobility is one of the most obvious factors to be influenced by dam construction. The replacement of forests by an aquatic environment may have both positive and negative effects. The disruption of silt transport may have adverse consequences for the agricultural industry downstream from the dam because farmlands benefit from nutrients carried by a river.

In addition to environmental effects, hydroelectric dam construction generally has social and cultural consequences. People living in areas that are flooded by dam construction need to be relocated. It is estimated that, to date, many tens of millions of people have been affected in this way. It is a matter not only of relocating people but also, in most cases, of dealing with cultural or historical sites and cemeteries. A recent example of social and cultural effects of hydroelectric power is the construction of the Three Gorges Dam in China. The project required the relocation of 1.24 million

Figure 11.25: Construction of the Three Gorges hydroelectric facility.

Zoonar GmbH/Alamy Stock Photo

residents. The relocation costs were approximately the same as the actual construction costs for the hydroelectric facility. In addition, the project affected approximately 1300 archeological sites. Although many were preserved and relocated, other known, as well as yet undiscovered, sites were destroyed.

A final point to consider in the implementation of hydroelectric power is safety. A large dam can fail with serious consequences. Dams may be constructed for a number of reasons other than for the production of hydroelectricity, such as water storage, irrigation, flood control, etc., and many dams are considered multipurpose. All have at least some risk of failure. The severity of such a failure is a function of many variables, such as the volume of water impounded by the dam, the population distribution living downstream, and the warning time before failure. The most notable dam failure was the Banqiao Reservoir Dam and 61 other associated dams in China in 1975 during Typhoon Nina. The Banqiao Dam was constructed on the Ru River in the 1950s. Excessive rain associated with the typhoon increased water flow and levels in the river and ultimately caused a series of dam failures. The Banqaio Dam, at a height of 118 m, was the largest of these. A design that did not include an adequate number of sluice gates to release excess water was a factor in the failure of the dam. Several other dams were intentionally destroyed in an attempt to control water levels. The flooding caused by these failures caused 26,000 deaths, and the resulting epidemics and famine in the region led to an estimated additional 145,000 deaths. Although such catastrophic events are rare, there have been more than 30 dam failures in the past century that have resulted in more than 100 fatalities. A well-known dam failure that resulted in 11 deaths was the Teton Dam in Idaho. The dam had existed for less than a year before failure resulted from poor design and unstable geological conditions in the area. Figure 11.26 shows the reservoir emptying through the failed dam within a few hours of the catastrophic event.

Figure 11.26: Water flows through an opening in the Teton Dam in Idaho in 1976.

Bettmann/Contributor/Getty Images

Example 11.6

If hydroelectric power is considered carbon-free, estimate the annual savings in CO_2 emissions for the Three Gorges Dam in 2008 compared to the same power generation by natural gas–fired stations operating at a Carnot efficiency of 40%?

Solution

The total electricity generated by the Three Gorges Dam during 2008 is given in Energy Extra 11.2 as

$$(7.4 \times 10^{10} \text{ kWh}) \times (3.6 \times 10^6 \text{ MJ/kWh}) = 2.66 \times 10^{17} \text{ MJ.}$$

From Table 3.2, the energy content of methane is 55.5 MJ/kg. At 40% efficiency, this corresponds to 22.2 MJ/kg.

Thus, the amount of methane needed to equal the hydroelectric production will be

$$(2.66 \times 10^{17} \text{ MJ})/(22.2 \text{ MJ/kg}) = 1.20 \times 10^{16} \text{ kg.}$$

From the combustion reaction in equation (1.15),

$$CH_4 + 2O_2 \rightarrow CO_2 + 2H_2O,$$

we see that 1 mole of CH_4 (12 + 4) = 16 g produces 1 mole (12 + 2 × 16) = 44 g of CO_2.

Thus the combustion of 1.20×10^{16} kg of methane will produce

$$(1.20 \times 10^{16} \text{ kg}) \times (44/16) = 3.30 \times 10^{16} \text{ kg of } CO_2.$$

Despite notable dam failures, hydroelectric power is, on average, a relatively safe option overall, as indicated in Table 6.8. However, dam failure, although unlikely, has the greatest risk for large-scale disaster of any of the anthropogenic causes presented in Figure 6.32.

11.8 Summary

In this chapter, the basic physics of the potential and kinetic energy associated with water were presented. The potential energy associated with water as a result of the gravitational interaction is given as $E = mgh$, where m is the mass, g is the gravitational acceleration, and h is the height. The kinetic energy of moving water is analogous to that of moving air, as discussed in the previous chapter, and is related to the cube of the velocity.

This chapter also reviewed the various designs of turbines. Different runner geometries are appropriate for different applications on the basis of flow rate and pressure. Generally, Kaplan turbines are most suitable for low head systems, whereas Francis turbines are most suitable for high head systems.

High head hydroelectric systems are typically constructed using a dam to create a large reservoir. Generating capacity can exceed 20 GW_e, compared with about 1 GW_e for a typical coal-fired or nuclear generating station.

Low head and run-of-the-river systems typically do not incorporate a reservoir to create a significant head but primarily utilize the kinetic energy associated with the flow of the river. These systems typically have a much smaller capacity than high head systems, often in the 10–100 MW_e range.

The availability and use of hydroelectric power were discussed. Hydroelectric power comprises about 16% of the world's electricity generation. It is a major source of energy and, together with nuclear power (which has a similar share), contributes the majority of the world's non–fossil fuel electricity. Although much of the practical hydroelectric capacity in North America and Europe has already been developed, extensive new development continues in many other regions of the world. Hydroelectric power has some very positive attributes. It is low cost, and the facilities are relatively low maintenance. It comprises a significant enough resource that it can make a major contribution to the world's electrical needs. Thus far, the safety record of hydroelectric facilities has been quite good, and the average overall risk is probably lower than for most large-scale generating techniques. Small-scale hydroelectricity, specifically run-of-the-river facilities, has relatively low environmental impact

On the negative side, hydroelectric power, at least large-scale high head installations, may not be as environmentally neutral as other alternative energy technologies, particularly in warmer climates. Hydroelectric installations may contribute to greenhouse gas emissions in addition to having adverse effects on agriculture and wildlife. The longevity of hydroelectric dams is related to geographical conditions, and, in the long term, some hydroelectric resources may have limited renewability. Finally, there is always a risk of catastrophic dam failure, which carries the possibility of substantial human causalities.

Problems

11-1 One cubic meter of water at 20°C is dropped from a vertical height of 100 m. If the change in potential energy is converted into heat with an efficiency of 100% and this thermal energy is used to heat the water, what is its final temperature?

11-2 The Three Gorges Dam is rated at 22.5 GW_e maximum output. If its generating efficiency is 90% and its head is 81 m, what is the rate of flow (in kg/s) of water through its turbines at maximum output?

11-3 Water falls from a head of 100 m at a rate of 1 m^3/s, and this energy is converted into electrical energy at an efficiency of 90%. How many typical single-family homes can this provide with electricity?

11-4 Consider the run-of-the-river system described in Example 11.3. What is the most reasonable turbine runner design for such a system and why?

11-5 Use the total annual U.S. hydroelectric production and the information in Chapter 2 to confirm the percentage of U.S. electricity that comes from hydroelectric facilities as shown in Table 11.2.

11-6 The reservoir created by the Three Gorges Dam has an area of 1045 km^2. It is anticipated that, in the long term, the capacity factor of this generating facility will be around 0.5. Consider an alternative situation where the land area utilized by the Three Gorges reservoir was utilized instead for a photovoltaic array with an average efficiency of 20%. Would one of these situations be significantly more advantageous than the other in terms of energy production?

11-7 A reservoir is 1 km wide and 10 km long and has an average depth of 100 m. Every hour, 0.1% of the reservoir's volume drops through a vertical height of

100 m and passes through turbines to produce electricity with an efficiency of 92%. What is the electrical power output of this facility?

11-8 The term *micro hydro* refers to very small-scale hydroelectric power and includes installations that can supplement other sources of electricity for single-family residences. Typical efficiencies are less than commercial hydroelectric installations and may be in the range of 75%. A homeowner plans to install a micro hydro generator on a stream that has an average flow of 1200 L/min and a vertical drop of 5 m. What is the average electric power that can be expected from this facility? Should this project be seriously considered if the construction cost is $3000?

11-9 A rectangular reservoir is 2 km wide by 10 km long. It is 50 m deep and drains through a 93% efficient hydroelectric facility with an average head of 100 m.

(a) Calculate the total electrical energy that would be produced by the water in the reservoir.

(b) What is the value of the electricity that is generated at rate of $0.10 per kWh?

(c) If the electricity is used by a city with a population of 100,000 with a per-capita power requirement of 1 kWh_e, how long would the electricity last?

11-10 A 1.2 m-diameter penstock carries water from a reservoir with a 150-m head to a 90% efficient hydroelectric generating facility that produces 10 MW_e.

(a) What is the flow rate of the water in the penstock in mass per unit time?

(b) What is the velocity of the water in the penstock?

11-11 The total flow of the Niagara River is 6.0×10^3 m^3/s. Using the height of Niagara Falls as given in the text and the fact that a maximum of 75% of the flow can be utilized for hydroelectric power outside tourist season, calculate the maximum available electric power if the generators are 85% efficient.

11-12 A swimming pool with dimensions 8 m wide by 14 m long with an average depth of 2.6 m is filled from a lake at a level of 50 m below the pool using an electric pump with an efficiency of 85%. How much does the electricity used to fill the pool cost at a rate of $0.11 per kWh?

11-13 If electricity produced by the Three Gorges hydroelectric station is sold at a rate of $0.06 per kWh, what is the total revenue for 2008 based on actual power generated?

11-14 If a high head hydroelectric facility is considered to be carbon-free, what is the annual reduction in CO_2 emissions for a 1000 MW_e hydroelectric station compared to a 1000 MW_e coal-fired station operating at 38% efficiency? Assume both stations have a 70% capacity factor.

11-15 Hydroelectric power accounts for about 6% of the primary energy used for electricity generation in the United States.

(a) If all of the viable undeveloped hydroelectric resources in the United States were implemented, what fraction of U.S. electricity needs would hydroelectric power satisfy?

(b) Approximately how many 1000 MW_e coal-fired generating stations could be eliminated by existing and potential hydroelectric facilities? Assume a 60% capacity factor for all facilities.

(c) What would be the annual reduction in CO_2 emissions if hydroelectric power is considered carbon-free?

11-16 A public utility is planning to construct a high head hydroelectric facility. The approximate cost for large facilities is $3000 USD per kW_e installed. The utility is considering two options, a 600 MW_e facility that would operate at a 70% capacity factor or an 800 MW_e facility that would operate at a 65% capacity factor. If electricity is sold for $0.13 per kWh, estimate the payback period for the additional infrastructure costs for the larger facility. Ignore the cost recovery factor and operating costs.

11-17 A homeowner in an area with substantial elevation changes and a large annual rainfall (1.54 m) plans to supplement electricity needs of 1.1 kW average over the year with a hydroelectric installation. An open tank of 15 m in diameter collects precipitation that falls on its surface. Periodically the tank is drained through a penstock to a hydroelectric generator at an elevation of 100 m below the tank. Discuss the merits of this approach.

11-18 Water flows through a 2-m-diameter penstock at a velocity of 2.5 m/s. It passes through an 87% efficient turbine/generator. What is the power output?

11-19 The Amazon River has a flow rate of 2.09×10^5 m^3/s (the largest in the world) at an average velocity of 0.7 m/s. If 10% of this power could be extracted with a run-of-the-river hydroelectric facility, what would be the output in MW_e?

11-20 A Kaplan turbine runner has a diameter of 5 m, and the turbine/generator assembly has an efficiency of 80%.

(a) What is the velocity of water required to produce an output of 1 MW_e?
(b) What is the flow rate in kg/s of water through the turbine?

Bibliography

G. Boyle (Ed.). *Renewable Energy.* Oxford University Press, Oxford (2004).

R. Hinrichs and M. Kleinbach. *Energy: Its Use and the Environment* (5th ed.). Brooks-Cole, Belmont, CA (2012).

M. Kaltschmitt, W. Streicher, and A. Wiese (Eds.). *Renewable Energy: Technology, Economics and Environment.* Springer, Berlin (2007).

R. A. Ristinen and J. J. Kraushaar. *Energy and the Environment.* Wiley, Hoboken (2006).

C. Simeons. *Hydropower: The Use of Water as an Alternative Source of Energy.* Pergamon Press, Oxford (1980).

N. Smith. "The origins of the water turbine." *Scientific American* **242** (1980): 138.

J. Tester, E. Drake, M. Driscoll, M. W. Golay, and W. A. Peters. *Sustainable Energy: Choosing Among Options.* MIT Press, Cambridge, MA (2006).

C. C. Warnick. *Hydropower Engineering.* Prentice-Hall, Englewood Cliffs, NJ (1984).

World Energy Council. *World Energy Resources 2016*, available online at https://www.worldenergy.org/publications/2016/world-energy-resources-2016/

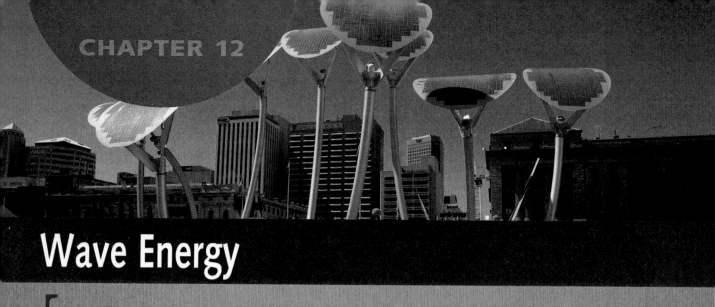

Wave Energy

Adrian Lyon/Alamy Stock Photo

Learning Objectives: After reading the material in Chapter 12, you should understand:

- The availability of wave energy worldwide.
- The relationship between wave properties and energy.
- The types of wave energy devices.

- The design of oscillating water columns and the properties of the Wells turbine.
- The use of floating and pitching devices for wave energy generation.
- The use of wave-focusing devices.

12.1 Introduction

Energy can be obtained from the ocean either from temperature or salinity gradients or from the movement of water. The former is discussed in Chapter 14. The latter may be from tides, ocean currents, or waves. Tidal energy, which is the result of the gravitational interaction, is discussed in the next chapter. Ocean currents may have components that result from temperature gradients, winds, and/or tides. The possibility of obtaining energy from currents that arise primarily from tidal motion is discussed in the next chapter. This chapter discusses energy from ocean waves.

Ocean waves are primarily the result of the interaction between wind and the sea surface. Like wind energy, wave energy is primarily a manifestation of solar energy. Waves occur everywhere in the oceans, but some locations on earth are more likely to have larger waves than others. Figure 12.1 shows the average wave power available at various locations worldwide. Typically, the larger wave energies are over 50 kW/m averaged over the year. Areas with wave energies much below this value are probably not economically viable. For a region with an average wave energy of 50 kW/m, a facility that extracts the energy from 20 m of wave front could theoretically have an average output of as much as 1 MW. The more industrialized regions where wave power is attractive are the southern coast of Australia and the Atlantic coast of Norway, the United Kingdom, Ireland, and Portugal. Much of the activity in wave energy development occurs in these countries, although there are also significant activities elsewhere, such as in Canada, the United States, China, Denmark, and Japan.

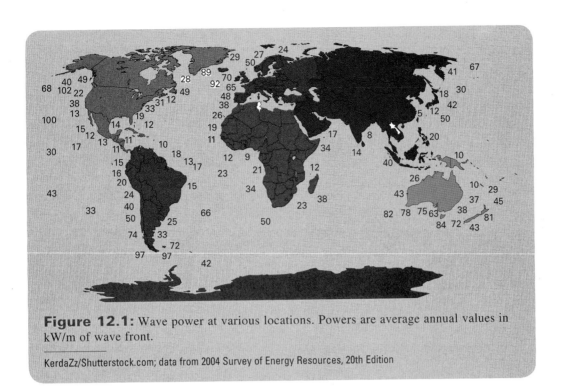

Figure 12.1: Wave power at various locations. Powers are average annual values in kW/m of wave front.

KerdaZz/Shutterstock.com; data from 2004 Survey of Energy Resources, 20th Edition

This chapter begins with a review of the basic physics of wave energy. This is followed by an overview of some of the technologies that have been developed to extract useful energy from waves and a summary of the utilization of these technologies.

12.2 Energy from Waves

A schematic of an ocean wave in cross section is shown in Figure 12.2 on the next page. The wave amplitude is A, and the wavelength is λ. A water wave has energy associated with it for two reasons: (1) It is moving, so it carries kinetic energy, and (2) the surface of the water is not flat, so that potential energy is associated with the raising of some of the water above the flat surface and the lowering of some water below the surface. It is easiest to consider the potential energy in detail.

As illustrated in Figure 12.2, the wave is formed by raising the surface on the left side of the figure and depressing the surface on the right side. If the wave is modeled as an ideal sine wave, it can be created by merely flipping the water in the darker blue portion of the wave above the zero level to form the lighter blue portion. The area of the portion of the wave above the zero level is found by integrating a sine wave of wavelength λ and an amplitude over half a wavelength. This area is $A\lambda/\pi$. Thus, the mass of this half wave per unit length in the direction orthogonal to the plane of the figure (and the direction of propagation of the wave) is

$$\frac{m}{l} = \frac{\rho A\lambda}{\pi},$$

(12.1)

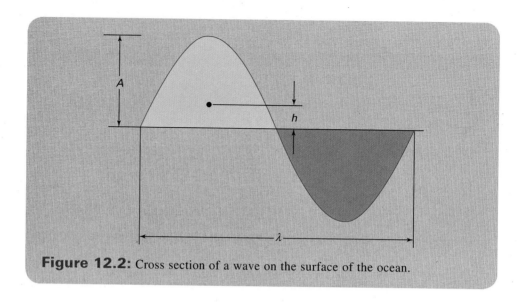

Figure 12.2: Cross section of a wave on the surface of the ocean.

where ρ is the water density. The center of gravity of this half wave is a distance h above the zero level (Figure 12.2). From some simple geometry, h can be calculated to be

$$h = \frac{\pi A}{8}.$$ (12.2)

Thus the formation of the wave from a flat surface by flipping over the lower portion of the wave to form the upper portion of the wave corresponds to moving a mass, as given in equation (12.1), vertically upward by a distance of $\Delta h = 2h$. The potential energy (per unit length of the wave front) associated with this process is

$$\frac{E}{l} = \frac{m}{l}g\Delta h = \left(\frac{\rho A \lambda g}{\pi}\right)\left(\frac{\pi A}{4}\right) = \frac{1}{4}\rho A^2 \lambda g.$$ (12.3)

It can be determined (although not very easily) that the kinetic energy of a water wave per unit length is equal to its potential energy per unit length. Thus the total energy associated with the wave is equal to twice that given in equation (12.3):

$$\frac{E}{l} = \frac{A^2 \lambda \rho g}{2}.$$ (12.4)

This is the energy content of one full wavelength of the wave. If a device is utilized to convert this wave energy into a usable form (e.g., electricity), then the energy content of one wavelength of the wave is incident on the device during one period of the wave. It can be shown that the period of a water wave, T, is related to its wavelength by the expression

$$\lambda = \frac{gT^2}{2\pi}.$$ (12.5)

Substituting this expression into equation (12.4) gives

$$\frac{E}{l} = \frac{A^2 g^2 \rho T^2}{4\pi}.$$ (12.6)

Since we are primarily concerned with the power generated, which is the energy per unit time, then the left-hand side of equation (12.6) is divided by the wave period to give the power per unit length of the wave front as

$$\frac{P}{l} = \frac{A^2 g^2 \rho T}{4\pi}. \tag{12.7}$$

Note that in this expression the amplitude A, as shown in Figure 12.2, is one-half of the crest to trough height of the wave, H, i.e. $H = 2A$. Equation (12.7) is, therefore, often written in terms of the wave height as

$$\frac{P}{l} = \frac{H^2 g^2 \rho T}{16\pi}. \tag{12.8}$$

Using the density of seawater ($\rho = 1025$ kg/m^3) and appropriate values for the constants in equation (12.8), the power in kilowatts per meter is related to the wave height in meters, and the period in seconds and is expressed as

$$\frac{P}{l} = [1.96 \text{ kW/(m}^3\text{·s)}]H^2 T. \tag{12.9}$$

Example 12.1

An ocean wave has a height of 2 m and a period of 10 seconds. What is the power available in 1 km of wave front?

Solution
From equation (12.9), the power in kilowatts per meter of wave front is

$$\frac{P}{l} = [1.96 \text{ kW/(m}^3\text{·s)}] \times (2 \text{ m})^2 \times 10 \text{ s} = 78.4 \text{ kW/m}.$$

For 1 km of wave front, the total power available is

$$P = 78.4 \text{ kW/m} \times (1000 \text{ m/km}) = 78.4 \text{ MW}.$$

Equation (12.5) also gives the wave's velocity. As it travels a distance of λ in time T, its velocity is

$$v = \frac{\lambda}{T} = \frac{gT}{2\pi}. \tag{12.10}$$

Example 12.2

An ocean wave has a period of 8 seconds. What are its velocity and wavelength?

Solution
From equation (12.10), the velocity is given in terms of the period as

$$v = \frac{gT}{2\pi} = \frac{9.8 \text{ m/s}^2 \times 10 \text{ s}}{2 \times 3.14} = 15.6 \text{ m/s}.$$

From equation (12.5), the wavelength is

$$\lambda = \frac{gT^2}{2\pi} = \frac{9.8 \text{ m/s}^2 \times (10 \text{ s})^2}{2 \times 3.14} = 156 \text{ m}.$$

12.3 Wave Power Devices

Devices to extract energy from ocean waves can be located onshore, where they obtain energy from breaking waves, or they can be offshore and extract energy from waves as they propagate in the ocean. From a practical standpoint, facilities are best located on the shore or at least not far offshore. Facilities near shore can be connected via transmission cables, as is the case for offshore wind farms, and this has been found to be fairly practical (see the information later in the chapter about the Pelamis). Connection of onshore facilities is straightforward. Facilities that are far offshore need an appropriate mechanism for transporting the generated electricity back to shore. Energy storage mechanisms, such as batteries (Chapter 19) or hydrogen (Chapter 20), are either expensive or have low efficiency. Unfortunately, the efficiency of wave power devices is fairly low, and the actual (electrical) energy output is substantially less than the theoretical wave energy. For an offshore device, the situation is somewhat like that for a wind turbine; to extract all of the kinetic energy from a wave, the wave has to be stopped. For an onshore device, the problem is that waves lose energy as they approach shore due to interaction with the sea floor. At a water depth of 20 m, a typical wave has lost over 60% of the energy it had in deep water.

There are three main approaches to the development of devices for harnessing wave power: oscillating water columns are primarily onshore devices while floats and pitching devices, as well as wave focusing devices, are typically located offshore. These three technologies are discussed in detail below.

12.3a Oscillating Water Columns (OWCs)

Oscillating water columns are devices that are typically permanently attached to the shoreline or anchored close to shore. Onshore units have been constructed in Scotland, and near-shore units have been developed in Australia (Figures 12.3 and 12.4, respectively).

Figure 12.3: Illustration of LIMPET the 250 kW oscillating water column constructed by Wavegen in Scotland.

Claus Lunau/Bonnier Publications/Science Source

Figure 12.4: Oscillating water column developed by Energetech in Australia that is anchored near shore, rated at 500 kW.

NOAA

The basic design of the OWC is shown in Figure 12.5. The details of the generator assembly are shown in Figure 12.6 on the next page. The device is positioned onshore or near-shore, and, because waves are incident on the device, the level of the water in the wave chamber rises. When the height of the water column rises, then air is pushed out through a turbine, which drives a generator to produce electricity. When the water level in the wave chamber falls, then air is pulled in through the turbine. A typical wind turbine is designed so that, if air flows in one direction, the turbine rotates in one direction, and, if the air flows in the opposite direction, the turbine rotates in the opposite direction. This situation is acceptable for a traditional wind turbine because, if the direction of the wind changes, then this is a slow process, and the wind turbine can be rotated to ensure that the turbine always faces into the wind. The problem with the OWC is that the direction of air flow changes with the periodicity of the ocean waves, which might be about 10 seconds.

The practical solution to this problem is to design a turbine that always rotates in the same direction regardless of the direction of the air flow. There are two approaches

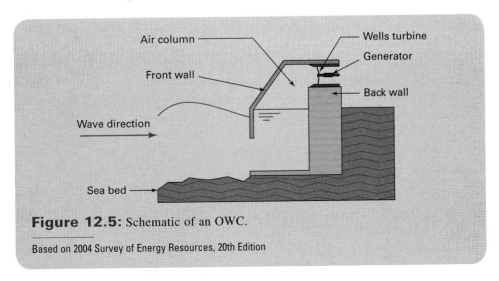

Figure 12.5: Schematic of an OWC.

Based on 2004 Survey of Energy Resources, 20th Edition

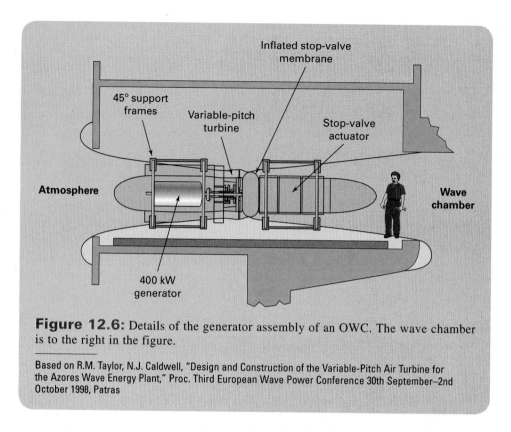

Figure 12.6: Details of the generator assembly of an OWC. The wave chamber is to the right in the figure.

Based on R.M. Taylor, N.J. Caldwell, "Design and Construction of the Variable-Pitch Air Turbine for the Azores Wave Energy Plant," Proc. Third European Wave Power Conference 30th September–2nd October 1998, Patras

to the design of such a device: the variable pitch turbine and the Wells turbine. *Wind turbines* are often variable pitch devices where the blade can be rotated on its axis (the axis along the length of the blade) in order to change its angle relative to the wind. In a more extreme case, the angle of the blade can change enough to accommodate changes in the direction of the airflow. An example of a variable pitch turbine for application in an OWC is shown in Figure 12.7. Although this type of turbine is the most efficient option, energy is required to change the pitch of the blades every time the direction of airflow changes. Also, the device is somewhat complex, with correspondingly high production costs and risk of mechanical failure.

A simpler, less expensive, and potentially more reliable alternative is the *Wells turbine*, which was developed by Alan Wells of Queen's University, Belfast, in the 1980s. The trade-off is a potentially lower efficiency. In a Wells turbine (Figure 12.8 on the page 396), the blades are fixed and symmetric (somewhat almond-shaped in cross section, as shown in the figure), so that they appear the same from either air flow direction. The net resulting force on the blade causes a rotation in the same direction regardless of the airflow direction.

12.3b **Floats and Pitching Devices**

Devices of several different geometries fall into this category, and two characteristic designs are described in the section. These are anchored near shore, and the electricity that is generated can be transferred to shore through appropriate power lines. The most common of these devices is the *Pelamis* (pronounced pel-AH-mis), which is named after a genus of sea snake (because of its geometry). (*Note:* The genus Pelamis contains a single species, *P. platura*, which is highly venomous and inhabits

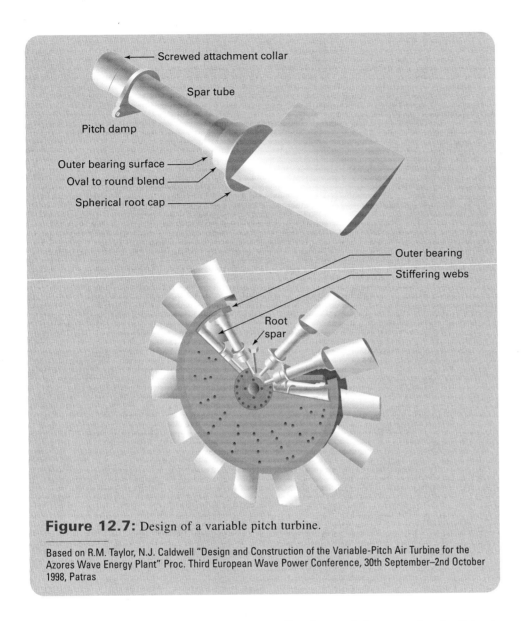

Figure 12.7: Design of a variable pitch turbine.

Based on R.M. Taylor, N.J. Caldwell "Design and Construction of the Variable-Pitch Air Turbine for the Azores Wave Energy Plant" Proc. Third European Wave Power Conference, 30th September–2nd October 1998, Patras

coastal regions in much of the Pacific and Indian Oceans.) An example of a Pelamis (the energy generator) is shown in Figure 12.9 on the next page. These devices have been developed in Scotland and Portugal. They are typically about 3.5 m in diameter and 120 m long and are made in sections (typically four) that are connected by hinges. They are anchored in water that is typically about 50 m deep and are allowed to rotate so that they face in the direction of the waves. As the waves pass the sections of the Pelamis, it flexes at the hinges, giving rise to a snake-like appearance. Hydraulic cylinders associated with the hinges are used to pump hydraulic fluid through hydraulic turbines, which drive electric generators. Typical Pelamis generators are rated at 750-kW$_e$ output.

Other devices that have been tested in recent years are buoys that ride up and down in the waves as shown in Figure 12.10 on the page 397. These typically generate electricity either by pumping fluid through a turbine that is connected to a generator or by

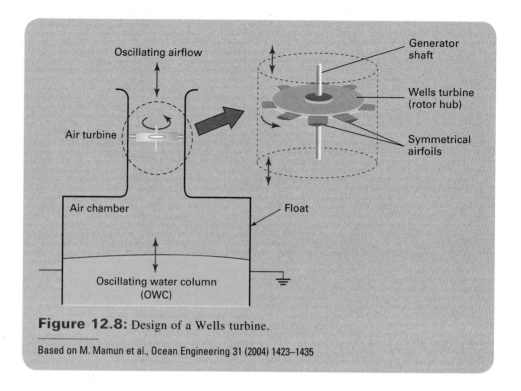

Figure 12.8: Design of a Wells turbine.

Based on M. Mamun et al., Ocean Engineering 31 (2004) 1423–1435

mechanical means that use the wave motion to turn an electric generator. Figures 12.11 and 12.12 show a prototype wave buoy produced by NEMOS GmbH in Germany. Rather than coupling to only the vertical movement of the buoy, a system of cables, as shown in Figure 12.12, harvests energy from both vertical and horizontal buoy movement to drive a generator positioned on a nearby tower. Towers of offshore wind turbines are ideal for this purpose; they will facilitate power transmission to shore by

Figure 12.9: Pelamis prototype machine on-site at the European Marine Energy Centre off Orkney.

Figure 12.10: Experimental wave energy buoy off Kaneohe Bay, Oahu, Hawaii. Portions of the device above and below the surface are shown.

David Fleetham/Alamy Stock Photo

Figure 12.11: Prototype NEMOS wave power buoy. The tower supporting the electric generator can be seen to the right of the photograph.

Courtesy of NEMOS GmbH

Figure 12.12: Schematic diagram of a NEMOS waver power buoy showing underwater footings and cables.

Courtesy of NEMOS GmbH

ENERGY EXTRA 12.1
The design of a Pelamis

The Pelamis has been one of the most successful devices for harvesting wave energy. Its efficiency in converting wave energy into electricity depends on its ability to couple the wave motion to the mechanical motion of device. This coupling is optimized when the frequency response of the Pelamis (that is, its resonant frequency) is close to the frequency of the waves. Although the Pelamis can be designed so that its resonant frequency is near the average wave frequency, it is necessary to adjust the Pelamis frequency response

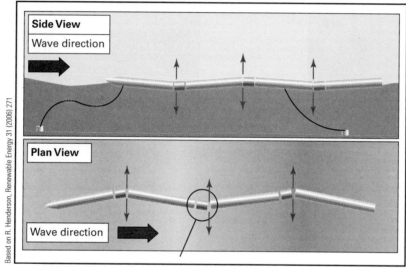

Based on R. Henderson, Renewable Energy 31 (2006) 271

Movement of a Pelamis in the vertical (top) and horizontal (bottom) directions.

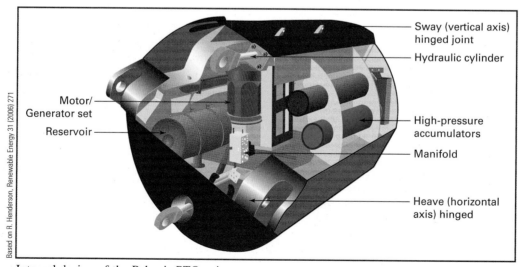

Based on R. Henderson, Renewable Energy 31 (2006) 271

Internal design of the Pelamis PTO unit.

Continued on page 399

Energy Extra 12.1 continued

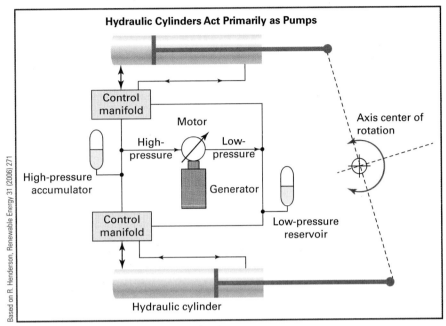

Based on R. Henderson, Renewable Energy 31 (2006) 271

Hydraulic Cylinders Act Primarily as Pumps

Control manifold

Motor

High-pressure

Low-pressure

High-pressure accumulator

Generator

Axis center of rotation

Control manifold

Low-pressure reservoir

Hydraulic cylinder

Schematic of the PTO unit in a Pelamis.

to account for variations in the wave frequency. It is also necessary to vary the response of the unit in a cyclic way through the wave cycle in order to maximize energy transfer. Because the Pelamis generates electricity by using hydraulic cylinders to pump fluid through a turbine, the resonant frequency of the device can be adjusted by controlling the hydraulic pressure in the system and thereby controlling the resistance of the joints. A detailed picture of the motion of a Pelamis in waves is shown in the first diagram. The sections of the Pelamis flex in both the vertical and horizontal directions, as shown in the figure.

These sections are connected by shorter elements, called the power take-off (PTO) mechanisms (see the figure), that contain the hydraulic systems and allow the sections to flex in both directions. The design of these elements has recently been reviewed by Henderson, "Design, simulation, and testing

of a novel hydraulic power take-off system for the Pelamis wave energy converter" [*Renewable Energy* **31** (2006): 271].

A simplified schematic of the hydraulic control system is shown in the above figure. The control manifolds ensure that optimal resistance to wave motion is provided by the hydraulic cylinders and maximize the efficiency of energy transfer. High-pressure accumulators store fluid at high pressure and provide a smooth flow of fluid to the hydraulic motor to even out any cyclic variability and random anomalies in electricity generation.

Topic for Discussion
The Pelamis is constructed in four hinged sections and has a total length of 120 m. Discuss the relevance of this geometry and size.

using existing infrastructure and provide a system in which wave and wind power will complement each other by evening out the intermittent nature of both energy sources.

12.3c **Wave-Focusing Devices**

The Wave Dragon (Figure 12.13) is a wave-focusing device. The device is moored offshore, typically a few kilometers from land. Two arms act as curved reflectors to gather waves, which are directed to the sloping ramp at the far end of the device (Figure 12.14).

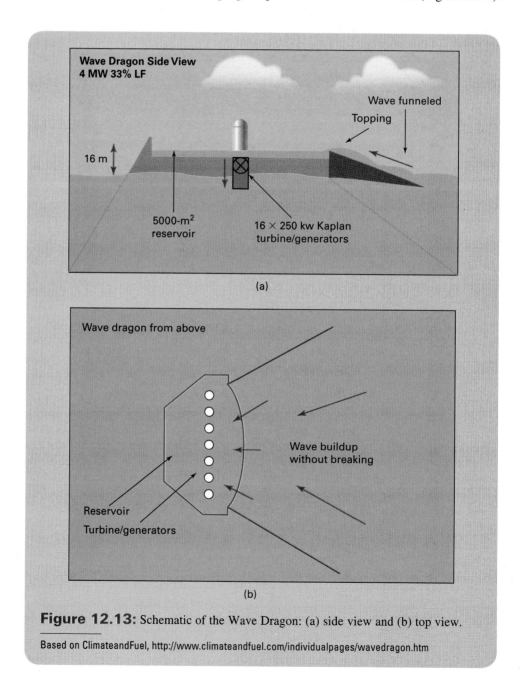

Figure 12.13: Schematic of the Wave Dragon: (a) side view and (b) top view.

Based on ClimateandFuel, http://www.climateandfuel.com/individualpages/wavedragon.htm

Figure 12.14: Photograph of the Wave Dragon.

Erik Friis-Madsen/Wave Dragon

Example 12.3

What is the total average wave power available in Portugal? The country has a continental coastline of about 940 km. How does the annual wave energy compare with an estimate of the national energy needs?

Solution

Figure 12.1 shows an average wave power in the region around Portugal of 48 kW/m. For a total coastline of 940 km, the total power is

$$P = (48 \text{ kW/m}) \times (1000 \text{ m/km}) \times (944 \text{ km}) = 4.53 \times 10^7 \text{ kW}.$$

The per-year energy is

$$E = (4.53 \times 10^7 \text{ kW}) \times (8760 \text{ h/y}) = 3.97 \times 10^{11} \text{ kWh},$$

or

$$E = (3.9 \times 10^{11} \text{ kWh}) \times (3.6 \times 10^6 \text{ J/kWh}) = 1.43 \times 10^{18} \text{ J}.$$

Figure 2.5 shows an average per-capita energy use of around 150 GJ for most southern European countries. Portugal, with a population of 10.6 million, has a total energy requirement of about

$$(10.6 \times 10^6) \times (150 \times 10^9 \text{ J}) = 1.59 \times 10^{18} \text{ J}.$$

These values are very similar, although it is important to note, as just discussed, that probably only a small fraction of the available wave energy is practical for utilization.

The water collected in the reservoir runs back into the sea through turbines, which are used to run generators. Several prototype devices have been constructed off the coasts of the United Kingdom, Portugal, and Denmark. Because the device can be large, it harvests energy from a substantial length of wave front. Prototype and planned devices have outputs in the range of tens of megawatts electrical for moderate sea states.

12.4 Wave Energy Resources

At present all wave power devices are prototypes or have had limited commercialization. The future development of this energy resource requires a consideration of the extent of wave energy. Certainly, it is a renewable resource with a lifetime that is, for all practical purposes, infinite. However, as shown in Figure 12.1, it is a resource that is best exploited in a limited number of locations around the world. It is estimated that the average deepwater wave power worldwide is in the range of 10^{12}–10^{13} W. This is similar to, or somewhat less than the current world power requirements. Since all of this wave power cannot be exploited, it cannot meet our total needs. An ambitious effort to utilize wave power with current or foreseeable technology could make up to perhaps a few percent of the available wave power commercially viable. Because wave energy resources occur in specific geographical areas, it is a resource that can be locally significant. An analysis of wave power places production costs in the 5–10 cents per kilowatt-hour range, which, according to Table 2.6, make it reasonably competitive with other options.

Like wind power, wave power is subject to daily and seasonal fluctuations, as well as random fluctuations due to changing meteorological conditions. Thus, this is best used as a source of energy in conjunction with storage systems and/or other energy resources, as discussed in Chapter 18.

Wave power has minimal environmental impact. The Wells turbines used in most onshore OWCs tend to be noisy and need to be acoustically insulated, particularly in populated areas. Offshore wave farms, consisting of devices like the Pelamis, wave buoys or the Wave Dragon must be cognizant of navigational considerations. Aesthetically, wave farms are probably less conspicuous than offshore wind farms.

12.5 Summary

Wave energy is an extensive resource, comparable to our total energy needs. However, only a small fraction of wave energy can be extracted in a viable manner. This chapter described the geographic distribution of wave resources that may be of commercial interest. If viable resources are fully utilized, wave energy could account for a few percent of our energy requirements. Because waves are variable and somewhat unpredictable, wave energy is most appropriately utilized in conjunction with other resources and a suitable energy storage system.

This chapter considered the basic physics of the power available from ocean waves and demonstrated that the total power per unit length (l) of a wave is $\dfrac{P}{l} = \dfrac{H^2 g^2 \rho T}{16\pi}$, where H is the wave height, and T is the wave period. The dependence of power on the square of the wave height indicates that the majority of energy available occurs during periods of above-average wave conditions.

The chapter described the three current technologies for harnessing wave energy: oscillating water columns, floats and pitching devices, and wave-focusing devices. The oscillating water column uses wave motion to force air through a turbine to drive a generator to produce electricity. The most common turbine for this purpose is the bidirectional Wells turbine.

Floating and pitching devices, such as the Pelamis, have probably been the most extensively developed of the wave energy–harvesting technologies. These devices use the mechanical movement of the wave to operate turbine/generators to produce electricity. Wave-focusing devices, such as the Wave Dragon, use ramps to convert the kinetic energy of waves into gravitiational potential energy, which can then be used to drive turbines.

Although the technology for the conversion of wave energy to electricity is relatively straightforward, the development of commercial devices is still in its early stages. Portugal and Scotland have been the most active countries in developing this resource. In the long term, wave energy may be economically competitive with other alternative energy technologies. Overall, its environmental impact is relatively low, and safety and security are not significant concerns.

Problems

12-1 North Carolina has some of the highest average wave energies in the United States along its roughly 500 km of coastline (Figure 12.1). Would it be reasonable for North Carolina to utilize wave energy for all of its electricity? Assume that a total annual per-capita electricity requirement of 15 MWh$_e$ and that 5% of the coastline is used for wave electricity generation at an average efficiency of 20%.

12-2 A small boat with two occupants and a 15-kW outboard motor (total mass = 500 kg) moves up and down in a wave with a height of 1 m and a period of 10 seconds. Calculate the ratio of the average wave power lifting the boat from the trough of the wave to the crest of the wave to the power available from the motor.

12-3 An ocean wave has a velocity of 8.5 m/s. What are its period and wavelength?

12-4 An ocean wave has a height of 2.5 m and a period of 10 s. What is the total power available from 20 m of wave front for this wave?

12-5 The average wave power available in Hawaii (Figure 12.1) is 100 kW/m. For waves with a period of 10 s, what is the average wave height?

12-6 In a storm, waves may have a height of 12 m and a period of 15 s. Compare the power per meter of wave front to the average values shown in Figure 12.1.

12-7 In deepwater, tsunamis have a relatively small height (typically 2 m) and a very long period (typically 30 minutes). The amplitude becomes larger as they pile up when they reach shallow water. For a tsunami in deep water, calculate the energy per meter of wave front (in J/m). Compare this with the energy per meter of width of a 2-m wide, 1500-kg automobile traveling at 120 km/h. This comparison emphasizes the damage that can be caused by a tsunami.

12-8 A wave travels at 10 m/s and has a height of 2 m. If it is incident on a wave energy device that generates electricity from wave energy with an average efficiency of 20%, how large would the device have to be to generate 1 MW$_e$?

12-9 A seiche is a standing wave that occurs due to resonant conditions on the surface of lakes and enclosed ocean bays. It is the result of surface perturbations caused by phenomena such as seismic activities or weather conditions. The period of a seiche is approximated by the expression

$$T = \frac{2L}{\sqrt{gh}},$$

where L is the length of the resonant enclosed region, g is the gravitational acceleration, and h is the mean water depth. Lake Ontario is particularly susceptible to seiches. Calculate the period of a seiche along the length of Lake Ontario.

12-10 A section of ocean shoreline has incident waves with a constant period of 15 s. During one day the waves have a height of 2 m for 12 hours, 3 m for 8 hours, and 5 m for 4 hours.

(a) What is the total energy available per km of coast?
(b) During which period of different wave heights is the greatest energy available?
(c) What is the average wave power per km of coast during the day?

12-11 Most of Greenland's electricity comes from hydroelectric facilities. These produce a total of about 300 million kWh annually. If wave power on Greenland's southeast coast was developed, how many kilometers of coast would be required at a wave-power-generation efficiency of 25%, to replace the hydroelectric generation?

12-12 An ocean wave has a height of 5 m and a period of 12 s.

(a) Calculate the wavelength of this wave.
(b) Calculate the power per unit wave front for this wave.
(c) Calculate the energy per unit wave front for one period of this wave.

12-13 An ocean wave has a wavelength of 120 m and a period of 16 s.

(a) Calculate the wave's velocity.
(b) If the amplitude of the wave is 1.5 m, calculate the energy per km of wave front in one period.

12-14 A tsunami in deep water has a period of 5.7×10^2 s.

   (a) What is its velocity?
   (b) What is its wavelength?
   (c) If the tsunami has a height of 2.5 m, what is its power per km of wave front?

12-15 (a) If all other factors are the same, which is the single most important parameter in maximizing the energy per unit length of wave front for one period of the wave: wavelength, wave height, or wave velocity?

   (b) What is the effect of doubling each of these parameters on the wave energy per period?

12-16 If a tsunami in deep water has a period of 10 minutes, what would the wave height have to be to provide the same power in 1 km of wave front as the output of a 1000 MW$_e$ coal-fired generating station? This result emphasizes the fact that a tsunami in deep water is actually difficult to detect in comparison to normal ocean waves with wavelengths of hundreds of meters and wave heights of several meters.

12-17 An ocean wave has a velocity of 20 m/s.

   (a) What is its period?
   (b) What is its wavelength?
   (c) If the energy per meter of wave front is 1 MJ per period, what is the wave height?

12-18 Hawaii has about 1200 km of coast line and a population of about 1.4 million.

   (a) If Hawaii's per-capita energy consumption is similar to the rest of the United States, calculate the total annual primary energy use in this state.

   (b) Compare the results from part (a) with the total annual wave energy available in Hawaii. What fraction of the total wave energy would satisfy the primary energy needs?

12-19 An ocean wave has a period of 10 s and a height of 2 m in deep water.

   (a) What is its wavelength?
   (b) As it approaches shore, its period decreases to 6 s. What is its wavelength at this time?
   (c) For the situation in part (b), if wave energy per wavelength is conserved, what is the new wave height?

12-20 The LIMPET wave-generating facility in Scotland has a width of 21 m. Estimate its efficiency.

Bibliography

G. Boyle (Ed.). *Renewable Energy.* Oxford University Press, Oxford (2004).

R. Cohen. "Energy from the ocean." *Philos. Trans. Royal Soc. London* **A307** (1982): 405.

A. Goldin. *Oceans of Energy—Reservoir of Power for the Future.* Harcourt, Brace, Jovanovich, New York (1980).

M. Knott. "Power from the waves." *New Scientist* **179** (2003): 2473.

M. McCormick. *Ocean Wave Energy Conversion.* Wiley, New York (1981).

T. R. Penney and D. Bharathan. "Power from the sea." *Scientific American* **256** (1987): 86.

D. Ross. *Energy from the Waves.* Pergamon Press, Oxford (1979).

R. J. Seymour. *Ocean Energy Recovery: The State of the Art.* American Society of Civil Engineers, New York (1992).

World Energy Council. *World Energy Resources 2016*, available online at https://www.worldenergy.org/publications/2016/world-energy-resources-2016/

Tidal Energy

Learning Objectives: After reading the material in Chapter 13, you should understand:

- The reasons for tidal motion and resonance effects in enclosed basins.
- The energy associated with tidal movement.
- The design of barrage systems.
- The availability and utilization of tidal energy based on barrage systems.
- Environmental and other factors that affect the viability of barrage systems.
- The design of tidal current energy systems.
- Experimental and commercial tidal current systems.

13.1 Introduction

In addition to waves, energy is associated with the movement of water in the ocean because of tides. Tidal motion is much more predictable than wave activity, and the realization that the energy associated with this movement could be harnessed dates back to well before 1000 CE. In medieval Europe, barrages were constructed to trap tidal water and subsequently release it through waterwheels. Like hydromechanical energy harnessed from flowing rivers, this tidal energy was used to perform tasks such as grinding grain. The interest in using tidal energy to generate electricity on a commercial scale originated in the 1960s as part of the growing desire to develop nonfossil fuel energy sources. The first (and still functional) commercial tidal electricity generating facility grew out of this interest and opened in 1966. Although tides exist everywhere in the world's oceans, a sufficiently large tidal range is necessary to make tidal energy practical. These conditions exist only in a few locations on earth. This is particularly true in the United States where tidal energy opportunities are very scarce. Nonetheless, interest in tidal energy as a alternative to fossil fuels has grown, and there are now several new tidal energy initiatives in North America and elsewhere. This chapter reviews the historical, scientific and engineering aspects of tidal power development and the progress that has been made in this area in recent years.

13.2 Energy from the Tides

The ocean tides are the result of the gravitational interaction of the ocean's water with the sun and the moon. This source of energy, along with nuclear and geothermal, are the only energy sources available that are not in some way a direct manifestation of solar radiation. Tidal energy may be available in the form of gravitational potential energy resulting from the raising and lowering of the ocean water level, or it may be kinetic energy associated with currents flowing through channels as a result of changing water levels. There has been increased interest during the past half century or so in the development of tidal power. It has, however, been used since the eleventh century in England and France, where the use of watermills, powered by tidal water motion paralleled the use of traditional waterwheels utilizing flowing rivers.

The tides arise primarily from the gravitational force acting between the oceans and the moon. Water on the side of the earth facing the moon is closer to the moon than the water on the other side of the earth, and is therefore more strongly attracted to the moon. This causes the water to form a bulge toward the moon. The water on the side of the earth away from the moon is less strongly attracted to the moon and bulges in the opposite direction. This behavior is illustrated in Figure 13.1. As the earth rotates, this tidal bulge moves around the earth with a period of half the *lunar day* (that is the time it takes the earth to return to the same position relative to the moon). This effect gives rise to the *diurnal period* of the tides, which is about 12 hours and 25 minutes.

Tidal behavior is complicated by the presence of the sun, which also has a gravitational effect on the oceans. This effect is significant, although smaller than that of the moon. When the moon is in the new phase or full phase, the earth, moon, and sun align in a single line (Figure 13.1). In this case, the solar tide adds to the lunar tide, giving rise to an increased tidal range. This situation is referred to as a *spring tide*. When the moon is in its quarter phase, it lies at right angles to the line between the earth and the sun (Figure 13.1). In this case (known as a *neap tide*), the lunar and solar effects do not

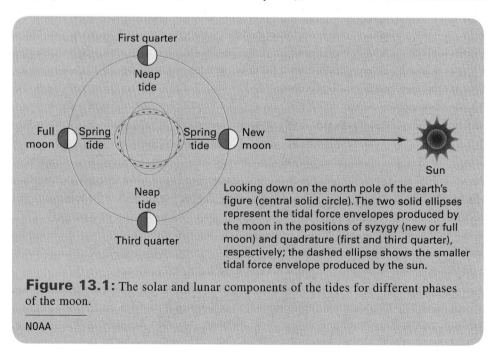

Looking down on the north pole of the earth's figure (central solid circle). The two solid ellipses represent the tidal force envelopes produced by the moon in the positions of syzygy (new or full moon) and quadrature (first and third quarter), respectively; the dashed ellipse shows the smaller tidal force envelope produced by the sun.

Figure 13.1: The solar and lunar components of the tides for different phases of the moon.

NOAA

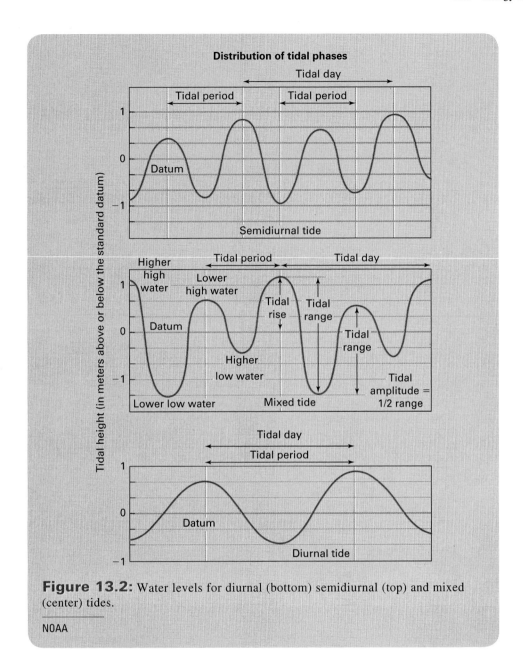

Figure 13.2: Water levels for diurnal (bottom) semidiurnal (top) and mixed (center) tides.

NOAA

work together and the tidal range is not as great. During a neap tide, the earth's oceans would experience tidal bulges four times during the lunar day. If the amplitude of the lunar tidal bulge and the amplitude of the solar tidal bulge are similar, the periodicity of the tide would be one-fourth of the lunar day, or about 6 hours and 13 minutes. This is referred to as a *semidiurnal tide*. In most cases, tides are a mixture of diurnal and semidiurnal, as shown in Figure 13.2. The exact details of the diurnal and semidiurnal components of the tide at a particular location depend on the relative positions of the sun and the moon and on the specific coastal geography.

    Interest in harnessing the energy of the tides for the production of electricity has, until fairly recently, dealt with the utilization of the gravitational potential energy

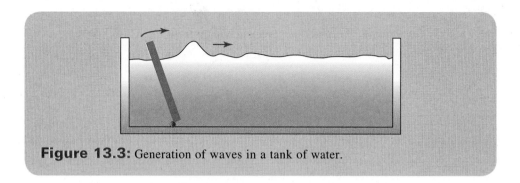

Figure 13.3: Generation of waves in a tank of water.

associated with the rising and falling water level. This approach has been primarily aimed at using enclosed basins where resonance conditions (as described later in the chapter) make the tidal range particularly large. Particular sites around the world that satisfy this condition are described in detail in the next section. The tidal range in an enclosed basin results from the tidal period and the geometry of the basin. Increased tidal range occurs as a result of resonance effects. This behavior can be illustrated using the properties of water in a tank with a mechanism (a paddle) for making waves. If a wave is produced as shown in Figure 13.3, it will travel to the far end of the tank and reflect back to the paddle. The period of time it takes to travel this distance depends on the properties of the water and the geometry (e.g., length) of the tank and is referred to as the *resonant period*. If a second wave is created just as the first wave is returning to the paddle, the wave amplitudes will add together. Thus, if waves are produced at the resonant period of the tank, then a resonance effect occurs, and the wave amplitude is very large. If the resonant period of an ocean basin is near the tidal period, then this effect gives rise to an increased tidal amplitude. This is what happens, for example, in the Bay of Fundy where the resonant period of the bay is about 13.3 hours compared to the tidal period of about 12.4 hours, leading to a very large tidal range of about 13 m. These extreme tides are illustrated in Figure 13.4.

Figure 13.4: Low tide at Parrsboro, Nova Scotia, on the Bay of Fundy, illustrating the extreme tide conditions experienced at this location.

Richard A. Dunlap

The potential energy associated with the tide can be calculated. If a basin has an area, A, and a tidal range, h, then the volume, V, of water that cycles through the basin every tidal period, T, is

$$V = 2Ah. \tag{13.1}$$

The factor of 2 accounts for the water entering the basin during the rising tide and leaving the basin during the falling tide and is appropriate for bi-directional generation. The mass of this water is given in terms of its density, ρ, as

$$m = 2Ah\rho. \tag{13.2}$$

The potential energy available as a result of the tidal movement is

$$E = 2Ahg\rho\left[\frac{h}{2}\right] = Ah^2g\rho, \tag{13.3}$$

where $h/2$ is the change in height of the center of mass of the water in the basin between high tide and low tide, and g is the gravitational acceleration. If this energy is available over the tidal period, it will correspond to an average power of

$$P = \frac{Ah^2g\rho}{T}. \tag{13.4}$$

It is seen in this equation that having a large area is important for the production of power, but having a large tidal range is even more important.

Example 13.1

Calculate the average tidal power that is available from a basin that is 10 km wide and 25 km long if the tidal range is 10 m.

Solution

The average available power is given by equation (13.4) as

$$P = \frac{(10 \times 25 \times 10^6\,\text{m}^2) \times (10\,\text{m})^2 \times (9.8\,\text{m/s}^2) \times (1025\,\text{kg/m}^3)}{(12.4\,\text{h} \times 3600\,\text{s/h})} = 5.6\,\text{GW},$$

equivalent to several large coal-fired or nuclear power plants (Chapters 3 and 6).

For reasons to be described, however, only about a quarter of this is actually accessible; that is a capacity factor of about 25%.

13.3 Barrage Systems

Barrage systems make use of tidal energy in the manner just described. The barrage is basically a dam that closes off a basin (sometimes called a *headpond*). *Sluice gates* in the barrage are doors that can be opened or closed to allow the water to enter or leave the basin. In a simple depiction (Figure 13.5 on the next page), the sluice gates are opened

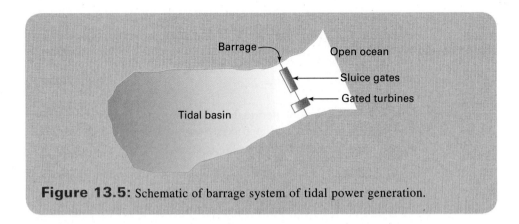

Figure 13.5: Schematic of barrage system of tidal power generation.

as the tide is rising to allow the basin to fill from the ocean. When the tide is high and the basin is filled, the sluice gates are closed. At this point, the turbine gates are closed, and the water level in the ocean drops as the tide goes out. When the tide is low, the turbine gates are opened to allow the water to flow through the turbines and back into the ocean, thereby generating electricity. The output of the turbines depends on the head of water enclosed in the basin, and the turbines are designed so that the water flow rate through the turbines (and hence the electric power generated) is compatible with the timescale of the tidal period. A simplified diagram of the barrage system in cross section is shown in Figure 13.6.

The main problem with this simple scheme is that, as the water is flowing out of the basin, the water level in the ocean is rising due to the rising tide. Thus, as the tide rises, the effective head of the water in the basin is decreasing, and when the water levels become equal on the two sides of the barrage, no further power is generated.

One way of improving on this simple scheme is shown in Figure 13.7. This is known as the *ebb generation scheme*. In this scheme, the basin is allowed to fill, and then the turbine gates are opened when the tide in the ocean is still on the way down, and electricity is generated until the tide rises again to the level of the falling water in the basin. This scheme, shown in the figure, illustrates that electricity is generated from a point where the tide is about half low until it is half high again.

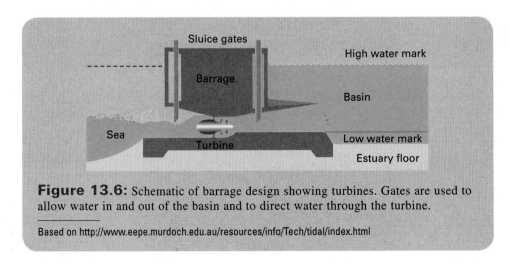

Figure 13.6: Schematic of barrage design showing turbines. Gates are used to allow water in and out of the basin and to direct water through the turbine.

Based on http://www.eepe.murdoch.edu.au/resources/info/Tech/tidal/index.html

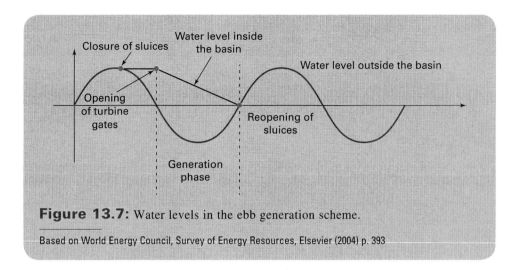

Figure 13.7: Water levels in the ebb generation scheme.

Based on World Energy Council, Survey of Energy Resources, Elsevier (2004) p. 393

An alternative approach is to close the sluice gates at low tide and allow the water to rise in the ocean before opening the turbine gates to allow the water into the basin. This situation (Figure 13.8) is referred to as the *flood generation scheme*. In general, the ebb generation scheme is more efficient than the flood generation scheme because, as a result of the typical shape of a basin, more water is contained in the top half than in the bottom half. Also, any rivers that flow into the basin will augment ebb generation but reduce flood generation. A simple analysis of Figures 13.7 and 13.8 provides a rough estimate of the capacity factor. Both illustrations show that power generation occurs over only one-half of the total tidal period; that is, water is flowing through the turbines only when the water level in the basin is changing linearly. In addition, it is seen that the total range of the water level inside the basin is one-half of the total range of the water level in the ocean. Thus, combining these two factors of one-half yields a typical capacity factor of 25%.

The implementation of barrage systems worldwide has been minimal. At present three major facilities are operating. The largest tidal power facility is the Sihwa Lake Tidal Power Station near Seoul, South Korea. This facility utilized an existing

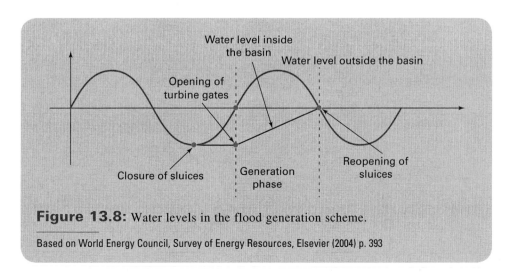

Figure 13.8: Water levels in the flood generation scheme.

Based on World Energy Council, Survey of Energy Resources, Elsevier (2004) p. 393

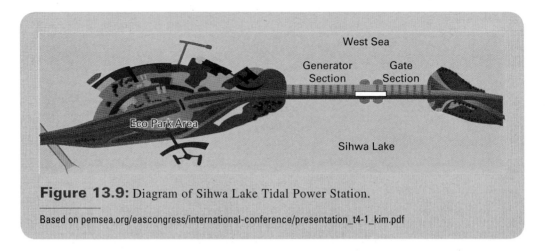

Figure 13.9: Diagram of Sihwa Lake Tidal Power Station.

Based on pemsea.org/eascongress/international-conference/presentation_t4-1_kim.pdf

12.7-km-long seawall that was constructed in 1994 to form a freshwater lake for agricultural purposes. By 2001, the water in Sihwa Lake became sufficiently polluted that it was no long useable for irrigation. The use of the seawall as a barrage for a tidal power station has allowed water to flow between Sihwa Lake and the ocean and has reduced pollution in the lake (which is now saltwater and is no longer used for agricultural irrigation). The tidal generating station became operational in 2012 and utilizes a flood generation scheme. The mean tidal range is 5.6 m, and ten turbines provide a total capacity of 254 MW_e at a capacity factor of 24.8%. Figure 13.9 shows a diagram of the facility and illustrates the relationship of the turbines and the sluice gates. Figure 13.10 shows the seawall with the generator section in the foreground.

The second largest operational tidal power plant is the Rance River Tidal Generating Station in Bretagne, France (Figure 13.11). Located on the estuary of the Rance River at its outflow into the English Channel between Dinard and Saint-Malo (Figure 13.12), it became operational in 1966 and has been in operation the longest of the currently used

Figure 13.10: Sihwa Lake Tidal Power Station with the generator section shown in the foreground.

Topic Images Inc./Getty Images

Figure 13.11: Barrage at the Rance River Tidal Generating Station in France.

Robert Estall photo agency/Alamy Stock Photo

tidal power facilities. The barrage is 750 m in length and accommodates a roadway across the river. It encloses a basin of 22.5 km² with a mean tidal range of about 8 m. The barrage contains 24 *bulb turbines* of the type shown schematically in Figure 13.13 on the next page. These are similar to Kaplan turbines (Figures 11.4 and 11.5). The maximum capacity of the station is 240 MW$_e$, but the actual average output is 68 MW$_e$, consistent with about a 25% capacity factor.

The third significant tidal power plant, at Annapolis Royal in Nova Scotia, Canada, became operational in 1984. This location is shown on the maps in Figures 13.14

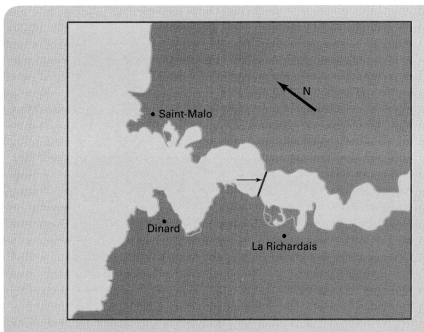

Figure 13.12: The location of the tidal barrage is shown by the small arrow.

Based on Google Earth

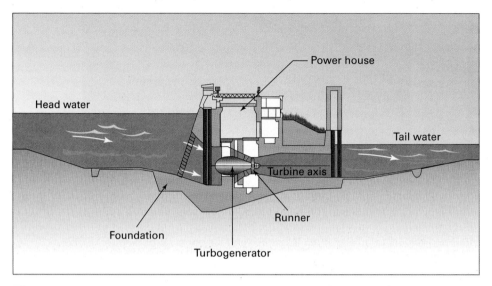

Figure 13.13: Bulb turbine used for a tidal power plant.

Based on Figure 1 in http://www.docstoc.com/docs/31700275/Electric-Power-Generation-Equipment
-Incorporating-Bulb-Turbine-generator---Patent-4289971 and http://nptel.iitm.ac.in/courses/Webcourse-contents
/IIT-KANPUR/machine/chapter_7/fluid2%5B1%5D.jpg

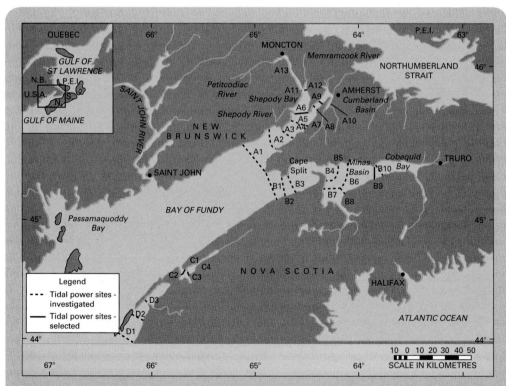

Figure 13.14: The Bay of Fundy showing the location of the tidal generating station
at Annapolis Royal (site C4) and other possible tidal power sites.

Atlaspix/Shutterstock.com, data from Google Maps

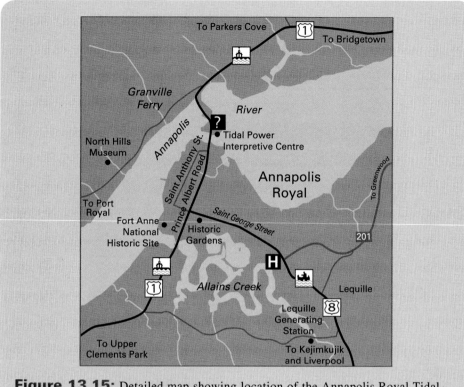

Figure 13.15: Detailed map showing location of the Annapolis Royal Tidal Generating Station on the Annapolis River.

and 13.15. The basin or headpond is formed upstream by the barrage. The facility is shown in Figure 13.16 on the next page, and the details of the sluice gates (which are large concrete doors that can be lowered and raised) are shown in Figure 13.17 on the next page. The barrage also serves as a causeway for vehicular traffic. The generating station has a maximum capacity of 18 MW_e, and typically generates electricity for about 5 hours during each tidal cycle. A *rim turbine* design (Figure 13.18 on the page 419) is used in this facility and reduces maintenance costs compared with bulb turbines.

At present, the fraction of tidal power resources that have been utilized is very small. Certainly there exists the possibility of implementing large-scale systems similar to that in France (or larger) at a number of locations worldwide. If tidal range and basin area are considered, then a dozen or more clearly identifiable sites are available for possible exploitation (Table 13.1 on the page 419). Even considering a capacity factor of about 25%, most of the sites listed in Table 13.1 would provide as much power as a major nuclear or fossil fuel generating station. In fact, some would provide substantially more. The Penzhinsk location has the potential to satisfy the electricity needs of a population of 6 million or more (at typical North American usage).

It is interesting to consider why development of these resources has not been pursued. The basic technology is straightforward, and in most ways, is similar to that used for hydroelectric power. However, large-scale tidal barrage systems are a substantial

Figure 13.16: Annapolis Royal, Nova Scotia tidal generating station. The photograph was taken from the north side of the Annapolis River looking southwest (see Figure 13.15). The headpond is in the foreground, and the exit to the ocean is behind the barrage.

Richard A. Dunlap

Figure 13.17: Sluice gates at the Annapolis Royal tidal power station.

Richard A. Dunlap

engineering challenge (as explained later in this chapter). Also, the longevity of tidal power resources and their environmental impact need to be investigated thoroughly.

Constructing a barrage across an ocean basin has some similar effects to constructing a major hydroelectric installation. The natural flow of water is disrupted. Consequences can include effects on navigation, fish and mammal movements, and the transportation of sediments. In the case of the Rance River station, locks have been constructed to allow small craft to pass from one side to the other, although changes in the marine ecology

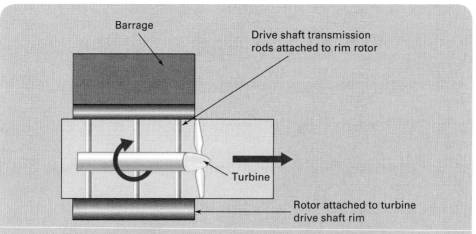

Figure 13.18: Schematic diagram of a rim turbine as used at the Annapolis Royal tidal power station. In this arrangement, the rotating turbine shaft is connected by rods to the generator, which is in the form of a ring outside the region that is exposed to the marine environment. This facilitates maintenance and reduces adverse environmental effects.

Based on Bright Hub Engineering, Marine Energy - Tidal Barrage, http://secure.brighthub.com /engineering/marine/articles/55520.aspx?image=58429

have been observed since the facility was constructed. In the case of the Sihwa Lake Tidal Power Station, an existing seawall was utilized as the barrage, and the water flow between Sihwa Lake and the ocean has helped to alleviate existing environmental problems. The effects on the transport of sedimentation are analogous to the collection of silt

Table 13.1: Possible sites for the development of tidal power using a barrage system. The table is not necessarily comprehensive but includes all major possibilities for barrage systems with a maximum capacity of 1000 MW or greater.

country	location	mean tidal range (m)	basin area (km²)	maximum capacity (MW)
Argentina	San Jose	5.8	778	5040
	Golfo Nuevo	3.7	2376	6570
	Santa Cruz	7.5	222	2420
	Rio Gallegos	7.5	177	1900
Australia	Secure Bay	7.0	140	1480
	Walcott Inlet	7.0	260	2800
Canada	Cobequid	12.4	240	5338
	Cumberland	10.9	90	1400
	Shepody	10.0	115	1800
India	Gulf of Khambhat	7.0	1970	7000
United Kingdom	Severn	7.0	520	8640
Russia	Mezen	6.7	2640	15,000
	Tuigar	6.8	1080	7800
	Penzhinsk	11.4	20,530	87,400

behind hydroelectric dams, as discussed in Chapter 11. The Bay of Fundy is an interesting case in this respect because the region is very muddy and sedimentation collection on the basin side of the barrage may substantially reduce the lifetime of the resource and contribute to ecological changes.

Perhaps the most important consideration for the implementation of a large-scale barrage system is the effect of the barrage on the characteristics of the tides themselves. For example, in the case of the Bay of Fundy, these effects have been modeled extensively. Figure 13.14 shows some possible sites for future tidal barrages in this area. In all cases, it is not the entire Bay of Fundy that would be barraged but relatively small inlets off the bay. The three locations given in Table 13.1 can be seen in the figure; Shepody (A6), Cumberland (A9), and Cobequid (B9). Constructing a barrage at any of these locations would decrease the effective length of the Bay of Fundy. As can be readily seen in the example shown in Figure 13.3, decreasing the length of the body of water decreases the time it takes a wave to propagate its length. Thus, barraging a portion of the Bay of Fundy would cause its resonant period to be decreased. Because the resonant period of the bay is currently 13.3 hours, decreasing this would bring it closer to the tidal period of 12.4 hours. This change would increase the resonance effect and the tidal range. The increased range can have significant effects, even very far from the Bay of Fundy. Computer models have shown that constructing a barrage across Cumberland Basin (site A8) would give rise to an increased tidal range of about 3 cm in Boston. Constructing a barrage across Cobequid Bay (site B9) would result in a corresponding 13-cm increase in tidal range in Boston. Even very moderate changes in mean ocean water level can have severe consequences for coastal environments and marine structures.

Finally, the length of the necessary barrage needs to be considered as an important factor in the economic viability of this approach. For example, a barrage across Cobequid Bay would be 8 km in length, while a barrage across the Severn Estuary in England would be 17 km in length. The Gulf of Khambhat in India would require a barrage 25 km long. Feasibility studies of the Severn barrage system have estimated the infrastructure cost to be around US$30 billion. Certainly these are major undertakings from the standpoints of both construction and maintenance.

Example 13.2

For a generation efficiency of 25%, calculate the total tidal energy and average power available during one tidal cycle for Cobequid Bay in Canada.

Solution
Use equation (13.3),

$$E = Ah^2 g\rho,$$

and the following values:

$A = 240 \text{ km}^2 = 2.4 \times 10^8 \text{ m}^2.$
$h = 12.4 \text{ m}.$
$g = 9.8 \text{ m/s}^2.$
$\rho = 1025 \text{ kg/m}^3.$

Continued on page 421

Example 13.2 continued

The total energy is

$$E = (2.4 \times 10^8 \text{ m}^2) \times (12.4 \text{ m})^2 \times (9.8 \text{ m/s}^2) \times (1025 \text{ kg/m}^3) = 3.7 \times 10^{14} \text{ J}.$$

At an efficiency of 25%, this provides $(0.25) \times 3.7 \times 10^{14} = 9.3 \times 10^{13}$ J. The average power is found by dividing by the tidal period as

$$\frac{9.3 \times 10^{13} \text{J}}{12.4 \text{h} \times 3600 \text{s/h}} = 2.083 \times 10^9 \text{W} = 2083 \text{MW}.$$

For a variety of reasons, barrage systems are less appealing than they may have been in the past. The focus on tidal energy is now aimed at systems that do not involve barrages. Some of these approaches are discussed in the next section.

13.4 Nonbarrage Tidal Power Systems

Several approaches to the utilization of tidal power that do not require a barrage system have been investigated. The more significant options include tidal lagoons, which are artificial enclosures that utilize the same principles as barrage systems, and underwater turbines and tidal fences, which make use of the kinetic energy associated with tidal currents.

13.4a Tidal Lagoons

A tidal lagoon is an artificial enclosure located up to a couple of kilometers offshore in a region where there is a large tidal range. Although, in principle, the operation of a tidal lagoon is similar to that of a barrage system, it avoids many of the potential environmental consequences that result from blocking off an ocean basin with a barrage. It does not significantly affect the resonant period of the basin, and adverse effects due to sedimentation are minimized. Also, navigation and aquatic animal movements are not greatly influenced. Bidirectional flow generation (i.e., generation during both ebb and flood cycles) can be much more easily implemented than in traditional barrage systems. This ability is largely due to the more regular shape of the lagoon compared with a natural basin. Typically, the design also eliminates the need for separate sluice and turbine gates. The water levels in the ocean and inside the lagoon, as a function of tidal period, are shown in Figure 13.19 on the next page, where periods of ebb and flood generation are illustrated. Bidirectional generation is expected to give about a 35% capacity factor and is predicted to be a cost-effective improvement over single-direction ebb generation. This improvement in capacity factor can be roughly seen by combining part of the ebb generation period in Figure 13.7 with part of the flood generation period in Figure 13.8. An illustration of a typical (proposed) tidal lagoon installation is shown in Figure 13.20 on the page 423, where possible turbine locations allowing energy generation during water flow into or out of the lagoon are indicated. The calculated power output during the tidal cycle is shown in Figure 13.21 on the page 423. A prototype tidal lagoon is in the planning stages at Swansea off the coast of Wales. The mean annual tidal range in this region is 6.3 m, and the proposed facility

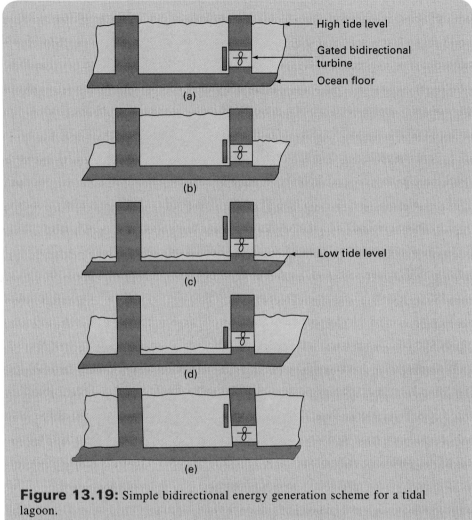

Figure 13.19: Simple bidirectional energy generation scheme for a tidal lagoon.
(a) High tide–turbine gates closed.
(b) Falling tide–head created inside lagoon. Turbine gates opened when tide is about half way down to generate electricity.
(c) Low tide–turbine gates closed as tide begins to rise.
(d) Rising tide–head created outside lagoon. Turbine gates opened when tide is about half way up to generate electricity.
(e) High tide–turbine gates closed when tide is high to return to diagram (a).

would have a lagoon area of about 5 km^2. Twenty-four 2.5-MW$_e$ turbines would give a maximum output of 60 MW$_e$. A larger (300-MW$_e$) tidal lagoon system is in the planning stages in China.

13.4b Underwater Turbines

Underwater turbines make use of the kinetic energy associated with tidal currents and rely more on current velocity and volume than actual tidal range. The same technology is applicable to the harvesting of energy associated with any type of ocean current. The basic

Figure 13.20: Illustration of typical proposed tidal lagoon power generation system.

© Aquaret (coordinated by AquaTT), www.aquaret.com

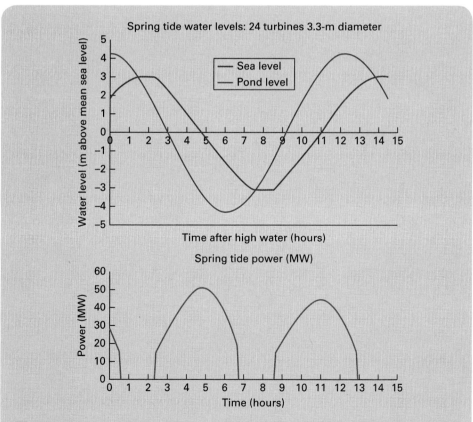

Figure 13.21: Water levels and predicted power output of Swansea Tidal Lagoon.

Based on Tidal Lagoon Power Generation Scheme in Swansea Bay. April 2006. A report on behalf of the Department of Trade and Industry and the Welsh Development Agency

design follows along the lines of the design of a wind turbine. Analogous to equation (10.3) for wind energy, the power per unit area (P/A) generated by a water turbine is given as

$$\frac{P}{A} = \frac{1}{2}C\rho v^3, \tag{13.5}$$

where C is referred to as the coefficient of performance and is a measure of the turbine efficiency. Typical C values are in the range of 0.3 to 0.4. ρ is the water density, and v is the current velocity. Current velocities are typically less than wind velocities, and velocity appears as v^3 in the equation. However, the difference between the density of air (1.204 kg/m^3) and the density of seawater (1025 kg/m^3) is the most significant difference between the analysis of the energy content of currents and wind. Even relatively small (5- to 10-m-diameter) water turbines are rated in the megawatt range for moderate tidal currents.

Example 13.3

A basin of area 12 km^2 with a tidal range of 11 m drains through an opening 200 m wide. What is the average water velocity during the falling tide cycle?

Solution
The total volume of tidal water in the basin is

$$(12 \text{ km}^2 \times 10^6 \text{ m}^2/\text{km}^2) \times 11 \text{ m} = 1.32 \times 10^8 \text{ m}^3.$$

If this water drains from the basin over the period of the falling tide (6.2 hours), then the average flow rate is

$$\frac{1.32 \times 10^8 \text{ m}^3}{6.2 \text{ h} \times 3600 \text{ s/h}} = 5.9 \times 10^3 \text{ m}^3/\text{s}.$$

The area of the tidal opening is

$$200 \text{ m} \times 11 \text{ m} = 2.2 \times 10^3 \text{ m}^2.$$

The average velocity is therefore

$$\frac{5.9 \times 10^3 \text{ m}^3/\text{s}}{2.2 \times 10^3 \text{ m}^2} = 2.7 \text{ m/s}.$$

Example 13.4

For a coefficient of performance of 0.35, calculate the diameter of an underwater turbine that would be required to produce an output of 5 MW in a current of 3.5 m/s.

Solution
Rearranging equation (13.5) to solve for the turbine area gives

$$A = \frac{2P}{C\rho v^3}$$

Continued on page 425

Example 13.4 continued

Using values of

$$P = 5 \times 10^6 \, \text{W},$$

$$C = 0.35,$$

$$\rho = 1025 \, \text{kg/m}^3, \text{ and}$$

$$v = 3.5 \, \text{m/s}$$

gives an area of

$$\frac{2 \times 5 \times 10^6 \, \text{W}}{0.35 \times 1025 \, \text{kg/m}^3 \times (3.5 \, \text{m/s})^3} = 650 \, \text{m}^2,$$

giving a diameter of

$$d = (4 \times 650 \, \text{m}^2/\pi)^{1/2} = 29 \, \text{m}.$$

A number of devices along these lines have been tested and are expected to have minimal environmental impact. The simplest, perhaps, are bladed devices that are similar to modern wind turbines (Figures 13.22 to 13.24). With these devices, it is straightforward to track the direction of the current, as is done for the wind direction with wind turbines by rotating the rotor assemblies about a vertical axis to ensure that the blades always face into the direction of the current. Other turbine geometries that are similar to rim turbines (Figure 13.25 on the page 427) are also being investigated. It is possible that *shrouded turbines*—that is, turbines with an enclosure to direct water toward the blades—may be effective for water devices. In a suitable location, a number of individual tidal turbines of, say, 1-MW output could be combined into a tidal energy farm, much as wind farms have been implemented.

Figure 13.22: Underwater turbine constructed by Seagen at Strangford Lough. Turbine blades are in their operating position below the water and the support tower is shown above the surface. Compare with Figure 13.23.

Figure 13.23: Illustration of the Seagen underwater turbine with rotors raised above the water's surface for maintenance.

Bloomberg via Getty Images

Figure 13.24: Atlantis' AR 1000 turbine.

Courtesy of Atlantis Resources Corporation

There has been significant activity in the development of tidal energy in the United Kingdom, Australia, and Canada, and several tidal turbines have been constructed or are in the planning stages. The facility shown in Figure 13.22 is located off the coast of Northern Ireland and has operated commercially since 2008 and is rated at 1.2-MW$_e$ output. In Canada a 2.0 MW tidal turbine has recently installed off

Figure 13.25: The rotor for the OpenHydro system being transported. Note the workers in red to the right of the rotor for scale.

THE CANADIAN PRESS/Andrew Vaughan

Cape Sharp in the Bay of Fundy near Parrsboro, Nova Scotia (Figure 13.26). A second 2 MW turbine is expected to be installed in the same area in the near future. Smaller experimental devices have also been tested on Canada's Pacific coast off the west coast of Vancouver Island.

Figure 13.26: The 16-m-diameter tidal turbine produced by OpenHydro and rated at 2 MW being deployed in the Bay of Fundy, November 2016.

THE CANADIAN PRESS/Andrew Vaughan

The Seagen tidal turbine (Figures 13.22 and 13.23) was installed in the 500-m-wide mouth of Strangford Lough in Northern Ireland (see the following map) in 2008. The twin rotor turbine was constructed by Marine Current Turbines (MCT) at a cost of about £8.5 million and has a total rated output of 1.2 MW. Each of the 16-m-diameter rotors is attached to a drive system with a mass of 27 t at the end of a transverse arm. The arm assembly is mounted on a 3-m-diameter tower, which is attached to the seabed by a single piling. The total device has a mass of about 1000 t.

During the maximum tidal current of about 3.6 m/s, the rotors rotate at about 14 rpm, corresponding to a tip speed of about 12 m/s. Because of the difference in density of air and water, the tidal turbine is most efficient at a tip speed of about one-third that of

a typical modern wind turbine (Chapter 10). The rotor assembly can be rotated on a vertical axis around the tower by 180 degrees so that the turbine is equally functional for both tidal current directions. Because of the predictable nature of the tides, this 1.2-MW tidal current generator actually produces about the same total energy output integrated over time as a wind turbine rated at 2.5 MW.

During normal operation (Figure 13.22), the water surface is about 3 m above the tips of the rotors, even at low tide. Thus, the turbine has no effect on navigation for small boats. The twin rotor assembly can be raised on the central tower (Figure 13.23), making maintenance straightforward. Studies have suggested that Seagen has minimal impact on marine animals because of the relatively low speed of the rotors.

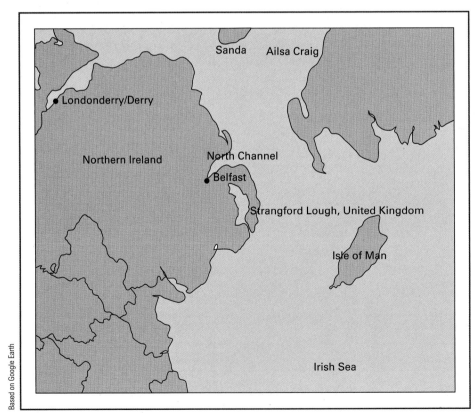

Location of Strangford Lough in Northern Ireland.

Continued on page 429

Energy Extra 13.1 continued

Although the Seagen turbine was decommissioned in 2017 it was the first successful commercial tidal current generator and generated electricity that fulfill the needs of over 1000 homes. It has provided significant information about the operation of tidal turbines and has led to the planning of further tidal current initiatives in the U.K. over the next few years including a proposed tidal current farm off the coast of Scotland. It has also renewed interest in this technology in other parts of the world.

Topic for Discussion

A public utility determines its cost ($/kWh) for generating electricity at a particular facility by dividing the total cost of the facility over its lifetime by the total number of kilowatt-hours the facility produces. The former includes capital infrastructure costs, operation and maintenance costs, fuel costs, and financing costs (if any). The latter is based on rated capacity and the overall capacity factor. This analysis is well-established for coal-fired stations or nuclear plants, for example. Discuss how this analysis might be the same or different for a tidal turbine farm and where there may be different certainties and uncertainties in the various factors.

13.4c **Tidal Fences**

Water turbines are horizontal-axis machines. Like the Darrieus rotor used for wind power generation, water turbines can also be vertical-axis machines. An advantage of this type of design over the horizontal-axis machine is that the gearbox and generator can easily be located above water, where it is much less susceptible to adverse environmental conditions. It is anticipated that a number of such turbines could be combined to span the mouth of an estuary or basin. Such a construction, referred to as a *tidal fence*, is shown in Figure 13.27. Although prototype vertical-axis machines have

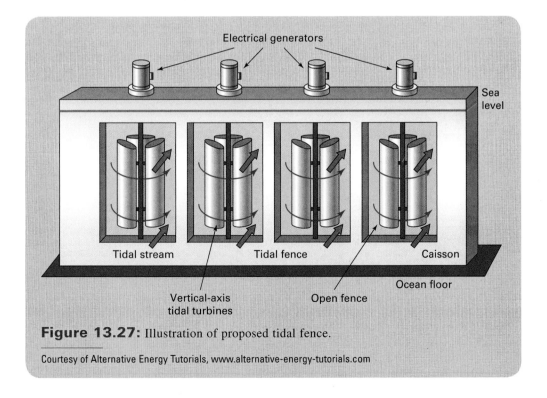

Figure 13.27: Illustration of proposed tidal fence.

Courtesy of Alternative Energy Tutorials, www.alternative-energy-tutorials.com

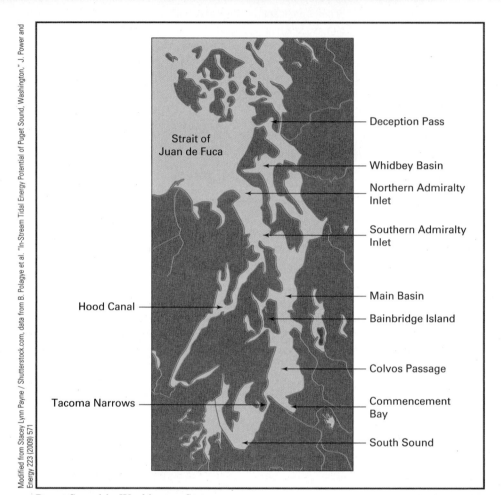

Modified from Stacey Lynn Payne / Shutterstock.com, data from B. Polagye et al. "In-Stream Tidal Energy Potential of Puget Sound, Washington," J. Power and Energy 223 (2009) 571

Puget Sound in Washington State.

While the most significant tidal energy resources worldwide are located outside the United States, a few sites within the United States are under consideration for development of this energy resource. The three locations that have attracted the most interest are Puget Sound in Washington State, Cook Inlet in Alaska, and the Bay of Fundy in Maine.

Puget Sound is a complex estuarine system that connects to the Pacific Ocean through the Strait of Juan de Fuca, as shown in the map above. Several organizations have investigated possible tidal generation in Puget Sound. Tacoma Power has looked at installing an in-stream turbine at Tacoma Narrows but has concluded that this is not commercially viable at the present. The U.S. Navy has shown interest in tidal power in Puget Sound, but their investigations are at the very early stages. The most advanced investigations are by the Snohomish County Public Utility District which has considered sites at Admiralty Inlet and Deception Pass (where tidal currents can exceed 4 m/s) and it plans to deploy two underwater turbines in the near future.

Continued on page 431

Energy Extra 13.2 continued

Archival photograph by Mr. Steve Nicklas, NGS/RSD

The tidal bore on the Turnagain Arm of Cook Inlet.

Cook Inlet has the second largest tidal range (after the Bay of Fundy) in North America. It has been estimated that 90% of the tidal energy resources in the United States occur in Alaska. The Turnagain Arm on Cook Inlet is one of only a few dozen locations worldwide where a tidal bore occurs. A tidal bore is a wave-like structure that is formed in a tidal river or narrow inlet by the leading edge of the incoming tide (see photograph). Ocean Renewable Power Company is in the early stages of projects at two locations in Cook Inlet, one near Anchorage and one near the town of Nikiski. If these projects move to commercialization, the electricity generated will help to offset Alaska's dependence on electricity generated from local fossil fuel resources.

The most advanced tidal energy efforts in the United States are in Maine. Maine borders the Bay of Fundy, along with New Brunswick and Nova Scotia, Canada, and experiences tidal ranges that can be as much as 8 m. Since 2006, pilot projects run by Ocean Renewable Power Corporation have been in place in Cobscook Bay near the Maine–New Brunswick border. Additional collaborative projects are underway in collaboration with the Canadian company, Fundy Tidal Power, to utilize the energy potential of this border region.

Topic for Discussion

Would it be reasonable to expect that a tidal energy farm consisting of 1-MW$_e$ turbines located in Cook Inlet could provide all the necessary electric power for Anchorage, Alaska? How many turbines would be needed?

been built, full-scale implementation of a tidal fence is still in the future. The Dalupiri Passage in the Philippines is a possible site for such development and could have a maximum capacity of 2200 MW$_e$. Capacity factors can be as high as about 50% for such installations as power generation is bidirectional, and it is not necessary to wait for a tidal basin to fill and/or empty before generation begins. The Severn Estuary (or Mouth of Severn) in England (Figure 13.28) is another site where a tidal fence may be a viable alternative to a traditional barrage system. The tidal fence can also serve as a causeway for vehicles. Environmental and navigational concerns for tidal fences

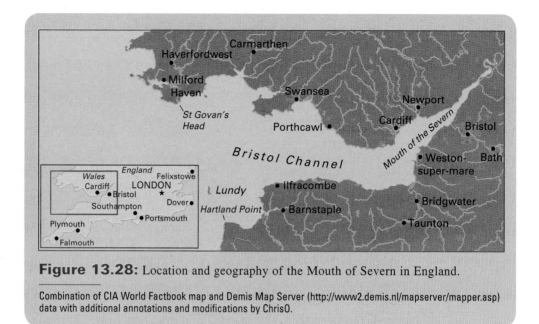

Figure 13.28: Location and geography of the Mouth of Severn in England.

Combination of CIA World Factbook map and Demis Map Server (http://www2.demis.nl/mapserver/mapper.asp) data with additional annotations and modifications by ChrisO.

may be more serious than for horizontal-axis water turbines, either isolated or in tidal power farms, and careful consideration of any possible adverse environmental impacts is necessary.

If properly implemented, tidal power can be a renewable resource with low environmental impact. Although the tidal energy in the oceans is substantial, the number of locations where it can be utilized as a viable energy source is somewhat limited. However, new, developing technologies have met with success and may lead the way to further implementation of this energy source.

13.5 Summary

The tides in the earth's oceans are the result of the combined gravitational forces from the sun and the moon. This chapter has reviewed the properties of tidal energy and has discussed the possible approaches to harnessing this energy. Although the tidal amplitude in most locations on earth is relatively small, the tidal range becomes amplified in certain enclosed basins as a result of resonance effects. These resonance effects are caused by the similarity of the tidal period and the natural oscillation period of the basin. In these areas, the possibility of utilizing tidal energy is the greatest. This chapter showed that, analogous to hydroelectric power from rivers, both the potential and kinetic energy associated with the tides can be utilized.

A barrage system uses the potential energy associated with the rising and falling tide to generate electricity. For example, at high tide, the water inside an enclosed basin is trapped by a gate. At low tide, when the sea level outside the basin has dropped, the water inside the basin can be released and can drive turbine/generators to produce electricity. Two small barrage systems, one in France and one in Canada, have been used successfully for many years for the commercial generation of electricity.

Unfortunately, barrage systems can have significant undesirable environmental impact. Manipulating the flow of tidal water can affect the tidal range over a very extended region and can also alter the life cycles of biological organisms in the area. Also, silt deposits can adversely affect the functioning of a barrage system.

This chapter also described tidal current generating systems, which have been more actively pursued in recent years. This approach is analogous to the run-of-the-river approach to hydroelectric power. The kinetic energy associated with tidal movement is harnessed using underwater turbines. This type of system minimizes the impact on the environment. Experimental devices have been tested, and there has been some limited commercial utilization of this technology in Canada and the United Kingdom. The tidal current approach also increases the number of possible locations that can be developed. The harvesting of tidal current energy focuses on the development of relatively small individual devices, rather like the approach toward wind energy, which can then be incorporated into tidal energy farms. Tidal energy development in the future is likely to have fairly local, rather than global, influences on energy production. This is because, like wave energy, it is confined to a fraction of the coastal areas worldwide.

Problems

13-1 Calculate the total kinetic energy (in MJ and in kWh) of a 1-m^3 parcel of seawater moving with a velocity of 1 m/s.

13-2 What is the total tidal energy available (at 100% efficiency) during the falling tide from a basin of area 100 km^2 with a tidal range of 8 m?

13-3 Using the tidal range and basin area for Penzhinsk inlet given in the chapter, estimate the monetary value of the electrical energy that can be generated (at 25% efficiency) during one falling tide cycle. Assume electricity has a value of $0.10/kWh.

13-4 A water turbine 15 m in diameter is placed in a channel with a tidal current moving with a velocity of 3.5 m/s. Estimate the power produced by the turbine.

13-5 What is the power available from an underwater turbine with a coefficient of performance of 0.4 that is 10 m in diameter in a current of seawater traveling at a velocity of 2.0 m/s?

13-6 For a tidal current with a velocity of 1.5 m/s in a channel that is 0.5 km wide and 28 m deep (on average), calculate the mass of water per unit time moving through the channel.

13-7 Calculate the average current velocity that is necessary for an 8-m-diameter water turbine with a coefficient of performance of 0.3 to generate 1 MW.

13-8 Consider the four proposed tidal barrage generating facilities for Argentina as described in Table 13.1. Calculate the average power available and compare with the capacity as given in the table. Comment on any differences between these quantities.

13-9 A 10-m-diameter tidal turbine is placed in an inlet and operates with a coefficient of performance of 0.3. During a tidal period of 12.4 hours, the tidal current in the inlet varies from 3.5 m/s in one direction to 3.5 m/s in the opposite direction. If it is assumed that the change in tidal current velocity is linear in time, what is the total energy output of the turbine during one tidal period.

13-10 The velocity variation of a tidal current through a tidal cycle is more accurately described as sinusoidal (rather than linear as assumed in Problem 13.9). Reconsider this previous problem using a sinusoidal variation. How does this new assumption affect the result?

13-11 It has been suggested that tidal turbines could be used to generate power from ocean currents. The Gulf Stream, for example, has a constant velocity of about 1.5 m/s. Discuss the advantages or disadvantages of this approach compared to the use of turbines in a tidal current that varies from −3.0 m/s (in one direction) to +3.0 m/s (in the other direction).

13-12 A tidal barrage system generates 15 MW_e from 85% efficient turbines/generators at the beginning of the power generation cycle. The mean head of water above the turbines is 5.7 m and the water flows through the turbines with a velocity of 6 m/s. What is the total turbine area?

13-13 Although the Gulf Stream is probably the best-known ocean current, the Antarctic Circumpolar Current is the largest ocean current. This current circulates around the continent of Antarctica and represents the transport of about 1.5×10^8 m³/s with a velocity of up to 1.1 m/s. Calculate the total power available from this current, and compare this result to the current world power use.

13-14 Based on a calculation of the total energy available and an assumed capacity factor of 0.24, calculate the total annual energy production for tidal generating stations at the three Canadian sites listed in Table 13.1. Estimate the number of homes that could be supplied with electricity from these three sites.

13-15 A tidal inlet has an area of 320 km² and a tidal range of 6.7 m.
   (a) Calculate the total energy available during one tidal cycle.
   (b) Calculate the average power available over one tidal cycle.
   (c) If the capacity factor is 0.27, what number of homes (on average) could be supplied with electricity if each home, on average, requires 1.3 kW?

13-16 In some tidal areas, such as the Bay of Fundy, some smaller basins could be used for tidal generation using barrage systems, while more open areas could be used for in-stream tidal turbines. Discuss how the variations in power output from these two approaches during the tidal cycle would (or would not) complement each other.

13-17 (a) A tidal turbine has an efficiency of 40%. In a tidal stream with a velocity of 7 m/s, what is the diameter of a turbine that is required to produce 1 MW_e?
   (b) A wind turbine has an efficiency of 40%. In a wind with a velocity of 7 m/s, what is the diameter of a turbine that is required to produce 1 MW_e?

(c) From the answers to parts (a) and (b), calculate the relative mass of water to air passing through the rotor per second.

13-18 Consider the possibility of implementing either tidal or wave energy in India.

   (a) From the maximum capacity for tidal generation in the Gulf of Khambhat as given in Table 13.1 and an estimated capacity factor of 0.24, calculate the total annual energy generated.

   (b) For wave energy at a capacity factor of 0.25, how many km of coastline on the south coast of India would have to be developed to generate the same annual energy as tidal energy in the Gulf of Khambhat as calculated in part (a)?

13-19 For the proposed tidal lagoon facility in Swansea (Section 13.4a), calculate the average tidal power available during the tidal cycle. Are the proposed turbines appropriate for this facility?

13-20 Estimate the average velocity of water leaving the Gulf of Khambhat during the falling tide cycle.

Bibliography

A. C. Baker. *Tidal Power.* Peter Peregrinus, London (1991).

G. Boyle (Ed.). *Renewable Energy.* Oxford University Press, Oxford (2004).

R. Charlier. *Tidal Energy.* Van Nostrand Reinhold, New York (1992).

R. Charlier. *Ocean Energies: Environmental, Economic, and Technological Aspects of Alternative Power Source.* Elsevier, New York (1993).

R. H. Clark. "Tidal power." *Annual Review of Energy* (1997): 2648.

R. H. Clark. "Tidal power." In *Wiley Encyclopedia of Energy Technology and the Environment,* Vol. 4. A. Bisio and S. Boots (Eds.). Wiley, New York (1995).

R. H. Clark (Ed.). *Tidal Power: Trends and Developments.* Thomas Telford, London (1992).

R. Cohen. "Energy from the ocean." *Philos. Trans. Royal Soc. London* **A307** (1982): 405.

M. W. Conley and G. R. Daborn (Eds.). *Energy Options for Atlantic Canada.* Formac Publishing, Halifax (1983).

A. Goldin. *Oceans of Energy—Reservoir of Power for the Future.* Harcourt, Brace, Jovanovich, New York (1980).

D. A. Greenberg. "Modeling tidal power." *Scientific American* **257** (1987): 128.

T. J. Hammons. "Tidal power." *Proc. IEEE,* **8** (1993): 419.

T. R. Penney and D. Bharathan. "Power from the sea." *Scientific American* **256** (1987): 86.

R. J. Seymour. *Ocean Energy Recovery: The State of the Art.* American Society of Civil Engineers, New York (1992).

World Energy Council. *World Energy Resources 2016*, available online at https://www.worldenergy.org/publications/2016/world-energy-resources-2016/

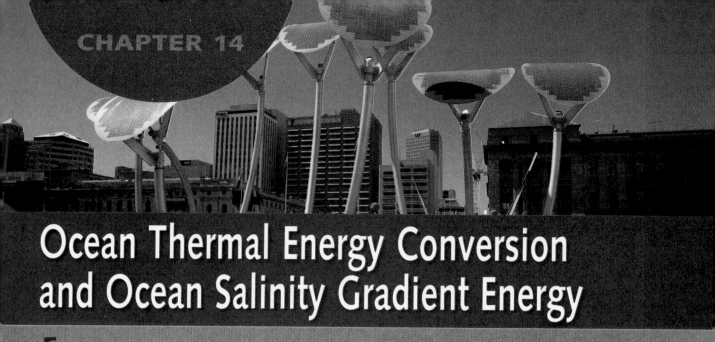

CHAPTER 14

Ocean Thermal Energy Conversion and Ocean Salinity Gradient Energy

Learning Objectives: After reading the material in Chapter 14, you should understand:

- The distribution of thermal energy in the oceans.
- The use of a heat engine to extract energy from the oceans and convert it into electricity.
- The design of the different types of OTEC systems.
- Experimental OTEC facilities and their performance.
- The basic principles of osmotic energy production.
- Experimental facilities for osmotic energy production.

14.1 Introduction

Several options for obtaining energy from the kinetic or potential energy associated with the movement of water in the oceans have been discussed in the past two chapters. Two additional methods of extracting energy from the oceans are discussed in this chapter. The first, *ocean thermal energy conversion* (OTEC), is a method of making use of the thermal energy in the ocean. It is basically a heat engine, like the heat engine used to convert the thermal energy of steam produced by burning fossil fuels or from nuclear reactions into mechanical energy, which can be used to generate electricity. Another option is to make use of energy associated with the chemical composition of seawater. The removal of salt from seawater to produce freshwater (i.e., desalination) requires energy. Conversely, energy can be extracted when seawater and freshwater are mixed. Energy obtained by this method is referred to as salinity gradient energy, or *osmotic energy*.

14.2 Basic Principles of Ocean Thermal Energy Conversion

The conversion of thermal energy in the ocean to mechanical energy depends on the availability of a cold reservoir into which excess heat can be transferred. This situation exists in the ocean because water temperature is a function of depth. In deep water, the depth dependence of the temperature is shown in Figure 14.1.

The figure shows that, below about 1000 m, the temperature is fairly constant at about 4°C. This is independent of the location on earth, the time of the year, and the air temperature. In temperate and colder regions, particularly in the winter, the temperature difference between deep water and surface water is relatively small. In tropical regions, particularly in the summer, the temperature difference between deep water and surface water is large. The OTEC process relies on a temperature gradient between the surface and the very deep water in the ocean and is suited only to tropical regions. Figure 14.2 on the next page shows the mean annular temperature difference between the ocean surface and water at 1000 m depth for various locations on earth. A temperature difference of at least 20°C (and preferably as large as possible) is necessary to make use of OTEC.

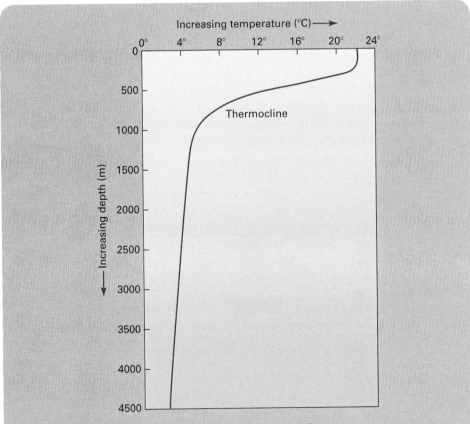

Figure 14.1: Typical depth variation of the ocean temperature in a tropical region.

Based on National Earth Science Teachers Association, http://www.windows2universe.org/earth/Water/temp.html

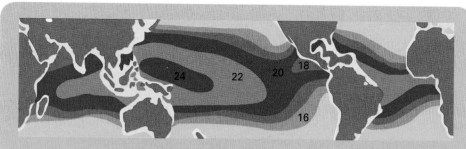

Figure 14.2: Temperature differences between the ocean surface and water at a depth of 1000 m measured in °C.

Based on World Energy Council, 2010 Survey of Energy Resources, http://www.worldenergy.org/documents/ser_2010_report_1.pdf

In an OTEC facility, cold water is pumped from deep in the ocean. Heat may then be transferred from the warm surface water to the cold water obtained from the ocean's depths. According to Figure 1.4, the warm surface water acts as the hot reservoir, and the cold water acts as the cold reservoir, and the transfer of heat allows for the extraction of mechanical energy. This process, as in all heat engines, is governed by the laws of thermodynamics, and the maximum efficiency that can be obtained is limited by the Carnot efficiency (Chapter 1):

$$\eta = 100\left(1 - \frac{T_c}{T_h}\right). \tag{14.1}$$

As in Chapter 1, it is essential that the temperatures used in this equation are expressed using an absolute temperature scale, such as Kelvin.

Example 14.1

Calculate the Carnot efficiency for an OTEC system where the warm surface water is at a temperature of 22°C, and the cold water from deep in the ocean is at a temperature of 4°C.

Solution

To calculate the thermodynamic efficiency, the temperatures must be expressed in Kelvin. Thus the temperatures of the hot and cold reservoirs are

$$T_h = 22°C + 273 = 295 \text{ K}$$

and

$$T_c = 4°C + 273 = 277 \text{ K},$$

respectively. The Carnot efficiency is then given from equation (14.1):

$$\eta = 100 \times \left(1 - \frac{277 \text{ K}}{295 \text{ K}}\right) = 6.1\%.$$

The thermal energy content of the warm water is substantial, and the efficiency of electricity generation from mechanical energy is quite good, as is typically the case for a generator. However, the low Carnot efficiency, as calculated in Example 14.1, is a factor that must be considered in the design and implementation of an OTEC system.

The possibility of extracting useful energy by the OTEC process was first proposed in 1881 by the French physicist Jacques Arsene d'Arsonval. The first operational OTEC system was constructed by Georges Claude in Mantanzas Bay, Cuba, in 1930. It produced 22 kW of electricity, but unfortunately, due to its low thermodynamic efficiency, it consumed more than that to operate. It is important to understand the operation of an OTEC plant, as described in the next section, in order to assess the possibilities of improving on this design.

14.3 OTEC System Design

14.3a Open Cycle Systems

There are basically three types of OTEC systems: *open-cycle systems, closed-cycle systems, and hybrid systems.* The open-cycle system (Figure 14.3) is the simplest, and uses seawater itself as the working fluid.

Warm water from near the ocean surface enters the system on the left side of the diagram and is vaporized in an evaporator by lowering the vapor pressure with a pump. The details of this process become evident in the phase diagram of water in Figure 14.4 on the next page. At a pressure of 1 atm, the transition line from the liquid (water) region to the vapor (steam) region (that is, the boiling point) occurs at 100°C. If the pressure is lowered, the transition line from liquid to vapor, as illustrated in the figure, shows that the boiling point occurs at a lower temperature. For a sufficiently low pressure, the water boils at (or below) room temperature. So when the warm seawater enters the evaporator (Figure 14.3), the pressure is lowered sufficiently to cause it to become vapor. This vapor

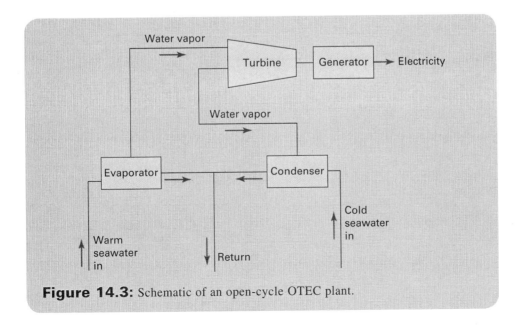

Figure 14.3: Schematic of an open-cycle OTEC plant.

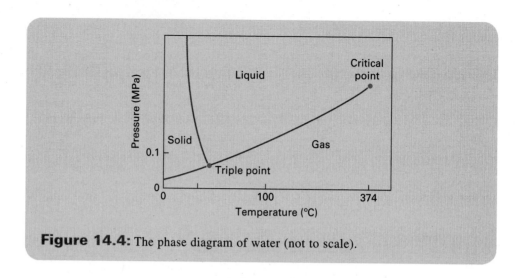

Figure 14.4: The phase diagram of water (not to scale).

drives a turbine (which, in turn, drives a generator to produce electricity) and then enters into a condenser where it is turned back into a liquid. The condenser is cooled by the cold seawater pumped from deep in the ocean, and this lowers the temperature of the vapor to a temperature below the liquid–vapor transition line. Generally, such systems utilize equal quantities of warm surface water and cold deep-ocean water. Ideally, the energy produced is determined by the energy extracted from the vapor by the turbine, which is related to the temperature difference between the evaporator and the condenser, and by the thermodynamic efficiency of the system, which is determined from the difference between the hot reservoir (warm surface water) and the cold reservoir (cold deep-ocean water). However, efficiencies are never ideal, and, in addition, pumps to bring water from deep in the ocean to the surface require substantial energy input.

A detailed diagram of the operation of an open-cycle OTEC is shown in Figure 14.5. As the figure shows, a by-product of energy production is desalinated water.

Example 14.2

Calculate the maximum power available from seawater that is cooled from 22°C to 4°C at a rate of 1 m³ per second. Note the specific heat of seawater is 3930 J/(kg·°C) and its density is 1025 kg/m³.

Solution

From equation (8.12) the relationship between the change of temperature, $(T_h - T_c)$, and heat, Q, is

$$Q = (T_h - T_c)Cm,$$

where C is the specific heat and m is the mass. Thus the energy available from cooling 1 m³ (or 1025 kg) of seawater is

$$Q = (22°C - 4°C) \times (3930 \text{ J/(kg·°C)}) \times (1025 \text{ kg}) = 72.5 \text{ MJ}.$$

If this energy is available during 1 second, then the power will be 72.5 MW.

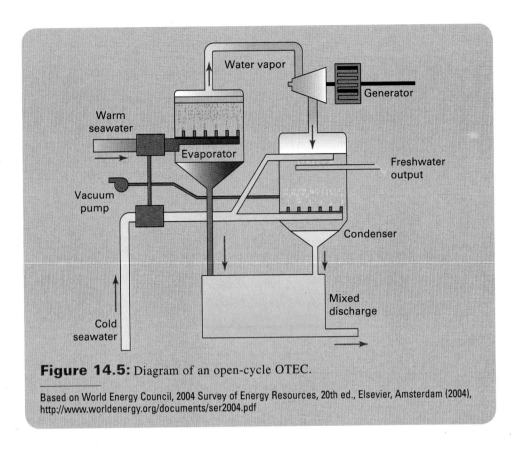

Figure 14.5: Diagram of an open-cycle OTEC.

Based on World Energy Council, 2004 Survey of Energy Resources, 20th ed., Elsevier, Amsterdam (2004), http://www.worldenergy.org/documents/ser2004.pdf

This is produced by the condensation of water vapor and can be used to provide freshwater, which is often a valuable resource in locations where OTEC plants may be viable.

14.3b Closed-Cycle Systems

A schematic diagram of a closed-cycle OTEC system is shown in Figure 14.6 on the next page. In this system, the working fluid is a substance with a phase diagram that is compatible with the temperatures and pressures present in the system. These substances, such as ammonia, are often used in commercial refrigerators. Figure 14.6 shows that the warm seawater from the ocean's surface is used to heat the working fluid, which is in a closed-system, to a temperature above its boiling point. The vaporized fluid is used to drive the turbine, after which it is condensed back into a liquid by cooling it with cold seawater pumped from deep in the ocean. A more detailed diagram of a closed-cycle OTEC is shown in Figure 14.7 on the next page. Closed-cycle systems have the disadvantage that leaks may release hazardous materials to the environment.

14.3c Hybrid Systems

The operation of a hybrid system (Figure 14.8 on the page 443) follows along the lines of the operation of a closed-cycle system. However, the warm seawater used to vaporize the working fluid is itself vaporized. The vaporized working fluid runs the turbine, as in the closed-cycle system, and the vaporized water is recondensed to produce desalinated water. Thus the system works as a closed-cycle system with the added value of producing freshwater.

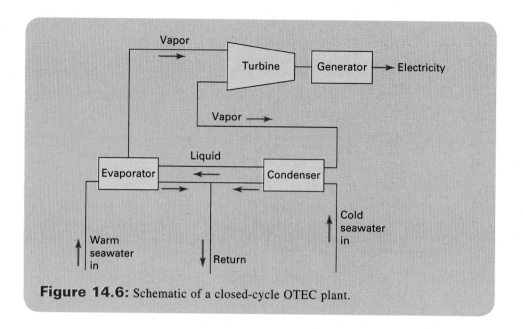

Figure 14.6: Schematic of a closed-cycle OTEC plant.

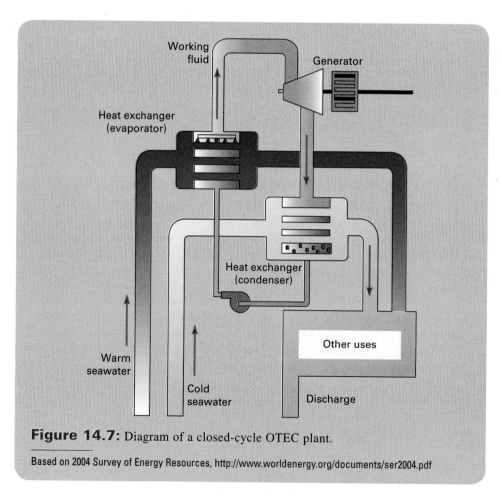

Figure 14.7: Diagram of a closed-cycle OTEC plant.

Based on 2004 Survey of Energy Resources, http://www.worldenergy.org/documents/ser2004.pdf

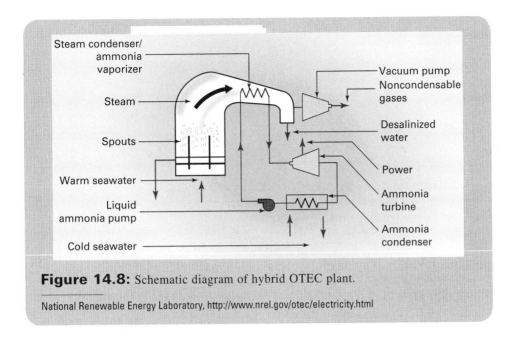

Figure 14.8: Schematic diagram of hybrid OTEC plant.

National Renewable Energy Laboratory, http://www.nrel.gov/otec/electricity.html

14.4 Physics of the Operation of an OTEC System

It might seem from Examples 14.1 and 14.2 that a simple analysis of an OTEC device might be obtained merely by multiplying the Carnot efficiency times the power available. This, however, does not provide an accurate picture of how the OTEC system works. The actual detailed operation of an OTEC device is quite complex. However, we can gain insight into the operation of the system and obtain a reasonable assessment of its operating characteristics using the following approach.

Consider a parcel of seawater that is cooled by an amount $(T_h - T_c)$. The energy that becomes available is

$$Q = mC(T_h - T_c), \tag{14.2}$$

where C is the specific heat. If water flowing at a rate of dm/dt is cooled from T_h to T_c, then the power available is

$$P = (dm/dt)C(T_h - T_c). \tag{14.3}$$

If we look, for example, at the details of a closed cycle OTEC system (Figure 14.6), we see that the total temperature drop between the warm seawater and the cold seawater occurs over several parts of the system. There is a temperature drop across the evaporator, there is a temperature drop across the turbine, and there is a temperature drop across the condenser. The sum of these temperature drops must be $(T_h - T_c)$. It can be shown that the system performance is optimized if the three temperature drops are

$$\Delta T(evaporator) = (T_h - T_c)/4.$$
$$\Delta T(turbine) = (T_h - T_c)/2. \tag{14.4}$$
$$\Delta T(condenser) = (T_h - T_c)/4.$$

The Carnot efficiency of the closed turbine cycle is, therefore,

$$\eta = (T_h - T_c)/2T_h. \tag{14.5}$$

The heat extracted by the evaporator will be

$$Q = mC(T_h - T_c)/4 \tag{14.6}$$

and the corresponding power will be

$$P = (dm/dt)C(T_h - T_c)/4, \tag{14.7}$$

where dm/dt is the flow rate in kg/s through the evaporator.

Combining equation (14.7) with the heat engine efficiency from equation (14.5) gives the thermodynamic power generated as

$$P = (dm/dt)C(T_h - T_c)^2/(8T_h). \tag{14.8}$$

This is sometimes written in terms of the total water flow rate, $(dm/dt)_{tot}$ through the evaporator and the condenser, combined (on the assumption that these are equal) as

$$P = (dm/dt)_{tot}C(T_h - T_c)^2/(16T_h). \tag{14.9}$$

The electrical power output will be the quantity in equation (14.9) times the efficiency of the turbine/generator, typically 85% to 90%.

The result in equation (14.9) emphasizes the importance of a large temperature difference between the warm reservoir and the cold reservoir, as this appears squared on the right-hand side of the equation. This result comes from the fact that the energy extracted from the seawater is proportional to the temperature difference and the efficiency of the heat engine is also proportional to the temperature difference.

It is important to realize that the power required to operate the facility must be subtracted from output as given by equation (14.9) in order to determine the net power gain. The most significant component of operating power is that required for the cold water pumps to bring seawater from deep in the ocean to the facility. Total operating power can account for more than half of the power generated.

Example 14.3

Consider an OTEC system with a warm water temperature of 24°C and a cold water temperature of 4°C and temperature drops are described by equation (14.4).

(a) Calculate the ideal Carnot efficiency.
(b) For a flow rate of 100 m³/s through the evaporator and a turbine/generator efficiency of 85%, calculate the expected electrical power.

Solution

(a) The total temperature drop will be

$$(24°C - 4°C) = 20°C,$$

Continued on page 445

Example 14.3 continued

and the temperatures of the inlet and outlet of the turbine will be

$$24°C - 20°C/4 = 19°C = 292 \text{ K.}$$
$$4°C + 20°C/4 = 9°C = 282 \text{ K.}$$

So the Carnot efficiency will be

$$\eta = 100 \times (1 - 282K/292K) = 3.4\%.$$

(b) From equation (14.8), the power output will be

$$P = \eta_{\text{gen}}(dm/dt)C(T_h - T_c)^2/(8T_h),$$

or

$$P = (0.85) \times (100 \text{ m}^3/\text{s}) \times (1025 \text{ kg/m}^3) \times (3930 \text{ J/(kg·K)})$$
$$\times (20 \text{ K})^2/(8 \times (24 + 273)\text{K})$$
$$= 57.6 \text{ MW}_e.$$

Example 14.4

Consider an OTEC system where the temperature drops across the evaporator, turbine, and condenser were all $(T_h - T_c)/3$. Show that this would yield a lower power output than the conditions given in equation (14.4).

Solution

Following the development in Section 14.4 in the text, we assume that

$$\Delta T(evaporator) = (T_h - T_c)/3.$$
$$\Delta T(turbine) = (T_h - T_c)/3.$$
$$\Delta T(condenser) = (T_h - T_c)/3.$$

The Carnot efficiency of the closed turbine cycle is, therefore,

$$\eta = (T_h - T_c)/3T_h.$$

The heat extracted by the evaporator will be

$$Q = mC(T_h - T_c)/3,$$

and the corresponding power will be

$$P = (dm/dt)C(T_h - T_c)/3,$$

where *dm/dt* is the flow rate in kg/s through the evaporator. Combining the calculated power with the heat engine efficiency from above gives the thermodynamic power generated as

$$P = (dm/dt)C(T_h - T_c)^2/(9T_h).$$

Comparing this result with equation (14.8) for the situation given by equation (14.4) we see that the output power is less.

14.5 Implementation of OTEC Systems

OTEC systems can be constructed in three types of locations: onshore, mounted to the ocean floor, or floating.

In the case of onshore installations, the facility is located on land or in very shallow water close to shore in a region where appropriate ocean thermal gradients exist. Warm surface water is supplied from the region near the facility, while a supply pipe carries cold water from deeper water offshore. This type of facility has clear advantages and disadvantages. On the plus side, the plant itself is easy to maintain. It is less susceptible to the adverse marine environment, and it is simple to connect the electric output to the power grid. In general, it is important to avoid mixing output water with the warm water supply. However, for land-based systems, the freshwater produced from the condensed working fluid can be readily used as a supply of quality drinking water, and the waste cooling water is useful for air conditioning. Because OTEC facilities must be built in tropical regions, this cold water is an added benefit. Alternatively, the discharge water could be carried offshore by pipe. On the negative side, locations must be chosen to minimize the length of pipe necessary to bring cold water from depths of up to 1000 m (or more) offshore. The supply pipe is a substantial component of the infrastructure cost and is subject to adverse environmental conditions, leading to potential maintenance costs. This is particularly true because the pipes would traverse the surf zone near shore and would be subject to extreme stresses during storms.

Bottom-mounted systems can be seated on the continental shelf at depths of up to 100 m or so. Locating the plant outside the surf zone minimizes stress on the supply pipes, and, if the plant is not far offshore, the electrical connection to the grid is convenient.

Floating systems minimize the length of pipes needed to supply cold water from the ocean depths. This, in principle, would improve the net efficiency by reducing the power consumed in pumping water. However, a floating system has several major disadvantages. Because it would be in deep water, mooring the facility and stabilizing it in potentially adverse weather conditions is not straightforward. Also, the electricity that is produced must be transported to shore. This transport would involve expensive cables that would be subject to adverse environmental conditions or the use of an energy storage mechanism [such as hydrogen (Chapter 20)], which would further reduce the already low efficiency. Finally, it is important to ensure that waste cooling water is pumped far enough away from the facility so that it does not cool the warm surface water, thereby reducing the thermodynamic efficiency.

Serious efforts to utilize ocean thermal energy began in the mid-1970s. In 1974, the National Energy Laboratory of Hawaii Authority (NELHA) was established at Keahole Point in Hawaii. One of their major initiatives has been to research the possibility of producing energy by OTEC. Prototype facilities have generally taken one of two approaches to developing OTEC power; either an onshore facility, which minimizes operational and maintenance costs, or portable ship-based plants. The latter optimizes efficiency without incurring substantial infrastructure costs.

In 1979, NELHA launched the *MiniOTEC*, a 50-kW$_e$ OTEC facility mounted on a barge. The barge was moored about 2 km off the coast of Hawaii near Keahole Point. The closed-cycle OTEC system produced a gross electric output of 52 kW$_e$ but a net output (after using electricity for pumps, etc.) of about 15 kW$_e$. This was sufficient to run the lights and electronic equipment on the ship.

The most extensive testing of OTEC occurred between 1992 and 1998 at the NELHA facility in Hawaii. A 250-kW$_e$ open-cycle OTEC plant (Figure 14.9) was

Figure 14.9: Land-based 250-kW open-cycle OTEC demonstration plant at Keahole Point, Hawaii.

U.S. Department of Energy

constructed on shore at Keahole Point. Water at a temperature of 6°C was pumped from a depth of 820 m at a rate of about 400 L/s (0.4 m^3/s = 6400 gal/min) through a 1-m-diameter pipe. In May 1993, 50 kW$_e$ of net power was produced by this plant. Although this facility is no longer operating as an OTEC plant to produce electricity, it does pump cold water on shore. This is used locally for air conditioning and, at peak operation, replaces the equivalent of traditional air conditioning requiring 200 kW of electricity.

Although Japan has no OTEC possibilities itself, Japanese researchers have been involved in OTEC activities on the island of Nauru in the Pacific Ocean. From October 1981 to September 1982 a closed-cycle OTEC plant, rated at 100 kW$_e$, was operated on the island. The net power production after operating pumps was 31.5 kW$_e$, which was used to supply electricity to a local school.

It has been estimated that globally, OTEC resources amount to about 10^{13} W. This is close to our total world power requirements. However, only a small fraction of this resource is economically viable or practical. Onshore or close-to-shore sites are the easiest to implement. There are probably a few hundred sites, at most, that fulfill the necessary criteria for exploitation. The principal locations with cold water resources within 10 km or so of land are summarized in Table 14.1 on the next page. Many of these locations are islands where the use of OTEC could provide a substantial contribution to local energy needs and reduce the need for energy importation.

Certainly this discussion illustrates the problems facing the development of OTEC as an energy resource. Thus far, prototype plants have produced a small net energy output. However, infrastructure and operating costs, at this time, make OTEC commercially unattractive. Technical difficulties in the efficient implementation of OTEC are significant. Fundamental problems that need to be dealt with include the intrinsically low efficiency of the process and the difficulties associated with the marine environment. These difficulties include the corrosive nature of seawater and

Table 14.1: Principal locations for the utilization of onshore ocean thermal energy conversion.

area	country	temperature difference (°C)
Africa	Sao Tome	22
Latin America/Caribbean	Barbados	22
	Cuba	22–24
	Dominica	22
	Dominican Republic	21–24
	Grenada	27
	Haiti	21–24
	Jamaica	22
	Saint Lucia	22
	Saint Vincent	22
	Trinidad/Tobago	22–24
	U.S. Virgin Islands	21–24
Indian Ocean/Pacific	Comoros	20–25
	Cook Islands	21–22
	Fiji	22–23
	Guam	24
	Kiribati	23–24
	Maldives	22
	Mauritius	20–21
	New Caledonia	20–21
	Philippines	22–24
	Samoa	22–23
	Seychelles	21–22
	Solomon Islands	23–24
	Vanuatu	22–23

the probability of biofouling in the system. The transport of very large quantities of seawater is also a challenge. A full-scale plant producing, say, 100 MW$_e$ would require pumping something in the order of 1000 m^3 of seawater per second through the system.

However, OTEC energy production does provide a number of advantages. It has relatively low environmental impact and produces freshwater as a by-product as well as cold water for air conditioning use. Aquaculture is also an added benefit of OTEC. Cold water from the ocean depths has a large concentration of organic nutrients that are depleted in the surface water by biological processes, and this water may be used effectively for aquaculture. Also, the low temperature of this water opens up the possibilities for aquaculture in tropical regions that include the growth of fish (e.g., salmon) and invertebrates (e.g., lobsters) that are normally native to temperate regions—rather the opposite of geothermal aquaculture discussed in the next chapter. In some ways, OTEC may be considered as a viable resource for reasons other than energy production, with any generated electricity a bonus.

Predictions for the utilization of OTEC have often been far from accurate. A forecast made in 1980 (Figure 14.10) illustrates the optimism with which this resource

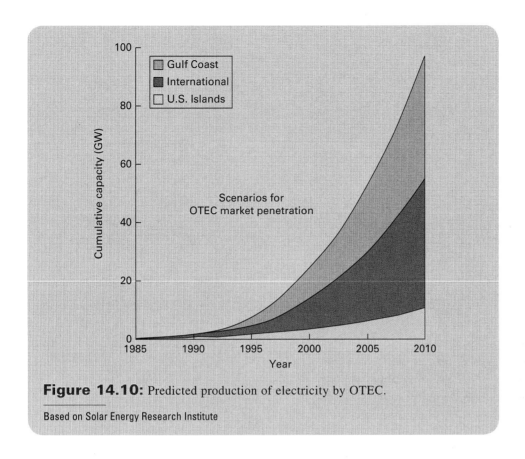

Figure 14.10: Predicted production of electricity by OTEC.

Based on Solar Energy Research Institute

was viewed. The actual cumulative capacity at a point close to the right-hand side of the graph is identically zero. Overall OTEC is an interesting, low-environmental-impact energy technology that unfortunately suffers from low efficiency. It is unclear whether the complex technological challenges facing OTEC can be overcome to make this resource economically viable.

14.6 Ocean Salinity Gradient Energy: Basic Principles

In 1784, the French physicist Jean-Antoine Nollet placed a pig's bladder filled with wine in a barrel of water. Over time, the bladder swelled and eventually burst. This behavior is the result of the osmotic pressures (Figure 14.11 on the next page) of freshwater and saltwater separated by a suitable membrane. In the figure, freshwater is placed in one side of a container and saltwater is placed in the other side of a container. The two sides are separated by a semipermeable membrane that allows water molecules to pass but not Na or Cl ions. A driving force tries to equalize the salt concentration on the two sides of the membrane, causing water to move from the fresh side to the salt side to dilute the saltwater, thereby increasing the height and pressure on the salt side of the membrane. The osmotic pressure is approximated by the Morse equation:

$$p = iRMT,$$ **(14.10)**

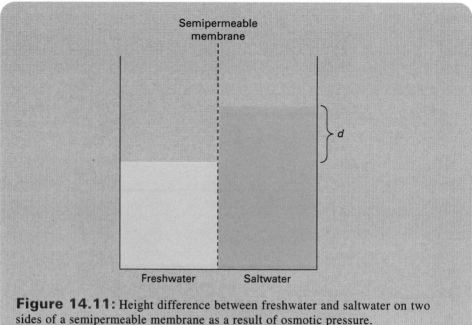

Figure 14.11: Height difference between freshwater and saltwater on two sides of a semipermeable membrane as a result of osmotic pressure.

where i is a constant (the dimensionless van 't Hoff factor) that depends on the solute ($i = 2.0$ for NaCl), R is the universal gas constant ($8.315 \, \text{J} \cdot \text{K}^{-1} \cdot \text{mol}^{-1}$), M is the molarity of the solution, and T is the absolute temperature. For seawater at, say, 10°C (i.e., 283 K) with a molarity of 600 mol/m³ (about average for the oceans), the osmotic pressure is

$$p = 2.0 \times (8.315 \, \text{J} \cdot \text{K}^{-1} \cdot \text{mol}^{-1}) \times (600 \, \text{mol/m}^3) \times (283 \, \text{K}) = 2.82 \times 10^6 \, \text{N/m}^2 \quad \textbf{(14.11)}$$

or about 28 atm. Thus, in Figure 14.11 the pressure difference pushes the level of the seawater above the freshwater by a distance (d in the figure), so that the pressure in the seawater at a height level with the top of the freshwater is $2.82 \times 10^6 \, \text{N/m}^2$. That means that the osmotic pressure is equal to the mass per unit area of the saltwater above the freshwater times the gravitational acceleration ($g = 9.8 \, \text{m/s}^2$), or

$$p = mg/A = \frac{dA\rho g}{A} = d\rho g, \quad \textbf{(14.12)}$$

where ρ is the density of water on the right side of the membrane.

The potential energy associated with the height difference calculated in Example 14.5 can be converted to kinetic energy and used to run turbines to generate electricity. In practice, only a small fraction of this energy could actually be extracted, because the membrane, which allows water (but not salt) to pass, will not withstand the osmotic pressure caused by the head (Figure 14.11). (This is why the pig's bladder burst.) A customary approach is to utilize the increased pressure on the saltwater side of the membrane to push water through a turbine to generate electricity. This approach to the production of useful energy is referred to as *pressure-retarded osmosis* (PRO).

Another method of producing electricity from the energy associated with salt concentration differences between freshwater and saltwater is *reverse electrodialysis* (RED). This method produces electricity directly, without first producing mechanical energy, by utilizing the charges associated with the dissolved ions in the water.

Example 14.5

Calculate the height difference between a column of freshwater and a column of seawater that results from osmotic pressure.

Solution

Equation (14.12) gives an expression for the height, d, as

$$d = \frac{p}{\rho g}.$$

Substituting equation (14.12) for the pressure gives

$$d = \frac{iRMT}{\rho g}.$$

We use the values $i = 2.0$ (for NaCl), $R = 8.315\ \text{J}\cdot\text{K}^{-1}\cdot\text{mol}^{-1}$, and $\rho = 1025\ \text{kg/m}^3$, and $g = 9.8\ \text{m/s}^2$. The molarity of seawater is given $M = 600\ \text{mol/m}^3$, and we assume a temperature of $10°\text{C} + 273 = 283\ \text{K}$. The height is then calculated to be

$$d = \frac{2.0 \times (8.315\ \text{J}\cdot\text{K}^{-1}\cdot\text{mol}^{-1}) \times (600\ \text{mol}\cdot\text{m}^{-3}) \times (283\ \text{K})}{(1025\ \text{kg}\cdot\text{m}^{-3}) \times (9.8\ \text{m}\cdot\text{s}^{-2})} = 281\ \text{m}.$$

Note: The units in this expression reduce to $\text{J}/(\text{kg}\cdot\text{m}\cdot\text{s}^{-2}) = \text{m}$.

A schematic diagram of an electrochemical cell that can be used for this purpose is illustrated in Figure 14.12. Pairs of membranes (+ and − in the figure) separate the negative (Cl⁻) and positive (Na⁺) ions, respectively, in a salt solution. Freshwater and saltwater are circulated through the device. Cl⁻ ions diffuse in one

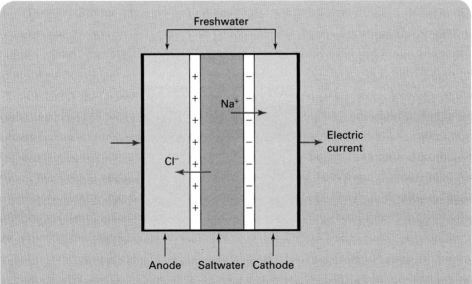

Figure 14.12: Schematic of reverse electrodialysis production of electricity.

direction (from the saltwater to the freshwater), and this flow of charged ions gives rise to a current in an external circuit as shown. To increase the voltage produced, several RED cells can be connected in series, or the cell can have several layers of alternating salt and freshwater channels.

14.7 Applications of Ocean Salinity Gradient Energy

The utilization of salinity gradient energy, sometimes referred to as osmotic energy, is possible at locations where both saltwater and freshwater are available in large quantities and in close proximity. This situation occurs most commonly at the outflow of rivers into the ocean, where an estimated 1 MW_e per m^3/s of river flow is available (assuming a typical 40% conversion efficiency). The world potential for viable power from salinity gradients at river outflows is estimated at about 150 GW_e. About half of this capacity is from the world's 50 largest rivers. Inland salt lakes are also a resource that could be utilized. These are highly saline and have a large osmotic pressure relative to freshwater. For example, the Dead Sea has a salinity that corresponds to a head of over 5000 m.

Although it is in the early stages of development, salinity gradient research is active in several places. The Netherlands has a research program to investigate electricity generation by RED. Freshwater that collects behind dykes is normally pumped to the sea, and this proximity of freshwater and saltwater (rather than at river outflows) makes salinity gradient power of interest. In Russia, a prototype RED plant at Vladivostok has been operational for several years. Norway has the capacity for about 3 GW_e of salinity gradient power from rivers flowing into the ocean and has the most advanced program to utilize this resource. Since 2009, a fully functional PRO facility has been operating outside of Oslo, Norway. Until fairly recently, the cost of osmotic membranes has been prohibitive. These membranes have a finite lifetime and must be replaced. Recent developments in membrane production have done much to alleviate this problem.

Generally, power produced from salinity gradients is environmentally friendly. However, there are some environmental concerns. The mixing of saltwater and freshwater produces brackish water. This happens naturally at the outflow of rivers, but if this process is altered by the utilization of salinity gradient power, then the environmental consequences of these changes must be considered. In PRO methods, the membranes tend to get clogged with impurities and particulate matter from the water. To optimize the lifetime of these membranes and to make salinity gradient power as economical as possible, the membranes must be cleaned. The chemicals used to clean membranes are potentially toxic and can have environmental consequences. Polyethylene (the material commonly used for plastic grocery bags and the standard material used for membranes) is derived from petroleum and is a by-product of the oil refining process. Thus, at present, salinity gradient energy seems to be dependent on the existence of fossil fuels, and the disposal of used polyethylene will contribute (although in a small way) to the production of greenhouse gases.

On the positive side, both OTEC and salinity gradient power do not suffer from a major drawback that affects most other sources of renewable ocean energy. That is, variations in time, either periodic, as for tidal energy, or somewhat less predictable, as for wave energy.

ENERGY EXTRA 14.1
Pressure-retarded osmosis in Norway

Pressure-retarded osmosis generating station in Norway.

The world's first fully functional osmotic power facility opened in 2009 in Tofle, on the Oslo Fjord in Norway (see the above photograph). The facility utilizes the pressure-retarded osmosis method to produce electricity and was constructed by the Norwegian energy company Statkraft at a cost of $7–8 million. The facility is located where a river flows into the ocean, providing a source of both freshwater and saltwater.

The Statkraft facility is designed for a total electrical output of 2 to 4 kW (on average, about enough for two or three single-family homes) using 2000 m³ of membrane. It is intended as a three-year study to assess the viability of a large-scale facility. Scaling up such a facility is not necessarily straightforward. To make a large-scale commercial plant economically viable will require the improvement of membrane characteristics from its current output of about 1 W/m² to about 5 W/m². Even under these

conditions, a small 25-MW$_e$ plant (small compared to a coal or nuclear plant) would require 5×10^6 m² of membrane. This area is about 10% of the current total world production of such membranes (largely for desalination facilities). The 25-MW$_e$ osmotic generating station would be about the size of a football stadium and would require a flow of 25 m³/s freshwater and 50 m³/s saltwater. Statkraft anticipates that the construction of a commercial osmotic power plant would take place around 2015–2017 and that this technology could be economically competitive with other alternative energy generation methods by around 2030.

The general design of a PRO generating station is shown in the following figure. Osmosis of freshwater across a membrane in order to equalize the salt concentration gradient, produces excess pressure of the seawater side of the membrane and this drives the turbine/generator to produce electricity.

Continued on page 454

Energy Extra 14.1 continued

Based on http://www.yalescientific.org/2013/02/the-elixir-of-life-generating-electricity -from-water/; http://pre.docdat.com/docs/index-219628.html; http://www.theengineer .co.uk/news/osmotic-power/302450.article

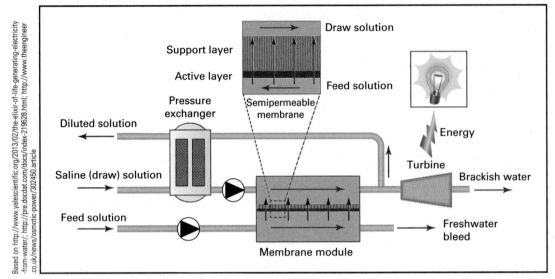

Schematic of pressure-retarded osmosis generating system.

Topic for Discussion

Consider the merits of the following energy generation proposal: An onshore OTEC facility brings in cold seawater and produces electricity. It also produces desalinated water. A PRO facility could be located next to the OTEC facility and could pump in additional seawater and use the desalinated water from the OTEC plant to produce more electricity.

14.8 Summary

The utilization of thermal gradients in the oceans for the production of energy requires a large temperature difference between the surface and deep water. This chapter overviewed the availability of such a situation in the world's oceans and the possibility of constructing an OTEC generating station. As the temperature of water deep in the ocean is relatively constant worldwide, a large gradient is characteristic of tropical regions where the surface temperature is high.

This chapter discussed the availability of a large thermal difference between surface and deep water as a mechanism for generating electricity using a heat engine. Warm surface water forms the hot reservoir, and cold water pumped from the oceans depths is the cold reservoir. The basic principle of operation follows the general design of a heat engine and allows for the transfer of thermal energy to produce mechanical energy, which can then be used to drive a generator to produce electricity.

The chapter reviewed the three possible designs of an OTEC system: open-cycle, closed-cycle, and hybrid systems. The open-cycle uses seawater as the working fluid in the heat engine. The closed-cycle system uses heat exchangers to transfer thermal energy from the seawater to a closed-system containing a working fluid such as ammonia. The hybrid system is similar to the closed-cycle system except that, in the process of transferring heat from the working fluid, the seawater is vaporized and freshwater, which can be used for domestic purposes, is produced.

The principles of OTEC have been known for well over a century, and experimental work has been conducted for over 80 years in the warmer regions of the world. The overall efficiency of these systems is very low because the temperature difference between the hot and cold reservoirs running the heat engine is very small (compared to the situation for a coal-fired or nuclear power plant). As a result of this low efficiency, the economic viability of OTEC-generated electricity is questionable.

This chapter also described another method of extracting energy from the oceans based on the salinity gradient that exists where freshwater from rivers mixes with seawater. This approach is based on the osmotic energy that results from two liquids with different solute concentrations. Two technologies allow for the utilization of osmotic energy: pressure-retarded osmosis (PRO) and reverse electrodialysis (RED). The chapter overviewed the basic principles of osmotic pressure and the methods for using osmotic pressure to drive a turbine/generator by the PRO technique. RED uses the charges associated with ions in solution to generate electricity directly in an electrolytic cell. The principles of salinity gradient energy also have been known for a number of years, but technical developments for its use as a source of electricity have not been undertaken until relatively recently. At present, a pilot PRO facility in Norway has shown encouraging results. While the cost of electricity generated by this method is still high, the hope is that new developments in membrane technology may make PRO- or RED-generated electricity a small but viable component of future energy production.

Problems

14-1 A 1 m^3-sparcel of seawater is cooled from 24°C to 8°C in 1.2 seconds. Calculate the average thermal power that is available during that time.

14-2 An OTEC system operates with a warm reservoir of 23°C and a cold reservoir of 5°C. Calculate the mass of water (vapor) flowing through the evaporator needed to generate 1 MWh of electricity. Assume a 90% conversion from mechanical to electrical energy.

14-3 An ideal OTEC system has a flow rate of 100 m^3/s of warm seawater. The warm reservoir is at 26°C, and a cold reservoir is at 6°C. For a conversion efficiency for mechanical to electrical energy of 90%, what is the facility's output in MW_e?

14-4 It has been speculated that an OTEC facility could use wave energy to offset its low efficiency by providing electrical energy for operating the plant. Where would the most advantageous location(s) be for such a facility? If the facility could intercept 200 m of wave front for your chosen location and had a total output of 100 MW_e, what fraction of this total output would be from waves? Assume the efficiency of wave-generated electricity to be 35%.

14-5 A 20 MW_e OTEC facility operates at a thermal efficiency of 2.4%. The warm water temperature is 22°C. Calculate the total flow rate of warm water in cubic meters per second (m^3/s). Assume that the conversion from mechanical to electrical energy is 86% efficient.

14-6 Make a plot of the height of a column of water (in meters) that the osmotic pressure between freshwater and saltwater can support as a function of salinity between 0 and 10% (weight of salt per weight of solution) of NaCl in water. Assume a constant density of 1025 kg/m^3.

14-7 The salinity (mostly NaCl) in some parts of the Great Salt Lake (Utah) is 27%; this is sometimes expressed as 270 parts per thousand, meaning 270 g of salt per liter of solution. Estimate the height of a column of water that can be supported by the osmotic pressure between Great Salt Lake water and freshwater. The density of a 27% salt solution is $\approx$ 1200 kg/m^3.

14-8 Consider the experiment of Jean-Antoine Nollet as described in the text, which illustrates the osmotic pressure between water and wine. Assume wine is 12% (by volume) of ethanol in water, and estimate, as in Example 14.4, the height of a column of water that the osmotic pressure between water and wine can support. Note that the van 't Hoff factor depends on the nature of the solvent and that a value of $i \approx 1.0$ is appropriate for ethanol in water.

14-9 In the Arctic, winter air temperatures can average around $-35°C$, while sea temperatures (under the ice) can typically be around $-2°C$. Consider the possibility of constructing an OTEC facility that uses the air as the cold reservoir and pumps water from the ocean to serve as the "warm" reservoir.

(a) Calculate the ideal efficiency of such a facility.
(b) Would there be constraints on the design of such a system (i.e., open system or closed system) and the working fluid?

14-10 An offshore OTEC facility operates with a warm reservoir temperature of 22°C and a cold reservoir temperature of 5°C.

(a) If 65% of the output power is required for operation of the facility (e.g., primarily water pumps), what is the net efficiency?
(b) In order to transport the electricity that has been generated to land, hydrogen is used as an energy storage mechanism. Hydrogen is produced by the electrolysis of water and is converted back to electricity using a fuel cell, with an overall efficiency of 40% (see Section 20.9). What is the net overall efficiency, and how does this compare with, for example, wind or solar photovoltaics?

14-11 Consider the possible design of an OTEC facility that has an electrical output comparable to an average nuclear power plant (e.g., 1 GW_e). Assume a turbine/generator efficiency of 90%.

(a) If the surface temperature is 23°C and the cold reservoir is at 4°C, what is the efficiency?

(b) What is the total flow rate of water needed?

(c) Compare the result of part (b) to the flow rate needed for a 1-GW$_e$ hydro-electric facility with a head of 200 m.

14-12 The Amazon River has a flow rate of 2.09×10^5 m^3/s at an average velocity of 0.7 m/s. Compare the total potential for osmotic energy and the total potential for run-of-the river hydroelectric energy (at 40% capacity factor) for the Amazon.

14-13 The Mississippi River, the largest of all rivers in the United States, has a flow rate of 1.8×10^4 m^3/s. If 10% of the osmotic energy potential of the Mississippi could be harvested, estimate the number of households that could be supplied with electricity.

14-14 An OTEC facility has a total flow rate of 500 m^3/s of water. The cold water from the deep ocean is at 5°C. During the year the surface water temperature varies from 19°C to 22°C. Plot the output as a function of warm water temperature over this range.

14-15 An OTEC facility operates with a cold water temperature of 6°C and has a total water flow rate of 50 m^3/s. If the operating energy (for pumps, etc.) is 10 MW$_e$, what is the minimum surface temperature needed to generate a net gain in electricity?

14-16 Estimate the salinity of the Dead Sea from the information in this chapter.

14-17 Compare the osmotic power available from freshwater flowing at a rate of 1000 m^3/s into the ocean (assuming an efficiency of 40%) and the thermal power available from water flowing at 1000 m^3/s through a heat exchanger and cooling by 5°C (assuming an efficiency of 3%).

14-18 (a) Calculate the energy available when a 10^6 kg-parcel of water is cooled by 4°C.

(b) If this energy is used (at an efficiency of 100%) to lift the parcel of water against gravity, what would be its elevation?

(c) How long would this energy satisfy the average per-capita primary energy needs in Norway?

14-19 (a) Calculate the gravitational potential energy associated with 1 m^3 of seawater that is lifted through a vertical distance of 1 km.

(b) Compare the result in part (a) with the total thermal energy difference between 1 m^3 of seawater at 4°C and 1 m^3 of seawater at 22°C.

14-20 If the temperature drops across the evaporator and condenser of an OTEC system are equal, prove that the relative temperature drops as given in equation (14.4) will maximize the power output.

Bibliography

W. H. Avery and C. Wu. *Renewable Energy from the Ocean: A Guide to OTEC.* Oxford University Press, New York (1994).

G. Boyle (Ed.). *Renewable Energy.* Oxford University Press, Oxford (2004).

R. Charlier. *Ocean Energies: Environmental, Economic, and Technological Aspects of Alternative Power Source.* Elsevier, New York (1993).

R. Cohen. "Energy from the ocean." *Philos. Trans. Royal Soc. London* **A307** (1982): 405.

A.V. da Rosa. *Fundamentals of Renewable Energy Processes,* (3rd ed.) Academic Press, Oxford (2013).

A. Goldin. *Oceans of Energy—Reservoir of Power for the Future.* Harcourt, Brace, Jovanovich, New York (1980).

B. K. Hodge. *Alternative Energy Systems and Applications.* Wiley, Hoboken (2010).

J. A. Kraushaar and R. A. Ristinen. *Energy and Problems of a Technical Society,* (2nd ed.) Wiley, New York (1993).

C. Ngô and J. B. Natowitz. *Our Energy Future: Resources, Alternatives and the Environment.* Wiley, New York (2009).

T. R. Penney and D. Bharathan. "Power from the sea." *Scientific American* **256** (1987): 86.

R. J. Seymour. *Ocean Energy Recovery: The State of the Art.* American Society of Civil Engineers, New York (1992).

World Energy Council. *World Energy Resources 2016*, available online at https://www.worldenergy.org/publications/2016/world-energy-resources-2016/

Geothermal Energy

Learning Objectives: After reading the material in Chapter 15, you should understand:

- The origins of geothermal energy.
- The types and distribution of geothermal energy resources.
- The direct use of geothermal heat.
- The types of geothermal electric generating facilities.
- The use of geothermal energy worldwide.
- The sustainability and environmental consequences of geothermal energy use.

15.1 Introduction

Geothermal energy refers to energy that is extracted from the interior of the earth. It is most convenient to access this energy in locations where hot regions are close to the surface. This source of energy has been used for more than 2300 years. The oldest known use of geothermal energy was the utilization of hot springs for bathing purposes in China. The Roman Empire also used geothermally heated water for bathing in England in the first century CE. The first extensive use of geothermal energy for heating purposes occurred in France in the fourteenth century when the city of Chaudes-Aigues developed a district heating system. The first commercial geothermal electric generating station was constructed in Italy in 1911. Since then, the use of geothermal energy for electricity generation has grown worldwide. The current chapter reviews the properties of geothermal energy, how it can be used for electricity generation, and its prospects for the future.

15.2 Basics of Geothermal Energy

The earth consists of an outer layer called the *crust*, an intermediate layer called the *mantle*, and an inner layer called the *core*. The crust consists primarily of rock and is typically about 30 km thick but can range from about 3 km to about 60 km in different locations. The mantle is about 2900 km thick and consists of rock that, because of the

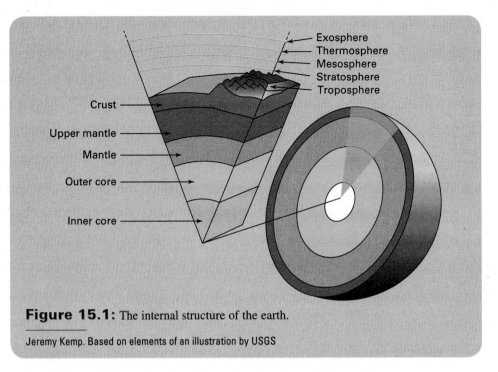

Figure 15.1: The internal structure of the earth.

Jeremy Kemp. Based on elements of an illustration by USGS

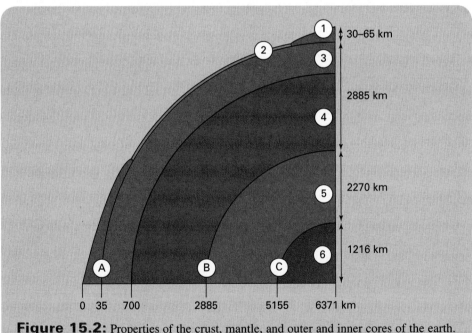

Figure 15.2: Properties of the crust, mantle, and outer and inner cores of the earth.
1. Continental crust; 2. Oceanic crust; 3. Upper Mantle; 4. Lower Mantle;
5. Outer Core; 6. Inner Core; A: Crust/Mantle Discontinuity (Mohorovicic Discontinuity);
B: Mantle/Core Discontinuity (Gutenberg Discontinuity);
C: Outer Core/Inner Core Discontinuity (Lehmann Discontinuity)

temperature and pressure present, is very plastic and made up largely of Mg, Fe, Al, Si, and O. The core is comprised primarily of Fe and is divided into an outer core (about 2300 km thick), which is liquid, and an inner core (about 2400 km diameter), which is solid (Figures 15.1 and 15.2).

The interior of the earth is hot. This heat is believed to have three causes:

1. The primordial heat associated with the material that condensed to form the earth during the early history of the solar system.
2. The decay of radioactive nuclides inside the core.
3. Friction in the liquid outer core that results from the tidal forces between the earth's core and the moon and sun.

Large planets, like Jupiter, cool slowly, and their interior temperature is dominated by primordial heat. The earth has lost a larger fraction of its primordial heat than Jupiter, and its interior temperature is dominated largely by the rate of radioactive decay of nuclides of uranium, thorium, and potassium. It is estimated that approximately 75% of the heat in the interior of the earth comes from this source. The heat from the interior of the earth, flowing outward (toward the cooler crust), amounts to about 0.087 W/m^2 at the surface. This is small (less than 0.1%) of the heat flow into the earth from solar radiation (average of 168 W/m^2; see equation 8.3).

Example 15.1

Calculate the total geothermal heat flow through the surface of the earth.

Solution

The heat flow from the earth's interior averages 0.087 W/m^2. The total surface area of the earth is $4\pi r^2$, where the mean radius of the earth is $r = 6371$ km $= 6.371 \times 10^6$ m (Figure 15.2). Thus the total heat flow is

$$4 \times 3.14 \times (6.371 \times 10^6 \, \text{m})^2 \times (0.087 \, \text{W/m}^2) = 4.4 \times 10^{13} \, \text{W}.$$

The crust of the earth consists of separate regions referred to as *tectonic plates*. The heat transfer from the hot interior of the earth toward the crust generates convection currents in the mantle. These convection currents drive the movement of tectonic plates, referred to as *continental drift*. Near the middle of the oceans, the oceanic plates move apart, allowing molten rock (*magma*) from deep within the mantle to push upward and to form a rift. At continental boundaries, plates collide. As this occurs, one plate (the oceanic plate) is pushed below the other (or *subducted*). This causes the subducted plate to heat (because it is pushed into hotter regions of the earth) and sends plumes of magma up toward the surface.

The possibility of utilizing geothermal energy is the greatest at or near plate boundaries because the thermal gradient below the surface is the largest in these regions. Figure 15.3 on the next page shows the tectonic plates on the surface of the earth. The locations of volcanoes are also shown in the figure and, in most cases, are seen to fall on or near plate boundaries. These regions correspond to the subduction of oceanic plates by continental plates, as along the Pacific coast of North and South America or at midoceanic ridges, as in Iceland (Figure 15.4 on the page 463). The most active regions of the earth occur along the edges of the Pacific Plate, where the greatest number of volcanoes occurs. For this reason, the edge of the Pacific Plate is sometimes referred to as the *Ring of Fire*. This ring offers the greatest potential for the use of geothermal energy (although other active regions, e.g., in the Caribbean and in Iceland, have important resources as well).

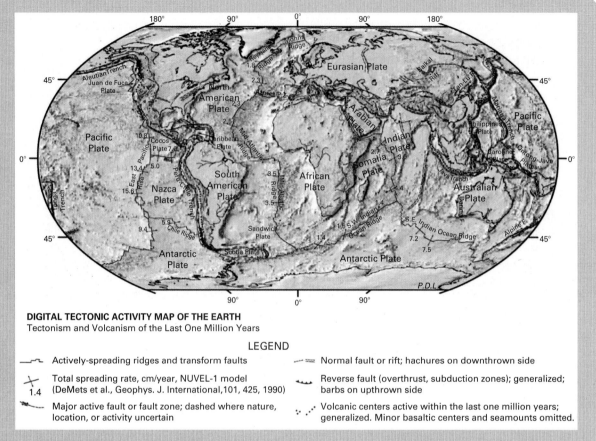

DIGITAL TECTONIC ACTIVITY MAP OF THE EARTH
Tectonism and Volcanism of the Last One Million Years

LEGEND

Actively-spreading ridges and transform faults	Normal fault or rift; hachures on downthrown side
Total spreading rate, cm/year, NUVEL-1 model 1.4 (DeMets et al., Geophys. J. International,101, 425, 1990)	Reverse fault (overthrust, subduction zones); generalized; barbs on upthrown side
Major active fault or fault zone; dashed where nature, location, or activity uncertain	Volcanic centers active within the last one million years; generalized. Minor basaltic centers and seamounts omitted.

Figure 15.3: The earth's tectonic plates. The locations of volcanoes that have erupted during historic times are shown by the red dots.

NASA

The temperature inside the earth increases as a function of depth. The temperature gradient in normal cases is in the range of 17–30°C per km. In active regions, the temperature gradient can be much greater. The thermal gradient is a measure of the usefulness of geothermal energy in a particular region. The measured temperature gradient for different regions of the United States is shown in Figure 15.5 and suggests that the western parts of the country are the most suitable for the development of geothermal energy resources.

Geothermal resources are roughly divided into six categories:

1. Normal geothermal gradient.
2. Hot dry rock.
3. Hot water reservoirs.
4. Natural steam reservoirs.
5. Geopressurized regions.
6. Molten magma.

The possibilities of utilizing these resources are now discussed.

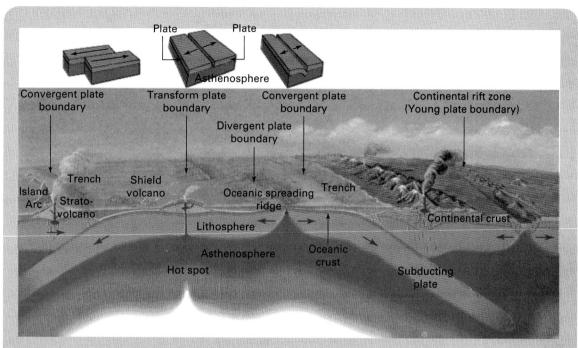

Figure 15.4: Movement of tectonic plates resulting in the formation of hot regions close to the surface of the earth at midoceanic ridges and at continental shelf boundaries.

U.S. Geological Survey

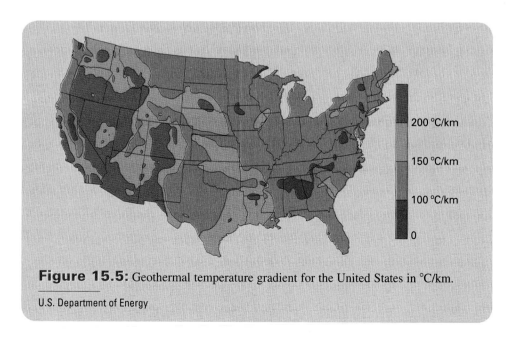

Figure 15.5: Geothermal temperature gradient for the United States in °C/km.

U.S. Department of Energy

15.2a **Normal Geothermal Gradient**

Even in regions where the normal geothermal gradient exists, the rock becomes hot enough to provide useful energy if one drills deep enough. For a gradient of 30°C/km, a well drilled to a depth of 6 km will access a temperature of about 200°C. A working fluid (e.g., water) can be injected into the well and returned at an elevated temperature. Although it is technologically feasible to use this as a method of generating electricity, this approach is not economically viable at this time. However, the heat capacity associated with the soil and rock near the surface of the earth can be exploited using a geothermal heat pump (Chapter 17).

15.2b **Hot Dry Rock**

Hot dry rock refers to a resource that is basically the same as the normal geothermal gradient except that the temperature increases at a greater rate (i.e., more than about 40°C/km). This feature increases the possibility of utilizing geothermal energy and is a resource that is estimated to be capable of providing up to 200 GW of thermal energy in the United States alone. The basic approach would be to drill a well into the region of elevated temperature and then, using hydraulic or explosive techniques, to fracture the rock to form a reservoir into which a working fluid (water) could be injected to artificially replicate the situation found in hot water reservoirs, as discussed later in this chapter. Thus far, the technology to implement this approach is still in its early stages, and its economic viability is uncertain.

15.2c **Hot Water Reservoirs**

Surface water from rain, melting snow, and other sources penetrates the earth's surface in regions where the crust has faults and cracks. If this occurs in a region where there is an anomalously large geothermal gradient, then the water can be heated underground to form hot water and/or steam. If the water or steam has a clear path back to the surface, it can appear, as it does in many locations, as a hot spring or geyser. If the water or steam gets trapped between layers of impermeable rock, then it can form an underground geothermal reservoir. When the deposit is mostly hot water under pressure, the resource is referred to as a *hot water reservoir*. These underground sources of hot water are accessed by drilling into the region. Wells of anywhere from 100 m up to a few kilometers deep are used. Hot water reservoirs are in common use for direct heating and can be used for the generation of electricity, although for the latter purpose, natural steam deposits are preferred.

15.2d **Natural Steam Reservoirs**

When little or no liquid water is associated with a deposit, it is referred to as a *natural steam reservoir* or sometimes a *dry steam deposit*. Because the temperatures involved are quite high, these deposits are the preferred source of geothermal energy for electricity generation. Unfortunately, such resources are somewhat rare, and only two sizable natural steam deposits have been identified: The Geysers field in California and Larderello in Italy. Hot water reservoirs and natural steam reservoirs constitute the vast majority of geothermal resources that have been exploited commercially.

15.2e **Geopressurized Regions**

Geopressurized resources are reservoirs of pressurized hot water containing dissolved methane gas. They typically have temperatures in the range of 90–200°C and occur at depths of 3–6 km. The only extensive deposits of this type that have been identified occur along the Gulf of Mexico coast. This resource can provide energy by three mechanisms: heat associated with the water, combustion of the methane, and direct use of the pressure to drive turbines. Test facilities were constructed in the 1980s and 1990s in the Gulf region. Thermal, as well as chemical energy, was extracted, although the mechanical energy associated with the pressure was not utilized. It was concluded that extraction of all forms of available energy was not economically feasible at that time. It is probable that some aspects of the energy associated with geopressurized regions will be developed in the future in conjunction with natural gas recovery.

15.2f **Molten Magma**

Regions where molten magma appears at or near the surface of the earth (e.g., active volcanoes) are too unstable and unpredictable to provide a reliable and safe means of extracting geothermal energy.

15.3 **Direct Use of Geothermal Energy**

Geothermal energy can be used in two ways: (1) direct use of the heat for space heating and similar applications or (2) use in a heat engine (e.g., turbine/generator) to produce electricity. The heat associated with geothermal resources has been utilized directly in several ways (Figure 15.6), and the more important ones will be described in some detail.

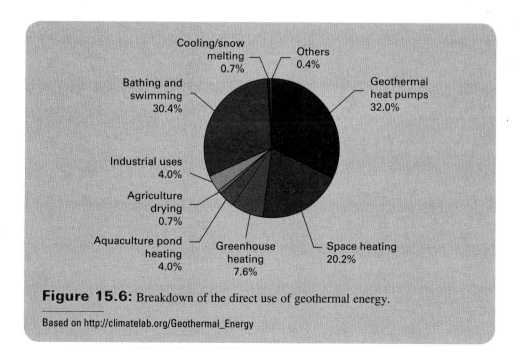

Figure 15.6: Breakdown of the direct use of geothermal energy.

Based on http://climatelab.org/Geothermal_Energy

ENERGY EXTRA 15.1
Geothermal aquaculture

Carol Bond/Alamy Stock Photo

Geothermal power station and Huka Prawn Park, Taupo, New Zealand.

Aquatic farming in ponds heated by geothermal energy has increased the growth rate and diversity of animals that can be raised. Fish, shellfish, and reptiles have been produced by aquafarmers in geothermal regions, often in locations where they might otherwise not survive.

Geothermal aquaculture can nicely complement other geothermal energy uses. In locations where the temperature of the geothermal resource is not high enough to provide an acceptable thermodynamic efficiency for electricity generation, aquaculture is still a possibility. Also, in locations where geothermal energy is used for electric generation or direct heating, the runoff from such systems is still at sufficiently high temperatures to be put to use in geothermal farms. Geothermal water can be mixed with cold water in a controlled manner to regulate the temperature in a pond. Circulation is typically necessary to avoid excessive thermal gradients.

The availability of heat from geothermal sources provides reasonably constant temperature ponds for aquafarming. Because most species of aquatic creatures grow fastest and remain healthiest in an environment where the temperature remains within a certain range, productivity in geothermally heated aquatic farms can exceed that in unheated farms, and such facilities can function economically and without the carbon footprint of farms that utilize artificial heating methods.

Notable geothermal aquafarming activities in the United States exist in Arizona, California, Colorado, Idaho, and Nevada. Total heat utilization in the United States for geothermal aquaculture is in the range of a few million gigajoules per year. Commonly grown organisms are catfish, tilapia, trout, sturgeon, and freshwater prawns. Other activities include the growth of tropical fish for aquaria and tropical reptiles, such as alligators in Idaho.

In other locations worldwide, geothermal aquaculture is becoming more common. Some activities include raising eels in Slovakia, fish and shellfish (particularly abalone) in Iceland, fish in China, and eels and alligators in Japan. In addition to aquatic animals, aquatic plants that are useful for human or animal food can also be grown in geothermally heated ponds.

Topic for Discussion

Discuss the design of a system using hot geothermal water along with a cold water supply to regulate the temperature of a pond or tank for aquaculture.

15.3a **Bathing**

Perhaps the oldest use of geothermal energy is for bathing (or *balneology*). In regions where geothermally heated hot water appears at the surface, these so-called hot springs have been used since ancient times for bathing and therapy. Today, these resources form an important tourist industry in the United States, Japan, Mexico, New Zealand, and elsewhere.

15.3b **Space Heating**

Space heating is another traditional use of geothermal energy where hot water (typically at 60°C or hotter) from geothermal reservoirs is pumped through buildings to provide heating. After passing through a heat exchanger to heat the building, the water is reinjected into the geothermal reservoir for reheating. This may be done on an individual building basis but is often done collectively for large residential and commercial districts. The largest development of this type is in Reykjavik, Iceland, although the resource is used substantially in France, the United States, Turkey, Poland, and Hungary.

15.3c **Agriculture**

Along the lines of space heating, hot water from geothermal reservoirs can be used to provide heat to greenhouses to improve growing conditions. This approach is in common use in Italy and in Hungary, where it satisfies about 80% of the greenhouse heating needs.

15.3d **Industrial Uses**

Heat from geothermal sources has been used worldwide for a variety of applications that would otherwise use electric heat or heat from fuel combustion. Some industrial uses that have been found for geothermal heat are drying of food and wood products.

15.3e **Snow Melting**

Geothermally heated water is used in a variety of locations to melt snow and ice in the winter. A common approach is to pump hot water from geothermal reservoirs through pipes embedded in sidewalks and roads to keep them free from snow and ice in cold weather. A geothermally heated sidewalk is shown in Figure 15.7 on the next page.

15.3f **Geothermal Heat Pumps**

The temperature of the ground at a depth of about 3 m remains very constant throughout the year and ranges from about 10°C to about 16°C depending on the local climate. This very constant temperature reservoir can be utilized by a heat pump (Section 1.5) for space heating purposes. For traditional heat pumps, as discussed in more detail in Section 17.6, the outside air is typically used as the cold reservoir. However, geothermal

Figure 15.7: Construction of a sidewalk in Reykjavik, Iceland, utilizing heating pipes carrying geothermally heated water to prevent icing.

heat pump systems offer the advantage that the cold reservoir is at a much more stable temperature. This type of system utilizes the heat capacity of the earth as a heat source and does not depend on the availability of enhanced geothermal activity. As such, it can be implemented in any location.

15.4 Geothermal Electricity

Geothermal energy is converted into electricity by extracting hot geothermal fluid (water or steam or a mixture of the two) from a *production well* drilled into the geothermal reservoir. Hot water, extracted from the well, is converted into steam (as described in Section 15.4b). The steam is then used to run a turbine, which in turn drives a generator to produce electricity. The fluid is then injected back into the geothermal reservoir through an injection well in order to replenish the supply of water in the reservoir and to minimize the distribution of geothermal water on the earth's surface (Section 15.5). The general layout of a geothermal power plant is shown in Figure 15.8.

There are several methods by which the hot fluid is actually utilized to drive the turbine, and the choice of method depends to some extent on the nature of the geothermal resource. Several possibilities are now described.

15.4a Dry Steam Plants

A few geothermal reservoirs provide essentially pure steam with little hot water. This simplifies the utilization of geothermal fluids for driving a turbine. Basically, the steam can be input directly into the turbine as it comes out of the well (after passing through a filter to remove debris). After driving the turbine, the steam is condensed and reinjected

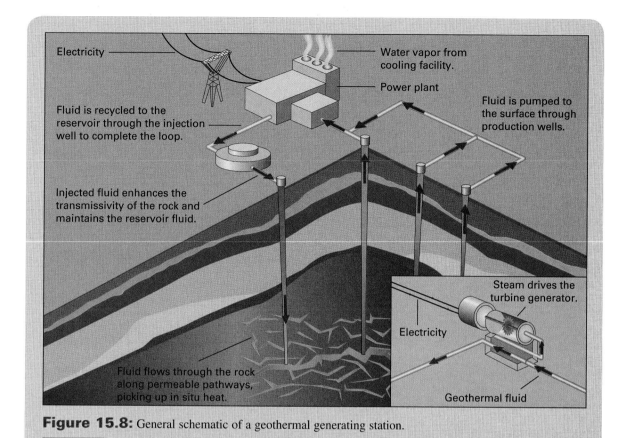

Figure 15.8: General schematic of a geothermal generating station.

U.S. Department of Energy, What is an Enhanced Geothermal System (EGS)?http://www1.eere.energy.gov/geothermal/pdfs/egs_basics.pdf

Example 15.2

Calculate the ideal Carnot efficiency of a geothermal generating station if the temperature of the geothermal fluid is 210°C and the temperature of the cold reservoir is 85°C.

Solution

From equation (1.34), the ideal Carnot efficiency of a heat engine is

$$\eta = 100\left(1 - \frac{T_c}{T_h}\right).$$

Converting temperatures to Kelvin gives

$$T_h = 210°C + 273 = 483 \text{ K}$$

and

$$T_c = 85°C + 273 = 358 \text{ K}.$$

The efficiency is therefore

$$\eta = 100 \times \left(1 - \frac{358 \text{ K}}{483 \text{ K}}\right) = 25.9\%.$$

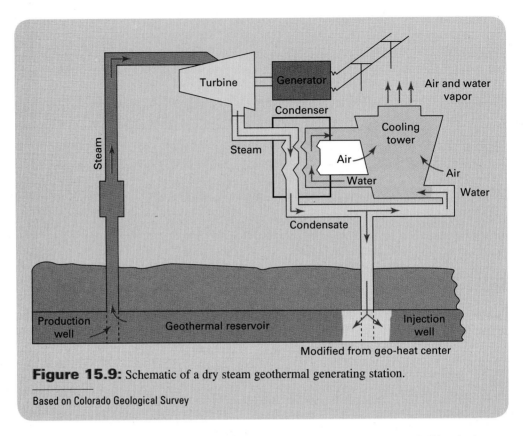

Figure 15.9: Schematic of a dry steam geothermal generating station.

Based on Colorado Geological Survey

into the reservoir (Figure 15.9). The Geysers reservoir in Northern California is currently the world's largest known dry steam resource. One of the 22 power plants associated with the Geysers resource is shown in Figure 15.10. Typical facilities of this type have capacities in the range of 50 MW_e.

Figure 15.10: A power generating station at The Geysers in California.

Bureau of Land Management, U.S. Department of Interior

15.4b **Flash Steam Plants**

Flash steam power plants are the most common type of geothermal generating plants in operation today. This is because most reservoirs contain hot water at an elevated pressure, and flash steam plants are the most convenient technology for utilizing this type of resource. When hot water under pressure is removed from its underground reservoir, the pressure drops, and, according to the water phase diagram in Figure 14.4, this lowering of the pressure causes the water to vaporize, producing steam. This steam is then used to run a turbine, and after going through a condenser, it is reinjected into the reservoir. A general schematic of a flash steam power plant is shown in Figure 15.11, and a photograph of the Krafla Geothermal Power Plant in Iceland is shown in Figure 15.12 on the next page. Geothermal fluid at various pressures is supplied from more than twenty production wells. Two flasher or steam separators (see Energy Extra 17.1) are utilized, one for high-pressure geothermal fluid and one for low-pressure geothermal fluid.

15.4c **Binary Power Plants**

In both the dry steam and flash steam generating systems, the hot water and/or steam from the geothermal reservoir are used to drive the turbine. In a binary cycle power plant (Figure 15.13 on the next page), the thermal energy of the hot water and/or steam is transferred to a second working fluid through a heat exchanger. This working fluid is in a closed-system and is vaporized in the heat exchanger.

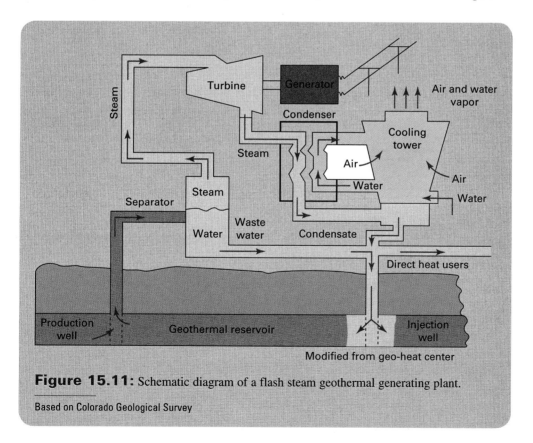

Figure 15.11: Schematic diagram of a flash steam geothermal generating plant.

Based on Colorado Geological Survey

Figure 15.12: Geothermal generating station at Krafla, Iceland. The station consists of two 30 MW$_e$ units.

Richard A. Dunlap

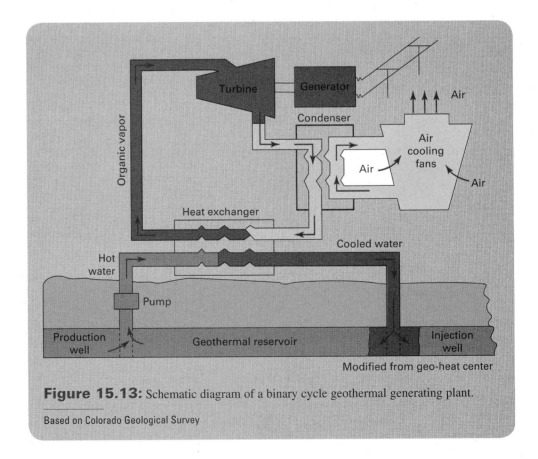

Figure 15.13: Schematic diagram of a binary cycle geothermal generating plant.

Based on Colorado Geological Survey

Figure 15.14: Binary cycle geothermal plant in Mammoth Lakes, California.

Visions of America, LLC/Alamy Stock Photo

The vapor is used to run the turbine and is then returned as a liquid to the heat exchanger. This system has the advantage that a working fluid with a boiling point that is lower than that of water can be used, thus allowing geothermal reservoirs with lower temperatures to be used for electricity generation. A binary cycle plant in Nevada is shown in Figure 15.14.

15.4d Hybrid Power Plants

Hybrid power plants combine a binary cycle power plant with another method of utilizing geothermal energy. Figure 15.15, for example, shows a combined flash steam and binary cycle plant. Flash steam is produced from geothermal hot water and is used to

Figure 15.15: Hybrid flash/binary cycle geothermal plant on the Big Island of Hawaii.

U.S. Department of Energy

run a turbine. The condensed steam (which is still quite hot) can be used in a binary cycle system to extract more of the heat content of the water.

15.5 Utilization of Geothermal Resources and Environmental Consequences

The installed capacity for geothermal energy use in various countries is summarized in Tables 15.1 and 15.2 for electricity generation and direct use, respectively. Capacity factors are typically about 0.7 for electricity generation and around 0.3 for direct use, but they can vary considerably, particularly for direct use where demand (e.g., for heat throughout the year) is not constant. The data for direct use are also subject to substantial uncertainty. Resources are utilized in specific locations where geothermal reservoirs occur. The nature of the resource is a major factor in the development of geothermal energy for electricity generation or direct use. The distribution of electricity-generating facilities that utilize geothermal energy is illustrated in Figure 15.16. This correlates well with the availability of geothermal energy, as shown in Figure 15.4, and the population density. The United States has the largest installed geothermal electric generating capacity. Roughly two-thirds of this capacity is in California where geothermal electricity accounts for about 4.5% of the state's

Table 15.1: Electricity generation capacity from geothermal sources as of 2015 from the world's principal producers and total world capacity.

country	installed capacity (MW$_e$)
United States	3567
Philippines	1930
Indonesia	1404
Mexico	1069
New Zealand	979
Italy	824
Iceland	665
Turkey	624
Kenya	607
Japan	533
El Salvador	204
Costa Rica	202
Nicaragua	155
rest of world	312
world total	**12,763**

Table 15.2: Capacity for direct use of geothermal power as of 2014 from the world's principal producers and total world capacity.	
country	**installed capacity (MW)**
China	17,870
United States	17,416
Sweden	5600
Turkey	2886
Germany	2849
France	2347
Japan	2186
Iceland	2040
Switzerland	1733
Finland	1560
Canada	1467
Norway	1300
Italy	1014
rest of world	9749
world total	**70,017**

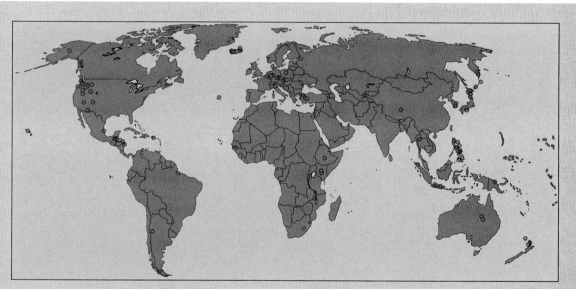

Figure 15.16: Distribution of geothermal electricity generating plants.

Pyty/Shutterstock.com. Data from http://www.thinkgeoenergy.com/thinkgeoenergy-updates-global-geothermal-power-plant-map/

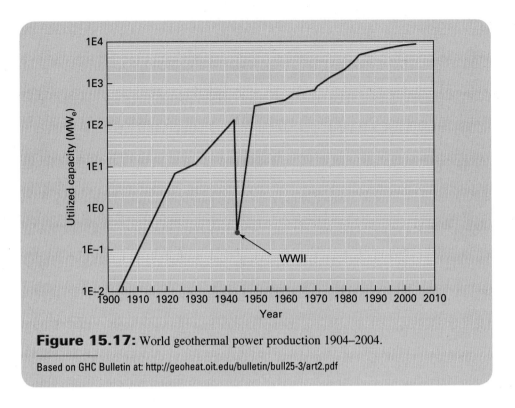

Figure 15.17: World geothermal power production 1904–2004.

Based on GHC Bulletin at: http://geoheat.oit.edu/bulletin/bull25-3/art2.pdf

total electric use. Although there have been increases in geothermal energy production in some developing countries over the past few years, the overall growth worldwide has been minimal since the late 1980s (Figure 15.17). Figure 15.18 shows the situation in the United States, where the geothermal generating capacity leveled off in

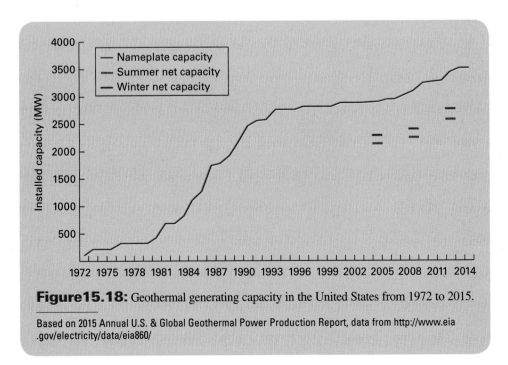

Figure 15.18: Geothermal generating capacity in the United States from 1972 to 2015.

Based on 2015 Annual U.S. & Global Geothermal Power Production Report, data from http://www.eia .gov/electricity/data/eia860/

scenario	1990	2010
Table 15.3: Prediction from about 1980 for growth in geothermal generating capacity in the United States in MW_e for different scenarios.		
business as usual	2930	18,870
national commitment	8270	60,900

the early 1990s, although there has been some additional growth since around 2010. Table 15.3 gives predictions from around 1980 for the U.S. geothermal generating capacity. While the actual growth between 1980 and 1990 is reasonably consistent with the business-as-usual scenario, the prediction for 2010 greatly overestimated the actual situation. This is perhaps due to the continued low prices of fossil fuels and the economic and environmental advantages of pursuing other renewable energy sources, such as wind.

Since the nuclear accident at the Fukushima reactor in Japan in 2011, there has been renewed interest in developing that country's geothermal resources for electricity production. At present, Japan has a geothermal electric capacity of 536 MW_e, or about one-half of the capacity of a medium-sized nuclear reactor. The Geothermal

Example 15.3

If the geothermal electricity capacity in the United States is used at a capacity factor of 60%, estimate the annual savings in CO_2 emissions compared to the same energy generated by coal-fired power plants operating at a Carnot efficiency of 38%.

Solution

From Table 15.1, the installed capacity in the United States is 3567 MW_e. At a capacity factor of 70%, this represents a total annual energy production of

$$(3567 \text{ MW}_e) \times (0.60) \times (8760 \text{ h/y}) = 18.7 \times 10^6 \text{ MWh/y}.$$

The energy content of bituminous coal is 31 MJ/kg
or

$$(31 \text{ MJ/kg})/(3600 \text{ MJ/MWh}) = 8.6 \times 10^{-3} \text{ MWh/kg}.$$

Thus at 38% efficiency the geothermal energy production is equivalent to

$$(18.7 \times 10^6 \text{ MWh/y})/[(8.6 \times 10^{-3} \text{ MWh/kg}) \times (0.38)] = 5.72 \times 10^9 \text{ kg/y}.$$

If we approximate the combustion of coal by the combustion of carbon

$$C + O_2 \rightarrow CO_2,$$

then 1 kg of coal will produce

$$(1 \text{ kg}) \times (12 \text{ g/mol} + 32 \text{ g/mol})/(12 \text{ g/mol}) = 3.67 \text{ kg of } CO_2 \text{ per kg coal}.$$

So the geothermal generation represents

$$(5.72 \times 10^9 \text{ kg/y}) \times (3.67) = 2.10 \times 10^{10} \text{ kg } CO_2.$$

> ### Example 15.4
>
> Iceland has a population of 323,000. Compare its per-capita installed capacity for geothermal heat to the average per-capita residential home heating requirements. Assume an average residential volume of 150 m^3 per person. Note that Reykjavik, Iceland has 5010 heating degree days (°C) per year.
>
> ### Solution
>
> From Table 15.2 the installed capacity for direct use of geothermal heat in Iceland is 2040 MW. This represents a per-capita capacity of
>
> $$(2040 \text{ MW})/(323,000) = 6.32 \text{ kW per capita.}$$
>
> From Section 8.5 the annual residential heating requirement is 67 kJ/m^3 per degree day (°C). This gives a total annual heating requirement per capita of
>
> $$(67 \text{ kJ/m}^3 \cdot {}^\circ\text{C}) \times (150 \text{ m}^3) \times (5010{}^\circ\text{C}) = 50.4 \text{ GJ}$$
>
> for an average power requirement of
>
> $$(5.04 \times 10^{10} \text{ J/y})/(3.15 \times 10^7 \text{ s/y}) = 1.60 \text{ kW per capita.}$$
>
> So Iceland's installed per-capita geothermal heat capacity substantially exceeds its estimated residential heating requirement. In fact, the geothermal heat capacity in Iceland provides nearly all residential, commercial, and industrial heating needs.

Research Society of Japan has estimated that 1.5 to 2.4 GW$_e$ of Japan's 23 GW$_e$ geothermal potential could be developed within the next 40 years. These numbers can be viewed in the context of Japan's current nuclear generating capacity of about 47 GW$_e$.

The total worldwide geothermal capacity is difficult to assess, and estimates have varied greatly. In an analysis of the future use of geothermal energy, it is important to consider the lifetime of these resources. Although geothermal energy is considered a renewable resource and one may have the impression that its lifetime is more or less infinite, this is not generally the case. Because of the thermal gradient in the earth, heat is constantly flowing from the inside toward the surface. When heat is removed from a reservoir beneath the surface, it is being removed faster than it is being replenished from deeper within the earth. Reinjecting water into the reservoir provides a source of fluid to carry the heat, but it does not help replace the extracted heat. If we stop removing heat from a reservoir, then the temperature increases due to the heat flowing from the interior. This process is, however, very slow and occurs on a timescale of probably hundreds of thousands of years. How long the geothermal energy in a particular reservoir will last depends on how much thermal energy is in the reservoir and how fast it is removed. In general, lifetimes can be as short as a few years, or as long as a few hundred.

The environmental impact of geothermal power is generally minimal, but some concerns should be mentioned. The water or steam extracted from a geothermal reservoir contains a number of impurities. Among these are dissolved gases such as carbon dioxide, nitrogen compounds, and sulfur compounds. These can be released to the atmosphere, particularly in dry steam and flash steam plants (less in binary cycle plants). These emissions typically amount to about 5% of those released by a coal-fired generating station (per unit energy produced), and appropriate methods of removing at least some of these pollutants can be implemented. Water from underground reservoirs often contains anomalously high levels of toxic elements, such as arsenic, mercury, lead, antimony, and the like. Environmental consequences that could result from the release of these toxins are minimized by the common practice of reinjecting geothermal fluids back into the deposit. An undesirable side effect of this practice, however, is the possibility of introducing geological instabilities, and these effects may preclude the possibility of future development of certain geothermal resources.

A final point in geothermal energy's favor is land utilization. Compared with renewable energy sources that have a very low energy density on the earth's surface, such as solar or wind, or compared with coal, which requires substantial mining activities, geothermal energy utilizes land area very effectively. Per unit of power produced, geothermally generated electricity occupies only about 15–30% of the land area compared with most other energy sources.

15.6 Summary

This chapter described the origins of geothermal energy. This energy is believed to be associated with three factors: the primordial heat from the material that formed the earth, radioactive decay in the earth's interior, and tidal friction in the earth's molten core. Geothermal energy resources may be categorized in a number of ways. The major types of geothermal energy resources are the normal geothermal gradient, hot dry rock, hot water reservoirs, natural steam reservoirs, geopressurized regions, and molten magma. Hot water reservoirs and natural steam reservoirs are the most useful for wide-scale utilization of geothermal energy.

This chapter presented the ways in which geothermal energy resources may be utilized. These uses fall into two general categories: direct use of the thermal energy and conversion of the thermal energy into electricity. Typical direct uses of geothermal energy include bathing, space heating, agriculture, and aquaculture. Hot water reservoirs are typically the most suitable for these applications.

Geothermal electricity-generating facilities extract energy in the form of hot water or steam from a production well. This fluid is used to generate electricity using a turbine/generator and is then reintroduced into the reservoir through an injection well. Hot dry steam is the most convenient resource for this application and is the most commonly found in the geothermal areas of California. Most geothermal electricity generation worldwide uses hot water as a source of geothermal energy.

The United States has the greatest utilization of geothermal energy resources both for direct use and for electricity generation. While some further development is possible in the United States, the greatest potential for additional geothermal energy use is probably in countries, such as Japan, that lie in geothermally active areas but that have not been as active in pursuing the use of this resource.

As noted in the chapter, geothermal resources are not indefinitely renewable because utilization on any reasonable commercial scale for electric generation depletes the resource faster than it is replenished by heat flowing outward from the earth's interior. Lifetimes of geothermal resources depend on the nature of the resource and the manner of its use. Typically, with extensive use, the lifetime of a resource may be in the order of a century.

Problems

15-1   Calculate the ideal Carnot efficiency for a turbine operating with a hot reservoir at a temperature of 225°C and a cold reservoir at a temperature of 75°C.

15-2   Assuming that the geothermal heat flux in the United States is (on average) typical of that which occurs worldwide, calculate the fraction of total U.S. primary energy use that could be provided by geothermal energy if this energy could be utilized in its entirety.

15-3   Consider the total energy use for a typical North American single-family home, as discussed in Chapter 2. If the house has a footprint of 160 m², what fraction of its energy needs could be met by the average geothermal heat flow from the earth's interior?

15-4   Heat is extracted from hot rock at a temperature of 250°C and used to produce electricity with the ideal Carnot efficiency. If the temperature of the cold reservoir is 75°C, what mass of rock would be needed to yield 1 GWh$_e$? See Chapter 8 for the thermal properties of typical rock, and assume a generator efficiency of 90%.

15-5   Heat is extracted from geothermal pressurized water at a temperature of 225°C with the ideal Carnot efficiency. If the cold reservoir is at 80°C and the generator efficiency is 90%, what is the flow rate of geothermal water (in m³/s) needed to yield 20 MW$_e$ output?

15-6   Assume that geothermal heat transfer (at least near the surface of the earth) is by conduction through the crust rocks. For an average geothermal heat flow of 0.087 W/m² and typical thermal gradient of 100°C/km, calculate the thermal conductivity of the rock. Compare with the known thermal conductivities of similar materials given in Chapter 8.

15-7   It is clear that geothermal generating stations extract energy from a resource more rapidly than it is replenished from the interior of the earth. Assume that a resource has 10 EJ of energy that can be extracted economically. If this resource is used to generate 250 MW$_e$ of electricity, how long will the resource last? It is necessary to know the efficiency of electric generation. Assume an efficiency of one-third the ideal Carnot efficiency (Example 15.2). This is typical of the actual average operational efficiency of a geothermal generating station.

15-8   A geothermal electrical generating station has an actual efficiency of one-third of the ideal Carnot efficiency. For a cold reservoir temperature of 70°C, calculate and plot the efficiency of the station as a function of the geothermal fluid temperature from 100°C to 300°C. If a minimum efficiency of 5% is necessary to make the utilization of the resource economical, what minimum geothermal fluid temperature is needed?

15-9 Consider the possibility of extracting power from a flow of water of 1000 L/s for the four following scenarios and calculate the power output. In all cases, assume a turbine/generator efficiency of 85%.

    (a) Geothermal water at 200°C with a cold reservoir of 40°C.

    (b) Hot water at 24°C supplied to an OTEC facility where the cold water is at 4°C and the indicated flow is the flow through the evaporator.

    (c) Water falling through a head of 100 m in a high head hydroelectric installation.

    (d) Water flowing through a run-of-the-river hydroelectric installation at a velocity of 5 m/s.

15-10 Assuming an average geothermal heat flux of 0.087 W/m^2 and an average per-capita power use of 10 kW, what is the maximum population density that could be accommodated if geothermal heat is used at an average efficiency of 50% to satisfy all primary energy requirements? Are there countries that (on the average) meet this condition? Give examples.

15-11 For a cold reservoir temperature of 60°C, plot the Carnot efficiency of a heat pump operating on geothermal fluids as a function of the geothermal fluid temperature from 120°C to 400°C.

15-12 Estimate the world geothermal power production for the year 2025

    (a) If the growth continues exponentially at the actual rate since the mid-1980s.

    (b) If the growth continues exponentially since the mid-1980s at the actual rate between 1970 and 1985.

    (c) If the growth continues exponentially since 1920 at the actual rate between 1905 and 1920.

    (d) Discuss the feasibility of these three scenarios.

15-13 A geothermal resource has a life expectancy of 100 years if an average power of 50 MW$_e$ is generated at two-thirds of the ideal Carnot efficiency. If the reservoir temperature is pressurized water at 300°C and the cold reservoir temperature is 75°C, what is the total energy available in the reservoir?

15-14 If a typical home uses 8×10^{10} J of heat per year, how many homes could be heated using the hot dry-rock resources in the United States if they were fully implemented?

15-15 Consider a geopressurized deposit with water at a temperature of 200°C and dissolved methane with a molar fraction of 10^{-5}. The heat content of the water is used to generate electricity in a heat engine with one-half of the Carnot efficiency and a cold reservoir of 50°C, and the chemical energy content of the methane is used by burning the methane and generating electricity with a heat engine with an overall efficiency of 25%. What is the energy produced by m^3 of resource by each of these methods?

15-16 As indicated in the chapter, geothermal heat comes primarily from the radioactive decay of uranium, thorium, and potassium. The isotopes of these elements that contribute to the radioactivity and their half-lives are: ^{235}U (7.1×10^8 years), ^{238}U (4.5×10^9 years), ^{232}Th (1.4×10^{10} years), and ^{40}K (1.4×10^9 years). Assume that these four nuclides existed in equal quantities at the time the earth was formed 4.54×10^9 years ago.

(a) What are the relative proportions of each nuclide at present?

(b) What are the relative contributions of each nuclide to the total radioactive decay in the earth?

15-17 In recent years, annual geothermal electricity generation in the United States has been about 16.8×10^6 MWh_e.

(a) What is the average capacity factor?

(b) If geothermal electricity is considered to be carbon free, what is the reduction in the mass of emitted CO_2 compared to the equivalent generation of electricity by typical coal-fired generating stations?

15-18 Granite has a typical thermal conductivity of 1.9 W/(m·°C).

(a) What is the thermal conductivity in the units used in Chapter 8, $(W \cdot cm)/(m^2 \cdot °C)$?

(b) Calculate the power per km^2 transferred through granite in the presence of a thermal gradient of 20°C per km.

15-19 A geothermal generating facility operates at the ideal Carnot efficiency and utilizes a pressurized hot water resource at 240°C.

(a) Calculate the Carnot efficiency as a function of the temperature of the cold reservoir from 40°C to 90°C.

(b) The waste heat from electricity generation may be utilized for space heating, and the facility may, therefore, supply a community with both electricity and heat. This approach is referred to as cogeneration and is discussed in more detail in Chapter 17. Using the relative proportions (averaged over the year) for residential electricity and heat requirements as presented in Chapter 2, is there an approach to regulating the cold reservoir temperature that would make the best use of the geothermal resources?

15-20 If the average geothermal electricity generation in the United States amounts to 16.8×10^6 MWh, what fraction of the total electrical energy generation in this country does this represent?

Bibliography

H. Armstead. *Geothermal Energy*, (2nd ed.) E&FN Spon, London (1983).

G. Boyle (Ed.). *Renewable Energy*. Oxford University Press, Oxford (2004).

M. Dickson and M. Fanelli. *Geothermal Energy: Utilization and Technology*. Earthscan Publishers, London (2005).

R. Harrison, N. D. Mortimer, and O. B. Smarason. *Geothermal Heating: A Handbook of Engineering Economics*. Pergamon Press, Oxford (1990).

B. K. Hodge. *Alternative Energy Systems and Applications*. Wiley, Hoboken (2010).

M. Kaltschmitt, W. Streicher, and A. Wiese (Eds.). *Renewable Energy: Technology, Economics and Environment*. Springer, Berlin (2007).

R. A. Ristinen and J. J. Kraushaar. *Energy and the Environment*. Wiley, Hoboken (2006).

World Energy Council. *World Energy Resources 2016*, available online at https://www.worldenergy.org/publications/2016/world-energy-resources-2016/

Biomass Energy

16.1 Introduction

Biomass energy refers to energy extracted from recently grown biological matter. It is renewable (as compared with fossil fuels) because, as it is used, new material can be grown to replace it. Biofuels are the oldest source of energy used by humans; the use of wood for heating and cooking dates back to prehistoric times. Because of their renewability, biofuels are also a source of energy that has generated substantial interest in recent years. On a global scale, biofuels are currently the largest contributor to renewable energy. However, to be utilized in a renewable manner and remain carbon neutral, new biomaterial must be grown to replace material that has been used in order to sequester the carbon that has been produced. Biofuels may be materials, such as wood, that are utilized directly in the form in which they are found in nature. They may also be materials that are produced by processing naturally grown materials into a form of fuel that is more readily utilized (e.g., the production of ethanol from plant matter). Also, biofuels include societal waste products (e.g., municipal waste), which are either burned directly or processed to produce a more convenient form of fuel. In this chapter, biofuels are placed in four basic categories: wood, bioalcohols, biodiesel, and municipal waste.

16.2 Wood

Wood has long been used as a source of energy, and until the late 1800s, was the major source of energy worldwide, after which it was replaced (for some time at least) by coal. Over the years, wood has also been an important source of material for industry and building (and continues in this role). In North America, more than 70% of wood that is harvested is used for industry (largely the paper industry) and building construction. Less than 30% is used as a source of energy. In Bangladesh, 98% of the harvested wood is used for energy. Today one-third of world's population (mostly in developing countries) relies on wood as a major source of energy. This discussion is confined to the use of wood as a source of energy.

Wood that is used for the production of energy is used in one of three forms:

1. *Firewood*: Wood that is used directly.
2. *Charcoal*: Primarily carbon that is produced by heating wood.
3. *Black liquor*: A combustible oil-like substance that is a by-product of the pulp and paper industry.

The combustion of wood is basically the oxidation of carbon or hydrocarbons and is, from a chemical standpoint, very much like the combustion of a fossil fuel. The energy content of wood is typically about 14 MJ/kg compared with about 31 MJ/kg for bituminous coal. The main difference between burning wood and burning coal is that trees (or other living organic matter) sequester carbon dioxide so that the CO_2 released by the combustion of wood can be eliminated from the environment if the trees that are burned are replaced by an equal number of new trees. In this way, wood can be used as a renewable resource that does not have a net contribution to the greenhouse gas in the environment. In this case, it is referred to as being *carbon neutral*. Unfortunately, much of the wood used for both energy production and industrial purposes is not used in a renewable way, and the resulting deforestation means that the greenhouse gases released into the environment by the burning, decomposition, or decay of wood are not sequestered by the growth of new trees. Efforts exist in many areas to alleviate this problem (Figure 16.1).

Figure 16.1: Sign promoting reforestation activities in New Brunswick, Canada.

Richard A. Dunlap

Example 16.1

How many kilograms of wood have the same energy as one tonne of bituminous coal?

Solution

From Appendix IV 1 tonne of bituminous coal has an energy content of

$$(3.10 \times 10^7 \text{ J/kg}) \times (1000 \text{ kg/t}) = 3.10 \times 10^{10} \text{ J/t}.$$

From Appendix IV, 1 kg of wood has an energy equivalence of 1.4×10^7 J. Thus $(3.10 \times 10^{10} \text{ J/t})/(1.4 \times 10^7 \text{ J/kg}) = 2.21 \times 10^3$ kg wood is equivalent to 1 tonne of coal.

Much of the wood that is used as a source of energy is used on a residential basis, rather than for generating electric power for the grid. Difficulties with the widespread use of wood for commercial energy production stem largely from problems with processing and transportation costs. A 1500-MW$_e$ wood-fired generating station operates in Sweden, and there is a 500-MW$_e$ facility in Austria. A 3.96-GW$_e$ coal-fired generating station in the United Kingdom has recently been partially converted to a wood-burning facility using fuel imported from the United States and Canada. A number of smaller wood-fired generating stations exist worldwide, including a few in North America. Even so, wood, primarily as a heating fuel, remains a major contributor to renewable energy, as illustrated by a breakdown of renewable energy in the United States in 2015 (Figure 16.2).

Although the use of wood in a renewable way is carbon neutral, the burning of this resource is not without other environmental consequences. Wood contains much less sulfur than coal, so SO$_x$ emissions are not a concern. However, burning wood

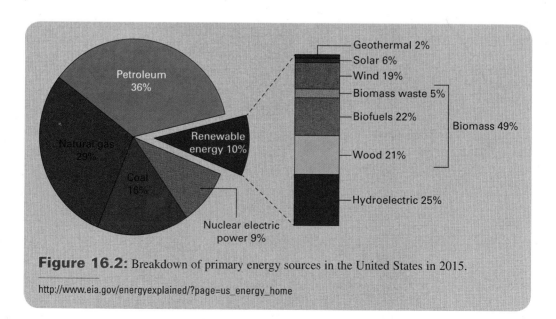

Figure 16.2: Breakdown of primary energy sources in the United States in 2015.

http://www.eia.gov/energyexplained/?page=us_energy_home

produces substantial quantities of NO_x and particulate matter. Because wood use is not generally a large-scale commercial operation, control of emissions is more difficult. Also, the combustion of wood releases benzo(a)pyrene, a known carcinogen.

The consequences of leaving dead trees to decay is also a concern, and it has been suggested that clearing deadwood from forests for fuel use is the lesser of two evils. Decaying wood emits carbon mostly in the form of CH_4, while burning wood emits carbon mostly in the form of CO_2. Although both CH_4 and CO_2 are greenhouse gases (Table 4.6), methane is about 25 times more effective at absorbing infrared radiation than carbon dioxide.

16.3 Ethanol Production

Light hydrocarbons such as methane and ethane are gaseous at room temperature (Table 3.2), and hydrated light hydrocarbons are typically liquid at room temperature. Some of these are alcohols, such as methanol, ethanol, and the like, which are represented by the formula $C_nH_{2n+1}OH$. The properties of the light alcohols are shown in Table 16.1. These alcohols may be produced as a by-product of the distillation of petroleum or by the fermentation of sugar containing biological materials. Bioalcohols can be used as a source of energy and, because they are liquids at room temperature, provide a convenient replacement for petroleum-based transportation fuels such as gasoline. This discussion focuses on ethanol because this has been shown to be the most practical in terms of ease of production and suitability for direct replacement of gasoline in internal combustion engines. Methanol has found some use as a fuel for fuel cells and is discussed further in Chapter 20.

Currently about 95% of all ethanol is bioethanol; the remainder is produced from petroleum. Ethanol is the alcohol found in alcoholic beverages and is also used in various industrial processes. The majority of ethanol produced at present is used for the production of energy. Energy from bioalcohol is a manifestation of solar energy. Light from the sun produces glucose ($C_6H_{12}O_6$) by means of photosynthesis using water and CO_2. The process is

$$6CO_2 + 6H_2O + \text{light energy} \rightarrow C_6H_{12}O_6 + 6O_2. \qquad \textbf{(16.1)}$$

Living organisms can modify glucose to form fructose (same chemical formula as glucose but different structure), bond glucose together in strings to form starch or cellulose, or bond glucose to fructose to form sucrose (normal table sugar). The

			molecular mass (g/mol)	density (g/cm³)	boiling point (°C)	heat of combustion (MJ/L)
n	name	formula				
1	methanol	CH_3OH	32.04	0.792	64.7	17.9
2	ethanol	C_2H_5OH	46.07	0.789	78.4	23.5
3	propanol	C_3H_7OH	60.10	0.785	82.3	26.3
4	butanol	C_4H_9OH	74.12	0.810	117.7	29.7

Table 16.1: Properties of the light alcohols of the composition $C_nH_{2n+1}OH$. Heat of combustion is the HHV (Chapter 1).

Example 16.2

Calculate the mass of CO_2 (in kilograms) produced by the combustion of 1 kg of methanol.

Solution

Methanol combines with oxygen to yield CO_2, H_2O, and energy. Balancing the chemical formula shows that the reaction is

$$2CH_4O + 3O_2 \rightarrow 2CO_2 + 4H_2O + \text{energy.}$$

Therefore, 2 moles of methanol yields 2 moles of carbon dioxide, or a 1:1 molar ratio. Using approximate molar weights of methanol (32 g/mol) and carbon dioxide (44 g/mol) means that 1 g of methanol produces (1 g) $\times$ (44 g/mol)/ (32 g/mol) = 1.37 g of CO_2. Thus 1 kg of methanol produces 1.37 kg of carbon dioxide.

fermentation of simple glucose produces ethanol, along with CO_2 and heat, according to the process

$$C_6H_{12}O_6 \rightarrow 2C_2H_5OH + 2CO_2 + \text{heat.} \qquad \textbf{(16.2)}$$

The combustion of ethanol produces CO_2 and water along with heat:

$$C_2H_5OH + 3O_2 \rightarrow 2CO_2 + 3H_2O + \text{heat.} \qquad \textbf{(16.3)}$$

The photosynthesis process is inherently inefficient (perhaps about 2%), meaning that the production of electricity by burning ethanol in a heat engine has an overall efficiency that is considerably lower than that achieved by photovoltaics. However, as the technologies involved in these two approaches are quite different, a more detailed analysis of biofuel use is needed.

The steps in the commercial production of ethanol from organic matter are: fermentation, distillation and dehydration. The details of these are as follow:

16.3a Fermentation

Traditional fermentation processes convert simple sugars (including glucose, fructose, sucrose, and starches) to ethanol according to equation (16.2). More complex processes (discussed later in the chapter) are required to break down cellulose prior to fermentation. Some plant material (e.g., sugarcane) contains a significant proportion of simple sugar, whereas others, such as corn, contain much less. For example, the current production of ethanol from corn utilizes only about 50% of the dry corn kernel [although the remainder of the plant may be used for other purposes (e.g., livestock feed, etc.)].

16.3b Distillation

The distillation of the fermentation product is normally required to remove water from the ethanol. Traditional distillation techniques yield an *azeotropic* mixture of ethanol with about 4% water. Although this mixture can, in principle, be used directly to produce energy by combustion, the presence of water makes it immiscible with gasoline, limiting its use for fuels that are a gasoline-ethanol mixture (see Section 16.3d). It is therefore generally necessary to use dehydration methods to remove the remaining water.

16.3c Dehydration

Traditional dehydration techniques mix, say, benzene with the azeotropic ethanol-water mixture produced by distillation. The water preferentially mixes with the benzene, and the ethanol can be separated. While this technique allows for the extraction of high-purity dehydrated ethanol, the waste material contains benzene, a known carcinogen. To eliminate the health and environmental hazards associated with carcinogenic materials, new benzene-free techniques have been developed that use molecular sieves to absorb water and allow the ethanol to be extracted. These techniques are now in common use and can take the place of both the distillation and dehydration processes.

16.3d Use of Ethanol

Ethanol can be used in its pure form as a fuel in an internal combustion engine, or it may be blended with gasoline in various proportions. Fuel mixtures are designated by the percentage (by volume) of ethanol where, for example, E10 represents a mixture of 10% ethanol and 90% gasoline. The mixing of ethanol with gasoline as a transportation fuel dates back to the 1920s. It became popular in the 1970s during the oil crisis when E10 fuel was commonly marketed under the name of *gasohol*. Normal gasoline internal combustion engines can use ethanol-gasoline mixtures up to about E10, or in some cases slightly higher, without any modification. In some countries, low-ethanol blends have been mandated by government regulation (e.g., E10 in Sweden and E5 in India). As of 2008, nine states in the United States have mandated the use of E10. In other states, the addition of ethanol to gasoline is a common practice, and ethanol in some amount is present in more than 60% of the gasoline sold in the United States. In some cases (e.g., see Figure 16.3 on the next page), this is indicated on the gasoline pump.

The use of higher percentages of ethanol in gasoline requires minor modifications to the engine design. Most modifications are related to the increased reactivity of polymers to ethanol-containing fuels and require the replacement of some plastic components in the fuel delivery system. Typical modifications to an engine that will allow the use of E85 cost less than US$100. However, retrofitting older engines can be problematic because the ethanol dissolves accumulated organic deposits in the fuel system (i.e., those that are relatively insoluble in gasoline) and causes clogging in the engine. Many vehicles sold in North America (referred to as *flex fuel* vehicles) come with the ability to utilize ethanol mixtures up to E85 (Figure 16.4 on the next page). In considering fuels with higher ethanol content, it is important to realize the lower energy content (per unit volume) of ethanol compared with gasoline when analyzing fuel consumption, fuel cost, and driving range. Ethanol (Table 16.1) has an energy content of 23.5 MJ/L, whereas gasoline has 34.8 MJ/L, although increased engine efficiency may partially offset these differences. The utilization of ethanol-containing fuels in the gasoline engines of recreational vehicles has been somewhat controversial, particularly for marine engines. In addition to the potential clogging problems, the presence of ethanol in fuel stored in a humid marine environment can promote the condensation of

Figure 16.3: Gasoline pump in Maine indicating 10% ethanol content (E10).

Richard A. Dunlap

Figure 16.4: Nameplate on vehicle designed to run on gasoline with up to 85% ethanol (E85).

Richard A. Dunlap

moisture from the air, resulting in performance loss and mechanical difficulties. This problem is often exacerbated by the fact that boaters often replace fuel less frequently in their boats than drivers do in their automobiles.

Example 16.3

For vehicles with the same range, calculate the fuel tank volume and equivalent fuel mass for vehicles using methanol and ethanol compared to a gasoline-powered vehicle with a fuel tank volume of 70 L. Assume that the engine efficiencies are the same for all three fuels. (*Note:* The density of gasoline depends on the exact ratio of the different hydrocarbons present but has an average value of about 0.72 kg/L.)

Solution

If the energy content of gasoline is 34.8 MJ/L, a 70-L tank would have a total energy content of 34.8 MJ/L $\times$ 70 L = 2436 MJ. Using the energy content per unit volume given in Table 16.1 for methanol and ethanol (17.9 MJ/L and 23.5 MJ/L, respectively), the volumes of these two biofuels necessary to produce the same energy as 70 L of gasoline are

$$\text{Methanol:} \qquad \frac{2436 \text{ MJ}}{17.9 \text{ MJ/L}} = 136 \text{ L}$$

and

$$\text{Ethanol:} \qquad \frac{2436 \text{ MJ}}{23.5 \text{ MJ/L}} = 104 \text{ L.}$$

Using the densities from Table 16.1, the masses of these two biofuels are

$$\text{Methanol:} \qquad (136 \text{ L}) \times (0.792 \text{ kg/L}) = 108 \text{ kg}$$

and

$$\text{Ethanol:} \qquad (104 \text{ L}) \times (0.789 \text{ kg/L}) = 82 \text{ kg.}$$

By comparison, 70 L of gasoline have a mass of (70 L) $\times$ (0.72 kg/L) = 50 kg. This shows that biofuel-powered vehicles must accommodate both larger volumes and masses of fuel.

The breakdown of world ethanol production is shown in Table 16.2 on the next page. Currently, about 90% of ethanol production is for fuel purposes. The increase in fuel ethanol production worldwide since the mid-1970s is shown in Figure 16.5 on the next page. From Table 16.2, it is seen that almost 70% of the ethanol produced worldwide is produced in either the United States or Brazil. It is interesting to compare the approaches taken to fuel ethanol production in these two countries.

In the United States, virtually all fuel ethanol production is produced from corn. As previously explained, only a fraction of the corn plant can be utilized in ethanol production using current technology. Cellulosic ethanol production (i.e., the use of plant cellulose rather than just the fermentation of sugars) would provide a substantial increase in opportunities for ethanol production in the United States and elsewhere by allowing for greater utilization of plants like corn and also allowing for the utilization

country	2007	2008	2009	2010	2011	2012	2013	2014
United States	24,685	35,238	41,405	50,338	52,799	50,346	50,346	54,131
Brazil	18,999	24,499	24,900	26,203	21,096	21,111	23,723	23,432
Europe	2158	2778	3937	4577	4421	4463	5190	5470
China	1840	1900	2052	2052	2101	2101	2635	2404
Canada	799	901	1102	1351	1749	1700	1980	1931
rest of world	1192	1473	3460	3729	2642	2847	4815	5640
world total	**49,676**	**66,790**	**76,855**	**88,241**	**84,808**	**82,567**	**88,688**	**93,007**

Table 16.2: Ethanol production in recent years (10^6 L).

Based on data from Renewable Fuels Association. http://www.ethanolrfa.org/resources/industry/statistics/#1454098996479-8715d404-e546

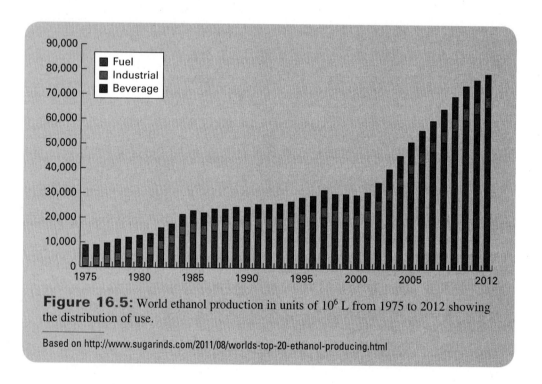

Figure 16.5: World ethanol production in units of 10^6 L from 1975 to 2012 showing the distribution of use.

Based on http://www.sugarinds.com/2011/08/worlds-top-20-ethanol-producing.html

of other (low-sugar-containing) plant species such as grasses. At present, this is not technologically viable but may be developed at some point in the future. The utilization of sugarcane as a source material for ethanol production has recently been initiated in some states with warmer climates (i.e., Louisiana, Hawaii, Texas, and Florida), but in most parts of the country, sugarcane is not a viable crop. Some of the properties of ethanol production in the United States are summarized in Table 16.3.

In Brazil, sugarcane is used almost exclusively for ethanol production. Brazil has seriously pursued the use of ethanol as a fuel for more than 40 years; some of the characteristics of their program are summarized in Table 16.3. The improvement in ethanol production per area of farmland over the years in Brazil is illustrated in Figure 16.6. While normal U.S. passenger vehicles typically tolerate up to 10% ethanol in gasoline

Table 16.3: A comparison of ethanol production and use in the United States and Brazil.

property	United States	Brazil	units
major crop	corn	sugar cane	
total ethanol production (2014)	54,131	23,432	10^6 L
total arable land (excludes Alaska)	27.0	35.5	10^5 km^2
land used for ethanol production	1.0	0.36	10^5 km^2
percent arable land used for ethanol production	3.7	1.0	%
productivity	3.8–4.0	6.8–8.0	10^5 L/km^2
energy balance	~1.4	~9.2	
carbon payback period	93	17	years
ethanol fueling stations*	2326	35,017	
percent fueling stations selling ethanol	1.0	100	%
ratio of fuel ethanol to gasoline used	~1.0	~0.1	

* U.S as of 2010, Brazil as of 2007

mixtures and flex fuel vehicles can use up to E85, vehicles in Brazil are designed to operate on a much wider variety of fuels. Most run on any mixture up to E100. In fact, flex fuel vehicles in Brazil can utilize hydrated ethanol (that is the azeotrope consisting of about 96% ethanol with 4% water), thus eliminating the need for additional dehydration after distillation.

A comparison of the information in Table 16.3 for the United States and Brazil shows that, although the total ethanol production in the two countries is similar, the program in the United States is a minor addition to gasoline use, while in Brazil it fulfils the major portion of transportation fuel needs. This comparison is clear from the fraction of fueling stations in the two countries that sell ethanol-rich fuel mixtures: 1% in the United States and 100% in Brazil. It is also seen in the table that U.S. ethanol

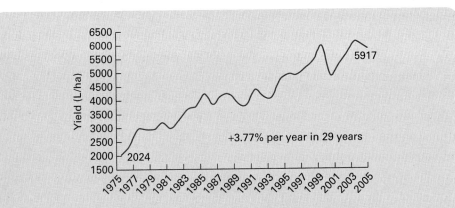

Figure 16.6: Increase in ethanol productivity in Brazil, 1975–2005.

José Goldemberg, "The Brazilian biofuels industry," Biotechnology for Biofuels 2008, 1:6. © 2008 Goldemberg; licensee BioMed Central Ltd.

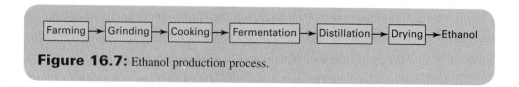

Figure 16.7: Ethanol production process.

production utilizes a larger fraction of available arable land than is the case in Brazil. To understand why ethanol production and use seem to be much more successful in Brazil than in the United States, it is important to look at some of the details of the ethanol production process. As described, plant material is grown, ground, and fermented and the resulting ethanol is distilled and possibly dehydrated (Figure 16.7). At present, each of these steps, at least in the United States, is accomplished by machines that are fueled by fossil fuels or utilize electricity generated largely by burning coal. This will continue to be the situation until ethanol production becomes self-sustaining. Even so, energy input is required to produce ethanol, and the ratio of energy input to the energy content of the ethanol produced is the important factor. In the United States, this ratio is about 1.4:1, meaning that 1 J of energy input produces ethanol with an energy content of 1.4 J; that is a gain of 0.4 J of energy. Although there is a net energy gain in the process, the numbers indicate that the 0.4 J of energy gained is expensive in terms of both energy and dollars. In Brazil, the ratio is about 9.2:1, meaning that 1 J of energy input into ethanol production yields a net gain of 8.2 J—clearly a much more advantageous situation.

Several factors contribute to this situation:

1. Brazil has a climate that is much more conducive to plant growth than much of the United States has. Thus, plants grow faster and can be grown over a greater fraction of the year.
2. Because of the climate, sugarcane can be readily grown in Brazil. This is a much more efficient source of ethanol, as illustrated by the volume-per-unit-farming-area values shown in Table 16.3.
3. Brazil has invested considerable effort into making their ethanol production process as efficient as possible (Figure 16.6).
4. The U.S. agricultural and ethanol production technologies are more mechanized and therefore require greater energy input than is the case in Brazil.

There is also the question of the overall contribution to greenhouse gas production by various energy production methods. In the United States, large-scale ethanol utilization would require the development of new farmland and the development of new farming infrastructure (e.g., tilling of new soil, etc.). The carbon payback period in the United States, that is, the time required to compensate for the carbon released to the atmosphere during the development of ethanol production (recall the carbon payback period for hydroelectric discussed in Energy Extra 11.2), has been estimated to be about 93 years. A similar period for Brazil is about 17 years. This difference results from the greater difficulty in growing crops with a greater ethanol return in a more temperate climate coupled with the greater degree of mechanization (and associated carbon emissions) in the U.S. agricultural system.

A final point that needs to be considered is the impact of fuel ethanol production on food production. The extensive use of farmland for the production of plant matter for ethanol production would clearly impact the ability to produce food crops. Such food crops are essential both for direct human consumption and for use as feedstock for animals, although some use of waste material from ethanol production from corn for livestock feed is possible. The extent of the conflict between ethanol production and food production in the United States is emphasized by the fact that, if all corn currently grown in the United States were used for ethanol production, it would replace only about 12% of the gasoline used. Thus, the large-scale use of ethanol will require considerable new farming activity or a decrease in food productivity or (most likely) both. A decrease in food production will lead to a decrease in food supply and an increase in prices. Agricultural methods that are the most effective at maximizing ethanol production will likely lead to a decline in soil fertility, the increased use of pesticides and fertilizers (which require energy for their production), increased deforestation (to make additional land available), and a reduction in the availability of water for irrigation. Thus, in addition to the reallocation of existing farmland, increased ethanol production can have adverse indirect consequences on food production and the environment. Cellulosic ethanol production will increase the productivity of ethanol from corn and open up the possibility of producing ethanol from other crops. This could include the use of grasses that can be grown on land that is less desirable for food production. Switchgrass (*Panicum virgatum*) is a rapidly growing grass that thrives in most parts of the United States and in a wide variety of habitats. The cellulosic ethanol industry is in its very early stages of development, and further technological advances, as well as careful integration into the agricultural system, will be needed to benefit fully from this approach.

Overall, a comparison of the Brazilian and American approaches to ethanol production emphasizes the importance of climate, as well as the development of efficient production technologies. It also makes it clear that, from a climatic and geographical perspective, ethanol production is not a viable energy alternative in all parts of the world.

16.4 Biodiesel

Biodiesel fuel is comprised of short chain alkyl esters and is similar in most ways to traditional petroleum-derived diesel fuel. It is made by the transesterification of vegetable oils or animal fats. Biodiesel is distinguished from unprocessed vegetable oil (often called straight vegetable oil, SVO), which is sometimes used as a fuel. The latter may be nonfood-grade vegetable oil or waste vegetable oil from the food industry. Although SVO may be an attractive alternative to diesel fuel, engines require significant modifications to utilize this fuel. Biodiesel, on the other hand, can readily replace petroleum-derived diesel fuel, much as ethanol can be used to replace gasoline. Minor engine modifications, as for ethanol use, are required because the natural rubber components often used in diesel engines may degrade upon exposure to biodiesel. Also, engines that have previously run on petroleum diesel typically have deposits left behind from the fuel, and these are soluble in biodiesel fuel, possibly leading to clogging of filters. In addition to use in vehicles, biodiesel is a direct replacement for domestic or commercial heating fuel, subject to the solubility issues described. It may also find use as a fuel for trains or aircraft.

Biodiesel may be blended with petroleum diesel, and the resulting fuels are designated by the percentage of biodiesel (e.g., B5 for 5% biodiesel, 95% petroleum diesel- etc.), up to B100 for 100% biodiesel. B99 is a common fuel in some parts of the world. The 1% petroleum diesel is added to biodiesel to retard mold growth. The compatibility of biodiesel blends with unmodified diesel engines is somewhat unclear, although blends up to B5 are generally felt to be acceptable.

While waste vegetable oil and waste animal fat may seem like an attractive source material from which to make biodiesel, it is unlikely to be a major component in the production of this fuel. The collection of waste oil may be inconvenient or difficult, as well as being expensive. The major problem, however, is availability. In the United States, approximately 1.8×10^{11} L of diesel fuel are used for transportation and heating annually. Estimated primary production of vegetable oil is about 1.3×10^{10} L per year, and the production of animal fat is about half that amount. Thus, waste from the food industry could account for, at most, about 10% of the diesel needs in the United States. It is therefore necessary to utilize predominately source material that is grown for the specific purpose of producing biodiesel. In the United States, most current biodiesel production uses oil extracted from soy. Supplying the U.S. needs for diesel fuel with biodiesel produced from soy would require about 2×10^6 km^2 of farmland. This is the majority of all arable land available (Table 16.3). Clearly this is not a viable situation; biodiesel would replace food production (and ethanol production as well). Other sources of materials for biodiesel need to be found if this energy source is going to make a significant contribution to our energy needs. Table 16.4 shows the productivity for various plants that can be used for biodiesel production. Productivity from all common terrestrial oil-producing plants that are compatible with a temperate climate is similar. Tropical plants (palm and coconut) are somewhat more productive. However, algae is clearly the best choice for producing biodiesel by far. An additional feature of algae is that it can be grown in marine environments and in ponds on land that is not otherwise suitable for farming. Total U.S. diesel needs could be satisfied by algae grown in an area of about 100,000 km^2, or about the land area of Iceland or South Korea. Certainly future developments in biodiesel production must consider algae as an important possibility.

In recent years, biodiesel production has increased substantially as Figure 16.8 indicates. European countries, Germany in particular, have been very active in developing biodiesel.

The question of the environmental impact of biodiesel use is important. Compared with petroleum-derived diesel, biodiesel produces less sulfur compounds, less hydrocarbon emission, and fewer particulates. However, it produces somewhat more NO$_x$. Clearly, the combustion of a hydrocarbon releases carbon to the atmosphere in the form of CO$_2$. Although carbon is absorbed from the atmosphere by the growth of

Table 16.4: Typical annual productivity of biodiesel from different plant materials per unit farming area.	
	annual biodiesel production
plant	10^3 L/km^2
algae	1700
palm oil	475
coconut	215
rapeseed	95
soy	55–91
peanut	84
sunflower	77

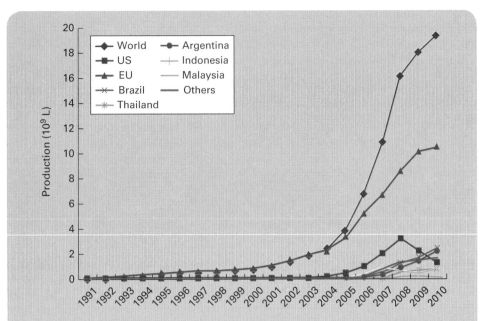

Figure 16.8: Growth of biodiesel production worldwide and by major producers from 1991 to 2010.

Based on Guyomard Hervé, Forslund Agneta and Dronne Yves (2011). Biofuels and World Agricultural Markets: Outlook for 2020 and 2050, Economic Effects of Biofuel Production, Dr. Marco Aurelio Dos Santos Bernardes (Ed.), InTech, DOI: 10.5772/20581. Available from: http://www.intechopen.com/books/economic -effects-of-biofuel-production/biofuels-and-world-agricultural-markets-outlook-for-2020-and-2050

Example 16.4

The actual ocean coastline of a country is a somewhat ambiguous quantity, depending on the scale on which harbors, islands, estuaries, and other features are included. For example, the coastline of the continental United States has been estimated at a minimum of about 8000 km or a maximum of about 80,000 km, if all small-scale features are included. Assuming, from a practical standpoint, that a maximum of 16,000 km of the U.S. coastline to a distance of 100 m offshore is utilized for algae production to make biodiesel, what fraction of the total U.S. diesel needs could be met?

Solution

The total available area for algae production is

$$(1.6 \times 10^4 \text{ km}) \times (0.1 \text{ km}) = 1600 \text{ km}^2.$$

The annual productivity of biodiesel from algae is given in Table 16.3 as 1.7×10^6 L/km^2. The available coastal area would allow for the production of

$$(1600 \text{ km}^2) \times (1.7 \times 10^6 \text{ L/km}^2) = 2.7 \times 10^9 \text{ L of biodiesel per year.}$$

If the total diesel use in the United States is 1.8×10^{11} L, the amount that can be produced from algae grown under the conditions specified in this example would amount to $(2.7 \times 10^9 \text{ L})/(1.8 \times 10^{11} \text{ L}) = 0.015$, or 1.5% of the requirement. This calculation emphasizes the magnitude of the agricultural task of producing biofuels.

ENERGY EXTRA 16.1
Environmental effects of biofuels

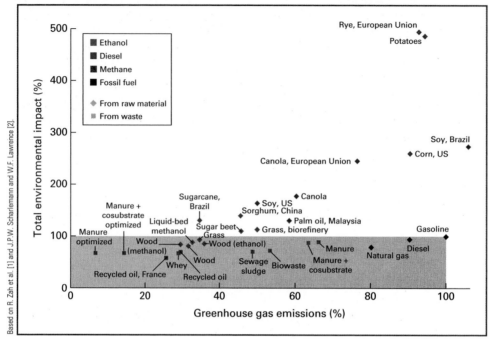

Greenhouse gas emissions and total environmental impact factors for some biofuels.

As illustrated by the comparison in the text between ethanol production from Brazilian sugarcane and from U.S. corn, it is clear that all biofuels do not offer the same degree of carbon reduction relative to fossil fuels. While Brazilian sugarcane would seem to provide substantial benefits as an alternative fuel, corn produced in the United States would appear to have only a marginal environmental advantage over petroleum. The question of the environmental impact of biofuels, however, is a much more complex problem than merely a comparison of the relative carbon footprint. A recent analysis by Zah and colleagues[1] has provided a quantitative analysis of the overall environmental consequences of the use of various biofuels by establishing a single quantitative measure of environmental impact. Scharlemann and Laurance[2] have provided further clarification of this approach.

One of the factors that contribute to the overall environmental impact of a particular fuel assesment is the effect on the indigenous ecosystem. For example, deforestation caused by the destruction of carbon-sequestering rain forests in Brazil to make land available for sugar cane production contributes to an increase in greenhouse gas emissions that will negate some of the positive benefits of ethanol use. Another factor concerns the use of crops that rely heavily on nitrogen-based fertilizers. The nitrous oxide released by such agricultural practices is a significant greenhouse gas (Chapter 4).

Zah and colleagues have established a single quantitative measure of the overall environmental consequences of a particular biofuel, but other factors are difficult to quantify. For example, the displacement of food crops with, say, corn for ethanol production will likely affect market prices for certain crops, in turn, possibly altering the approach to agricultural land utilization. However, Zah and colleagues'

Continued on page 499

Based on R. Zah et al. [1] and J.P.W. Scharlemann and W.F. Lawrence [2].

Energy Extra 16.1 continued

analysis is a huge step forward in quantifying the benefits of biofuels.

A summary of Zah and colleagues' analysis of biofuels is shown in the figure. All data are normalized to the values for gasoline; to be environmentally advantageous to gasoline, a fuel must have both of these factors less than 100%. Very nearly all biofuels are acceptable in terms of greenhouse gas emissions. However, consistent with the discussion in the text, it is seen that Brazilian sugarcane does quite well in this area, whereas U.S. corn is fairly marginal. A large fraction of biofuels fail to meet the criterion for total environmental impact. Those that fail include both Brazilian sugarcane and U.S. corn (which is one of the worst performers in this respect). The analysis by Zah and colleagues did not include crops, such as switch grass, which may, at some point in time, provide a suitable material for cellulosic ethanol production.

A careful consideration of the type of analysis present by Zah and colleagues is crucial in establishing policies for biofuel production.

1. R. Zah et al., *Ökobilanz von Energieprodukten: Ökologische Bewertung von Biotreibstoffen* (Empa, St. Gallen, Switzerland, 2007).
2. J. P. W. Scharlemann and W. F. Laurance, How green are biofuels?" *Science* **319** (2008): 43–44.

Topic for Discussion

Discuss the relative merits (in the United States) of the following two possibilities:

- Converting land used for growing corn as a food crop to land for growing corn for ethanol production.
- Erecting wind turbines in corn fields (grown for food) and using as much of the land as possible for dual purposes.

plant material (such as soy), a detailed and reliable evaluation of the net carbon contribution from biodiesel use is difficult at best. As with ethanol use, many factors must be considered, including the carbon payback period for infrastructure development, and this depends greatly on crop selection and production methods.

16.5 Biogas

The term biogas refers to a mixture of gases produced by the decomposition of organic matter in an oxygen-free environment. Biogas may be produced by the digestion of feedstock by anaerobic organisms or by the fermentation of biodegradable materials. Feedstock can be materials such as sewage, manure, agricultural waste, or organic municipal waste. The use of municipal waste will be discussed in more detail in the following section.

Biogas consists primarily of methane with a smaller amount of carbon dioxide. It may also contain small concentrations of nitrogen, hydrogen, hydrogen sulfide, and oxygen. The gas may be treated to increase the methane content in order to improve its quality for energy production. As biogas is similar in composition to natural gas, its applications are also similar, and it has found use as a heating fuel, a transportation fuel, and for the generation of electricity.

A diagram of a typical biogas facility is illustrated in Figure 16.9 on the next page. This facility is based on the use of manure and agricultural waste. The biogas feedstock is anaerobically digested, and the resulting gas is treated to increase its methane content. The resulting fuel is used for transportation fuel and to operate a motor/generator for the production of electricity. Excess exhaust heat from the motor is used

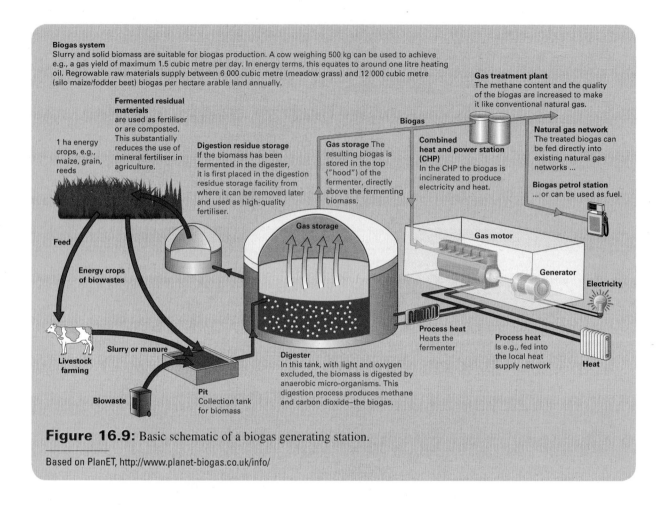

Figure 16.9: Basic schematic of a biogas generating station.

Based on PlanET, http://www.planet-biogas.co.uk/info/

for direct heating purposes. This use of excess heat, which is referred to as combined heat and power or cogeneration, is discussed in more detail in Section 17.3. Figure 16.10 shows a biogas generating station in Germany where the digesters are seen on the right and the generating facility is shown on the left in the image.

India and China, as well as several European countries, have been active in developing the use of biogas. At present, biogas use is the most extensive in Germany, where agricultural waste and manure (Figure 16.9) have commonly been used as feedstock.

16.6 Municipal Solid Waste

Humans produce considerable quantities of waste. This includes household waste, as well as commercial and industrial waste. Municipal solid waste refers to waste from all sources but does not include chemically or biologically hazardous waste or radioactive waste because these must be disposed of by special means according to the hazards they pose. It also does not include sewage and other nonsolid waste. Municipal solid waste does include durable goods such as old appliances and furniture, as well as nondurable goods, such as newspapers and food waste. The most appropriate method

Figure 16.10: Biogas production facility in Germany.

RikoBest/Shutterstock.com

of disposing of municipal solid waste depends on the nature of the waste. Methods include recycling (as is appropriate for metals, glass, and plastics), composting (as is appropriate for biodegradable materials), landfilling, and waste-to-energy incineration (as is appropriate for combustible materials).

The amount of waste produced per person has increased over the years. An example for the United States is shown in Figure 16.11 on the next page. Per capita, the United States has one of the highest rates of waste production. Other similarly industrialized countries typically produce less waste per capita; for example, values for Canada and New Zealand are about half those shown for the United States in Figure 16.11. Over the years, the ways in which waste has been treated have changed. While energy recovery from waste has increased over the years, landfilling (disposal) is still the predominant means of waste disposal in the United States, followed by recovery (recycling), as Figure 16.12 on the next page illustrates. Overall, waste management is an extensive and complex problem; however, the use of waste for the production of energy is relevant to this discussion.

The utilization of municipal solid waste to produce energy can be approached in several ways. Noncombustible materials, such as glass and metal, can first be removed, and the remaining combustible material can be burned directly to produce heat, which can then be used to generate steam to run a turbine and generator. This is the most common approach in the United States. Another simple approach is to collect gas from the decomposition of organic material at landfill sites (mostly methane) and burn it to produce heat and subsequently electricity. A more sophisticated approach is to shred the combustible material to produce refuse-derived fuel (RDF). RDF material can then be formed into pellets for combustion, or it can be heated to produce gas (i.e.,

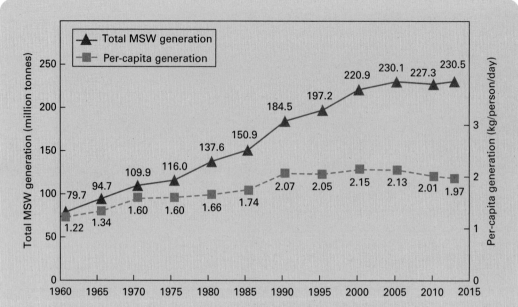

Figure 16.11: Municipal solid waste production in the United States from 1960 to 2013.

U.S. Environmental Protection Agency, Wastes - Non-Hazardous Waste - Municipal Solid Waste, https://archive
.epa.gov/epawaste/nonhaz/municipal/web/html/index.html

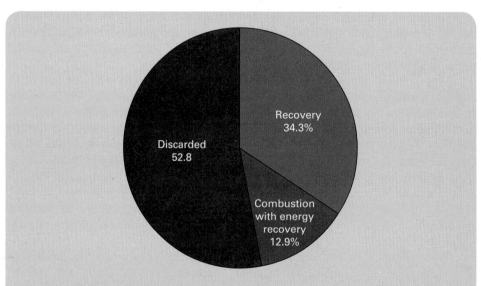

Figure 16.12: Relative importance of different municipal solid waste disposal
methods in the United States in 2013.

U.S. Energy Information Administration, http://www.eia.gov/energyexplained/index.cfm/data/index
.cfm?page=biomass_waste_to_energy

gasification). At low cooking temperatures, mostly methane is given off, whereas at higher temperatures the total gas production increases, and the gas is largely hydrogen. Methane can be used for the thermal generation of electricity, and hydrogen is suitable for combustion or fuel cell utilization (Chapter 20).

Society produces substantial quantities of waste, and there are benefits to the proper disposal of this waste, both from an energy standpoint and an environmental standpoint. It is important, however, to realize that the possible contribution of RDF to our overall energy needs is quite small.

Example 16.5

Estimate the percentage of the total energy needs of the United States that could be provided by MSW if the total content of this waste is 10 MJ/kg and it is utilized with an efficiency of 30%.

Solution

From Figure 16.9, the daily average per-capita waste production in the United States is about 2 kg. This has an energy content of 20 MJ per day or 7.3 GJ per year. From Figure 2.5, the per-capita primary energy use in the United States is about 330 GJ per year. Thus, RDF can supply a maximum fraction of $7.3/330 = 0.022$, or about 2% of the energy needs. This is a fairly optimistic estimate because it does not take into account the energy required to collect and process the MSW.

The exact effect of municipal solid waste energy generation on carbon emissions is difficult to determine. All combustion emits greenhouse gases, primarily CO_2. However, the extent to which municipal solid waste combustion contributes to the net carbon release depends on many factors, principal among them being the content of the waste. Burning organic matter, such as scrap wood or paper, which originates from plant material, is generally beneficial. Although the burning process produces CO_2, placing the same material in a landfill in which it decomposes will produce substantial quantities of methane, which is a more effective greenhouse gas than CO_2. However, municipal solid waste also contains substantial material (e.g., plastics) that is derived from petroleum products. This is not renewable and, when burned, contributes to greenhouse gases in the same way as the burning of fossil fuels. Determining the best way in which to minimize the impact of this component of municipal solid waste (that is, combustion, recycling, or landfill) is a complex problem.

The question of pollution is also difficult to analyze. Most combustible components of municipal solid waste are hydrocarbon-based materials, such as wood scraps and paper, and these release pollution into the atmosphere during burning. Typical pollutants include particulate matter, unburned hydrocarbons, and nitrogen compounds. However, some material may also contain toxic components such as lead, mercury, and cadmium. It is important from an environmental standpoint to remove these materials prior to combustion in order to avoid their release into the atmosphere or into groundwater supplies.

The overall effectiveness in producing energy from municipal waste, compared to the benefits of recycling and/or composting, needs to be established.

16.7 Summary

Wood is the most traditional biomass-based energy, and has been a component of our energy production for many years. As discussed in this chapter, a portion of the use of wood may be considered carbon neutral (or at least low carbon) because it includes an awareness of the renewability of the resource. This, however, is not true of wood use in many parts of the world where this resource is used to produce energy at the expense of deforestation.

This chapter described the straightforward method of producing ethanol from organic matter that contains glucose. Ethanol production has been successful in Brazil but less so in countries with more temperate climates. The economic and environmental factors related to ethanol production in, for example, the United States are not simple to evaluate. The relatively high cost from both an economic and an energy standpoint, coupled with the agricultural demands on farm land, make the wide-scale conversion from fossil fuels to biofuels for transportation use a difficult task. Certainly, the development of efficient cellulosic ethanol production methods would make the situation much clearer, although much research is necessary to make this a reality.

Biodiesel is derived from vegetable oils or animal fats and is a direct replacement for petroleum-based diesel fuel, either alone or as a mixture. The production of biodiesel is subject to considerations similar to those of ethanol production. The availability of arable land resources is a major concern, and current technologies may be viable only in regions with tropical climates. The ability to utilize plant material grown in an aquatic environment or on land that is otherwise not agriculturally useful would be a significant factor.

The chapter showed that the environmental consequences of biofuel utilization are not straightforward to assess. At present, fossil fuel energy is a major component of the energy used to produce biofuels, and as a result, the net reduction of greenhouse gas emissions may be less than perceived. Also, the interrelationship of agriculture for biofuel production, and agriculture for food production, needs to be carefully analyzed.

The use of waste material (e.g., municipal solid waste or agricultural by-products) for the production, of heat/electricity or for conversion into appropriate transportation fuels, may form a small component of our future energy mix. Although it is not always obvious that significant net economic and environmental benefits are associated with the utilization of refuse-derived fuels, the alternatives for waste management may be less attractive. The recovery of some of the energy content of waste material has advantages over the mere use of energy for waste collection and processing, followed by landfill greenhouse gas emissions, without the energy benefits.

Problems

16-1 How many tonnes of wood per day would be needed for a wood-fired generating station that produces 1000 MW_e continuously (typical of a coal-fired facility) at 35% efficiency?

16-2 Calculate the mass of CO_2 (in kilograms) produced by the combustion of 1 kg of ethanol.

16-3 Calculate the mass of CO_2 (in kilograms) per MJ of energy produced for the hydrocarbons with $n = 1$ through 4 shown in Table 16.1.

16-4 Following Example 16.3, estimate the necessary fuel tank volume and fuel mass for a vehicle operating on E85 as compared to a gasoline vehicle in order to achieve the same driving range.

16-5 Photosynthesis is a very inefficient process for converting sunlight into usable chemical energy. While the theoretical efficiency is 25%, the actual efficiency is affected by the wavelength distribution of the light, the absorptance of the plant matter, and various other factors and may typically be in the range of about 1%. Using the energy content of wood as a guide and the average solar insolation of 168 W/m^2 (Chapter 8), how much land area would be needed to fulfill an average person's energy needs (in the United States)? Assume that biomass energy content can be converted into end user energy with the same efficiency as current energy sources (Chapter 2).

16-6 A large maple tree collects sunlight over an area that is 8 m in diameter. The average solar radiation (Chapter 8) is 168 W/m^2. The tree grows for 8 months of the year and is dormant for 4 months of the year. After 10 years, the mass of the tree has increased by 540 kg. What is the efficiency of converting sunlight into chemical energy?

16-7 Using information from Figure 16.11 and an energy content of MSW of 10 MJ/kg, estimate the number of 35% efficient 1000-MW_e coal-fired power plants that could be eliminated if 50% of MSW were burned to produce electricity.

16-8 Two methods are used to generate 1 GWh_e of electricity: (1) burning coal with a generation efficiency of 30% and (2) burning MSW with an efficiency of 25%. For MSW generation, assume an energy content of 10 MJ/kg and a total net energy requirement of 5000 MJ/tonne for collection and transportation, which is provided by burning fossil fuels at 20% efficiency. The MSW consists of 50% carbon-neutral organic material and 50% fossil fuel–derived material. Compare the relative contributions to greenhouse gas emissions for these two methods. Ignore any effects of possible carbon sequestration.

16-9 Consider a dense forest with an average solar insolation of 200 W/m^2 and a photosynthesis process with an efficiency of 1.0 %. Assume that the energy content of wood is exclusively the result of carbon that comes from CO_2 sequestered from the atmosphere. What is the mass of CO_2 sequestered per km^2 of forest each year?

16-10 Compare the relative amounts of CO_2 produced per MJ of energy from the combustion of methane and methanol.

16-11 The average wood content of forested areas in the United States is 250 tonnes per hectare. Each year forest fires consume approximately 4×10^5 hectares of forests in the United States.

(a) What is the total energy content of the wood burned in these forest fires?

(b) Estimate how many 1-GW_e coal-fired power plants operating at a capacity factor of 70% could be replaced by the combustion of this wood?

(c) Assuming that wood is 50% carbon, what is the mass of CO_2 emitted annually from these forest fires?

16-12 A homeowner uses 3000 L of oil annually for heat, using an 85%-efficient furnace. If the heating system is converted to a wood furnace operating at 70% efficiency, what mass of wood is required annually to provide the same heating value?

16-13 In North America each person produces approximately 2.0 kg of municipal solid waste (MSW) per day. If this MSW is converted into electricity with an efficiency of 38%, what fraction of a person's average electric power use of 0.6 kW could be provided?

16-14 The average per-capita rate of primary energy use in Canada is approximately 13 kW. Spruce trees grow at an average rate of approximately 0.005 kg per m^2 of forest per day and have an average energy content of 1.7×10^7 J/kg.

(a) If all energy in Canada were generated from spruce trees with the same average efficiency as present energy sources, what is the area of spruce forest needed to satisfy one person's energy needs?

(b) What is the area of spruce forest needed to provide all of Canada's population with energy? What fraction of the total land area does this represent?

16-15 Two vehicles implement different approaches to using ethanol as a fuel. One vehicle burns the ethanol in an internal combustion engine with an efficiency of 18% to convert the chemical energy in the ethanol to mechanical energy supplied to the vehicle's wheels. The second vehicle uses electricity to charge batteries to run electric motors to drive the wheels. The electricity is produced by burning the ethanol in a heat engine to drive a generator with a net efficiency of 38%. The batteries/electric motors are 89% efficient.

(a) Compare the relative amounts of ethanol needed for the two vehicles to travel the same distance.

(b) Compare the relative net CO_2 emissions for the two vehicles.

16-16 Methanol is a convenient fuel, because it is a liquid at standard temperature and pressure. Methanol may be produced by first reacting carbon with water as in equation (3.4) to produce hydrogen and carbon monoxide. The carbon monoxide may be reacted with water to produce carbon dioxide and hydrogen as in equation (3.5). Methanol (along with water) is produced by combining CO_2 with hydrogen.

(a) Write the equations for this series of reactions and show that excess CO_2 is produced.

(b) What is the ratio of the mass of the CO_2 produced by methanol production to the mass of the CO_2 produced by subsequently burning the methanol?

16-17 Estimate the total mass of fuelwood used in the United States per year.

16-18 What fraction of arable land in the United States would need to be utilized if biodiesel produced from sunflowers replaced all of the country's diesel requirements? Repeat this calculation for biodiesel produced from algae.

16-19 A 1-GW$_e$ wood-burning generating station operates at a Carnot efficiency of 30% and has a capacity factor of 75%. What is the mass of wood required annually? What is the volume in m^3 of this wood?

16-20 The light alcohols in Table 16.1 are referred to as monohydric alcohols and have the formula $C_nH_{2n+1}OH$. Common heavier monohydric alcohols are pentanol ($n = 5$) and cetyl alcohol ($n = 16$).

(a) Write the formula for the combustion of pentanol and cetyl alcohol.

(b) Calculate the mass of CO_2 produced per kg of alcohol for these two materials.

(c) Locate values for the heat of combustion for these two compounds and calculate the mass of CO_2 produced per MJ of energy. Comment on the trend that is shown in these results.

Bibliography

G. Boyle (Ed.). *Renewable Energy.* Oxford University Press, Oxford (2004).

R. C. Brown. *Biorenewable Resources: Engineering New Products from Agriculture.* Iowa State Press, Ames (2003).

H. S. Geller. "Ethanol fuel from sugar cane in Brazil." *Annual Review of Energy* **10** (1985) 135–164.

R. Hinrichs and M. Kleinbach. *Energy: Its Use and the Environment.* Brooks-Cole, Belmont, CA (2002).

B. K. Hodge. *Alternative Energy Systems and Applications.* Wiley, Hoboken, NJ (2010).

A. Nag. *Biofuels Refining and Performance.* McGraw-Hill, New York (2008).

L. Olsson (Ed.). *Biofuels (Advances in Biochemical Engineering/Biotechnology).* Springer-Verlag, Berlin (2007).

G. Pahl. *Biodiesel: Growing a New Energy Economy.* Chelsea Green Publishing, White River Junction, VT (2005).

D. Pimentel and T. Patzek. "Ethanol production using corn, switchgrass, and wood; Biodiesel production using soybean and sunflower." *Natural Resources Research* **14** (2005): 65.

P. Quack. *Energy from Biomass: A Review of Combustion and Gasification Technology.* World Bank, New York (1999).

V. Quaschning. *Renewable Energy and Climate Change.* Wiley, West Sussex, United Kingdom (2010).

R. A. Ristinen and J. J. Kraushaar. *Energy and the Environment.* Wiley, Hoboken, NJ (2006).

J. Tester, E. Drake, M. Driscoll, M. W. Golay, and W. A. Peters. *Sustainable Energy: Choosing Among Options.* MIT Press, Cambridge, MA (2006).

F. M. Vanek and L. D. Albright. *Energy Systems Engineering: Evaluation and Implementation.* McGraw Hill, New York (2008).

PART V

Energy Conservation, Energy Storage, and Transportation

In this section of the book, mechanisms for energy conservation, energy storage, and the ways in which energy can be used for transportation purposes are described. Energy conservation must go hand in hand with the development of sustainable energy development. The reduction of energy needs and the efficient use of the energy produced will help to maximize the effectiveness of any energy strategy. Energy conservation efforts may be based on improved energy utilization technologies, but they may also be based on changes in how we view energy use. Options for energy conservation may be things that we can do as individuals with the products that we purchase, or with our energy use in our homes and for transportation. Energy conservation may also be viewed on a much larger scale in terms of the development of new energy-efficient products and the implementation of proactive national energy policies. These factors are considered in Chapter 17.

Energy utilization is efficient and effective only if the energy is available when and where it is needed. The need for energy is not always readily compatible with the way in which it is produced, particularly for many of the nontraditional energy technologies. Wind energy is available only when the wind blows, and solar energy is available only during the daytime. Our energy needs, however, are not necessarily coupled to the behavior of nature. Most alternative energy technologies produce electricity. This is readily distributed on the electric grid for residential or commercial use, but appropriate technologies are needed to use electricity for transportation.

The availability of energy when and where it is needed depends on the implementation of appropriate energy storage mechanisms, as well as an effective distribution technology. These aspects of efficient energy use are often not given the serious consideration that they deserve. Inevitably, the storage of energy involves the conversion of one type of energy to another (and generally back again), and these conversions are always less than 100% efficient. The most appropriate energy storage mechanism for a particular application depends on a number of factors: the form of end-use energy, the amount of energy that needs to be stored, the rate at which energy needs to be stored and extracted (i.e., the power), and any space or weight constraints (i.e., portability).

Technologies that are potentially viable for large-scale energy storage to even out supply and demand variations on the electric grid are considered. For the most part, these systems are not portable, and size and weight considerations are secondary. Reliability, efficiency of energy storage, and conversion, as well as economy of infrastructure and operation, are generally the first concerns (Chapter 18).

Finally, in Chapters 19 and 20, technologies for portable energy storage are discussed. These technologies are appropriate for application to transportation. High-energy-density storage technologies, which mean small and lightweight devices, are a crucial factor for providing energy for vehicles. The two major competing approaches to this problem, batteries and hydrogen, are discussed in detail.

The photograph at the beginning of this part of the text shows a Tesla roadster being charged. This battery electric vehicle is produced by Tesla Motors in California and was first sold in 2008. It is the first production electric vehicle to utilize Li-ion batteries and the first to have a range of more than 320 km. ■

Jim West/Alamy Stock Photo

Energy Conservation

Learning Objectives: After reading the material in Chapter 17, you should understand:

- How government energy policies deal with conservation matters.
- How combined electricity and heat production can make the best use of resources.
- Approaches to electricity distribution and how the smart grid can be used to integrate different energy sources and regulate the use of electricity.
- Energy conservation in the community through the use of LED streetlights.

- The efficient operation of residential HVAC systems.
- The application of heat pumps to space heating needs.
- Reducing heat transfer in and out of buildings.
- High-efficiency lighting technologies.
- Vehicle fuel efficiency and government standards.
- The viability of hybrid vehicles as a means of fossil fuel conservation.

17.1 Introduction

While most of this text has dealt with methods of producing energy (i.e., extracting energy from our environment and converting it to a form that suits our needs), it is also important to make efficient use of the energy that we produce. Energy resources are limited, and the more efficiently they are utilized, the more available they will be in the future. The *conservation of energy* (Chapter 1) is a basic principle of physics. The term *conservation of energy*, or, more commonly, *energy conservation*, can also refer to actions taken to best minimize the quantity of energy that we use to fulfill our needs. Actions to minimize our energy use can occur on several levels, from the scale of the individual to a global scale. For example, we can control energy conservation measures: turning lights off when they are not being used or driving more fuel-efficient vehicles. On a somewhat larger scale, energy can be conserved by public utilities by improving the efficiency of electricity-generating stations or by cities by using LEDs (light-emitting diodes) for street lighting. On a national scale, government energy policies may include guidelines that help to promote efficient utilization of energy in addition to directing the methods by which energy is produced. Clearly, the topic of energy conservation is very diverse and may be approached in a number of ways. In many ways, individuals control their own energy use. They can make choices in their lifestyles and in how they satisfy their transportation needs that can contribute to more effective utilization of energy resources. Individuals

Jim West/Alamy Stock Photo

and organizations may, to the limit of their ability, influence corporate and/or government policy in order to optimize energy use on a local, regional, national, or even global scale.

In this chapter, the topic of energy conservation is approached from several viewpoints, beginning with the ways in which the efficient energy use can be promoted by policies and incentives at the national or international levels. Energy conservation measures that are applicable on a regional or local level are then presented. Finally, the things we can do as individuals to conserve the energy that we use in our own lives are discussed.

17.2 Approaches to Energy Conservation

Environmental concerns are integrated with energy production methods and conservation methods. Utilizing energy production technologies that are less harmful to the environment has a positive impact, although all energy technologies have some impact, as discussed throughout this text, and the severity of this impact is not always obvious. Using less energy has a positive environmental impact, whether the energy comes from traditional fossil fuel resources or from alternative technologies, and this emphasizes the need to actively pursue appropriate approaches to conservation in addition to developing new resources.

Actions to conserve energy may be made by individuals, businesses, or governments at the national, regional, state/provincial, or local levels. As individuals, our approach to energy conservation includes things that we can do in our own home, the ways that we satisfy our transportation needs, and the ways we interact with local, regional, or national representatives and government agencies or organizations to promote energy conservation measures. Business can contribute to energy conservation efforts through proactive approaches to reducing energy use, the utilization of suppliers who are energy conscious, utilizing environmentally aware manufacturing processes, and producing energy-efficient products. In the case of governments, the development of energy policies, as well as funding and incentives to other levels of government, industry, business, and/or individuals for the implementation of conservative approaches to energy use, can have a positive effect on the efficient utilization of energy resources.

Motivations for energy conservation may include concerns for the environment, long-term energy security and independence, but ultimately a significant component of the motivation for energy conservation is economic; if we save energy, we save money. The most appropriate approach to energy conservation varies considerably in different parts of the world. The availability of energy resources, climate, and the distribution of energy requirements between, for example, transportation, electricity, heating, industry, and so on, are all important factors in determining the best approach to energy conservation. Economy, social factors, and politics also play important roles in implementing conservation measures. In most countries, a national energy policy guides the use of energy resources and typically involves a consideration of such factors as:

- The assessment of energy needs.
- The availability of energy resources.
- The possibility of energy self-sufficiency.
- The environmental consequences of energy use.
- The promotion of energy-efficient products.
- The promotion of energy conservation measures.

- Interaction with energy-related activities at state/provincial and/or municipal levels.
- The development of mechanisms to implement energy policy, including incentives, subsidies, and the like.

Such policies involve various aspects of energy conservation, as well as the development of energy resources. Although it is difficult to generalize, some examples of the approach to energy conservation from diverse locations will be shown in this chapter to help provide insight into the ways of dealing with this issue.

17.2a **Energy Conservation in the United States**

Energy policy in the United Sates is largely implemented by the Department of Energy (DOE). The U.S. DOE divides energy use in the United States into four broad categories: residential, commercial, industrial, and transportation. According to DOE statistics, these sectors account for 21%, 17%, 33%, and 28%, respectively, of the primary energy use nationwide. The approaches to energy use and conservation in these four sectors are now outlined.

Heating and/or cooling accounts for the largest component of residential energy use, about 42% on average. About 13% of energy use is for water heating, about 10% for lighting, and the remainder mostly for electronics and appliances. The exact breakdown of residential energy use varies considerably in different regions of the United States and is directly related to climate. In colder regions, heating is a major component of energy use, while in warm regions, air conditioning is a substantial factor. In regions with a moderate climate, lighting and appliances may account for the largest component.

Historically, changes in the overall energy consumption are determined by two competing factors. Energy consumption is reduced by the use of more efficient appliances and measures to reduce energy loss, whereas energy consumption is increased by greater energy needs of consumers.

To promote more efficient energy use, the Department of Energy, as authorized by Congress, established the 1987 National Appliance Energy Conservation Act. This Act specifies minimum efficiencies for residential energy consuming devices, including appliances, electronics, and heating and cooling equipment. Regulations are upgraded on the basis of a yearly assessment of what is technologically feasible and economically justifiable. Exceptionally efficient devices are awarded the ENERGY STAR approval (Energy Extra 17.3). Because many appliances consume substantial electricity while in their standby mode, improvements in energy consumption when the appliance is in off mode are an important consideration for energy conservation. The ENERGY STAR program includes not only devices that consume energy but also construction materials and construction techniques that are a factor in determining energy needs.

Consumer energy demands have increased as a result of lifestyle changes. This has been particularly significant since the 1970s with increases in average home size, in the utilization of electronic devices, and in the prevalence of central air conditioning systems. These have all put a greater strain on energy requirements in the United States.

Both federal and state governments have promoted more efficient residential energy use (Section 17.6) by offering tax credits for home improvements that decrease energy consumption. In many areas, public utilities also offer subsidies for improvements leading to more efficient energy use. The possibility of the future implementation of a smart grid (Section 17.4) will require input from both utilities and various levels of government. A step in this direction may be the use of real-time energy (e.g., electricity) monitors to provide consumers with feedback on their energy use and to promote energy-conscious behavior.

The commercial sector includes retail businesses, restaurants, schools, and other facilities. Much of the energy use in the commercial sector follows along the lines of the residential sector in its energy use for heating, cooling, hot water, and lighting. However, the commercial sector typically uses a larger fraction of energy for lighting (up to about 25%) compared to the residential sector. As a result, fluorescent lighting is used for nearly all lighting in this sector. Centralized control of building energy use, such as programmed heating and lighting systems, provides an effective means for the implementation of energy conservation measures. Efficient energy use in the commercial sector is strongly motivated by economic factors. In most areas, municipal building codes for new commercial structures require designs that comply with energy efficiency standards. U.S. government policy specifies that federal government buildings must conform to certain energy efficiency requirements. Fifteen percent of existing buildings were required to meet minimum energy requirements by 2015, and all new federal government buildings must be *zero net energy* (*ZNE*) buildings by 2030. *ZNE building* is a term used somewhat ambiguously, and usage in North America is often different from its usage elsewhere. In the strict sense, a ZNE building is a building in which all required energy is produced on-site from renewable sources such as solar or wind. The official definition of a ZNE building used in U.S. government policy is a building whose construction and operation contribute zero net greenhouse gas emissions. This is sometimes referred to as a zero-net-carbon building.

Energy conservation in the industrial sector, which includes all manufacturing, mining, construction, and agricultural activities, is motivated by both environmental and economic issues. The steel industry and the paper industry are major energy users and have made substantial reductions in their energy use in the past 30–40 years. An important factor in this reduction has been the utilization of cogeneration (Section 17.3). Agricultural activities have become more energy efficient in recent years, and, as seen in Chapter 16, this increased efficiency will be observed directly as a reduction in the ratio of primary energy input to the caloric content of the food produced. In many regions, the energy used in the treatment and distribution of freshwater to consumers is a significant fraction of total energy use. Thus, the availability of freshwater is a long-term environmental concern and an integral component of our future energy concerns.

The total energy use in the industrial sector in the United States has decreased in recent years. This is, in part, due to more energy-efficient technologies but reflects changes in the types of industrial activities in the United States and the increase in the outsourcing of many manufacturing processes.

Transportation (Section 17.8) is a major component of energy use. In the United States, 65% of transportation energy use is for gasoline vehicles (mostly privately owned vehicles), 20% for diesel vehicles (mostly commercial trucks, trains, and ships), and 15% for aircraft. U.S. government energy policy has dealt with the improvement in fuel consumption by vehicles since the 1970s. These actions have primarily been through federal, or in some cases, state regulations (California most notably) aimed at automobile manufacturers. As a result of the energy crisis in the early 1970s, the U.S. government implemented the *CAFE* (*Corporate Average Fuel Economy*) policy in 1975 to pressure automobile manufacturers to improve the fuel efficiency of their vehicles. In 1978, the *Gas Guzzler Tax* was introduced to penalize drivers of vehicles that consumed excessive amounts of fuel. Although the tax is still in effect, few vehicles are subject to the tax as a result of general improvements in the fuel efficiency of passenger vehicles.

One approach to reducing the environmental impact of gasoline-powered vehicles has been the introduction of flex fuel vehicles. These vehicles (Chapter 16) can operate on various mixtures of gasoline and ethanol, typically up to 85% ethanol (E85) in the United States, although in other countries flex fuel vehicles may be able to use up to 100% ethanol or various mixtures of gasoline and methanol. Over 49 million flex fuel vehicles are on the roads worldwide. Most are in Brazil, but the United States is next with about 35% of the total. However, despite the substantial number of flex fuel vehicles on the roads in the United States, the availability of flex fuels (e.g., E85) is minimal. Many owners of flex fuel vehicles are not aware that their vehicles can run on high ethanol-gasoline mixtures, and so more than 90% of these vehicles are operated exclusively on gasoline. Part of the motivation for automakers to produce these vehicles [which are only minimally different than non–flex fuel vehicles (Chapter 16)] has been a government fuel economy credit for each flex fuel vehicle sold (even if owners run them on gasoline). This credit helps manufacturers meet CAFE guidelines without improving fuel efficiency.

Other activities that have promoted improved efficiency of transportation have included subsidized public transportation. Such subsidies are typically provided at the municipal level. Also, at the municipal or state level, the ride-sharing or carpooling of private vehicles has been encouraged by designating special lanes in urban areas for vehicles with more than a certain number of occupants.

Incentives are available for purchasers of plug-in hybrids (Section 17.8) and pure battery electric vehicles, BEVs (Chapter 19). These are in the form of tax credits and/or rebates and are offered by the federal government and a number of state governments.

17.2b Energy Conservation in Canada

According to the Constitution of Canada, provinces and territories have jurisdiction over natural resources, which include energy resources in the form of both nonrenewable resources such as oil and natural gas and renewable energy such as wind, solar, and tidal. In terms of energy, the federal government deals largely with interprovincial and international matters. Most of the initiatives for energy development and many of the programs dealing with energy conservation come at the provincial level. As presented in Chapter 2, Canada has among the highest per-capita energy consumption in the world. This is due to factors such as GDP, climate, industrial diversity, and population density. Although Canada is a net energy exporter (producing more energy than it consumes), conservation is an important concern. Heating needs (due to the climate) and transportation (due to the large land area) are of particular interest, and a number of initiatives and incentives are available in these areas.

A number of incentive programs for energy conservation vary from province to province depending on the specific energy needs and concerns. Many of these are aimed specifically at the home owner and include actions such as:

- Distributing free compact fluorescent lamps (Section 17.7) at major building supply and department stores.
- Free removal and a rebate for homeowners who replace old (energy-inefficient) appliances with new energy-efficient ones.
- Free energy-efficient upgrades such as blown-in insulation, energy-efficient lighting, programmable thermostats, and air leak reduction for low income families.
- Provincial and federal rebates for energy-efficient upgrades such as ENERGY STAR windows or improved insulation.

- Rebates for the development of residential renewable energy, such as solar collectors or residential wind turbines.
- Rebates for supplementing or modifying heating systems to use lower greenhouse gas emission options.
- Rebates for new home construction that meets certain energy efficiency standards.

The Office of Energy Efficiency of Natural Resources Canada offers a number of programs dealing with energy conservation that are aimed at business and industry. These include distribution of information on energy efficiency of appliances and equipment, guidelines for energy-efficient construction, and cost sharing programs to undertake energy assessments and retrofits for existing structures to improve energy efficiency.

Following the energy crisis of the early 1970s, the Canadian government introduced measures to improve the fuel efficiency of vehicles in that country. In 1975, the Joint Government–Industry Fuel Consumption Program (FCP) was introduced. Since that time, Transport Canada has collected fuel consumption data for vehicles, and this information is made available through Natural Resources Canada. The purpose of this program was to encourage the introduction of fuel-efficient vehicles in Canada and to promote public awareness of energy conservation. Annual voluntary CAFE guidelines, the equivalent of the U.S. CAFE, were established to encourage manufacturers and importers of automobiles and light-duty trucks to strive for improved fuel efficiency. This program was ended in 2010 with the introduction of new *Passenger and Light Truck Greenhouse Gas Emission Regulations*, which form part of the *Canadian Environmental Protection Act*. Motor vehicle manufacturers now submit data on vehicles sold in the country to Environment Canada.

Flex fuel vehicles are readily available in Canada, and, in 2014, about 1.6 million such vehicles were on the road. Because only three service stations that were open to the public in Canada in 2012 sold E85, virtually all flex fuel vehicles in that country operated exclusively on gasoline.

At the provincial and municipal levels, carpool lanes are common in a number of locations. The governments of Ontario and Quebec have initiated a rebate program for purchasers of plug-in hybrids and BEVs. Ontario also issues Green Licence Plates to BEVs and plug-in hybrids, which allow these vehicles to travel in carpool lanes regardless of the number of occupants.

17.2c Energy Conservation in the European Union

A common energy policy for the European Union (EU) was approved by the European Council in 2007. Key points of this policy are:

- Reducing greenhouse gas emissions.
- The development of renewable energy technologies.
- Energy conservation.
- Development of advanced nuclear reactors.
- Carbon sequestration.

While conservation is a specific goal of the policy, substantial emphasis in the EU has been placed on the reduction of greenhouse gas emissions, and conservation is an important component of this reduction. The EU energy policy is implemented through the *Strategic Energy Technology (SET) Plan*, which includes initiatives in wind energy, solar energy, bioenergy, carbon sequestration, smart grid, and nuclear fission. The plan coordinates EU programs with national programs of the member states. Much of the energy conservation effort in the European Union falls under the *SAVE (Specific Actions for Vigorous Energy Efficiency)* Programme. Originally, SAVE was established in 1991 and ran until 1995. It was followed by an updated version, SAVE II, from 1996 to 2002, and then SAVE III, now part of *IEE (Intelligent Energy–Europe)*. These programs deal specifically with energy conservation in the residential, commercial, and industrial sectors.

Various approaches to residential energy conservation in the EU parallel efforts in North America. The goal of these programs is both to reduce greenhouse gas emissions and to conserve energy resources. Long-term goals include requirements for all new homes to be ZNE buildings, meaning zero-net-carbon buildings, and the installation of real-time electricity monitors in all homes.

In the commercial sector, building efficiency is an important concern. Common methods for the implementation of efficiency standards for buildings in the EU are that:

- Requirements that all large new buildings must meet certain efficiency standards.
- Large existing buildings must undergo major renovations and comply with minimum requirements.
- All large buildings require energy certification before sale.
- Furnaces, boilers, and air conditioning systems must have regular inspections to ensure efficient operation.

EU energy policy includes a voluntary agreement with the *ACEA (Association des Constructeurs Européens d'Automobiles*, or, in English, *European Automobile Manufacturers' Association)* to achieve specified reductions in CO_2 emissions from automobiles. Improved fuel efficiency is an important component in greenhouse gas reduction.

Most EU countries have incentives for the use of biofuels for transportation. Flex fuel vehicles, typically operating on E85, are promoted in many locations, with E85 fueling stations becoming more common. Incentives in terms of reduced taxes on fuel cost or vehicle taxes help to offset the lower energy content of E85 compared with gasoline. Germany, in particular, has placed an emphasis on biodiesel vehicles. Incentives include a fuel tax waiver on biogenetic fuels. Because biodiesel is a 100% nonfossil fuel, there are no fuel taxes, whereas for ethanol-gasoline mixtures, the tax is prorated by the concentration of fossil fuel present.

The EU is also a member of *International Partnership for Energy Efficiency Cooperation (IPEEC)* along with Australia, Brazil, Canada, China, India, Mexico, Russia, South Korea, and the United States. This organization enhances global cooperation on energy policies with a focus on energy efficiency. Many of the initiatives are led by one of the member states. Some of these initiatives are as follows:

- *IPEEI (Improving Policies through Energy Efficiency Indicators)*, led by France, develops and implements methodologies to assess energy efficiency.
- *SBN (Sustainable Buildings Network)*, led by Germany, investigates policies dealing with building efficiency. Guidelines for new construction, as well as approaches to the upgrade of existing buildings, are being explored.
- *WEACT (Worldwide Energy Efficiency Action through Capacity Building and Training)*, led by Italy, facilitates policy creation and implementation related to energy efficiency in developing countries.

17.2d **Energy Conservation in India**

Energy use is growing rapidly in India as a result of its significant economic growth. At present, more than 30% of the population has no access to electricity, and this percentage is greater than 50% in rural areas. Thus, the needs that must be addressed by energy policy in India are more complex than in many other nations. Currently, about 70% of India's energy generation is from fossil fuel sources, and expanding efforts in alternative renewable sources is a major priority. In fact, there is active development in rural electrification, using renewable energy sources such as hydroelectric energy, wind energy, and solar energy. Due to the need for more energy generation in India, conservation of the available energy is an important component of energy policy. Several government agencies are involved in conservation efforts.

The *Petroleum Conservation Research Association* (*PCRA*) is an Indian government body established in 1977 for the purpose of promoting energy efficiency, specifically in the area of fossil fuel conservation measures. Media campaigns to promote awareness of the need for energy efficiency to conserve resources and reduce environmental pollution have been undertaken in recent years and have been effective in bringing information about these issues to the public. The association deals with energy conservation in four broad areas of national importance: residential (or domestic) energy use, industrial energy use, agricultural energy use, and transportation.

Public education is a major focus of the PCRA in the residential sector and involves programs aimed at promoting efficient energy use in the home. The PCRA is also involved in the development of fuel-efficient stoves, the use of energy-efficient lighting by compact fluorescent lamps, and the introduction of alternative renewable energy sources. In the industrial sector, the PCRA has been active in conducting energy audits for industries and providing recommendations for more efficient energy use. Operational adjustments that can save energy without capital investment are an important focus of these activities. Interactions with industry include seminars, clinics, and workshops, as well as follow-ups to assess actual energy savings realized. Agricultural initiatives include educational programs for farmers to promote the proper maintenance of equipment, such as tractors and pumps, to optimize energy efficiency. Finally, the transportation sector is the focus of driver education programs to promote fuel-efficient driving habits.

In 2001, the Indian Parliament passed the *Energy Conservation Act*. This Act places constraints on large energy consumers to adhere to energy utilization guidelines, requires that new construction follow energy-efficient building codes and requires that appliances follow energy-efficient guidelines. As part of the implementation of this Act, the *Bureau of Energy Efficiency* (*BEE*) was established with the mandate to develop programs to increase the efficiency of energy use in India. Several focus sectors for BEE have been identified:

- Replacement of incandescent lamps with compact fluorescent lamps (Section 17.7).
- Marketing of high-efficiency appliances and providing consumers with energy consumption information.
- Replacing municipal streetlights with LED lamps (Section 17.5).
- Improving energy efficiency in municipal water treatment and distribution systems.
- Establishment of voluntary energy efficiency standards for new building construction.
- Replacement of inefficient agricultural equipment with energy-conserving devices.

The BEE has been instrumental in coordinating a wide variety of conservation efforts in India and has established systems and procedures to monitor the effectiveness of implemented measures.

In order to reduce dependence on fossil fuels, India has been active in pursuing biofuels and nuclear energy. These activities include a major effort to produce biodiesel derived from jatropha plants for transportation purposes. The area around the railroad lines between Mumbai and Delhi is used to grow jatropha, and the train itself operates on a biodiesel mixture containing jatropha oil. India also has one of the most ambitious long-term nuclear development plans in the world as described in Energy Extra 6.4. Numerous projects producing and using biodiesel from this source have been undertaken. The area around the railroad lines between Mumbai and Delhi is used to grow Jotropha, and the train itself operates on a biodiesel mixture containing Jotropha oil. This program has been a major success for biodiesel development in India.

India is also a member IPEEC (Section 17.2c). One of India's roles in this organization is to lead the *AEEFM* (*Assessment of Energy Efficiency Finance Mechanisms*) initiative. This program identifies and documents methods to overcome barriers to the successful financing of energy efficiency projects worldwide. This includes assessing effective mechanisms for utilizing financing to undertake energy efficiency programs and how tax initiatives and subsidies can best be used to benefit these activities.

17.3 Cogeneration

All methods of producing electricity that involve a heat engine generate excess heat in accordance with the thermodynamic efficiency of converting thermal energy into mechanical energy. These methods include thermal generation using nuclear fuel, coal, and natural gas. On the one hand, this excess heat is an undesirable consequence of heat engines because energy is not converted into the desired form and because the excess heat must be removed. Another approach to this situation may be to view the excess heat as a resource. The heat may be used locally to provide heating and/or hot water to the community. This approach is referred to as *cogeneration*, or *combined heat and power* (*CHP*). As efficiencies of generating stations are typically less than about 40%, the majority of energy produced is available for use as heat rather than as electricity. A major difficulty in implementing CHP is to match the demand for electricity and the demand for heat. Electricity may be widely distributed via the grid, but thermal energy must, for practical reasons, be utilized locally. A cogeneration plant may be primarily designed to meet the base load requirements for electricity or the base load requirements for heat/hot water. In either case, other sources of energy may be required, particularly in periods of high demand for one or the other resource.

Although it might seem that the utilization of any excess heat for a useful purpose might be more beneficial than merely dumping it into the environment, a certain degree of efficiency must be achieved to justify (economically) the necessary infrastructure and maintenance costs. Distribution of electricity via connection to the electrical grid is relatively straightforward, but the distribution of heat and hot water requires a system of insulated pipes.

Cogeneration has been utilized more effectively for natural gas than for coal-fired plants, but it has also been successfully implemented for stations burning a variety of alternative fuels, including wood and agricultural waste. The greater need for heat in cooler climates has made the implementation of cogeneration popular and often economical in such regions. In fact, it has become an important component of the energy economy in Denmark, the Netherlands, and Finland. In warmer climates, the utilization of excess heat from power plants may provide heat/hot water, as well as cold water

Figure 17.1: Cornell University cogeneration plant.

Jon Reis/www.jonreis.com

for air conditioning purposes via an *absorptive chiller*. Such an approach to the use of energy from a thermal generating station is referred to as *trigeneration*.

Cogeneration has been used very effectively where a large amount of heat is used in a fairly small area. This is the case, for example, in Manhattan where approximately 100,000 buildings receive heat through cogeneration. Cogeneration is also commonly used in decentralized facilities such as businesses, manufacturing facilities, and hospitals. The needs of those small communities can be met through small generating stations. The natural gas–fired combustion turbine cogeneration facility at Cornell University is an example of this approach (Figure 17.1). The 30-MW generating facility provides the majority of energy needs on campus. The general design of such a system is illustrated in Figure 17.2 on the next page.

Example 17.1

A natural gas-fired generating station, rated at 200 MW_e with an efficiency of 39%, operates with a capacity factor of 68%. Hot water for heating use in the local community is produced by cogeneration. Calculate the heat energy available per year.

Solution

At the rated output, the total power produced is

$$\frac{200 \text{ MW}_e}{0.39} = 513 \text{ MW}.$$

Thus, the excess heat produced is

$$513 \text{ MW} - 200 \text{ MW} = 313 \text{ MW}.$$

At a capacity factor of 68%, the total energy produced per year is

$$313 \text{ MW} \times 0.68 \times 3.15 \times 10^7 \text{ s/y} = 6.7 \times 10^9 \text{ MJ}.$$

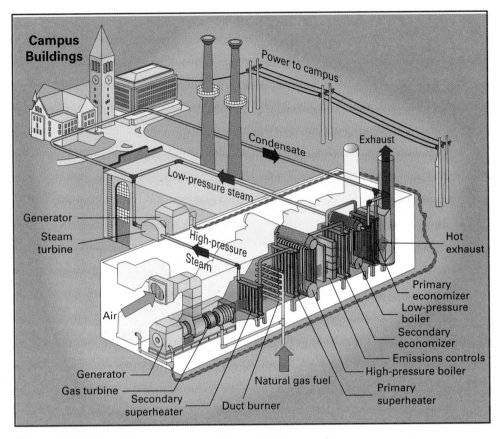

Figure 17.2: Schematic diagram of Cornell University cogeneration plant.

Based on Cornell University, Combined Heat and Power Plant, http://energyandsustainability.fs.cornell.edu/util
/heating/production/cep.cfm

Overall, cogeneration can, under the right circumstances, be an effective method of utilizing waste energy and improving the efficiency of our energy use. It has been more extensively used in Europe (particularly Northern and Eastern Europe) than in North America, but its use has begun to become more widespread in recent years.

ENERGY EXTRA 17.1
Geothermal cogeneration in Iceland

As described in Chapter 15, Iceland has substantial geothermal resources and has made extensive use of these resources for both direct heating and electricity generation. In fact, about 65% of all primary energy use in Iceland is geothermal. It provides about 26% of the country's electricity and 90% of the country's domestic heat and hot water. Except for fossil fuels used for

transportation (which accounts for about 15% of primary energy use), virtually all of the rest of Iceland's primary energy use comes from hydroelectric.

While the term cogeneration most commonly refers to the production of electricity and thermal energy from a fossil fuel–fired facility, Iceland has very effectively combined these two aspects of

Continued on page 521

Energy Extra 17.1 continued

Geothermal generating stations in Iceland. The small Bjarnarflag and Húsavik Stations provide some district heating and hot water to the local area.

name	location	electrical capacity (MW$_e$)	thermal capacity (MW$_{th}$)
Hellisheiði Power Station	Hveragerði	303	133
Reykjanes Power Station	Grindavík	100	0
Nesjavellir Geothermal Power Station	Þingvellir	120	300
Svartsengi Power Station	Grindavík	75	150
Krafla Power Station	Reykjahlíð	60	0
Bjarnarflag Power Station	Reykjahlíð	3	unknown
Húsavik Power Station	Húsavik	2	unknown

geothermal energy production. There are seven geothermal generating stations in Iceland, three of which , Hellisheiði, Nesjavellir and Nesjavellir, are major facilities that provide combined heat and power (CHP).

A simple cogeneration system is illustrated in the diagram. Borehole fluid, which is a mixture of steam and hot water, passes through a steam separator. The steam is directed into a turbine that drives a generator to produce electricity for distribution on the grid. Hot water from the exhaust of the turbine transfers its thermal energy to ground water for district heating use. The borehole fluid contains a variety of dissolved minerals and is not suitable for direct heating use. The hot water from the steam

separator transfers its heat to ground water in a heat exchanger, and this heated ground water is combined with the heated ground water from the turbine. This heated water is further processed and distributed for heating and hot water use.

The district heating system in Reykjavik is the largest such system in the world, carrying over 60 million cubic meters of hot water annually. Hot water at about 85 °C is stored in six large tanks, each with a capacity of 4,000 m^3, for distribution throughout the city. The storage facility, referred to as Perlan (Icelandic for "the pearl") sits atop a hill 61 m above sea level. The dome above the six tanks, as shown in the figure, houses a restaurant and gift shops and revolves once every two hours.

Hellisheiði Power Station in Hveragerði, Iceland.

Continued on page 522

Energy Extra 17.1 continued

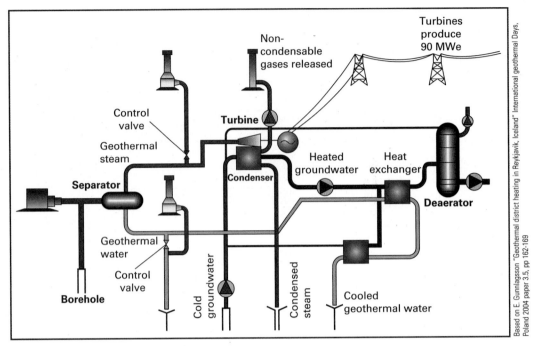

Simplified diagram of a geothermal CHP system.

While geothermal cogeneration is used elsewhere in the world, outside of Iceland it is a relatively small component of national primary energy requirements. Iceland, with its extensive geothermal resources and small population (about 330,000), is an ideal location for making use of an approach to cogeneration with low environmental impact.

Steam separators at Hellisheiði Power Station in Hveragerði, Iceland.

Continued on page 523

Energy Extra 17.1 continued

Perlan thermal storage tanks on Öskjuhlíð Hill in Reykjavík, Iceland.

Topic for Discussion

Iceland has the highest per-capita energy use in the world at 709 GJ per capita per year (see Figure 2.4 for comparison). This high energy use comes in part from a high standard of living and a cold climate but is mostly the result of substantial levels of energy-intensive industry (primarily aluminum mining and smelting) compared to the small population. Discuss the per-capita reduction in carbon emissions that results from Iceland's energy policies in comparison with the per-capita carbon emissions from other countries (see Section 4.6).

17.4 Smart Grid

The first electric distribution grids were established in the 1890s. Over the next 70 years or so, the electric grid system expanded and grew into the large-scale system that effectively supplied our electric needs in the 1960s and 1970s through a one-way interconnected system. During this time, the tendency was to construct large, centralized generating facilities (i.e., 1 GW_e and larger) in geographically appropriate locations (e.g., hydroelectric dams on rivers, nuclear stations near cooling water supplies, coal plants near railway connections, etc.). One-way transmission of electricity to users via an interconnected grid minimized the possibility for large-scale interruptions.

In recent years, the desire to integrate alternative energy sources into the grid has made the previous approach to electricity distribution less effective than it had previously been. This is because many alternative energy facilities are smaller (e.g., tens-of-MW_e wind farms), are intermittent in their output (e.g., wind, solar, tidal, etc.), and require the integration of energy storage systems (e.g., pumped hydroelectric, batteries, etc.). As many alternative energy resources, e.g., solar, wind, and tidal, are in locations that are remote from the regions of greatest power use, the expansion of the existing grid system for the transmission of power is an important aspect of an improved and well-integrated grid system. There is also a move for users to become suppliers by selling excess electricity generated by wind and/or solar back to the utility, as well as an incentive to shift usage to off-peak hours by real-time metering. The so-called *Smart Grid* would deal with these issues to make electricity distribution and use more efficient. Additional features of a smart grid would be increased reliability and security.

Some of the challenges to moving to smart grid technology are the installation of the necessary infrastructure and dealing with public concerns over technological changes. In the first instance, each household would have to be fitted with a smart meter for two-way, real-time monitoring and control of electrical use, along with the necessary integration of power generation and distribution systems. Public concerns include security of information on the Internet, infrastructure costs, and loss of control of personal energy use. In principle, utilities (or governments) would be able to regulate electricity use to even out peak use periods by, say, requiring the turn-off of unused appliances, scheduling of electric vehicle charging, and so on. Rapidly changing appliance technologies requires the implementation of suitable standards to ensure long-term compatibility. Public utilities tend to be conservative in their approach to technology, aiming for reliability rather than some degree of uncertainty, and they view economic factors as an important consideration when implementing changes. Government incentives and subsidies may be important in promoting change.

In the long term, it has been estimated that smart grid technology could save the United States more than $100 billion over the next 20 years as a result of improved efficiency of energy utilization. Smart grid technology is already making its way into the consumer market. The first full implementation of smart grid technology was in Italy in 2005 by the utility Enel S.p.A. (*Ente Nazionale per l'Energia eLettrica, Società per Azioni*). Enel designed and manufactured its own smart meters and developed its own computer control systems. It was estimated that the payback period for the infrastructure was about four years.

In North America, Austin, Texas, has been working on smart grid technology since 2003. Smart meters have been gradually introduced into the system, and integration of the smart grid is continuing. Boulder, Colorado, began implementing smart grid technology in 2008. Hydro-One in Ontario, Canada, is in the process of introducing smart grid technology to over 1 million customers. In 2009, however, electricity regulators in Massachusetts rejected plans to deploy smart grid technology out of concerns for lack of protection for low-income customers.

Smart grid technology is certainly an important component of a future electrical system that includes the effective integration of nontraditional energy generation technologies and energy storage methods. As evidenced by previous efforts, however, the implementation of this approach involves technical innovation but also must deal with political and economic challenges in order to function efficiently.

17.5 Energy Conservation in the Community—LED Streetlights

On a municipal level, many approaches can be taken to reduce energy consumption. Because up to 40% of municipal energy budgets are used for lighting streets, this is an area where significant energy savings can be realized. It is also an area where a clear, well-defined, and cost-effective approach can be taken. The replacement of traditional sodium vapor or metal halide streetlights with high-efficiency light-emitting diode (LED) lamps typically reduces energy consumption by around 60%.

The physics of the operation of an LED follows closely from the discussion of photovoltaics in Chapter 9. A photovoltaic cell is a semiconducting junction (diode) that produces an electric current in response to incident light. Electron-hole pairs are formed in the depletion region (Figure 9.20) as a result of energy deposited there by photons. The electron-hole pairs contribute to a current that results from the electric field across the device. A light-emitting diode operates in basically just the opposite manner. A voltage is applied across the junction, causing electron-hole pairs to combine and produce photons. Electrons in the n-type region on one side of the junction (Figure 9.20), can move toward the junction and recombine with holes from the p-type region that are moving in the opposite direction. The energy that results from this recombination produces photons. The energy of the photons is governed by the difference in the energy of the electrons near the bottom of the conduction band and the energy of the holes near the top of the valence band. The energy is therefore determined by the size of the energy gap and is fairly narrowly defined. Semiconducting materials with different energy gaps give rise to photons of different energies with small-energy gap materials, producing small-energy, long-wavelength light toward the red end of the spectrum and large-energy gap materials producing large-energy, short-wavelength light toward the blue end of the spectrum. Table 9.1 gives a general idea of the energy gaps associated with different common semiconductors. Figure 1.1, along with the relationship between energy (in eV) and wavelength (in nm),

$$E = \frac{1240 \, \text{eV} \cdot \text{nm}}{\lambda},$$
(17.1)

provides an idea of the colors of the light that can be produced. Early LEDs produced red light, whereas modern devices can be tuned by adjusting the composition of the semiconductors, and hence the energy gap, to produce virtually any color desired.

For lighting purposes, it is desirable (for good color rendition) to have a broad spectrum of wavelengths giving rise to light that, to the human eye, appears white. There are basically two ways of doing this using single-color LEDs. Perhaps the obvious way is to combine three LEDs producing light of the primary colors (red, green, and blue) and combining these into white light (Figure 17.3). The human eye perceives

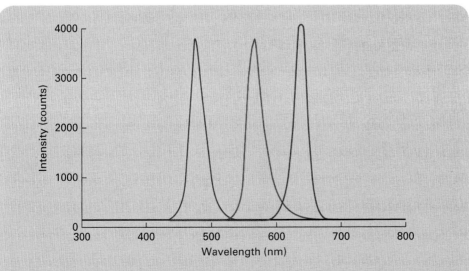

Figure 17.3: Spectrum of a three-color LED producing white light.

Based on The Midlands Dahlia Society, Propagation 2011 - What do LEDs provide? http://www.dahlia-mds.co.uk/Topics/Propagation_2011_5.htm

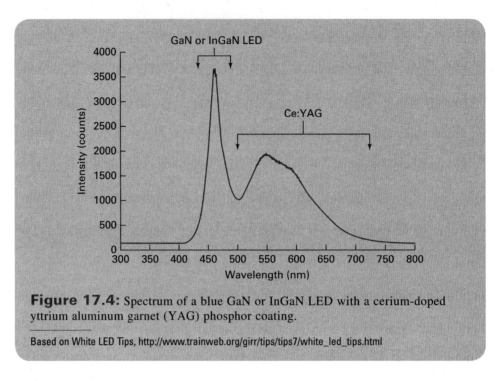

Figure 17.4: Spectrum of a blue GaN or InGaN LED with a cerium-doped yttrium aluminum garnet (YAG) phosphor coating.

Based on White LED Tips, http://www.trainweb.org/girr/tips/tips7/white_led_tips.html

this as white light, even though the spectrum is quite different from the broad continuum of sunlight shown in Figure 8.1.

An alternative, less expensive, and more commonly used method is to begin with a blue LED, which produces photons near the short-wavelength (highest-energy) end of the optical spectrum. The blue light is incident on a phosphor, which absorbs the blue light and re-emits it over a broad spectrum of lower energies (or longer wavelengths). The spectrum of this device contains the original spectrum of the blue LED and the broad re-emitted spectrum from the phosphor (Figure 17.4).

An example of the design of an LED streetlight fixture is shown in Figure 17.5. The finned geometry on the top side of the light fixture provides a large surface area

Example 17.2

A 1-km section of roadway has one lighting fixture every 40 m. If 200-W sodium vapor lamps are replaced with 80-W LED lamps, and the lights are on an average of 12 h per day, what is the annual savings for this section of road if electricity costs $0.10 per kWh?

Solution

The 1-km section of road has 1000 m/40 m = 25 fixtures. The power savings are (200 W − 80 W) = 120 W per fixture for a total of (25) × (120 W) = 3 kW. During a 1-year period at 12 hours per day, this is

$$(3 \text{ kW}) \times (365 \text{ d/y}) \times (12 \text{ h/d}) = 13{,}140 \text{ kWh},$$

or a value of (13140 kWh) × (0.10 $/kWh) = $1314.

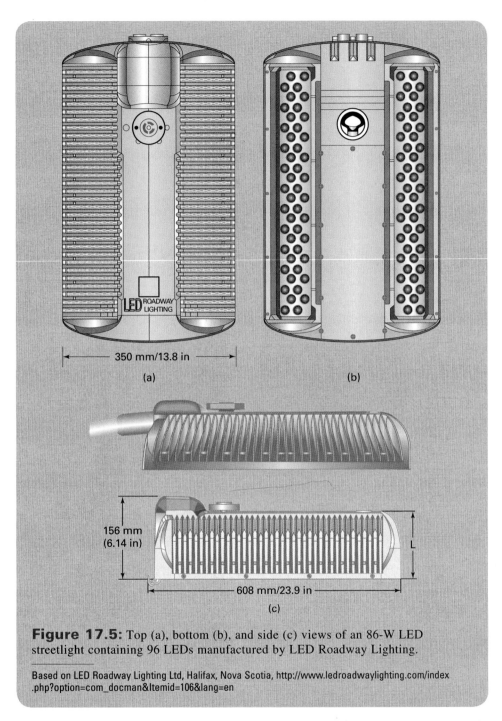

Figure 17.5: Top (a), bottom (b), and side (c) views of an 86-W LED streetlight containing 96 LEDs manufactured by LED Roadway Lighting.

Based on LED Roadway Lighting Ltd, Halifax, Nova Scotia, http://www.ledroadwaylighting.com/index .php?option=com_docman&Itemid=106&lang=en

to dissipate heat. This design of a heat sink is typical of that used on many electronic devices. An example of the effectiveness of LED streetlights is illustrated in Figure 17.6 on the next page, where 250-W high-pressure sodium fixtures consuming 275-W total (lamp and ballast) were replaced by 200-W LED streetlights (LED Roadway Lighting Ltd. model SAT-96M) leading to a reduction in energy consumption and increased public safety due to improved lighting conditions.

Figure 17.6: Conversion of streetlights on King William Street, Adelaide, Australia from (top) high-pressure sodium fixtures to (bottom) LED streetlights.

LED Roadway Lighting Ltd, Halifax, Nova Scotia

Overall, LED streetlights have several clear advantages over traditional sodium vapor or metal halide streetlights:

- Low power consumption (about one-fifth that of traditional lights for the same light output).
- Long lamp lifetime (about 12–15 years, or about 3 times that of traditional lamps).
- A payback period of about 3 years.
- Can be dimmed during conditions when full illumination is not needed.
- Minimize wastage light. Light is directed to road surface.
- Accurate color rendering.
- Elimination of toxic materials (e.g., Pb or Hg) found in many traditional streetlights.

Thus, the replacement of existing sodium or metal halide street lights with high-efficiency LED fixtures is a straightforward and cost-effective technique for lowering

ENERGY EXTRA 17.2
LEED certification

The Leadership in Energy and Environmental Design (LEED) certification system was developed in 1998 by the United States Green Building Council (USGBC). This system provides third-party verification that a building or group of buildings achieves a certain level of excellence in environmental performance in areas such as energy and water efficiency, CO_2 emissions, and indoor environmental quality. A second version (LEED NCv2.2) was released in 2005 and a third version (LEED NCv3) in 2009. In 2003, the Canada Green Building Council received approval from the USGBC to implement its own version of the LEED system. Thirty countries have now created LEED certification systems.

Five categories of constructions can apply for LEED certification:

1. *Green Building Design and Construction* includes new constructions and some major renovations of existing structures, for example, schools (as shown in the figure), retail businesses, and hospitals.
2. *Green Interior Design and Construction* includes the interior design of commercial buildings.
3. *Green Building Operations and Maintenance* deals with the operations and maintenance systems of existing buildings.
4. *Green Neighbor Development* deals with the development of environmental excellence in neighborhood design.
5. *Green Home Design and Construction* deals with residential homes.

Buildings are judged in a number of areas of environmental leadership and are awarded points in each category. The total number of points that can be achieved is 100 plus 10 bonus points. LEED accreditation is awarded according to the total score, and buildings are certified at one of the following levels:

- *Platinum certification:* 80 points and above
- *Gold certification:* 60–79 points
- *Silver certification:* 50–59 points
- *Certification:* 40–49 points

Studies have indicated that LEED-certified buildings are indeed more energy efficient and that workers in the buildings have increased productivity due to improved environmental conditions such as air quality, ventilation, temperature, and lighting. LEED-certified buildings have also been found to achieve higher rents, higher occupancy rates, and increased sale prices compared to their nongreen counterparts.

© Britton Images Photography

Dr. David Suzuki Public School in Windsor, Ontario; the first LEED Platinum–certified Education Building in Canada.

Continued on page 530

Energy Extra 17.2 continued

A directory of LEED-certified building projects in the United States is available online at http://www.usgbc.org/LEED/Project/CertifiedProjectList.aspx. For Canada, a directory is available at http://www.cagbc.org/Content/NavigationMenu/Programs/LEED/ProjectProfilesandStats/default.htm.

Topic for Discussion

Locate LEED-certified buildings in your area. Observe the building from the outside, or, if it is a public building or if access can be arranged, view the interior. What features do you see that have contributed to its LEED certification?

energy costs with a very reasonable payback period and for improving public safety. A number of installations are in place in municipalities across the United States, Canada, Eastern Europe, Australia, and Asia.

17.6 Home Heating and Cooling

A system that controls the climate in a building is often referred to as an *HVAC* (*heating, ventilation, and air conditioning*) *system*. The importance and design of the various components of such a system depend largely on the climate in the region. A single-family home may be connected to a *district system* (such as a district heating system supplied with steam or hot water from a cogeneration facility or geothermal resource) through a heat exchanger, or it may have a self-contained system within the home. The latter provides the homeowner the opportunity to make decisions concerning system design and operation that may have economic or environmental consequences. A home HVAC system may consist of individual distributed components, such as an electric heating system with baseboard heaters in the rooms or window air conditioners, or it may be a centralized facility, i.e., a furnace with a heat distribution system or central air conditioning. Although home HVAC systems can be quite varied in design, this discussion deals with some typical ways in which such systems operate, and how their efficiency can be optimized. We begin with systems that produce heat (i.e., furnaces) and then move on to systems that transfer heat into a building (i.e., a heat pump) and finally cooling (i.e., air conditioning) systems.

17.6a Furnace Efficiency

The term *furnace* is somewhat ambiguous. In North America, it is often used to refer to a centralized facility for producing heat for distribution to a building. It is also used, especially outside North America, to indicate an industrial furnace or kiln used for manufacturing processes. Sometimes the term furnace is used for a heater that distributes heat by circulating hot air, while the term *boiler* is used for a heater that distributes heat by circulating hot water or steam. Here we use a fairly generic definition of furnace as a centralized heater for providing heat to a building.

A furnace may produce heat by burning a combustible fuel or in some instances, through resistance heating, using an electric current. The fuel that is burned in a furnace may be natural gas, propane, oil, or in some cases an alternative fuel such as wood pellets. The efficient utilization of the energy content of the fuel may be viewed in two ways: (1) the efficient conversion of chemical energy into thermal energy and (2) the efficient distribution of the heat produced. The latter factor involves a control system to ensure that heat is provided when and where it is needed. A cost-effective means for doing this is with programmable thermostats that can provide heat at times and in rooms that are compatible with the occupants' needs.

The efficiency with which the chemical energy of the fuel can be converted into heat depends on the design and operational parameters of the furnace. Furnaces generally fall into two categories: *noncondensing* and *condensing*. The combustion of a hydrocarbon produces a number of by-products, principal among these are CO_2 and H_2O (in the form of water vapor). It is desirable if the heat produced during combustion is transferred to the fluid (typically air, water, or steam) that distributes the heat throughout the building. However, in practice, some of the heat produced is carried away by the by-products or exhaust gas of the combustion. If the exhaust gases, sometimes called flue gases, carry away substantial thermal energy, then they need to be vented through a heat-resistant chimney. This is traditionally the design of a heating system. Modern, high-efficiency furnaces are condensing gas furnaces (Figure 17.7). These furnaces extract enough of the thermal energy from the exhaust gases for use in the building in order that these exhaust gases are sufficiently cool and the water vapor in the gases condenses into a liquid. In such cases, the exhaust gases can be at a temperature of around 50°C and can be vented by means of a pipe (which is often plastic) directed through a wall or the roof of the house. Typically, furnaces with

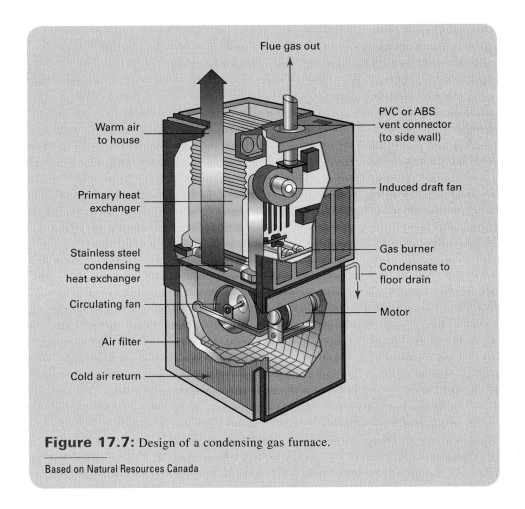

Figure 17.7: Design of a condensing gas furnace.

Based on Natural Resources Canada

ENERGY EXTRA 17.3
ENERGY STAR

In 1992, the United States Environmental Protection Agency (EPA) established the *ENERGY STAR* program to improve the efficiency of energy use and reduce greenhouse gases and to provide consumers with information about energy-efficient products. The program has subsequently been adopted by Canada, the European Union, Australia, Japan, New Zealand, and Taiwan. Consumers are most commonly aware of the ENERGY STAR program by the logo, shown in the figure, that appears on electronic devices (such as televisions and computers) and appliances (such as refrigerators and dishwashers). The logo indicates that the device has exceeded the specified level of energy efficiency in the ENERGY STAR testing program. In the United States, the criteria for ENERGY STAR approval are set for each product by either the Environmental Protection Agency or the United States Department of Energy.

The ENERGY STAR program, however, is much more diverse than just electronics and appliances. The ENERGY STAR rating also applies to construction materials, such as windows and furnaces. Upgrades to older homes can be made within the guidelines of the ENERGY STAR program, and newly constructed homes can earn the ENERGY STAR label if ENERGY STAR components and ENERGY STAR construction practices are utilized.

Commercial buildings such as hospitals, schools, and manufacturing plants can also be awarded ENERGY STAR approval. LEED and ENERGY STAR for buildings are complimentary approaches. ENERGY STAR rating deals more or less exclusively with energy consumption, whereas LEED takes a broader approach toward the environmental quality of the building. Buildings can achieve both LEED and ENERGY STAR certification. For existing buildings, the ENERGY STAR certification is a requirement for LEED approval.

U.S. Environmental Protection Agency

ENERGY STAR logo.

The ENERGY STAR program has promoted public awareness of the need for energy conservation and the corresponding benefits in greenhouse gas emission reduction. Consumer recognition of the ENERGY STAR logo has also motivated manufacturers to provide products that offer environmentally conscious options.

Topic for Discussion
Try to find similar appliances or electronic devices with and without the ENERGY STAR logo. Is a clear energy savings associated with the ENERGY STAR product? If so, make an estimate of the number of hours a typical user may use such a device and the annual electrical savings (in kilowatt-hours and dollars).

efficiencies of greater than about 90% are condensing gas furnaces. Such furnaces with efficiencies as high as 98% are available to the homeowner. Older furnaces may have efficiencies in the range of about 60%, and modern noncondensing furnaces typically have efficiencies of 80–85%. A homeowner who is installing a furnace and is concerned with efficient utilization of fuel should consider a condensing gas furnace.

ENERGY STAR-qualified furnaces are almost always condensing gas furnaces and offer an assurance of efficient operation.

Heat produced by a furnace may be used to heat air, which is circulated through ductwork to various locations in the building. Usually, a fan or blower is used to circulate the hot air (hence the common terminology applied to this design, *forced hot air heat*), and separate ducts return cool air to the furnace. It is important that ductwork be well sealed against air leaks and be appropriately insulated. Inferior ductwork that passes through unheated or minimally heated portions of the house (e.g., basements or attics) wastes energy. A furnace (often called a boiler) that heats water uses circulating pumps to distribute heat to the rooms of a house. Like air-carrying ductwork, pipes carrying hot water for heating must be appropriately insulated.

Example 17.3

If fuel oil costs $1.05 per liter, and a house requires net heat of 8.7×10^{10} J per year, calculate the annual savings that a homeowner could expect if an old 60% efficient system is replaced with a 94% efficient modern condensing gas furnace.

Solution

A 60% efficient furnace requires fuel with a total energy content of

$$\frac{8.7 \times 10^{10} \text{ J}}{0.6} = 1.45 \times 10^{11} \text{ J per year,}$$

whereas a 94% efficient furnace would require fuel with a total energy content of

$$\frac{8.7 \times 10^{10} \text{ J}}{0.94} = 9.26 \times 10^{10} \text{ J per year,}$$

for an energy difference of

$$(1.45 \times 10^{11} \text{ J} - 9.26 \times 10^{10} \text{ J}) \text{ per year} = 5.24 \times 10^{10} \text{ J per year.}$$

The energy content of fuel oil (about the same as crude oil) is given in Appendix IV as 3.85×10^7 J/L, so this energy difference translates to

$$\frac{5.24 \times 10^{10} \text{ J per year}}{3.85 \times 10^7 \text{ J/L}} = 1361 \text{ L/y.}$$

This has the monetary value of

$$(1361 \text{ L/y}) \times (\$1.05 \text{ per liter}) = \$1429 \text{ per year.}$$

17.6b Heat Pumps

The basic physics of a heat pump was presented in Section 1.5. The heat pump's operating principle can be incorporated into the design of a system that can use mechanical input to transfer heat from a cold reservoir (the outside of a building) to a hot reservoir (the inside of a building). The amount of heat (energy) transferred from the cold reservoir to the hot reservoir, Q_h, is related to the amount of work done, W, as

$$Q_h = COP \cdot W, \tag{17.2}$$

where the Carnot coefficient of performance (*COP*) is defined as

$$COP = \frac{1}{1 - \dfrac{T_c}{T_h}}. \tag{17.3}$$

T_h and T_c are the temperatures of the hot and cold reservoirs, respectively. (Recall that the temperatures in this equation must be expressed on an absolute temperature scale such as Kelvin.) To maintain a high coefficient of performance, the extraction of heat from the cold reservoir should not significantly lower the temperature of that reservoir. Thus, it is necessary that the total heat capacity of the cold reservoir be much larger than the heat capacity of the structure being heated. This is analogous to the situation for a thermal generating station where it is important that depositing waste heat from the facility into the cold reservoir does not raise the temperature of the reservoir. Raising the temperature of the cold reservoir lowers the ideal Carnot efficiency. For a home heat pump system, the obvious choices for an appropriate cold reservoir are the air outside the house (*air source heat pump system*) or the earth (*ground source heat pump system*). Where it is available, a water-based cold reservoir (a lake for example) can also be used. Home heat pump systems are sometimes designed to provide heating during the cold weather and to function as an air conditioner for cooling during the warm weather.

A schematic illustrating the operation of an air source heat pump is shown in Figure 17.8. This system utilizes the air outside the building as the cold reservoir and transfers heat by means of a working fluid. It has components outside the house to absorb heat from the environment and components inside the house to deposit the heat inside the building. The principle of operation is as follows: Fluid in the liquid phase

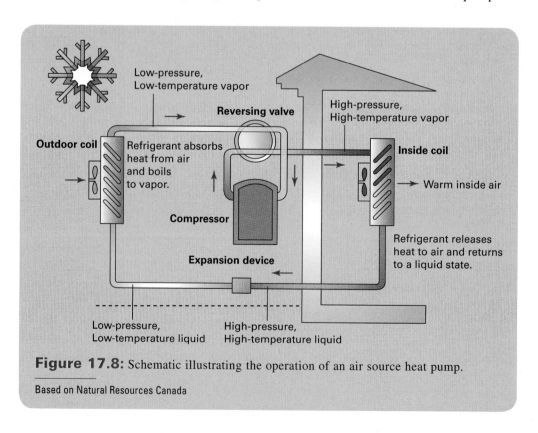

Figure 17.8: Schematic illustrating the operation of an air source heat pump.

Based on Natural Resources Canada

Figure 17.9: Exterior portion of a typical residential air source heat pump system.

Richard A. Dunlap

at low temperature and low pressure enters into coils outside the building, where it becomes a low-temperature, low-pressure vapor. This vapor is compressed by the compressor and becomes a high-temperature, high-pressure vapor and enters into the house. This vapor flows into coils inside the house, where it releases heat and becomes a high-temperature, high-pressure liquid. It then exits the building, where it passes through an expansion device to become a low-temperature, low-pressure vapor. It then flows into the cold outside the house and repeats the cycle. During this cycle, heat is extracted from the outside air and deposited inside the house. The heat released inside the house can be distributed via ductwork, as in the case of heat produced by a furnace. A typical exterior unit of an air source heat pump is illustrated in Figure 17.9.

Figure 17.10 on the next page shows the design of a ground source heat pump system. Water circulated through pipes in the ground transfers heat via a heat exchanger to a working fluid. The operation of the system is analogous to the air source system, and ultimately heat may be transferred to the air via a second heat exchanger for distribution within the house.

An analysis of the effectiveness of this heat transfer system is based on equations (17.2) and (17.3), along with a consideration of the heating requirements for the house, as presented in Chapter 8. Combining equations (17.2) and (17.3), the rate of heat transfer (or power) provided to the house can be expressed as

$$P = \frac{P_{\text{in}}}{1 - \dfrac{T_c}{T_h}}, \tag{17.4}$$

where P_{in} is the power input into the heat pump. T_c is the outside temperature, and T_h is the inside temperature. From equation (8.5), the rate of heat loss from a building is

$$P = \frac{A(T_h - T_c)}{R}, \tag{17.5}$$

where A is the surface area, and R is the R-value. For an actual building, heat loss through the various components (walls, windows, etc.) can be combined to yield the

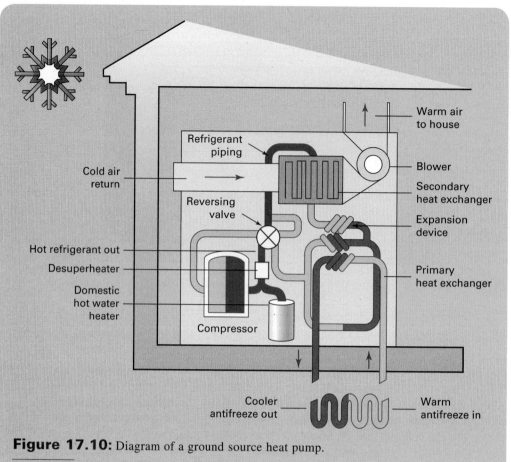

Cold air return →

Refrigerant piping

Warm air to house

Blower

Secondary heat exchanger

Reversing valve

Expansion device

Hot refrigerant out

Desuperheater

Domestic hot water heater

Primary heat exchanger

Compressor

Cooler antifreeze out

Warm antifreeze in

Figure 17.10: Diagram of a ground source heat pump.

Based on Heating and Cooling with a Heat Pump, Office of Energy Efficiency, Natural Resources Canada (Cat. No. M144-51/2004E)

Example 17.4

Calculate the power available from a heat pump when the input power is 4 kW and the inside and outside temperatures are 19°C and −3°C, respectively.

Solution

It is necessary to first express the temperatures in Kelvin:

$$T_c = -3°C + 273 = 270 \text{ K}$$

and

$$T_h = 19°C + 273 = 292 \text{ K}.$$

From equation (17.4), the power available is

$$P = \frac{4 \text{ kW}}{1 - \dfrac{270 \text{ K}}{292 \text{ K}}} = 53 \text{ kW}.$$

Example 17.5

A wall is 3 m by 6 m and has an R-value of 3.5 [in $(s \cdot m^2 \cdot °C)/J$]. A window in the wall with dimensions 1.0 m by 1.5 m has an R-value of 0.35 $(s \cdot m^2 \cdot °C)/J$. The inside and outside temperatures are 20°C and $-10°C$, respectively. Calculate the rate of heat loss through the wall in k/W.

Solution

The power loss is given by equation (17.5):

$$P = \frac{A(T_h - T_c)}{R}.$$

For the wall with a window, the loss through the wall and the loss through the window can be combined as

$$P = \frac{A_{wall}(T_h - T_c)}{R_{wall}} + \frac{A_{window}(T_h - T_c)}{R_{window}}.$$

The window has an area of $(1.0 \text{ m} \times 1.5 \text{ m}) = 1.5 \text{ m}^2$ and the wall (without the window) has an area of $(3 \text{ m} \times 6 \text{ m} - 1.5 \text{ m}^2) = 16.5 \text{ m}^2$. Using the appropriate R-values for the two components and $(T_h - T_c) = (20 - (-10))°C = 30°C$ gives the total power loss as

$$P = \frac{(16.5 \text{ m}^2) \times (30°C)}{3.5 \text{ } (s \cdot m^2 \cdot °C/J)} + \frac{(1.5 \text{ m}^2) \times (30°C)}{0.35 \text{ } (s \cdot m^2 \cdot °C/J)}$$

$$= 141 \text{ W} + 129 \text{ W} = 270 \text{ W}.$$

Interestingly, in this case the heat loss through the window is about the same as through the rest of the wall. This is not an uncommon situation for residential buildings.

---

total heat loss. To determine the effectiveness of a heat pump, the heat gain from equation (17.4) must be compared to the heat loss given by equation (17.5). The relationship between these two quantities depends on the design of the heat pump (i.e., the value of P_{in}) and the details of the building construction (i.e., the total areas and R-values of the building components). However, a general analysis of equation (17.4) shows that, as the value of T_c increases, the power increases, and it diverges as T_c approaches T_h (in this ideal treatment). An analysis of equation (17.5) shows that the power loss decreases linearly with increasing temperature and goes to zero as T_c approaches T_h. These characteristics are shown in Figure 17.11 on the next page. The point (i.e., temperature at which the heat pump becomes useful as temperature is increased) is the balance point. This simple analysis shows that below a certain temperature a heat pump is not useful. There are therefore limits to the climate conditions where heat pumps are economical, as well as constraints on the need for additional contributions to the heating requirements for a home (Section 17.6d).

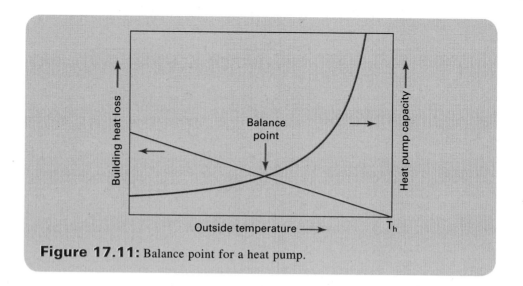

Figure 17.11: Balance point for a heat pump.

17.6c **Air Conditioning**

Centralized residential air conditioning systems are common in many locations with warmer climates. These systems generally consist of internal and external components. The general design and operation follow closely along the lines of an air source heat pump, except that heat is transferred from inside the building to outside the building. Figure 17.12

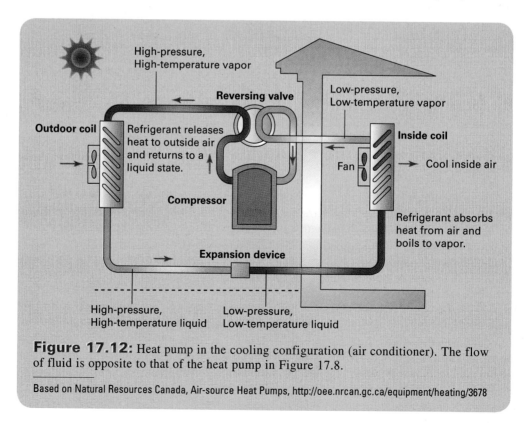

Figure 17.12: Heat pump in the cooling configuration (air conditioner). The flow of fluid is opposite to that of the heat pump in Figure 17.8.

Based on Natural Resources Canada, Air-source Heat Pumps, http://oee.nrcan.gc.ca/equipment/heating/3678

Figure 17.13: External portion of a residential central air conditioning system.

shows a schematic of an air conditioning system. The high-pressure and low-pressure sides are interchanged compared with the heat pump system. This is accompanied by a reversal of the flow of the working fluid that results from changing the direction of the compressor and the expansion valve. A typical external component of a residential central air conditioning is shown in Figure 17.13; the similarity with the heat pump shown in Figure 17.9 is obvious.

17.6d Integrated HVAC Systems

The design of an efficient residential HVAC depends substantially on the climatic conditions, the building design, and the needs of the occupants. In many regions, both heating and cooling are necessary, or are at least desirable, at different times during the year. The ways in which different approaches to residential HVAC systems can be integrated are very diverse. This section gives some common examples.

It is fairly straightforward to integrate forced hot air heating and central air conditioning (Figure 17.14 on the next page). Heat from a furnace is provided to the home in cold weather and heat is extracted through a heat exchanger and transferred to the outside by the air conditioning system when cooling is required.

In cooler climates, heating is generally a priority, and many alternative heating approaches can be beneficial and economical. Some approaches include heat pumps, as described, and solar thermal heating (Chapter 8). Figure 17.15 on the next page shows a system for combining heat extracted from the environment using a heat pump and an oil-fired forced hot air furnace. The coefficient of performance of the

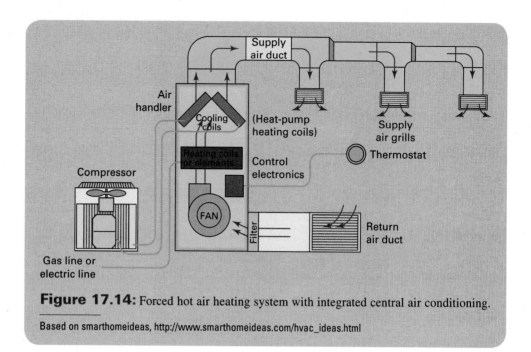

Figure 17.14: Forced hot air heating system with integrated central air conditioning.

Based on smarthomeideas, http://www.smarthomeideas.com/hvac_ideas.html

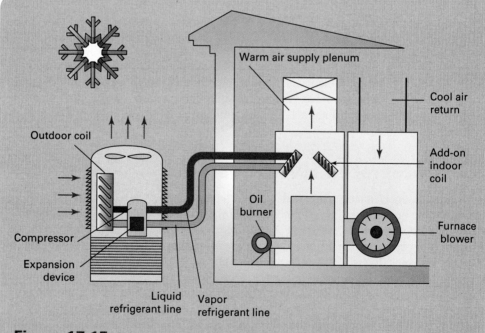

Figure 17.15: Schematic illustrating the operation of an air source heat pump in conjunction with an oil-fired furnace for residential heating.

Based on Heating and Cooling with a Heat Pump, Office of Energy Efficiency, Natural Resources Canada (Cat. No. M144-51/2004E)

heat pump decreases with decreasing temperature (Figure 17.11), and the heat loss from the building increases with decreasing temperature. Thus, as the outside temperature decreases, heating requirements increase, and the effectiveness of the heat pump diminishes. At some point, supplemental heating from another heating system (in this illustration, the oil-fired furnace) becomes necessary to maintain the interior temperature at the desired level. Another important factor that must be considered for efficient energy utilization is the coefficient of performance (*COP*) of the heat pump. The *COP* decreases with decreasing outside temperature, and if this approaches unity, then the energy required to run the heat pump, even in an ideal Carnot system, exceeds the energy gained. Thus, the effective control of the HVAC system is important for its most efficient operation.

The integration of a solar thermal heating system with a conventional fossil fuel–fired system (Figure 17.16) needs to deal with somewhat different requirements. Solar energy is available during the day and, as described in Chapter 8, must be stored during the night. This is often most convenient if water is used as the heating medium because it is also a convenient storage medium. Thus, Figure 17.16 shows a forced hot water heating system where heat is stored in a tank of water. Heat from the solar collector heats the water in the storage tank, but an auxiliary heating system, such as an oil- or natural gas–fired boiler provides additional heat as needed to maintain the proper storage water temperature, in accordance with the heating and hot water requirements of the home. Daily and seasonal variations in both the home energy requirements and the solar insolation need to be considered in the design of the system.

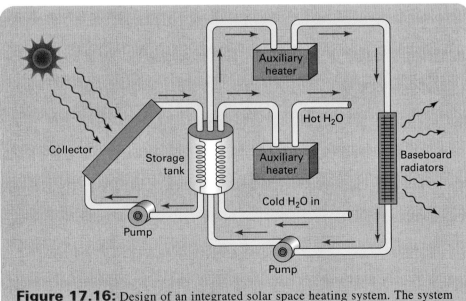

Figure 17.16: Design of an integrated solar space heating system. The system provides heating and hot water. Back-up heating is provided by an auxiliary system (e.g., fossil fuel-fired boiler).

Based on R. A. Hinrichs and M. Kleinbach, Energy, Its Use and the Environment, Brooks/Cole

17.6e Minimizing Heat Loss: Insulation, Windows, and Air Leaks

Although it is important to efficiently produce energy for space heating, it is also important to utilize that energy effectively. Thus, it is necessary to minimize heat losses from a building. In Chapter 8, the basic physics of heat transfer through walls and windows was presented in order to determine the heating needs for a house. In this section, these ideas are expanded to determine the most effective ways of reducing heat losses in a building and conserving energy used for space heating. Actions that are effective in minimizing heat flow from the warm (heated) interior of a building to the cold exterior in the wintertime are also effective at minimizing heat flow from the warm outside to the cool (air conditioned) interior of a building in the summertime. In this section, we focus on three important aspects of limiting heat loss from a building: insulation, windows, and eliminating air leaks.

Insulation is a material that occupies the space around the interior of the building, for example, inside the exterior walls. Its effectiveness is based on its ability to limit air circulation within the space. A reflective layer such as aluminum foil can also be effective in reducing radiative heat transfer. Insulation comes in various forms:

- *Blankets:* A flexible material such as fiberglass wool that comes in a roll.
- *Rigid insulation:* A fibrous or foam material that comes in sheets, such as closed-cell polystyrene foam, often referred to by its proprietary name, Styrofoam (Figure 17.17).
- *Foam insulation:* Sprayed on material that adheres to wall or roof surfaces, such as polyicynene or polyurethane.
- *Loose fill:* Loose fiber pellets such as vermiculite or particles of fiberglass that are blown into wall or ceiling spaces using pneumatic equipment.

The *R*-values per unit thickness given in Table 17.1 for some typical insulating materials. The *R*-value of insulation depends on the properties of the insulation

Figure 17.17: Rigid closed-cell polystyrene foam insulation Sheets are available in different thicknesses with a typical *R*-value of 0.35 (s · m^2 · °C)/J per cm of thickness.

Richard A. Dunlap

Table 17.1: Approximate *R*-values per unit thickness of some insulating materials.

material	R-value $(s \cdot m^2 \cdot °C)/J$ per cm thickness
vermiculite loose fill	0.15
fiberglass loose fill	0.21
polyicynene spray foam	0.25
fiberglass wool	0.26
polystyrene foam rigid sheet	0.35
polyurethane spray foam	0.42
polyisocyanurate spray foam	0.45

and is linearly proportional to thickness. The heat loss per unit area is, according to equation (17.5), inversely proportional to the insulation thickness. As discussed in Chapter 8, the *R*-values of interior and exterior walls need to be added to the *R*-values of the insulation to obtain the total *R*-value of the wall. From a practical standpoint, the values given in Table 17.1 should be used as guidelines for understanding the effects of insulation. However, in practice, building structure and construction techniques can often lead to actual *R*-values that differ substantially from those in the table. Insulation in walls fills the spaces between the wall studs; three factors related to the effectiveness of its installation are worthy of note.

1. The insulating properties of insulation are the result of its ability to minimize air transport. Compressing insulation such as fiberglass wool diminishes its insulating properties. Packing thicker insulation into a space than the space should accommodate reduces its effectiveness.

2. The insulation must fill the entire space between the studs. Air spaces between the edge of the insulation and the stud allow for heat loss.

3. The *R*-value for a wall is less than that of the insulation itself, because heat can be transported through the studs by conduction. This is referred to as a thermal bridge. It is a factor for wooden framed buildings but can be an even greater factor for metal framed buildings.

Considerations for insulation in attics are somewhat different because insulation can be laid or sprayed across ceiling joists, reducing concern for factors 2 and 3.

The construction of new buildings can accommodate insulation as appropriate for the climatological conditions in the area. The utilization of appropriate materials, as well as careful construction practices, optimizes the effectiveness of insulation.

In the case of upgrading insulation in existing buildings, access to spaces that need to be insulated may place restrictions on the available options. Adding additional insulation to existing insulation in attics is often fairly straightforward, although caution should be exercised to avoid compressing existing insulation under the weight of additional insulation and thereby reducing its effectiveness. Adding insulation to walls is somewhat more difficult because access to spaces that require insulation is often limited. In many cases, blown-in insulation is the only practical option. In addition to insulating wall and ceiling spaces, improving insulation on heating ducts and/or pipes carrying water for heating purposes through unheated spaces can be productive.

The economics of improving insulation for the purpose of reducing heat loss is often an important consideration. The payback period must make such improvements attractive for a homeowner. In general, a detailed analysis of the effects of improving insulation can be complex. However, Example 17.6 shows a very simple approach to understanding the results of adding insulation to an existing uninsulated building.

The significance of heat loss through windows has been emphasized in Example 17.5. The importance of ensuring that energy-efficient windows are utilized in new construction is obvious. The benefits of improving low-efficiency windows in existing buildings are also clear. Replacing older windows can be an effective means of reducing heating costs.

Example 17.6

Consider a simple house that is a cube 10 m on a side located near Portland, Oregon. It loses heat through the walls and ceiling but not through the floor. What is the heating requirement if the walls and ceiling are uninsulated and have an average R-value of 1.0 $(s \cdot m^2 \cdot °C)/J$? Repeat this calculation for an insulated house with an R-value for the walls and ceiling of 3.5 $(s \cdot m^2 \cdot °C)/J$. What is the annual savings in energy cost if heat costs $19.75 per GJ (typical of the price of oil or natural gas)?

Solution
From equation (17.5),

$$P = \frac{A(T_h - T_c)}{R}.$$

If A is in m^2, temperature is in °C, and the R-value is in $(s \cdot m^2 \cdot °C)/J$, then the power on the left-hand side is in W. If the right-hand side of the equation is multiplied by 86,400 s/d, then the left-hand side is in J/d. If the temperature on the right-hand side is expressed as the number of degree days per year, then the left-hand side gives the energy requirement in J/y.

For Portland, Oregon, Chapter 8 gives the number of degree days per year as 2426 in °C. The surface area of the walls and roof is 5×10 m $\times 10$ m $= 500$ m^2. Thus,

$$\text{energy/year} = (500 \text{ m}^2) \times (2426 \text{ °C d/y}) \times \frac{86,400 \text{ s/d}}{1.0 (s \cdot m^2 \cdot °C/J)} = 1.05 \times 10^{11} \text{ J/y}.$$

Repeating this calculation for an R-value of 20 gives

$$\text{energy/year} = (500 \text{ ft}^2) \times (2426 \text{ °C d/y}) \times \frac{86,400 \text{ s/d}}{3.5 (s \cdot m^2 \cdot °C/J)} = 0.30 \times 10^{11} \text{ J/y}.$$

The energy savings are $(1.05 - 0.30) \times 10^{11}$ J/y $= 7.5 \times 10^{10}$ J/y for a monetary value of

$$(75 \text{ GJ/y}) \times (\$19.75/\text{GJ}) = \$1481 \text{ per year}.$$

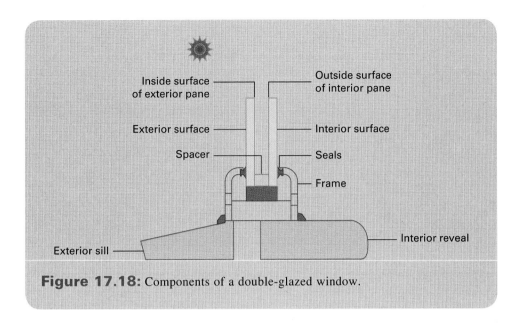

Figure 17.18: Components of a double-glazed window.

The purpose of a window is to allow light into a house while preventing (as necessary) the transfer of heat between the inside and outside. The simplest window is a single pane of glass (referred to as a single-glazed window). A double-glazed window consists of two panes of glass separated by an air space, and a triple-glazed window consists of three panes of glass, and so on. The primary purpose of the air space between the panes of glass is to reduce heat transfer from convection by preventing the circulation of air that comes in contact with the glass panes. The design of a typical double-glazed window is shown in Figure 17.18.

A typical single-glazed window has an R-value of about 0.18. The increase in R-value is approximately 0.18 per additional pane of glass. Thus, a basic double-glazed window (Figure 17.18) has R-value of about 0.36, and a triple-glazed window has an R-value of around 0.54. The R-value can be further increased by adding low-emissivity (low-e) coatings on the glass (see Energy Extra 8.1 for a description of the physics of optical coatings). Such a coating typically increases the R-value by about 0.18. Finally, the space between the panes of glass can be filled with a gas other than air. The most commonly used gas is argon. Argon has a higher viscosity and a lower thermal conductivity than air. The first property reduces heat transfer by convection because it reduces gas circulation in the space between the panes. The second property, obviously, reduces heat transfer by conduction. The net result is that replacing the air between the panes with argon increases the R-value by about 0.18.

From a practical standpoint, triple-glazed windows with argon fill and low-e coatings are about the practical limit for energy-efficient, high–R-value windows. These windows would have an R-value around 0.9 (0.54 for triple glazing, 0.18 for the argon, and 0.18 for the coating). Vacuum-glazed windows, where the gas is evacuated from the space between the panes have the highest R-values, up to around 2.2. However, these may not be practical for installations if wide temperature variations are expected because the vacuum seals may not be reliable under excessive thermal stress.

Example 17.7

Calculate the decrease in heat loss in kilowatt-hours per day when a 1.2 m × 2.1 m uncoated single-pane window is replaced with a double-glazed window with a low-e coating and argon fill (a relatively common configuration for current windows). The inside temperature is a constant 20°C, and the outside temperature is a constant −5°C.

Solution

From equation (17.5), the thermal power lost through the window is

$$P = \frac{A\Delta T}{R},$$

and the energy loss over a period of a day is

$$E = \frac{A\Delta T}{R} \times 86{,}400 \text{ s/d}.$$

If A is expressed in m², ΔT in °C, then the energy is given in joules. Thus for the single-pane window ($R = 0.18$),

$$E = (1.2 \text{ m} \times 2.1 \text{ m}) \times [20°\text{C} - (-5°\text{C})] \times \frac{86{,}400 \text{ s/d}}{0.18 \text{ s} \cdot \text{m}^2 \cdot °\text{C/J}}$$
$$= 30.2 \text{ MJ/d}.$$

For the double-glazed window with low-e and argon ($R = 0.72$),

$$E = (1.2 \text{ m} \times 2.1 \text{ m}) \times (20°\text{C} - (-5°\text{C})) \times \frac{86{,}400 \text{ s/d}}{0.72 \text{ s} \cdot \text{m}^2 \cdot °\text{C/J}}$$
$$= 7.6 \text{ MJ/d}.$$

The difference in energy loss is

$$30.2 \text{ MJ/d} - 7.6 \text{ MJ/d} = 22.6 \text{ MJ/d}.$$

Converting to kWh gives

$$\frac{22.6 \text{ MJ/d}}{3.6 \text{ MJ/kWh}} = 6.2 \text{ kWh/d}.$$

Example 17.8

Consider the situation in Example 17.7. If a house has 12 single-glazed windows that are replaced by low-e, argon-filled double-glazed windows, calculate the savings during 3 months of cold winter weather as typified in the example. Assume that heat is provided by electricity at a rate of $0.11 per kWh.

Solution

At the heat loss reduction given in Example 17.7, 6.2 kWh per day per window, the total savings in kilowatt-hours for 12 windows for 3 months are

$$(6.2 \text{ kWh/d}) \times (12) \times (90 \text{ d}) = 6696 \text{ kWh}.$$

At $0.11/kWh, the value of energy saved is

$$(6696 \text{ kWh}) \times (\$0.11/\text{kWh}) = \$737.$$

The actual net R-value for a window may be somewhat different (typically lower) than this simple approach would indicate. Heat transfer through the frame (Figure 17.18) can be important, particularly for windows with metal frames, and this additional heat transfer lowers the R-value. Nominally similar windows made by different manufacturers can have appreciably different actual performance.

The effectiveness of replacing lower R-value windows with higher R-value windows for conserving energy can be estimated on the basis of the physics of heat transfer as a function of R-value. The amount of energy that can be saved by upgrading windows is illustrated in Example 17.7, and the economics of this action are summarized in Example 17.8.

Heat loss through walls and windows is relatively straightforward to understand and quantify. Heat loss due to air leaks is due to convection and is more difficult to quantify. However, its importance is obvious; leave a window open in the wintertime in a cold climate, and the heat loss is apparent. Small spaces where air can leak into or out of a house can add up. These leaks include poor seals around windows and doors or between the house and the foundation (sill plate), but they also include places like dryer vents where outside air can get into the house. Figure 17.19 on the next page is an illustration of some of the common air leaks in a home. It is important, as in the case of considering insulation, to note that air leaks between the heated portion of the house and unheated spaces, such as attics, can be as significant as those to the outside. Because warm air is less dense than cold air, the warm air inside the house tends to leak out through openings in the higher portions of the house. These often lead into attic spaces (Figure 17.19). Lighting fixtures in the ceiling are a common source of air transport between the house and the attic. Also, air can leak into uninsulated interior walls through electrical outlets and switches or through spaces around plumbing and travel upward to the unheated

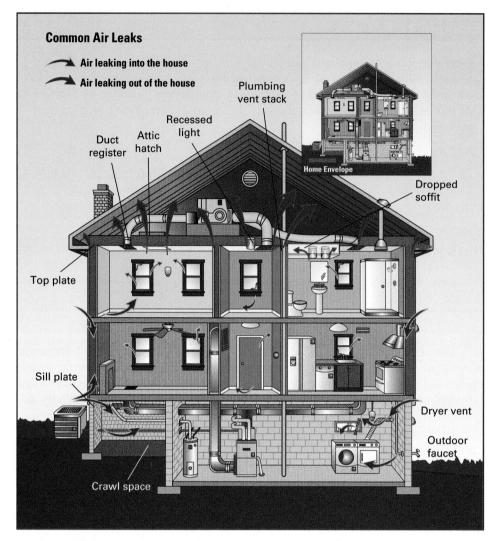

Figure 17.19: Common air leaks in a house.

Energy Star, Air Seal and Insulate with ENERGY STAR, http://www.energystar.gov/index.cfm?c=home_sealing .hm_improvement_sealing

spaces above, such as the attic (Figure 17.20). Thus sealing openings, even if they do not involve a direct connection to the outside, is important and can be accomplished with weather stripping or caulking as appropriate.

17.7 Residential Lighting

The efficient use of electricity for residential lighting involves personal actions such as turning lights off when they are not in use, as well as the analogous automated approach (i.e., the use of timers or photoelectric switches to control lighting). The most significant improvements in efficiency, however, are likely to come from the development of more efficient devices.

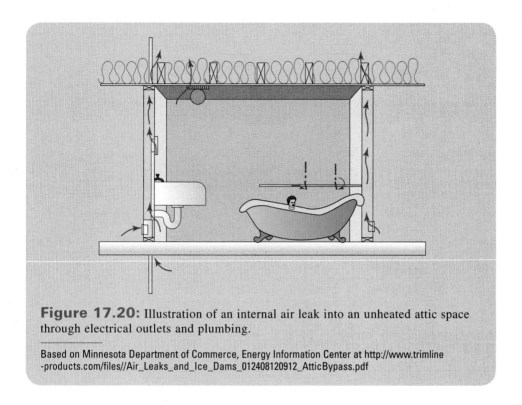

Figure 17.20: Illustration of an internal air leak into an unheated attic space through electrical outlets and plumbing.

Based on Minnesota Department of Commerce, Energy Information Center at http://www.trimline -products.com/files//Air_Leaks_and_Ice_Dams_012408120912_AtticBypass.pdf

Traditionally, room lighting consisted of incandescent bulbs or, where space and geometry allowed, fluorescent tubes. Fluorescent tubes have the advantage that they produce more light output for the same electrical energy input, typically about 5 times as much light output, compared with incandescent bulbs. The amount of light is measured in units of *lumens* (lm). A lumen is a measure of the amount of light over the visible portion of the electromagnetic spectrum and accounts for the actual light intensity, as well as the variations in the sensitivity of the human eye, over that range of wavelengths. Thus, it is an accurate measure of the brightness of the light as perceived by a person. Different devices that produce light output from an electrical input yield different amounts of light for the same power input, depending on the mechanism used for producing the light, as well as the wavelength distribution of the light produced. The light output (in lumens) for a given electrical input (in watts) is shown in Table 17.2 for some common light sources. It is clear that fluorescent lamps are substantially more efficient at converting electrical energy into light than incandescent bulbs. The energy that does not become light is given off as heat, clearly incandescent lights produce more heat than light.

Table 17.2: Typical light output per electrical input in lumens per watt (lm/W) for some light-producing devices.	
light source	output/input (lm/W)
incandescent bulb	15
compact fluorescent lamp (CFL)	60–70
light-emitting diode (LED)	100–150

Figure 17.21: Compact fluorescent lamps. Front left to right: 9-W spiral, 13-W spiral, 23-W spiral, 26-W spiral, and 9-W tube, with light output equivalent to incandescent bulb with ratings of 40 W, 60 W, 100 W, 120 W, and 40 W, respectively. Lamps shown have the standard North American Edison base.

Richard A. Dunlap

In recent years, compact fluorescent lamps (CFLs) have been developed. These are available in sizes that approximate the size of incandescent bulbs and are equipped with a standard base (either the Edison screw base, used in the United States and many other countries, or a bayonet base) so that they can be inserted into a standard lamp socket. Some examples of commercially available CFLs are shown in Figure 17.21. Two geometries are in common use: a spiral geometry, more common in North America, and a tube geometry, often used elsewhere. These have been promoted for their efficient production of light and their net positive contribution to the conservation of energy.

Example 17.9

A package for a 9-W CFL claims an expected lifetime of 8000 hours and a savings of $24.80 in energy costs. Confirm this claim.

Solution

A 9-W CFL has a light output equivalent to a 40-W incandescent bulb for a savings of 31 W. Over a lifetime of 8000 hours, this corresponds to a total energy savings of

$$(0.031 \text{ kW}) \times (8000 \text{ h}) = 248 \text{ kWh}.$$

If the cost of electricity $0.10 per kWh, then this corresponds to a savings of $24.80.

The principle of operation of the CFL is the same as for all fluorescent lamps. Electrons traveling through a gas interact with the atomic electrons of the gas atoms and cause them to be excited into higher energy levels; from there, they undergo transitions back to the ground state and thereby emit photons. Specifically, in a fluorescent lamp, the electrons are produced by the filaments at the ends of the tube, which are heated by electric current flowing through them. The filaments, referred to as cathodes because they emit electrons, are typically made of tungsten coated with a metallic oxide that readily gives off electrons. The tube contains mercury vapor at a very low pressure, and it is the electronic transitions in the mercury atoms that produce light. The photons that are produced by most electronic transitions in mercury are at fairly high energy and are in the ultraviolet portion of the electromagnetic spectrum (Figure 1.1) that is not visible to the human eye. To produce visible light, the ultraviolet photons are incident on a phosphor coating on the inside of the tube. The atoms in the phosphor coating absorb the ultraviolet photons and re-emit them at lower energies (in the visible portion of the spectrum) over a broad range of wavelengths. This is similar to how white light is produced by a blue LED (Section 17.5). The wavelength spectrum of the fluorescent lamp can be varied by using different combinations of phosphors. Classifications of fluorescent lamps such as cool white and daylight result from different combinations of phosphors.

Fluorescent lamps have an interesting electrical property. As the current flows through the mercury vapor, exciting electronic transitions in the gas atoms, the excited atoms actually make it easier for the current of electrons to flow. Thus as the current, I, increases, the resistance, R, decreases, making it easier for more current to flow. In the case where a constant voltage, V, is applied to the tube, the current increases in an uncontrolled manner according to Ohm's law:

$$I = \frac{V}{R}.$$

(17.6)

To avoid this situation, a ballast is utilized. A simple ballast is merely an inductor in series with the tube that limits the current flow in the AC circuit. CFLs typically use an integral electronic circuit built into the base (Figure 17.22) to fulfill this requirement.

Potential environmental hazards are associated with CFLs. Like other fluorescent lamps, CFLs contain mercury (Hg), which poses health risks. Government regulations (in the United States and most other locations) require that all mercury-containing products display the (Hg) symbol (Figure 17.23 on the next page). Each CFL manufactured at present contains 1–5 mg of mercury. Many locations have initiatives in place

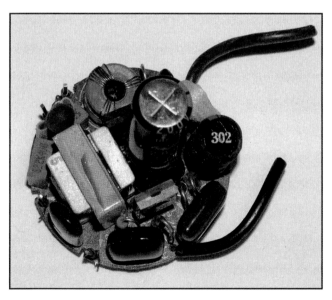

Figure 17.22: Internal electronic ballast from the base of a 13-W spiral compact fluorescent lamp.

Richard A. Dunlap

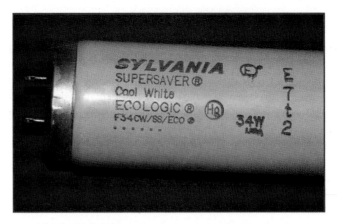

Figure 17.23: The mercury symbol, Hg inside a circle, on a fluorescent lightbulb.

Richard A. Dunlap

to divert CFLs from disposal in, say, landfills where the mercury can disperse in the environment. However, this practice is far from universal, and in most places discarded CFLs are likely to contribute to environmental mercury release. Although this may appear to be an undesirable situation, the U.S. Environmental Protection Agency has estimated that the mercury content of all CFLs sold in the United States during a year is only about 0.1% of the total annual U.S. anthropogenic mercury emissions.

A simple calculation provides information about the energy and economic advantages of CFLs. The difference in light output to produce the same light level in a room is a measure of the energy savings that can result for the replacement of incandescent bulbs with compact fluorescent lamps. However, this is not a complete picture of how CFLs affect energy utilization. A CFL mostly produces light and a relatively small amount of heat. By comparison, about 10% of the electrical energy used by an incandescent bulb is converted into light; the remainder of about 90% is converted into heat. Thus, most of the energy goes into heat, which may be a desirable or an undesirable by-product of the light production, depending on the situation. For example, in, say, Minnesota in the wintertime, the excess heat is not wasted energy but contributes to heating the interior of the building, thus reducing the energy necessary for heat by that amount. In this case, the total energy saved by converting incandescent bulbs to CFLs is not as great as one would expect. On the other hand, if we consider the situation, say, in Alabama in the summertime, the excess heat produced by an incandescent bulb is really waste heat and would typically have to be eliminated from the building by the air conditioning system. Thus, in this case, the incandescent bulb not only produces less light (per watt electrical input) but also puts additional energy demands on the cooling system, and the total energy saving for converting to CFLs is greater than expected.

As Table 17.2 indicates, compact fluorescent lamps are substantially more efficient at converting electrical energy into visible light than traditional incandescent bulbs. The table also shows that LEDs are even more efficient at producing light than CFLs. LEDs have become an important alternative to traditional lighting methods for roadway lighting. They have also recently become available as replacements for domestic room lighting. Due to their high efficiency, a 9-W LED lamp can provide as much light output as a 60-W incandescent bulb. An example of a commercially available LED lamp for room lighting is shown in Figure 17.24. The lifespan of a typical

Figure 17.24: A 10-W LED lamp with an Edison screw base.

Richard A. Dunlap

LED lamp exceeds that of a CFL; about 30,000 hours compared to about 8000 hours. Due to the substantially higher initial cost (at present), about a factor of 10 compared with CFLs, their use has not been as extensive, although prices have continued to decline, and this situation may change in the future.

17.8 Transportation

17.8a Fuel Economy

As illustrated in Figure 2.6, transportation accounts for nearly 30% of the energy use in the United States.and transportation accounts for over 70% of the energy obtained from petroleum. It is therefore an important concern to ensure that transportation technology makes efficient use of fuel. The problem of optimizing the energy efficiency of transportation is a complex one with many facets. One side of the problem deals with the prevalent mode of transportation. This varies considerably from one part of the world to another, from modes of transportation that are very efficient (i.e., bicycles) to moderately efficient (i.e., public transportation by bus or train) to the least efficient (personal automobiles). Globally, there has been a consistent shift from more efficient transportation modes to less efficient transportation modes with time. The move toward a transportation system that is dependent on automobiles occurred in the 1920s and 1930s in North America, in the 1950s and 1960s in much of Europe, and even later in many other parts of the developing world. Thus, a historical analysis of the use of energy for transportation is greatly influenced by economic, social, and cultural factors. Promoting more efficient use of transportation resources can help to conserve energy, and this might include encouraging consumers to drive less, ride bicycles, or take public transportation.

Another way of viewing the efficiency of transportation is to consider the trends in the efficiency of, say, the automobile over time. This approach deals primarily with the science and technology of transportation, rather than changes in human behavior and is the approach taken in the following discussion.

The fuel economy of a vehicle can be quantified in several different ways. The units used may be given as distance traveled per volume of fuel (commonly referred to as *fuel economy* or *fuel efficiency*) or as volume of fuel used per distance traveled (commonly referred to as *fuel consumption*). Fuel economy is commonly measured in kilometers per liter while fuel consumption is measured in liters per 100 kilometers (L/100 km). Different conventions are used in different countries around the world. There is an inverse relationship between fuel economy and fuel consumption given by

$$\frac{100}{(\text{L}/100\ \text{km})} = \text{km/L}. \tag{17.7}$$

This relationship is illustrated graphically in Figure 17.25.

In the 1960s and 1970s, it became increasingly apparent that the use of fossil fuels for transportation was a matter for concern. Concerns about the pollution that resulted from fossil fuel use (Chapter 4) were responsible for many governments establishing policies to deal with emissions from vehicles. Concerns also arose over the long-term availability of fossil fuel resources, and this provided motivation to improve the fuel efficiency of automobiles. Government regulations in a number of countries specified average fuel efficiency that manufacturers needed to achieve for their vehicles in that country [in the United States, this is the Corporate Average Fuel Economy (CAFE)], as discussed in Section 17.2. Usually vehicles were categorized as, say, passenger vehicles, light trucks, heavy trucks, and so on. Fulfilling fuel consumption regulations can be viewed in two ways: (1) preferentially marketing vehicles that meet fuel consumption guidelines or (2) improving vehicle technology to increase efficiency and

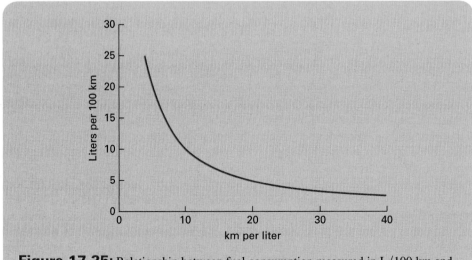

Figure 17.25: Relationship between fuel consumption measured in L/100 km and fuel efficiency measured km/L.

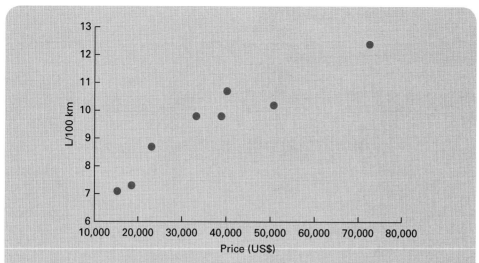

Figure 17.26: Fuel consumption in L/100 km average city/highway from U.S. Environmental Protection Agency (EPA) estimates as a function of model base price (in US$) for all nonhybrid passenger automobiles marketed in North America by Toyota (Toyota/Lexus) in 2016.

hence reduce fuel consumption. The former approach depends on the needs and desires of the consumer and is a function of factors like economy and geography, whereas the latter deals primary with vehicle technology.

Per-capita GDP plays an important role in what vehicles can be most successfully marketed, and this, in turn, is directly related to fuel economy. Generally, more expensive vehicles burn more fuel. Figure 17.26 shows the fuel consumption for all passenger vehicles manufactured by Toyota and marketed in North America as a function of their cost. People who purchase more expensive vehicles use more fuel.

Geographic and cultural factors also influence vehicle choice. People who live in urban (rather than rural) environments or in regions of higher population density may choose smaller (and typically more fuel-efficient) vehicles. For example, in the European Union, the requirement for average fuel consumption for 2012 passenger vehicles was 5.0 L/100 km, while in the United States the requirement for 2016 passenger vehicles was 6.6 L/100 km. While this is significant, it should be noted that the comparison of data from different countries can sometimes be misleading. The same vehicle may receive different fuel consumption ratings in different countries due to different testing procedures.

From a scientific and engineering standpoint, the development of new technologies that actually make an internal combustion engine vehicle more efficient in terms of converting the chemical energy of the fuel into mechanical energy is perhaps the most interesting. To understand where the most significant gains can be achieved, it is necessary to analyze where the energy losses actually occur. Based on studies by the U.S. Department of Energy, these losses are summarized for city and highway driving conditions in Figure 17.27 on the next page. As illustrated in

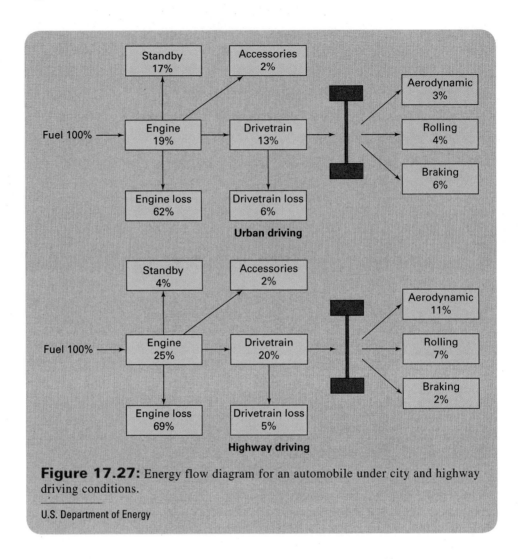

Figure 17.27: Energy flow diagram for an automobile under city and highway driving conditions.

U.S. Department of Energy

the figure, the typical overall efficiency for conversion of chemical energy of the fuel to energy delivered to the wheels for a gasoline-powered passenger vehicle for all driving conditions is about 17% (i.e., 13% for city and 20% for highway). This is an important reference point for comparison with other vehicle technologies, as discussed in the next two chapters.

Clearly, the greatest energy losses are from the engine. These losses are to some extent inevitable because of the basic thermodynamics of a heat engine. However, engine design and operation can optimize the efficiency of converting the chemical energy of the fuel into mechanical energy. The design of the drivetrain, control systems, and the overall geometry of the vehicle are some of the other factors that can make the vehicle more fuel efficient. Specifically, some of the design criteria that can be targeted are as follows:

- Improving cooling system efficiency (The use of efficient coolants can reduce the cooling system volume and lead to faster warmups and better temperature regulation.).

- Developing more efficient computer control of engine operating conditions, that is, operating temperature and fuel distribution.
- Using thinner and/or lower-friction engine oils and lubricants to reduce viscous drag on moving components.
- Increasing the number of gear ratios in the transmission or using a continuously variable drive system to best match the engine speed to the vehicle velocity.
- Improving the design of automatic transmissions to reduce slippage.
- Implementing more efficient technologies for shifting automatic transmissions at optimal points.
- Reducing weight by more efficient packaging of components and better utilization of occupant space.
- Reducing weight, particularly of moving parts, by utilizing advance materials.
- Utilizing more aerodynamic designs to reduce air resistance (The power required to overcome air resistance is proportional to the cube of the velocity, so this factor is particularly important at highway speeds.).
- Using low rolling resistance tires.
- Implementing methods for the efficient utilization of electrical energy for accessories.

Other approaches, including moving to new technologies for vehicle design such as gasoline-electric hybrids, are now discussed.

The actual benefits of actions to reduce fuel consumption are due, in some part, to improvements in vehicle technology. Studies by the U.S. Environmental Protection Agency (EPA) show a generally consistent improvement in fuel economy for passenger automobiles in the United States since the mid-1970s when government regulations began (Figure 17.28). There has been little change in fuel economy for heavy trucks over the past 40 years or so.

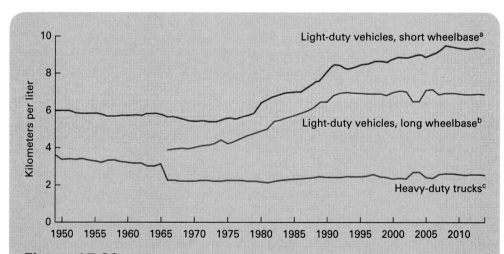

Figure 17.28: Fuel economy of vehicles in the United States 1949–2014. Passenger automobiles are typical of "Light-Duty Vehicles, Short Wheelbase."

U.S. Energy Information Administration, http://www.eia.gov/totalenergy/data/monthly/pdf/sec1_18.pdf

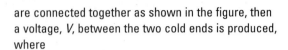

ENERGY EXTRA 17.4
Thermoelectric generators

In the present chapter a number of situations have been discussed in which energy is lost as heat. This occurs, for example, for furnaces (see Section 17.6a), where heat is carried away by exhaust gases, and for internal combustion engines, where heat is transferred to the cold reservoir of the heat engine. Thermoelectric generators are one way of potentially making use of this lost heat energy by generating electricity. The physics of these devices begins with a consideration of the electrical and thermal properties of materials.

Consider a bar of metal that is thermally connected to a hot reservoir at one end and a cold reservoir at the other end. Heat will be transported from the hot end to the cold end by conduction, as described in Chapter 8. Some of this heat will be carried by electrons flowing through the material. This flow of electrons will result in a potential difference between the two ends of the rod. This phenomenon is referred to as the Seebeck effect, and the potential difference may be expressed as

$$\Delta V = -S \Delta T$$

where S is the Seebeck coefficient, which is a quantity characteristic of the particular metal. The minus sign arises because the hot end of the rod will be at a higher potential than the cold end of the rod due to the negative charge of the electrons. If two dissimilar metals with different Seebeck coefficients, S_1 and S_2,

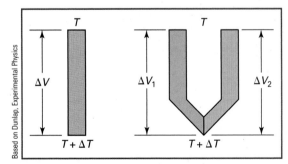

Seebeck effect in a rod of metal (left) and the creation of a thermocouple using two dissimilar metals (right).

are connected together as shown in the figure, then a voltage, V, between the two cold ends is produced, where

$$V = \Delta V_1 - \Delta V_2 = (S_2 - S_1)\Delta T.$$

The device described above is referred to as a thermocouple and is commonly used for measuring temperatures. In principle, it is also a device that can be used to convert temperature differences into electrical current. In practice, this device has low efficiency, because metals typically have high thermal conductivity, k, (see Section 8.3a) and high electrical conductivity, σ (i.e., they are good conductors). In general, the efficiency of a Seebeck device is related to its *figure of merit*, ZT, given by

$$ZT = \frac{\sigma S^2 T}{k}$$

and indicates that a good thermoelectric material should have a low thermal conductivity but a high electrical conductivity. It is also desirable to have a high Seebeck coefficient. Insulators typically have a low thermal conductivity but also a low electrical conductivity. Some semiconducting materials are the best compromise, as they have low thermal conductivity while maintaining reasonably high electrical conductivity.

A thermoelectric generator is a device analogous to the thermocouple but made from semiconducting materials: one piece of n-type material, with excess electrons, and one piece of p-type material, with excess holes (see Section 9.3) as the figure shows. The electrons (in the n-type material) and the holes in the p-type material) carry heat from the hot side of the device to the cold side of the device and produce the potential difference that drives the electrical current through the load.

Thermoelectric generators typically have efficiencies between about 4% and 7% for converting thermal energy to electrical energy, but despite this low efficiency, there are several potential

Continued on page 559

Energy Extra 17.4 continued

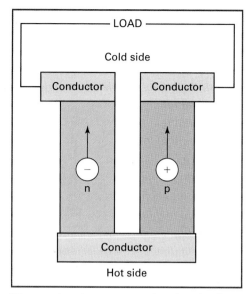

Thermoelectric generator showing the flow of electrons and holes from the hot side to the cold side.

applications for these devices. These applications basically fall into two categories:

- Use in remote locations where other energy harvesting methods are not appropriate

- Improvement of efficiency in devices with loose energy in the form of waste heat

In the first case, these devices have found use as a source of electrical power in space probes and in lighthouses above the Arctic Circle, where solar radiation is minimal. In these cases, the source of heat for the hot side of the device is often provided by the decay of a radioactive source.

Thermoelectric generators may also be utilized for the recovery of lost thermal energy from heating systems and industrial processes. However, in recent years their use for energy recovery from the exhaust of internal combustion engines in vehicles has been of interest. Several automobile manufacturers, including BMW in particular, have investigated this possibility. Electricity produced by the thermoelectric generator is used to help charge the vehicle's battery and to run electrical devices such as electronics, lights, and electric steering units. This reduces the load on the engine from the alternator and thus reduces fuel consumption. There are, however, drawbacks from this approach as well. Since exhaust gas must flow through a heat exchanger to extract thermal energy, there will be additional back pressure on the engine, and this can

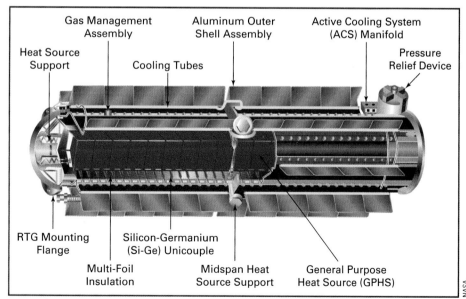

Thermoelectric generator module used by NASA on the Galileo, Ulysses, Cassini-Huygens, and New Horizon Space Probes.

Continued on page 560

Energy Extra 17.4 continued

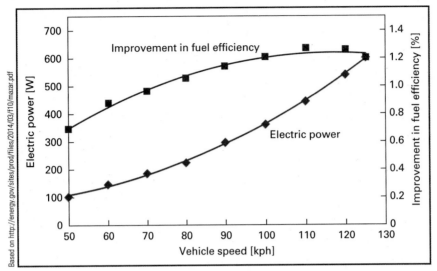

Test results for thermoelectric generator performance on a BMW X6.

adversely affect its performance. In order to maintain a suitable thermal difference across the device, fluid from the vehicle's cooling system is typically used to maintain the temperature of the cold side of the device. The weight of the thermoelectric generator system can be more than 50 kg for a passenger vehicle and more than 100 kg for an SUV or truck. This additional weight will increase fuel consumption. There is, therefore, a trade-off between the gains and losses of the system. Some results of the net gain of a thermoelectric generator are illustrated in the above figure.

Topic for Discussion

Benefits of the use of thermoelectric generators for the recovery of thermal energy from vehicle exhaust may be viewed as either environmental or economic. Make an estimate of the reduction in carbon emissions from a typical vehicle with a thermoelectric generator. If all internal combustion vehicles in the United States were equipped with thermoelectric generators, what would be the percent reduction in total carbon emissions? At present, the semiconducting materials that are used in thermoelectric devices tend to be fairly expensive, but research on the development of more economical materials is ongoing. What would be a reasonable expenditure for the addition of a thermoelectric generator to a passenger vehicle to yield an acceptable payback period based on improved fuel economy?

Recent and projected fuel economies in different countries worldwide is shown in Figure 17.29. The clear differences between Japan/European Union and the rest of the world are apparent in the figure.

In addition to the characteristics of a vehicle itself, the fuel economy that is actually realized also depends on drivers' habits. Energy-conserving driving behavior is promoted by many government energy policies (Section 17.2) and includes driving at moderate speeds on the highway, using cruise control, and minimizing idle time.

17.8b Hybrid Vehicles

Hybrid vehicles have become fairly common in recent years as an alternative to traditional gasoline-powered internal combustion engine vehicles. A traditional hybrid

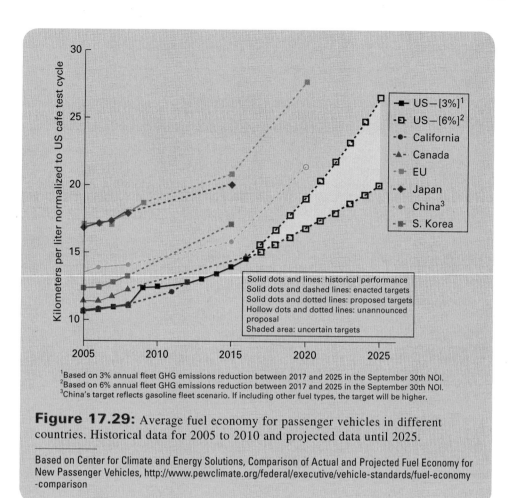

Solid dots and lines: historical performance
Solid dots and dashed lines: enacted targets
Solid dots and dotted lines: proposed targets
Hollow dots and dotted lines: unannounced proposal
Shaded area: uncertain targets

[1]Based on 3% annual fleet GHG emissions reduction between 2017 and 2025 in the September 30th NOI.
[2]Based on 6% annual fleet GHG emissions reduction between 2017 and 2025 in the September 30th NOI.
[3]China's target reflects gasoline fleet scenario. If including other fuel types, the target will be higher.

Figure 17.29: Average fuel economy for passenger vehicles in different countries. Historical data for 2005 to 2010 and projected data until 2025.

Based on Center for Climate and Energy Solutions, Comparison of Actual and Projected Fuel Economy for New Passenger Vehicles, http://www.pewclimate.org/federal/executive/vehicle-standards/fuel-economy -comparison

vehicle contains both an internal combustion engine and batteries that power an electric motor and is a step toward moving to a nonfossil fuel transportation technology. A gasoline engine is used to power the vehicle, and batteries drive an electric motor to provide supplemental power. The batteries (see the next chapter for more details on battery technology) are recharged by the gasoline engine and by energy obtained by regenerative braking. The first mass-produced hybrid was the Toyota Prius, which went on sale in Japan in 1997. The first hybrid to be sold in North America (beginning in 1999) was the Honda Insight (Figure 17.30 on the next page). Updated versions of both vehicles are currently on sale worldwide for a price of around US $20,000. Specifications for the first-generation Prius and Insight are given in Table 17.3 on the next page. The total power available at any time is somewhat less than the sum of the gasoline and electric outputs. These specifications suggest that these vehicles fall near the lower end of the performance scale for the majority of vehicles available in North America. In more recent years many family-sized passenger vehicles, such as the Honda Accord, and family-sized SUVs, such as the Volvo XC90, have become available in hybrid versions. Some high-end performance vehicles also use hybrid technology, primarily to increase total power to the wheels. For example, the gasoline engine in the Porsche 918 hybrid

Figure 17.30: A first-generation Honda Insight hybrid.

Richard A. Dunlap

(Figure 17.31) can provide a maximum of 447 kW while the two electric motors can provide an additional 205 kW.

As hybrids, like the Prius and the Insight, use gasoline as their primary fuel, they do not eliminate the dependency on fossil fuels, nor do they eliminate greenhouse gas emissions. They have improved fuel efficiency compared with similarly sized pure gasoline vehicles, although they may be more expensive due to the addition of batteries, an electric motor, and the associated systems. In a sense, they may be considered a transitional technology that helps to mitigate a problem (pollution, greenhouse gases, and dwindling fossil fuels) but cannot solve it.

Nearly all currently available hybrid vehicles fall under the category *parallel hybrids*, where the wheels are driven by both the gasoline engine and the electric motor. A more recent development is the *series hybrid vehicle*, where a gasoline engine is used exclusively to charge batteries and an electric motor is used exclusively to drive the vehicle's wheels. In this case, the gasoline engine can be operated at a constant speed and constant load, which provides greater efficiency than a gasoline engine that has to respond to fluctuations in driving conditions. Since the drive system in a series hybrid is essentially the same as in a pure electric vehicle and the gasoline engine is only used as a means of charging the battery, these vehicles are sometimes referred to as range-extended electric vehicles (or extended-range electric vehicles).

One of the earliest series hybrid vehicles was the Chevrolet Volt, as shown in Figure 17.32. The Chevrolet Volt is sold in North America, while the same vehicle is sold as the Holden Volt in Australia and New Zealand, the Vauxhall Ampera in the United Kingdom, and the Opel Ampera in the remainder of Europe. The Volt first went on sale in December 2010 and is still in production (as of 2017). A 1.4-L 4-cylinder gasoline engine is used to charge the batteries that power an electric motor to drive

vehicle	years	mass kg	gasoline power kW	electric power kW	fuel consumption L/100 km
Toyota Prius	1997–2001	~1200	43	30	5.6
Honda Insight	1999–2006	838	52	9.7	3.7

Table 17.3: Power output of the first-generation Toyota Prius and Honda Insight.

Figure 17.31: Porsche 918 hybrid.

eans/Shutterstock.com

the wheels. Under some circumstances the gasoline engine may also be used to drive the wheels directly, as in a traditional parallel hybrid design, so technically the Volt is a series/parallel hybrid and not a pure series hybrid. This configuration is sometimes called a *power-split hybrid* design; a similar approach has been used in the Toyota Prius since 1997. In the electric mode running on batteries only, the current version of the Volt has a range of 85 km. When the gasoline engine is utilized in conjunction with the batteries, the range is limited by the size of the gasoline fuel tank and is about 680 km.

The Fisker Karma (see Figure 17.33 on the next page) was a true series hybrid manufactured in Finland and sold from July 2011 to November 2012 (after which the company went bankrupt). The Karma utilized a General Motors 2.0 L turbo-charged 4-cylinder engine to charge the batteries. Its all-electric range is 51 km, and the total range using the batteries and gasoline engine is 370 km. The rights to the Karma design were purchased by the Chinese company Wanxiang, and an updated version of the Karma is now available under the name Karma Revero. Many of the Fisker Karma design features have been retained, including a photovoltaic roof

Figure 17.32: Chevrolet Volt (series/parallel hybrid).

6th Gear Advertising/Shutterstock.com

Figure 17.33: Fisker Karma series hybrid.

Fedor Selivanov/Shutterstock.com

(see Figure 17.34) that is able to provide an electric range of about 2.4 km on a one-day recharge. The current price of the Karma Revero is about $130,000 US.

A more affordable series hybrid, available since November 2013, is the BMW i3 with Range Extender (Figure 17.35), currently priced at about $46,000 US. The basic i3 is a battery electric vehicle (see Section 19.4 for more information on the electric BMW i3), and the Range Extender option adds a 647-cc 2-cylinder gasoline engine (used in the BMW C650GT Maxi-Scooter) to convert the vehicle to a series hybrid configuration. The gasoline engine increases the total range to about 300 km compared to a range of about 160 km for the pure electric version of the i3.

Nissan is currently developing a series hybrid system (called e-Power) that utilizes a 1.2-L 3-cylinder engine from the Nissan Micra. It is expected that Nissan will market a vehicle based on the e-Power system in the near future.

Figure 17.34: Solar roof on the Karma Revero.

ZUMA Press, Inc./Alamy Stock Photo

Figure 17.35: The BMW i3 is an electric vehicle with an optional gasoline engine that operates in a series hybrid configuration.

Art Konovalov/Shutterstock.com

Many hybrid vehicles now available are referred to as *plug-in hybrids*. This terminology refers to a hybrid gasoline/electric vehicle with batteries that can be charged by plugging the vehicle into the grid. These vehicles are also called *grid-connected hybrids* or *gas-optional hybrids*, since the batteries can be charged from the grid and the vehicle driven without the need for the gasoline engine. Since hybrids typically have battery packs with much smaller capacity than those in pure electric vehicles (see Chapter 19), the range in the gas-optional mode is also typically much less.

Although hybrids do not eliminate the need for fossil fuels, advances in battery technology and electrical components associated with the development of hybrid vehicles may have a direct benefit in the design of a new generation of battery electric vehicles and fuel cell vehicles, as described in Chapters 19 and 20.

17.9 Summary

Energy conservation is an essential component of a viable long-term plan to provide sustainable energy, and various aspects of its effective implementation are discussed in this chapter. Options for energy conservation are available on a variety of different scales, from the implementation of national energy policies by federal governments to what individuals can do in their own homes.

While scientific and technological advances can lead to improved conservation methods in the future, substantial technology already exists that can be implemented to aid in conservation efforts. As discussed in this chapter, the development of national and regional energy policies that include conservation are needed to provide guidelines for actions that promote the implementation of positive conservation measures. On a regional level, the most effective approaches to conservation are often related to factors such as climate, geography, and social traditions and economy. Regional and municipal governments, as well as public utilities, can do much to further conservation efforts in areas such as cogeneration, smart grid implementation, and street lighting.

Cogeneration refers to the direct use of excess heat from the thermal generation of electricity for space heating needs. This approach has been particularly effective for small generating facilities, which are appropriate for universities, large industrial plants, or isolated communities.

The implementation of a smart grid has been presented. The smart grid is an essential component for the efficient integration of renewable energy sources with tradition fossil fuel and nuclear generating facilities and for the effective regulation of electricity use in the community.

The principle of operation of LED lamps follows from the description of photovoltaic devices presented in Chapter 9. Whereas photovoltaic cells convert photons into a voltage, LEDs convert voltage into light. This chapter described the use of LEDs for municipal roadway lighting. Because municipal governments can pay as much as 40% of their budgets on street lighting, and LEDs lamps can provide a 60% savings in electricity consumption over traditional lighting, a net savings of up to a quarter of a municipality's budget can be realized by converting to a more energy-efficient lighting technology. Additional benefits include improved illumination and the elimination of the potentially toxic substances used in many traditional lamps.

This chapter also presented ways in which individuals can reduce home energy use by the use of efficient heating, ventilation, and air conditioning (HVAC) systems, the reduction of heat loss by efficient insulation, and the reduction of electricity consumption for lighting through the use of compact fluorescent lamps or LED lamps.

Finally, this chapter overviewed efforts to conserve energy associated with transportation. Government standards for vehicle fuel economy, present in most countries since the 1970s, have reduced the energy requirements of passenger vehicles and trucks and have provided a substantial reduction in the consumption of fossil fuels. This chapter also reviewed the viability of hybrid vehicles. Traditional parallel hybrid passenger vehicles offer improved fuel economy compared with their gasoline powered counterparts. More recently the availability of series hybrid vehicles is a step closer to the pure electric vehicles discussed in Chapter 19.

Problems

17-1 $R = 0.18$ windows in a house are replaced with $R = 0.52$ windows at a cost of $250 per m^2. Assume that the outside temperature is a constant 5°C [a reasonable approximation for a region corresponding to about 4300 degree days per year (°C)] and that heat costs $0.03 per MJ. How long will it take to recover the cost of the window replacement? (Do not include the cost recovery factor.)

17-2 The following table gives specifications for some 2012 automobiles sold in the United States.

automobile	average gasoline consumption (L/100 km)	base price (US$)
Honda Civic DX (automatic)	6.9	16,605
Honda Civic Hybrid	5.3	24,134
BMW 750i	13.1	84,300
BMW Active Hybrid 750i	11.7	97,000

If it is assumed that maintenance costs are the same for the gasoline and hybrid versions of the same vehicle, use the current price of gasoline in your area to determine how many miles would need to be driven for the better fuel economy of the hybrid to outweigh its higher purchase price. Repeat this calculation for the BMW.

17-3 Refer to Figure 2.6. What fraction of waste heat from electricity generation in the United States would be needed to satisfy all of the country's residential fossil fuel use (primarily for heating)?

17-4 A homeowner in Maine replaces six 60-W incandescent bulbs in a family room with equivalent CFLs. Using the following information, estimate the net annual energy savings (in dollars) for lighting and heating. The lamps are on an average of 3.5 hours per day, and electricity costs $0.105/kWh. The home is heated with an oil furnace at an efficiency of 87%, and home heating fuel costs $0.74 per liter. Heat is required 191 days per year, and there is no air conditioning.

17-5 In a local store, find the price of a 60-W incandescent bulb and CFL and LED bulbs with an equivalent light output. Based on a use of 4 hours per day and an electricity cost of $0.11/kWh, calculate the payback period for each of these bulbs compared to the incandescent.

17-6 Compare the overall efficiency of heating a house with oil at 85% efficiency and heating a house with a heat pump with a coefficient of performance of 6 using electricity generated by a heat engine at 35% efficiency.

17-7 Consider the heat losses through the four exterior walls of a house. The house is 9.0 m × 12.0 m, and the walls are 3.0 m high. There are 12 windows, each 1 m × 1.6 m. The walls are uninsulated and have an R-value of $R = 1.0$ and the windows are (uncoated) single pane. The homeowner has the option of either upgrading the windows to (uncoated) triple pane or introducing insulation into the walls to increase their R-value to $R = 3.5$. Which action will provide the greater benefit? In this problem, ignore doors and heat losses through the floor and roof.

17-8 Consider typical passenger automobiles in the United States. Calculate the reduction in CO_2 emissions (in tonnes) for a 2008 vehicle compared to a 1966 vehicle during the lifetime of the vehicle (assumed to be 250,000 km).

17-9 (a) A large North American city can have 100,000 street lamps. If these are typically 200-W lamps and electricity costs $0.08 per kWh, what is the annual electric bill if the lamps are on an average of 12 h per day?

   (b) What would be the capital cost to replace 100,000 lamps with 75-W LED lamps at a cost of $600 per lamp?

   (c) What would be the simple payback period (i.e., do not include capital recovery factor, interest, etc.) for this conversion?

   (d) Estimate the total CO_2 emission reduction over the payback period if the electricity is generated exclusively by coal-fired stations.

17-10 A heat pump operates with an outside temperature of −10°C and an inside temperature of +19°C. If the building requires 1.6 GJ of heat per day, what is the average power requirement in kW for the heat pump?

17-11 A house has 155 m² of exterior walls consisting of 2.9 cm of exterior wood, 15 cm of insulation, and 2.0 cm of interior wood.

    (a) On a winter day when the interior temperature is 20 °C and the exterior temperature is −2°C, what is the daily heat loss through the walls if they are insulated with fiberglass wool?

    (b) What is the percent reduction in heat loss if the fiberglass wool is replaced with polystyrene foam sheets?

17-12 A coal-fired cogeneration station provides electricity and heat to a community of 1000 homes.

    (a) Each home requires a constant 2.5 kW_e, and the generation of electricity is 40% efficient. If the cogeneration station maintains an output that is just sufficient to satisfy the electrical needs of the community, what is the available total power associated with the waste heat?

    (b) From part (a), what is the daily coal requirement in kg?

    (c) Each house has a volume of 450 m³ and maintains an inside temperature of 18.3°C. If 90% of the waste heat is available for heating purposes, what is the minimum (constant) outside temperature for which the cogeneration station can fulfil the heating needs of the community while just satisfying the electrical needs? See Chapter 8 for additional information on heating requirements.

17-13 For Figure 17.3, estimate the energy gaps for each of the three LEDs in the white LED package.

17-14 From equations (17.4) and (17.5), solve for T_c (i.e., the minimum temperature at which a heat pump can be effective) in terms of R, A, P_{in}, and T_h.

17-15 (a) A future homeowner is specifying construction materials for a new home in Edmonton, AB. The home is heated with electric heat at a cost of $0.11 per kWh. There are 200 m² of exterior walls consisting of 5 cm of wood with space for 15-cm-thick insulation. If fiberglass loose-fill insulation would cost $2500 and polystyrene foam sheet insulation would cost $4000, what is the payback period for the additional investment in polystyrene?

    (b) Repeat part (a) for the same home being built in San Francisco, CA.

    See Chapter 8 for additional useful information.

17-16 (a) From Figure 17.4, estimate the energy gap of a GaN or InGaN LED.

    (b) Are there other semiconducting materials that could be used (from an energy standpoint) to induce photon emission in a Ce:YAG phosphor?

17-17 (a) A 9-W LED lamp produces light at an intensity comparable to a 60-W incandescent bulb. If the lifetime of the bulb is 30,000 hours and electricity costs $0.12 per kWh, what is the total savings in energy costs over the life of the LED?

(b) If the LED is used an average of 5 hours per day, how long will it take to realize these savings? What is the average savings per day?

17-18 (a) An old, 68%-efficient natural gas furnace is used to heat a 400 m^3 home in Boston, MA. Estimate the annual savings in CO_2 emissions if the furnace is replaced with a 93%-efficient natural gas furnace?

(b) Estimate the annual CO_2 savings if the old furnace is replaced with an 80%-efficient coal-fired furnace.

17-19 Compare the primary energy requirement per km travelled for an average 1970 gasoline passenger vehicle in the United States to that for an average 2014 passenger vehicle.

17-20 One approach to cogeneration (electricity and heat) for a community is to design the facility to satisfy heating needs and to import or export electricity as needed. This approach is based on the philosophy that it is much easier to market excess electricity (or purchase it as needed) than to sell excess heat. Consider a typical natural gas–fired cogeneration facility that produces electricity at an efficiency of 40% and has 50% of the waste energy available for heating purposes. Discuss the validity of this approach for a northern community (near Anchorage, AK) compared to a southern community (near New Orleans, LA). Provide quantitative comparisons to support your evaluation. Are there other approaches to cogeneration that would be more reasonable for one, or both, of these locations?

Bibliography

B. Anderson and M. Riorden. *The New Solar Home Book*, (2nd ed.) Brick House Publishing, Andover, MA (1996).

B. Anderson and M. Wells. *Passive Solar Energy: The Homeowner's Guide to Natural Heating and Cooling*, (2nd ed.) Brick House Publishing, Andover, MA (1996).

A. K. Athienitis and M. Santamouris. *Thermal Analysis and Design of Passive Solar Buildings.* James & James, London (2002).

D. Chiras. *The Solar House: Passive Heating and Cooling.* Chelsea Green Publishing, White River Junction, VT (2002).

Natural Resources Canada. *Heating and Cooling with a Heat Pump.* Cat. No. M144-51/2004 (2004).

V. Quaschning. *Renewable Energy and Climate Change.* Wiley, Chichester (2010).

U.S. Department of Energy. *Insulation Fact Sheet.* DOE/CE-0180 (2008).

Energy Storage

Learning Objectives: After reading the material in Chapter 18, you should understand:

- The need for energy storage.
- Pumped hydroelectric storage and its use.
- The properties of compressed air.
- The use of compressed air for energy storage.
- Energy storage capabilities of flywheels and the relevance of materials properties.

- Properties of superconductors and the history of their development.
- The use of superconducting magnets for electrical storage.

18.1 Introduction

Storage is a major component of our energy use. Energy storage is necessary because it is sometimes inconvenient or impossible to produce energy when it is needed or where it is needed. The first instance may arise because of variations in demand compared with capacity (i.e., daily, weekly or seasonal fluctuations in electricity use) or inherently variable production methods (i.e., solar, wind, tidal, etc.). The second instance is most obvious in the case of energy needs for transportation. It must be realized that the storage of any form of energy often involves the conversion of one form of energy to another (and back again) and that any conversion is less than 100% efficient. So energy storage means energy loss, and this must be considered in the analysis of the viability of any particular method.

Chapter 8 presented some information about the storage of thermal energy. It is also important to consider other energy storage possibilities, particularly for the storage of electricity. Chapter 19 considers electrical storage using batteries, particularly with respect to electric vehicles, and Chapter 20 considers the possibility of using hydrogen as a method of storing energy. This chapter considers various other energy storage methods aimed at large-scale electricity storage for the grid.

Jim West/Alamy Stock Photo

18.2 Pumped Hydroelectric Power

The typical demand for electricity as a function of time throughout a week is shown in Figure 18.1. There are also longer-term seasonal variations. It is impractical to have generating facilities that can routinely provide power that can meet the peak demand. One approach is to have supplementary generating facilities that can be brought online quickly when the need arises. These might be combustion turbines used to supplement electricity generation by a coal-fired or nuclear power plant. Another approach is to store excess energy produced during periods of low demand and use this when demand exceeds capacity.

Pumped hydroelectric power is a well-established technology that is in common use by public utilities to account for variations in electricity demand. It was first used extensively in the 1930s and remains the most economical and practical method for large-scale electricity storage. The general design of a pumped hydroelectric facility is shown in Figure 18.2 on the next page. Excess electricity is used to pump water from the lower reservoir to the upper reservoir. In periods of high demand, the water is allowed to flow back into the lower reservoir to generate electricity by means of a

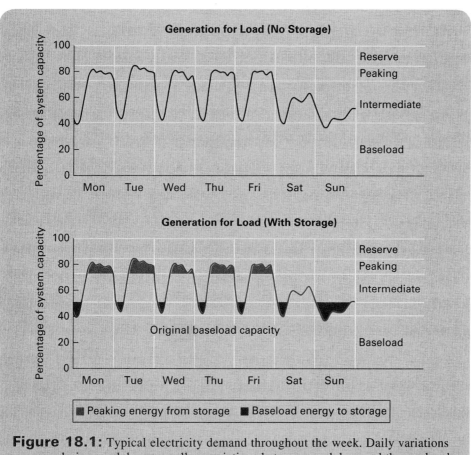

Figure 18.1: Typical electricity demand throughout the week. Daily variations are seen during weekdays as well as variations between weekdays and the weekend.

Based on National Academy of Sciences

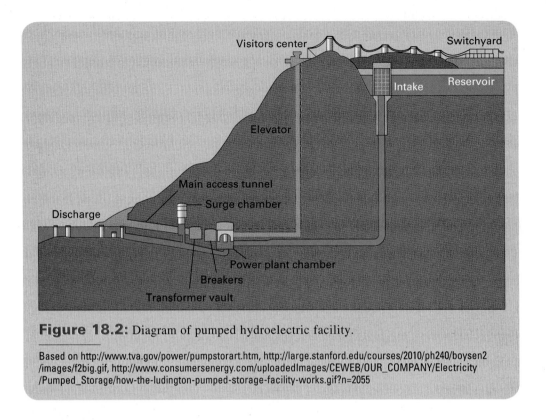

Figure 18.2: Diagram of pumped hydroelectric facility.

Based on http://www.tva.gov/power/pumpstorart.htm, http://large.stanford.edu/courses/2010/ph240/boysen2 /images/f2big.gif, http://www.consumersenergy.com/uploadedImages/CEWEB/OUR_COMPANY/Electricity /Pumped_Storage/how-the-ludington-pumped-storage-facility-works.gif?n=2055

turbine/generator. Most commonly, turbines such as a Francis turbine are used reversibly both as the pump and to drive the generator. The energy stored in water in the upper reservoir is given by

$$E = mgh, \tag{18.1}$$

where m is the mass of the water, g is the gravitational acceleration, and h is the average head. The power generated is

$$P = \frac{dE}{dt} = gh\frac{dm}{dt} = \rho gh\frac{dV}{dt}, \tag{18.2}$$

where ρ is the water density, and (dV/dt) is the flow rate (in volume of water per unit time). It might seem that there is a trade-off between head and flow rate and that the same power can be achieved in a high-flow low-head facility as in a low-flow, high-head facility. In practice high head is preferred because, to achieve a large enough flow in a low-head facility, the pipe may have to be of an impractical size. Also, to maintain power output for an extended period of time, the total volume of a low-head facility must be greater. Pumped hydroelectric facilities are therefore most practical in geographic locations that provide for sizable reservoirs that are relatively close together but furnish a sufficient head. Although this is most commonly achieved where only minor modifications of the natural geography are needed, it is also possible to utilize underground chambers (e.g., unused mines) for the lower reservoir. Although one might view pumped hydroelectric generation as related to normal hydroelectric generation, the geographic requirements are not necessarily compatible, and the electricity stored in a pumped hydroelectric facility need not be produced by any particular method or in particularly close proximity to the storage facility. In fact, pumped hydroelectric storage is most suitable as a means of electricity storage in conjunction with generating methods that can be naturally variable, such as solar or wind.

Example 18.1

A pumped hydroelectric storage reservoir has an area of 0.5 km^2 and an average depth of 10 m. The water flows to a lower reservoir with a head of 100 m. What is the total energy available in megawatt-hours?

Solution

The gravitational energy stored in the water is given by

$$E = mgh.$$

The total mass of the water stored in the upper reservoir is given by its volume and the density of water:

$$m = \rho V = (1000 \text{ kg/m}^3) \times (0.5 \times 10^6 \text{ m}^2) \times (10 \text{ m}) = 5 \times 10^9 \text{ kg.}$$

The energy stored is therefore

$$E = (5 \times 10^9 \text{ kg}) \times (9.8 \text{ m/s}^2) \times (100 \text{ m}) = 4.9 \times 10^{12} \text{ J.}$$

Converting to MWh gives

$$E = (4.9 \times 10^{12} \text{ J}) \times (2.78 \times 10^{-10} \text{ MWh/J}) = 1361 \text{ MWh.}$$

Pumped hydroelectric storage has fairly substantial infrastructure costs, but the maintenance and operating costs are quite low. Also, it is fairly low technology and quite reliable. It can be brought online quickly in response to demand variations. The lifetime of the storage is long, although not indefinite, because of natural seepage and evaporation. The efficiency is quite high, reaching about 80%. This is a result of the intrinsically high efficiency of converting the mechanical energy of falling water to electricity and the high efficiency of water pumps, although friction in the pipes decreases the efficiency by a small amount.

Example 18.2

For the pumped hydroelectric example in Example 18.1, how long could the energy stored in this facility provide electricity for a town of 50,000 people if the average power requirement for each person is 800 W$_e$? Assume a generator efficiency of 90%.

Solution

The town of 50,000 people uses a total of

$$(50,000) \times (800 \text{ W}_e) = 40 \text{ MW}_e.$$

At 90% efficiency the pumped hydroelectric facility can produce a total of

$$(1361 \text{ MWh}) \times (0.9) = 1225 \text{ MWh}_e.$$

The total time that the facility can supply the required electricity is

$$\frac{1225 \text{ MWh}_e}{40 \text{ MW}_e} = 30.6 \text{ h.}$$

Pumped hydroelectric is the most commonly used storage method for large-scale electricity storage for the grid. It is based on a well-established, reliable technology and is reasonably efficient. There is the potential for large energy storage capacity and readily available high power output. Although infrastructure costs can be high, there is relatively little maintenance.

The largest pumped hydroelectric facility in the world is the station in Bath County, Virginia, constructed between 1977 and 1985 at a cost of US $1.6 billion. The facility consists of two reservoirs separated by a vertical distance of 385 m. The lower reservoir has an area of 2.25 km^2, and the upper reservoir has an area of 1.07 km^2. The water flows between the reservoirs through three tunnels, each approximately 6 km in length. Each tunnel splits into two 5-m-diameter penstocks (for a total of six) before reaching the turbines. The map below shows the area around the facility and the locations of the upper and lower reservoirs. When demand exceeds the base load grid supply, water flows from the upper reservoir to the lower reservoir through six Francis-type turbines (at up to 850 m^3/s, generating a maximum of 2772 MW$_e$ for up to 11 hours. When electric demand is low, the generators function as electric motors, producing up to 480 MW to drive the turbines backward and pump water from the lower reservoir to the upper reservoir at a rate of 800 m^3/s.

A major advantage of pumped hydroelectric power to supplement base load generation compared to, say, bringing additional thermal generation online is response time. The Bath County station can bring 400 MW online just 6 minutes after start-up to add to base load capacity during periods of high demand.

Topic for Discussion

It may seem natural to think of pumped hydroelectric power in the context of normal hydroelectric power. However, pumped hydroelectric power is, in many ways, more relevant to other forms of electricity generation. Energy technologies such as wind, solar, tidal, and the like are the most variable in time, and most require an energy storage mechanism to best utilize their potential. Although it is not absolutely necessary for the facility that generates electricity and the facility that stores it to be in close proximity, it is convenient if they are. Discuss the ways in which the geographical requirements for various alternative energy technologies and the geographical requirements for pumped hydroelectricity are compatible or incompatible.

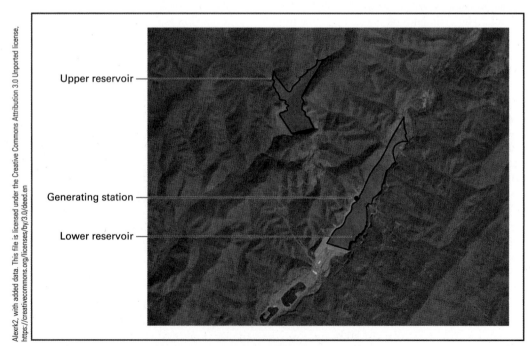

Upper reservoir

Generating station

Lower reservoir

Map of area around the Bath County pumped hydroelectric facility.

Table 18.1: Pumped hydroelectric storage in different regions of the world in 2015.

region	pumped hydroelectric capacity (MW$_e$)
Africa	1580
East Asia	57,999
Europe	50,949
Latin America and Caribbean	1004
Middle East and North Africa	1744
North America	2268
South and Central Asia	6146
Southeast Asia and Pacific	2425
world total	**144,465**

Based on data from U.S. Department of Energy, Global Energy Storage Database

Total pumped hydroelectricity capacity is in excess of 80 GW worldwide. This represents about 300 facilities that range from a few megawatts up to more than 2 GW. Table 18.1 summarizes the distribution of pumped hydroelectric storage capacity worldwide. These data indicate that pumped hydroelectric storage has been developed extensively in East Asia and Europe, where in the latter case it represents about 5.5% of the base load capacity. Less extensive development has taken place in North America, and in the United Stated pumped hydroelectric storage represents about 2.5% of base load capacity.

18.3 Compressed Air Energy Storage

Compressed air is another mechanism for storing energy. In one sense, it can be viewed much like pumped hydroelectric power, as a means for large-scale energy storage. It can also be viewed as a more portable power source that could compete in some ways with batteries or as a means of distributing power through pipelines that might be analogous to electricity. While current ideas concerning the use of compressed air fall primarily into the first two categories, compressed air was put to considerable use at the end of the 1800s as a means of distributing power. In the 1890s, Paris had more than 50 km of compressed air lines across the city that provided mechanical power for industrial applications, sewing machines, printing presses, and other machines.

For the purpose of storing energy in a compressed gas, it can be shown that (ideally) the energy required to change the pressure of a gas from P_i to P_f is

$$E/V = nRT \cdot \ln\left(\frac{P_{max}}{P_{min}}\right), \tag{18.3}$$

where n is the number of moles of gas, R is the ideal gas constant, and T is the absolute temperature (in Kelvin). Analogously, this equation gives the ideal amount of energy that can be extracted by reducing the pressure of the gas from P_{max} to P_{min}. Within certain limits of temperature and pressure, the energy stored in compressed air per unit volume can be approximated as

$$E/V = P_{max} \cdot \ln\left(\frac{P_{max}}{P_{min}}\right).$$ **(18.4)**

The derivation of these expressions assumes that the gas is compressed or expands isothermally, that is, without change in temperature. In this case (ideally), 100% of the energy input into compressing a gas can be extracted by allowing the gas to expand. Thus, a compressor can be used to compress gas, and a turbine can be used to convert the energy associated with the compressed gas back into mechanical or electrical energy, thereby using the gas as a mechanism for storing the energy. Unfortunately, in practice, when a gas is compressed, some of the energy that is input into the system goes into heating the gas (rather than compressing it). This kind of compression is called *adiabatic* and results in a loss of some of the energy. These losses are minimized if the gas is compressed and expands slowly, although some frictional losses are always associated with the compressor and turbine.

Compressed air is a reasonably practical method for large-scale energy storage that parallels pumped hydroelectric storage. A schematic of a typical system is shown in Figure 18.3. Heat exchangers are used to transfer heat away from the gas during compression and into the gas during expansion.

At present, two commercial facilities of this type are in operation: one in Huntorf, Germany, and one in Alabama. These are relatively small scale compared with pumped hydroelectric facilities and have capacities of 290 MW and 110 MW for the European and North American plants, respectively. Both facilities utilize natural underground caverns for gas storage.

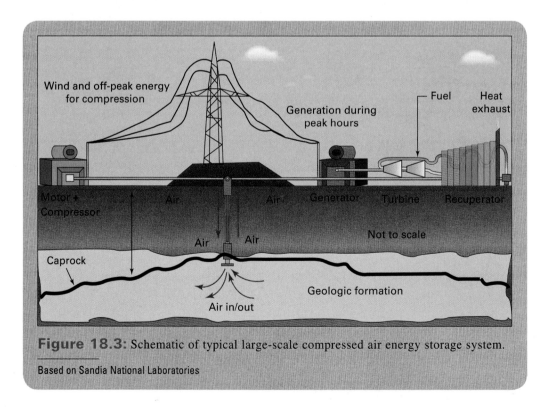

Figure 18.3: Schematic of typical large-scale compressed air energy storage system.

Based on Sandia National Laboratories

Example 18.3

Estimate the theoretical maximum energy that can be stored by compressed air in a $10 \times 10 \times 10$ m^3 chamber at a pressure of 5 MPa.

Solution

From equation (18.4),

$$\frac{E}{V} = P_{max} \cdot \ln\left(\frac{P_{max}}{P_{min}}\right).$$

Since standard atmospheric pressure is 0.1013 MPa, equation (18.4) gives the energy stored per unit volume (m^{-3}) as

$$\frac{E}{V} = 5 \text{ MPa} \cdot \ln\left(\frac{5 \text{ MPa}}{0.1013 \text{ MPa}}\right) = 19.5 \text{ MJ/m}^3.$$

For a volume of $10 \text{ m} \times 10 \text{ m} \times 10 \text{ m} = 10^3$ m^3, the total stored energy will be $(19.5 \text{ MJ/m}^3) \times (10^3 \text{ m}^3) = 19.5$ GJ.

The Huntorf facility was constructed in 1978 and has been in operation since then. An aerial photograph of this facility is shown in Figure 18.4. It utilizes two underground caverns with a total usable volume of 300,000 m^3 (equivalent to a cube about 70 m on a side). The caverns extend from 650 m underground at the top to 800 m at the bottom. The natural overburden of rock allows for a maximum operating pressure of 10^7 Pa. The construction of an above-ground storage tank of that volume and pressure capabilities would be impractical. The compressors require about 12 hours to fully compress the air within the caverns, and the turbines can produce up to 290 MW for up to 3 hours (total 870,000 kWh) as the air is released from the caverns.

The McIntosh facility in Alabama was constructed in 1991 and has improved efficiency over the Huntorf facility due to an improved heat recovery system. It uses a much larger underground cavern (about 5×10^6 m^3), and although it produces less power, it can supply power for up to 26 hours.

Figure 18.4: Huntorf compressed air energy storage facility.

Sandia National Laboratories and Strategen Consulting, LLC

18.3a **Implementation of Compressed Air Energy Storage**

Although both operational compressed air energy storage facilities have performed very well for many years, the implementation of this approach to energy storage requires the availability of fairly specific geological resources. There are proposals for the more widespread application of compressed air energy storage using underwater reservoirs. Large underwater plastic bags could be used for gas storage at relatively low pressures. Open-bottomed containers can also be used to store energy associated with the displacement of water by compressed air inside the container. Underwater approaches to compressed air energy storage are most appropriate for deep lakes and oceans where the depth of the water provides the pressure needed to compress the gas.

On a much smaller scale, compressed air has been considered as a possible energy carrier for vehicles. Compressed air from a high-pressure cylinder can be used to run a piston or turbine engine for vehicle propulsion. This approach has been utilized on a limited basis for well over 100 years. It is also nonpolluting, as long as the air is compressed using energy that is produced by nonpolluting methods.

Theoretically, if 1 m^3 of air is compressed into a cylinder with a volume of 5 L, it will have a pressure of 2×10^7 Pa and can store 530 kJ [according to equation (18. 4)]. From a practical standpoint, only about half of this can actually be achieved as usable energy to propel a vehicle. A suitable cylinder made of advanced materials (i.e., Kevlar or carbon fiber) would weigh about 2 kg, and 1 m^3 of air weighs about 1 kg. Thus a compressed air energy storage system has an energy density of about $(0.5 \times 530 \text{ kJ})/(2 \text{ kg} + 1 \text{ kg}) = 90$ kJ/kg. As a means of comparison with traditional internal combustion engines, the energy density of gasoline is 45 MJ/kg. At an efficiency of, say, 20% for the conversion of the chemical energy associated with gasoline to usable mechanical energy, gasoline, as an energy carrier, could provide about 10 MJ/kg of fuel (assuming the weight of the gas tank is relatively small). This comparison implies severe limitations for the range of a compressed air vehicle.

An analysis of the practicality of compressed air energy storage for transportation should also include a comparison with the more common alternative approach of using batteries to power an electric vehicle (Chapter 19). On the positive side, compressed air systems use straightforward technology involving nontoxic materials, and recharging the fuel supply in the vehicle can be done very quickly. On the negative side, the energy density that can be achieved in a compressed air system is comparable (at best) to rather low-tech batteries (i.e., lead acid) and is much less than that for currently available advanced batteries. Also, the safety of compressed air cylinders during accidents needs to be considered.

18.4 **Flywheels**

A *flywheel* is a mechanical device with a substantial moment of inertia. They have been used for many years as a method of evening out fluctuations in the rotational speed of a shaft that are caused by varying torque. This situation is common for, say, piston engines where the driving force (ignition of gasoline) is periodic. In recent years, interest has developed in using flywheels as a mechanism for the storage of energy.

Objects that are in motion have a kinetic energy associated with their movement given in terms of their mass, m, and velocity, v:

$$E = \frac{1}{2} mv^2. \tag{18.5}$$

Table 18.2: Constants, k, from equation 18.5 for different geometries.	
geometry	k
disk	1/2
ring	1
solid sphere	2/5
spherical shell	2/3

Objects that are rotating have a kinetic energy associated with their rotational motion given by

$$E = \frac{1}{2} I\omega^2, \qquad \textbf{(18.6)}$$

where I is the moment of inertia, and ω is the angular frequency of rotation. This is related to the frequency of rotation in revolutions per second, f, as

$$\omega = 2\pi f. \qquad \textbf{(18.7)}$$

The moment of inertia is related to the mass and the geometry of the object. For objects with axial (cylindrical) symmetry, the moment of inertia is related to the mass and radius of the object as

$$I = kmr^2. \qquad \textbf{(18.8)}$$

The constant k is related to the details of the geometry and is given for some common simple shapes in Table 18.2. These equations can be combined to give

$$E = 2\pi^2 kmr^2 f^2. \qquad \textbf{(18.9)}$$

The rotational energy of an object is greatest when the rotational frequency, the mass, and the radius are maximized and the geometry is optimized to give the largest value of k. Table 18.2 shows that the most advantageous geometry for a flywheel is a ring. This ensures that the mass is distributed as far away from the axis of rotation as possible. This has been a common design for flywheels used in the past to even out the load on a rotating shaft (Figure 18.5).

Figure 18.5: Steam engine with rim-loaded flywheel.

Richard A. Dunlap

Obviously from Example 18.4, a bicycle wheel would not make a very useful energy storage device, at least not at rotational frequencies that are typical for a bicycle wheel in normal use. We could, of course, rotate the wheel at a higher frequency, but there is a limit to how rapidly the wheel can rotate without sustaining damage. A more massive example might be a steel wheel on a freight train (which is approximately a solid disk with a typical mass of $m = 400$ kg and a typical diameter of $d = 0.85$ m (Example 1.3). The rotational energy content of such a wheel rotating at a frequency typical of a moving freight train would illuminate the 60-W bulb for about half an hour. A practical energy storage would require larger, more massive, and/or faster rotating flywheels.

Example 18.4

Consider an ideal bicycle wheel, all of whose mass is located in a thin ring at the edge with a diameter of 0.7 m and a mass of 1.2 kg. If the wheel has an initial rotational frequency of 2 revolutions per second (equivalent to a bicycle traveling at a speed of about 4 m/s), calculate the length of time the rotational energy content of the wheel could illuminate a 60-W lightbulb (at 100% conversion efficiency).

Solution

From equation (18.9),

$$E = 2\pi^2 kmr^2f^2.$$

Substitute the following values in the equation: $k = 1$ (as appropriate for a ring of mass given in Table 18.2), $m = 1.2$ kg, $r = 0.35$ m, and $f = 2$ s^{-1}. This gives

$$E = 2 \times (3.14)^2 \times (1) \times (1.2 \text{ kg}) \times (0.35 \text{ m})^2 \times (2 \text{ s}^{-1})^2 = 11.6 \text{ J}.$$

A 60-W lightbulb uses 60 J per second, so that 11.6 J will illuminate the bulb for

$$t = \frac{11.6 \text{ J}}{60 \text{ W}} \approx 0.2 \text{ s}.$$

From a practical standpoint, it is necessary to consider the strength of the material that the flywheel is made of. As the rotational speed of the wheel increases, the internal stresses that the wheel experiences also increase, and at some point the material fails. It turns out that optimizing k may result in unacceptable consequences in the strength of the flywheel because the spokes that support the heavy, loaded rim may not be able to withstand the stresses at high rotational speeds. The mechanical properties of the flywheel material are therefore of fundamental importance for the design. A simple quantitative approach to flywheel design considers the stress at the rim of a rotating circular object. Here the stress is the greatest because the actual velocity is the greatest. The rim stress can be expressed as

$$\sigma = \rho r^2 \omega^2, \tag{18.10}$$

where ρ is the density of the material. This equation may be combined with equations (18.6) and (18.8) to give

$$E = \frac{1}{2} km \left(\frac{\sigma}{\rho} \right). \tag{18.11}$$

Table 18.3: Densities, ρ, and tensile strengths, σ, and their ratios for some possible flywheel materials.

material	density (kg/m³)	tensile strength (N/m²)	σ/ρ (N · m/kg)
steel	7800	1.72×10^9	0.22×10^6
aluminum	2700	0.59×10^9	0.22×10^6
titanium	4500	1.22×10^9	0.27×10^6
fiberglass	2000	1.60×10^9	0.80×10^6
carbon fiber–reinforced polymer	1500	2.40×10^9	1.60×10^6

The maximum amount of energy that is possible is given by this expression when the stress σ is the breaking stress for the material (i.e., the tensile strength). Thus, for a flywheel of a given mass, m, and geometry (resulting in a value of k), the energy storage capability is optimized by maximizing (σ/ρ). The values of this quantity are given for some possible flywheel materials in Table 18.3. It is clear from this table that the use of advanced materials enables flywheels to store more energy than traditional metal designs.

From the numbers in the table and equation (18.11), it can be calculated that a practical-sized flywheel of a few hundred kilograms could store a maximum of about a hundred megajoules of energy (assuming a safety factor to prevent failure) at rotational rates of 50,000 to 60,000 rpm. If a motor/generator system is used for inputting and extracting energy, the efficiency should be quite good (typically 80–90%), and the flywheel could be used to provide, say, a few tens of kilowatts of electricity for a few hours. Although this is considerably less than pumped hydroelectric or compressed air systems could provide, the flywheel system is, relatively speaking, quite compact and may be useful for some specific applications. A number of flywheels can be combined to increase total energy storage capabilities. One possible use for such a storage system would be in smart grid systems (Chapter 17), which incorporate energy resources

Example 18.5

For a solid, cylindrical, steel flywheel with diameter equal to length, calculate the maximum energy stored for a wheel diameter of 1 m.

Solution

From equation (18.11), the maximum energy is

$$E = \frac{1}{2} km \left(\frac{\sigma}{\rho} \right).$$

The mass of a 1-m-diameter by 1-m-long flywheel made of steel ($\rho = 7800 \text{ kg/m}^3$) is

$$m = \pi r^2 l \rho = 3.14 \times (0.5 \text{ m})^2 \times (1 \text{ m}) \times (7800 \text{ kg/m}^3) = 6120 \text{ kg}.$$

Using the value of $\sigma/\rho = 0.22 \times 10^6$ Nm/kg for steel (from Table 18.3) and a value of $k = 0.5$ (from Table 18.2) for a solid flywheel, the maximum energy is

$$E = (0.5) \times (0.5) \times (6120 \text{ kg}) \times (0.22 \times 10^6 \text{ Nm/kg}) = 337 \text{ MJ}.$$

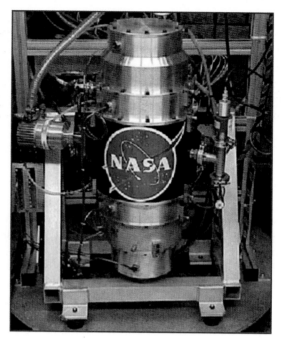

Figure 18.6: Small experimental high-speed flywheel constructed by NASA. The device is enclosed in a vacuum chamber to minimize air friction. The total height of the cylindrical housing is about 60 cm.

NASA

that are variable on a fairly short timescale (for example, a photovoltaic system that is susceptible to variations due to variable cloudiness). In such a case, a flywheel energy storage system could even out grid fluctuations. Another possible use of a flywheel energy storage system is in a vehicle where energy can be input into the flywheel during regenerative braking and extracted when needed. For automobiles, the gyroscopic effect associated with the rotating flywheel may introduce problems, but experimental applications in trains have met with limited success. A small experimental high-speed flywheel is shown in Figure 18.6.

A final important consideration in the design of a flywheel system is the minimization of frictional losses because of the high rotational speeds involved. Friction can occur in the bearings that support the flywheel and with the surrounding air. Practical high-speed flywheels that use magnetic bearings and are contained in a vacuum enclosure typically experience energy losses of about 1% per hour.

18.5 Superconducting Magnetic Energy Storage (SMES)

Electrical energy is transmitted as a current through a conductor. In a normal conductor, some resistance to the flow of current is always present, resulting in the loss of some of the electrical energy and the dissipation of heat. In a superconductor, the resistance is zero, and no losses in the energy are associated with the flow of current. Thus electrical energy could be stored (in principle, without loss) in a coil of superconducting wire. Electric current could be injected into the coil and would circulate indefinitely until it was needed. Superconducting magnetic energy storage (SMES) is in relatively common use in research laboratories as a mechanism to even out fluctuations in the line voltage resulting from variations in demand or to act more like a temporary backup for brief power outages. It is particularly useful when a user requires a highly stable source of power. In this sense, the SMES is analogous to the surge tank associated with a hydroelectric facility rather than actual pumped hydroelectric energy storage.

The energy stored in a current circulating in a coil is given by

$$E = \frac{1}{2}LI^2,$$ (18.12)

where L is the inductance of the coil, and I is the current. When L is measured in henries (H), and I is measured in amperes, then E is measured in joules. The inductance of a coil depends on its dimensions, the number of turns of wire, and the material in its core. For an air-core solenoid, a semiempirical relationship that gives a good approximation to the inductance is

$$L = frN^2,$$ (18.13)

where r is the mean radius, and N is the total number of turns of wire. The factor f is a geometry factor that, for r measured in meters, has a value of

$$f = \frac{3.9 \times 10^{-5}}{\left[9 + \dfrac{10l}{r}\right]}\text{J/(m} \cdot \text{A}^2),$$ (18.14)

Here, l is the length of the solenoid in meters.

Example 18.6

Consider a superconducting solenoid 1 m in diameter (0.5 m radius) and 1 m in length, consisting of 5000 turns of wire. Calculate the energy stored in the coil for a current of 1000 A.

Solution
Combining equations (18.13) and (18.14) gives the inductance of the coil as

$$L = \frac{3.9 \times 10^{-5}\text{J/(m} \cdot \text{A}^2)}{\left[9 + \dfrac{10l}{r}\right]}rN^2,$$

or for r and l in meters, the inductance in henries is

$$L = [3.9 \times 10^{-5}\,\text{J/(m} \cdot \text{A}^2)] \times (0.5\,\text{m}) \times \frac{5000^2}{9 + (10/0.5)} = 16.8\,\text{H}.$$

Using equation (18.12), the energy is found to be

$$E = \frac{1}{2}LI^2 = (0.5) \times (16.8\,\text{H}) \times (1000\,\text{A})^2 = 8.4\,\text{MJ}.$$

A superconducting coil, as described in Example 18.6, would be suitable for smoothing out power fluctuations for a single industrial or research user. A SMES facility that would compete with pumped hydroelectric storage for use by a public utility would need to be hundreds of meters or even kilometers in diameter.

A major difficulty in constructing a SMES facility is achieving the conditions that are necessary for making the coil superconducting. Superconducting materials exhibit normal electrical properties at high temperatures. If the temperature is lowered, they will, at some temperature, T_C (the *critical temperature*), undergo a transition to a superconducting regime. In the superconducting regime, the electrical resistivity goes to zero, and from a practical standpoint, the materials can be utilized for various applications possibly including energy storage. This phenomenon was first observed in mercury, which becomes superconducting below 4.2 K, by Heike Kammerling-Onnes in 1911. Since then, efforts have been made to find materials with as high a superconducting transition temperature as possible. Figure 18.7 shows the highest observed superconducting transition temperature as a function of year. Materials have now been discovered with T_C over 150 K. The rapid increase in observed T_C in the mid-1980s corresponds to the discovery of the so-called *high-temperature superconductors* (HTSCs) by Karl Alexander Müller and J. Georg Bednorz.

The recent discovery of superconductivity in H_2S under pressure at around 200 K, as seen in Figure 18.7, is both interesting and significant. Hydrogen sulfide is gas at room temperature and pressure. The fact that it becomes superconducting at low

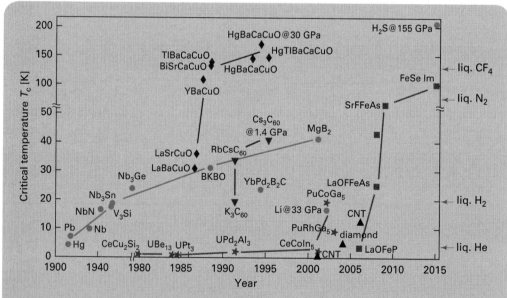

Figure 18.7: Progress in increasing T_C for superconductors as a function of time. The different symbols represent different classes of materials: carbon (▲), fullerenes [see Energy Extra 20.2] (▼), transition metal and alkali compounds (●), rare earth and actinide compounds (★), iron-based compounds (■), and perovskites (♦).

Based on Pia Jensen Ray. Figure 2.4 in Master's thesis, "Structural investigation of La(2-x)Sr(x)CuO(4+y) - Following staging as a function of temperature". Niels Bohr Institute, Faculty of Science, University of Copenhagen. Copenhagen, Denmark, November 2015. DOI:10.6084/m9.figshare.2075680.v2

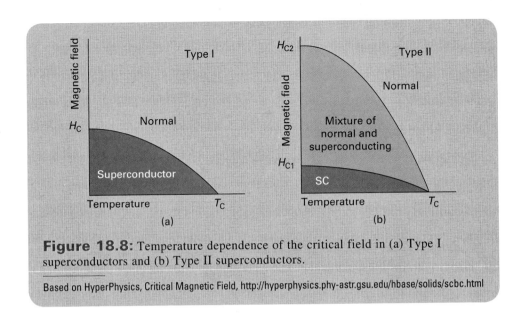

Figure 18.8: Temperature dependence of the critical field in (a) Type I superconductors and (b) Type II superconductors.

Based on HyperPhysics, Critical Magnetic Field, http://hyperphysics.phy-astr.gsu.edu/hbase/solids/scbc.html

temperature and elevated pressure suggests that other materials that are not conductors, or even solids, at ambient conditions may exhibit superconducting properties. It has been speculated that hydrogen and some simple hydrogen compounds may become superconducting near room temperature if sufficient pressure is applied.

Using a superconductor is not as simple as lowering its temperature below T_C. There is a close relationship between superconductivity and magnetism. In fact, the application of a magnetic field to a superconductor destroys its superconducting properties, making the material *normal* (from an electrical standpoint). In the simplest type of superconductors (Type I), a critical field at which the superconductivity suddenly and completely disappears can be defined. This critical field, H_C, is a function of temperature [Figure 18.8(a)]. A few superconducting materials (i.e., some elements) are Type I superconductors. However, most superconductors are Type II. As the magnetic field is increased in a Type II superconductor, a field is reached, H_{C1}, when portions of the sample become normal. However, other portions of the sample remain superconducting, and because an electric current takes the path of least resistance, the sample continues to carry current without resistance. As the field is increased, a point is reached, H_{C2}, when the entire sample becomes normal. At that point, the material does not conduct without resistance [Figure 18.8(b)]. Thus, at a given temperature, a Type I superconductor can be used at a field up to H_C, and a Type II superconductor can be used at a field up to H_{C2}.

A current flowing through a wire produces a magnetic field. Thus, there is a critical current, j_C, for which the current in a superconducting wire produces a field equal to the critical field, thereby driving the material normal. Although there is clearly a relationship between H_C and j_C, it is not straightforward. However, it is clear that superconductors just below their critical temperature cannot carry very much current and still remain superconducting. From a practical standpoint, it is customary to use superconductors as far below T_C as is possible. The maximum amount of current that can be carried is related to the value of the critical field at zero temperature, $H_C(0)$ for a Type I and $H_{C2}(0)$ for a Type II. To determine the usefulness of a superconductor, it is important to know the value of T_C and the value of the critical field.

Table 18.4: Properties of some superconducting materials. (T = tesla)				
material	type	T_C (K)	$H_C(0)$ (T)	$H_{C2}(0)$ (T)
Sn	I	3.7	0.03	—
Pd	I	7.2	0.08	—
Nb	II	9.3	—	0.4
NbTi	II	10	—	12
Nb_3Sn	II	18	—	25
$YBa_2Cu_3O_7$ (YBCO)	II	93	—	168
$Bi_2Sr_2Ca_2Cu_3O_{10}$ (BSCCO)	II	110	—	200

Some examples of the properties of superconductors are given in Table 18.4. The table shows that Type I superconductors have low values of T_C and low values of $H_C(0)$. This indicates that they are generally not very useful for carrying large amounts of current. However, some Type II superconductors (particularly the high-temperature superconductors) have fairly high values of T_C and $H_{C2}(0)$. Because j_C is not directly related to H_{C2}, it turns out that optimizing T_C and $H_{C2}(0)$ does not always optimize j_C. In this respect, YBCO, which was one of the earliest high-temperature superconductors to be discovered, seems to be in the lead.

While it may seem advantageous to utilize HTSC materials for the purpose of constructing SMES systems, the choice is not so clear. HTSC materials would minimize the refrigeration costs. However, for a large system, the increase in refrigeration costs between operating at, say, 80 K and operating at, say, 5 K, is not as substantial as it would be for a smaller system. In fact, even if HTSC materials are used, it may be advantageous to run the system at 5 K rather than at 80 K because the increase in critical field could outweigh the increase in refrigeration costs. However, as discussed in Chapter 2, HTSC materials are typically brittle and difficult to fabricate, so that capital costs and reliability may be more important factors in design viability than refrigeration costs.

SMES is not 100% efficient. Although the current circulates in the coil without loss, energy is required for refrigeration to maintain the coil at a temperature below the superconducting transition temperature of the wire. This is typically on the order of 0.1% of the energy stored in the coil per hour of storage. Also, losses are incurred when current is put into or taken out of the coil. Electric current carried in transmission lines is typically alternating current (AC), whereas the current stored in a superconducting magnet is direct current (DC). So to store electrical energy in a SMES and then use it, the AC must be converted to DC, and the DC must be converted back to AC for use. This AC-DC-AC conversion cycle is about 95% efficient and generally accounts for the major loss of energy in this system.

At present, small-scale SMES systems have been successful in providing grid stability (Figure 18.9). The world's largest (as of 2009) SMES system is at Sharp's Kameyama Plant in Japan and can provide up to 10 MW. Other similar or larger systems are in the testing stages. The practicality of large-scale systems that could compete with relatively low-tech options like pumped hydroelectric or compressed air has not been demonstrated.

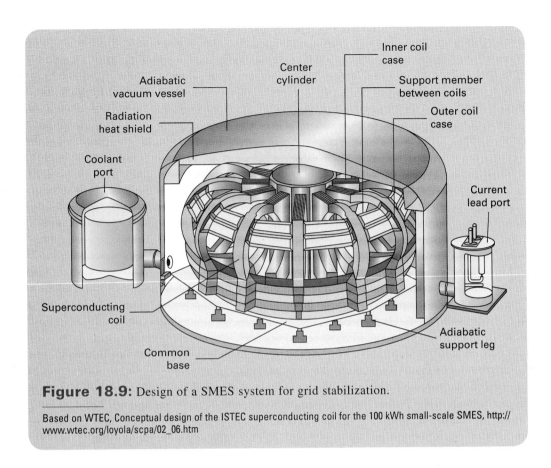

Figure 18.9: Design of a SMES system for grid stabilization.

Based on WTEC, Conceptual design of the ISTEC superconducting coil for the 100 kWh small-scale SMES, http://www.wtec.org/loyola/scpa/02_06.htm

18.6 Summary

Methods for energy storage are required because energy is not always available when and/or where it is needed. The former case results largely from variations in energy production capabilities and in energy demand and is particularly relevant when implementing renewable energy technologies. The latter case arises primarily because energy produced in centralized facilities, such as thermal generating plants, requires an appropriate storage mechanism for portable applications, such as vehicles. Several possible technologies for energy storage, that is, specifically electrical energy storage, were discussed in this chapter. These technologies do not include batteries and hydrogen; these are presented in Chapters 19 and 20, respectively.

The chapter discussed pumped hydroelectric storage, which is in common commercial use for large-scale electricity storage for the grid. This is a straightforward method where excess electricity generated in times of low demand is used to pump water into an elevated reservoir. When demand exceeds capacity, the reservoir is drained through turbines at a lower elevation to generate additional electricity. Although capital construction costs can be high, this method is low maintenance and cost-effective over the long term.

Energy storage using compressed air is also discussed. Excess electricity can also be used to compress air, which can be used to drive turbines in times of higher demand.

Two major commercial facilities of this type are in use, one in the United States and one in Germany. While this method of energy storage is potentially viable, requirements for an appropriate chamber are a major factor.

As pointed out in this chapter, other methods for large-scale energy storage, such as flywheels and superconducting magnet energy storage (SMES), are not at a technological point where they are viable. The chapter provided an introduction to the properties of flywheels and the factors that affect their energy storage capabilities. It was shown that, for a given mass and geometry, the maximum energy storage capability is directly proportional to the ratio of the tensile strength to the density (σ/ρ) of the flywheel material. Of commonly used materials, carbon fiber–reinforced polymer has the largest ratio and provides the best opportunity for high-density energy storage in a flywheel.

This chapter also reviewed the basics of superconductivity and the history of the development of superconducting materials. The physical principles of energy storage in a superconducting magnet have been presented.

At present, pumped hydroelectric storage is the most effective means of storing electricity for the grid. It has a high capacity and good efficiency and is reliable. Compressed air storage is also a viable means of electrical storage for the grid but is not as widespread as pumped hydroelectric, largely because of the physical requirements for a suitable location. It is not clear whether future technological developments for flywheel and SMES devices will make them competitive with the more traditional large-scale energy storage methods. However, SMES, at least, has found viable applications as a means of stabilizing power sources on a smaller scale.

Problems

18-1 A pumped hydroelectric storage reservoir has an area of 1 km^2 and an average depth of 30 m. The water flows to a lower reservoir with a head of 150 m.

(a) What is the total energy available in MWh$_e$? Assume a generator efficiency of 85%.

(b) What is the average power available (in megawatts) if the upper drains through the turbines over a period of 12 hours?

18-2 The total volume of the upper reservoir of a pumped hydroelectric facility flows through the turbines over a period of 8 hours. The upper reservoir has an area of 300,000 m^2 and an average depth of 15 m.

(a) What is the average flow rate (in cubic meters per second) between the upper and lower reservoirs?

(b) If the water between the two reservoirs flows through a (circular) penstock with a diameter of 7.5 m, what is the average velocity of the water?

18-3 Design a steel flywheel-based electricity backup system to provide 1 day of typical electricity use for a single-family home (no electric heat). Assume that 70% of the maximum theoretical energy storage capacity is available and can be converted to electricity at 90% efficiency.

18-4 Consider the possibility of constructing a vehicle that operates on compressed air. A practical vehicle requires about 0.7 MJ/km, and a practical high-pressure storage tank might store gas at a pressure of 7.5 MPa. Derive a relationship between range (in kilometers) and storage tank volume (in cubic meters). If a maximum practical tank volume is 1 m^3, what is the range of the vehicle?

18-5 The world's largest pumped hydroelectric facility in Bath County, Virginia, has a total energy storage capacity of about 30 GWh_e. Assuming the same conversion efficiency for a compressed air storage facility (which is optimistic), calculate the dimensions of a cubic chamber that, at a pressure of 5 MPa, has the same energy storage capacity as the Bath County station.

18-6 Consider solid cylindrical flywheels with the diameter equal to the length. For flywheels made of steel, aluminum, and carbon-reinforced polymer, calculate the diameter of a wheel that has a maximum energy storage capability of 10 MJ.

18-7 Energy is extracted from a flywheel with an initial rotational frequency of f_0 in such a way that the frequency decreases linearly with time. Describe the power provided as a function of time as the flywheel slows.

18-8 For large-scale electric storage for the grid, a capacity in the order of 1 GWh or greater is desirable. Consider the possibility of creating a SMES system that is 0.5 km in diameter, consisting of a 2-m-high coil with 20,000 turns of wire. What current is needed to store 1 GWh of energy?

18-9 The requirements for an ideal location for a pumped hydroelectric storage facility are not the same as those for a hydroelectric generating station. For pumped storage, a location for a reservoir at an elevation above a supply of water is required. It is convenient if the location is also compatible with the production of alternative energy facilities such as solar or wind, as these benefit greatly from the ability to store energy. Cape Breton Island in Nova Scotia, Canada is an ideal location to consider the possibility of combining wind power (as the region has good wind resources) with pumped hydroelectric storage (as the region consists of a plateau at an elevation of about 500 m above the Atlantic Ocean). Specify the design of a wind/pumped hydroelectric system on Cape Breton Island that could supply the electric needs of a population of 100,000. Consider a simple example where the per-capita electric use is a constant 0.7 kW and the wind velocity is constant during the day (12 hours), providing 55% of the rated turbine capacity, and zero at night. A fraction of the electricity generated during the day is used to pump sea water into a reservoir at an average elevation of 500 m, and this is used to provide electricity when there is no wind. Suggest design criteria for such a system, including at least the following:

- The number and capacity of wind turbines.
- The fraction of wind turbine output to be used to pump water to the reservoir.
- Capacity of hydroelectric turbines.
- Hydroelectric turbine and pumping efficiency.
- Size of the storage reservoir.

18-10 A figure skater with a mass of 55 kg spins at a rate of 4.5 rotations per second. What is the rotational energy of the skater? Approximate the skater as a cylinder with a diameter of 0.25 m and a length of 1.5 m. Express the answer in J and kWh.

18-11 Compare the following energies:

(a) The rotational energy of the earth (rotating on its axis)
(b) The kinetic energy of the earth (in its orbit around the sun)
(c) The rotational energy of the sun (rotating on its axis)
(d) The kinetic energy of the moon (in its orbit around the earth)

Consider all bodies as solid spherical objects with uniform density, and consider all orbits as circular. Masses, diameters, orbital dimensions, and rotational periods are readily available on the web.

18-12 When the Bath County Pumped Hydroelectric Facility is generating electricity at maximum output, what is the velocity of the water flowing through the penstocks?

18-13 (a) Assuming that the earth is a solid sphere of uniform density, calculate its moment of inertia and its rotational energy.
(b) The earth's rotation is slowing (i.e., the earth is losing rotational energy). The earth's rotational period (the length of the day) is increasing at a rate of 1.7×10^{-5} s per year. Calculate the annual rotational energy loss of the earth and compare this to the annual energy use by society.

18-14 A compressed air energy storage facility operates between a pressure of 5×10^5 Pa and a pressure of 7.5×10^6 Pa. How large would the compressed air chamber have to be to provide one day of electricity to a city of 100,000 where the average per-capita electric use is 0.8 kW_e? Assume a generating efficiency of 75%. Express your answer in cubic meters and the edge length in meters of a cubic chamber.

18-15 A superconducting magnetic energy storage coil has 10,000 turns of wire, a mean diameter of 10 m, and a length of 2 m. It carries a current of 2000 A.

(a) Calculate the energy storage capacity in MJ.
(b) A device with a larger capacity is required. Calculate the effect of doubling each of the following parameters while leaving the others the same.

(i) Length
(ii) Mean diameter
(iii) Current
(iv) Number of turns

18-16 A steel flywheel in the form of a solid disk has a radius of 0.2 m and a thickness of 0.1 m.

(a) What is the maximum rotational frequency that is compatible with the material characteristics of this flywheel?

(b) What is its energy storage capacity at maximum rotational frequency?

(c) If the flywheel is energized by a motor with a power output of 1 kW (at 100 % efficiency) how long would it take to increase the rotational frequency from zero to the maximum?

18-17 A small pumped hydroelectric facility has a mean head of 50 m. The facility has a maximum output of 500 kW_e and the water flows through a 1-m-diameter penstock.: The generating efficiency is 85%.

(a) What is the total flow rate at maximum capacity?

(b) What is the velocity of the water in the penstocks at maximum capacity?

18-18 It is common for homeowners to store thermal energy (e.g., from solar collectors) in a water tank for heating use at night. The ability to store one day's thermal energy needs can provide a guideline for a useful system. Consider the possibility of storing one day's electrical needs (e.g., from photovoltaic panels) in a home in the form of a tank of compressed air. Discuss the possibility and practicality of such an approach.

18-19 Other materials (in addition to those given in Table 18.3) have properties that are compatible with the construction of flywheels with high-energy storage capabilities (although these materials may not always be convenient in other respects for the construction of a flywheel). Calculate the maximum energy storage capabilities for a 1-kg disk-shaped flywheel made of the following materials, and compare these results with those for the materials in Table 18.3.

material	density (10^3 kg/m^3)	tensile strength (10^9 N/m^2)
bamboo	0.4	0.5
graphene	1.0	130.0
kevlar	1.44	3.76
zylon	5.8	1.56
boron	2.46	3.1

18-20 A pumped hydroelectric facility has an upper reservoir with an area of 2 km^2 and an average depth of 40 m. The average head is 200 m. The design requires that the upper reservoir can be filled from the lower reservoir in a period of 8 hours.

(a) What is the electrical power required by the pumps? Assume a pumping efficiency of 85%.

(b) What is the flow rate (in m^3/s) of water being pumped to the upper reservoir?

Bibliography

G. J. Aubrecht II. *Energy: Physical, Environmental, and Social Impact*, (3rd ed.) Pearson Prentice Hall, Upper Saddle River, NJ (2006).

J. A. Kraushaar and R. A. Ristinen. *Energy and Problems of a Technical Society*, (2nd ed.) Wiley, New York (1993).

C. Ngô and J. B. Natowitz. *Our Energy Future: Resources, Alternatives and the Environment.* Wiley, New York (2009).

A. Ter-Gazarian. *Energy Storage for Power Systems.* Peter Peregrinus, Stevenage, United Kingdom (1994).

Battery Electric Vehicles (BEVs)

Learning Objectives: After reading the material in Chapter 19, you should understand:

- The properties and applications of different types of batteries.
- The use of the Ragone plot to illustrate energy-power relationships for energy storage mechanisms.
- Energy requirements for vehicle propulsion.

- The historical development of electric vehicles and reasons for changes in their popularity.
- The advantages, disadvantages, and economic viability of electric vehicles.
- The utilization of supercapacitors for energy storage and their significance in electric vehicle design.

19.1 Introduction

Transportation places special demands on a source of energy; specifically, the source of energy must be portable. For this reason, petroleum-based products have been the overwhelming choice for transportation purposes. This stems from the fact that petroleum products, such as gasoline and diesel fuel, have a very high energy density (i.e., many megajoules per kilogram). In addition, there is a highly developed technology (i.e., the internal combustion engine) for converting the chemical energy stored in petroleum products into the mechanical energy needed to propel a vehicle, and for now, at least, petroleum is readily available and comparatively inexpensive. Many alternative energy sources are not appropriate for direct application to transportation needs. A major exception, perhaps, is biofuels, which are a direct replacement for petroleum and which can, more or less, make direct use of the existing distribution infrastructure and vehicle technology. As discussed in Chapter 16, these have been particularly successful in some parts of the world (e.g., Brazil). Although their implementation in other countries may not necessarily be straightforward, they may be one of the few viable nonfossil fuel options for aircraft. Other environmentally advantageous energy sources are not portable or suffer from very low energy density. The former is the case, for example, for nuclear, while the latter is the case, for example, for wind energy; although both of these may have applications for ocean transport. The implementation of alternative energy strategies for transportation, specifically road vehicles, requires a portable energy storage mechanism with sufficiently high energy density. While some of the methods discussed in the previous chapter (e.g., compressed air or flywheels) may be utilized, the most likely energy storage possibilities for vehicle use are batteries and hydrogen. Both of these serve as convenient mechanisms for storing the electrical energy produced by most alternative methods and are discussed in this and the next chapter.

19.2 Battery Types

A battery is an electrochemical device for storing electrical energy as chemical energy. It consists of one or more cells. Each cell consists of two half cells connected in series. One half cell contains the *anode* (or negative electrode) and an electrolyte, while the other half cell contains the *cathode* (or positive electrode) and an electrolyte. The two electrolytes may be either solid or liquid and may be common to the two half cells, or they may be separated by a membrane that is porous to the passage of certain ions. When a battery is charged, active atoms in the anode are ionized. The ions, thus formed, travel to the cathode through the electrolyte, and the electrons that are liberated travel to the cathode through an external circuit. At the cathode, the ions and electrons are recombined in the so-called *redox* (reduction-oxidation) reaction. When the battery is discharged, the ions travel from the cathode back to the anode through the electrolyte, while electrons flow from the cathode back to the anode through the external circuit and in the process can do useful work—heat a resistor, produce a magnetic field, turn a motor, and so on.

The voltage difference between the electrodes of a battery depends on the nature of the chemical reactions in the cell. For example, zinc-carbon batteries (by the nature of the chemical reactions present) produce about 1.5 V, whereas NiCd batteries produce about 1.2 V. Several cells can be connected together in series to produce larger voltages. During discharge, an ideal battery produces a constant voltage up to the point where all of the excess electrons and positive ions at the cathode have traveled through the external circuit and the electrolyte, respectively, and have recombined at the anode. Such an ideal battery would have no internal resistance. In practice, batteries do have internal resistance, and this resistance increases during discharge. As a result, the voltage produced by the battery decreases during the discharge. The exact details of the voltage curve during discharge depends on the battery chemistry and internal design.

Batteries are in common use in our everyday lives. They are prevalent in electronic devices and vary considerably in their design and electrical storage capacity. On the small end of the scale, common batteries include those used in watches. On the large end of the scale, common batteries include automobile batteries. The current chapter deals primarily with the use of batteries in electric vehicles.

Generally, batteries can be classified as *primary* batteries (nonrechargeable) or *secondary* batteries (rechargeable). Primary batteries are generally small, relatively inexpensive, and suitable for devices that require relatively little power (e.g., a watch or a smoke detector) and/or utilize power only on an intermittent basis (e.g., a flashlight). These batteries are designed to be replaced when they become discharged, and most comply with battery industry standards for size and voltage characteristics. The most common types of primary batteries are primary lithium (Li) cells (e.g., button cells used for watches, calculators, etc.), zinc-carbon cells, or alkaline cells (used for flashlights, radios, toys, etc.). Some characteristics of standard size zinc-carbon and alkaline cells are given in Table 19.1.

Table 19.1: Some properties of standard zinc-carbon and alkaline cells.

| size | diameter (cm) | length (cm) | volume (cm³) | Zn-C | | | alkaline | | |
				mass (g)	energy (mWh)	capacity (mAh)	mass (g)	energy (mWh)	capacity (mAh)
AAA	1.05	4.45	3.85	10	350	320	12	1410	1150
AA	1.45	5.05	8.34	19	650	591	23	2600	2122
C	2.62	5.00	27.0	48	2390	2172	66	9560	7800
D	3.42	6.15	56.5	98	5210	4733	135	20,830	17,000

It is clear from the table that the energy content of a battery depends on the battery chemistry and also on the size of the battery. The specific energy or energy density (i.e., energy per unit mass) depends on the battery chemistry, and amounts to about 55 Wh/kg for Zn-C batteries and 160 Wh/kg for alkaline batteries. To construct a viable battery electric vehicle (BEV), it is important to optimize the specific energy of the batteries in order to minimize the mass of the batteries. From a practical standpoint, primary batteries are not suitable for electric vehicles because they would have to be replaced at the end of their discharge. Secondary or rechargeable batteries are necessary for electric vehicles, and Table 19.2 on the next page gives the properties of some common secondary battery chemistries. To properly assess the suitability of certain chemistries, it is most relevant to consider factors such as the specific energy rather than the energy for a specific size battery.

ENERGY EXTRA 19.1
The cost of electricity

Most of the electricity purchased for personal use comes from a public utility. This is the electrical energy is used for lights, televisions, household appliances, and so on, and it typically costs about $0.12 per kWh. It is, however, sometimes convenient to use batteries as a portable source of electricity for flashlights, cameras, portable electronics, remote controls, and the like. It is interesting to consider the monetary cost of electricity that is used from such a source. Primary (i.e., nonrechargeable) AAA batteries are in common use, and the cost of the electrical energy obtained from such a battery is really the same as the cost of the battery itself. This is because, once the battery has been discharged, it is disposed of (or preferably recycled) and is not recharged. Table 19.1 shows that the energy content of a Zn-C AAA battery is 350 mWh, or 3.5×10^{-4} kWh. If a package of four AAA cells is purchased at a discount store for $1.00, then the cost of the electricity is $0.25/3.5 \times 10^{-4}$ kWh = $714 per kWh, or nearly 6000 times the cost charged by a public utility for the same amount of energy. To dry a typical load of laundry, a clothes dryer might use about 2.5 kWh of electricity. At a rate of $0.12 charged by a public utility, this amounts to a cost of about $0.30. If a clothes dryer ran on AAA batteries, the energy to dry a load of laundry would cost nearly $2000. While primary batteries are a convenient source of energy for portable electronic devices, the energy comes at a premium cost.

Topic for Discussion

Find prices at a local store for both Zn-C and alkaline AA and D cells. For the most objective comparison, consider the same name brand batteries in each category. In terms of economics, does size matter? Are alkaline batteries worth the extra cost?

Table 19.2: Properties of some secondary battery chemistries. Typical values are given. Specific energy is given in standard metric units (MJ/kg) and in traditional units often used for battery properties (Wh/kg), where 278 × (MJ/kg) = (Wh/kg). (NiMH = Nickel Metal Hydride)

chemistry	cell voltage (V)	specific energy (MJ/kg)	specific energy Wh/kg	specific power (W/kg)	self-discharge (%/month)	life (cycles)
Pb acid	2.1	0.13	36	100	4	600
Ni-Cd	1.2	0.22	56	150	20	1500
NiMH	1.2	0.28	78	800	20	1000
Li-ion	3.6	0.58	160	300	7	1200

The total energy contained in a battery (MJ) is related to the specific density (MJ/kg) and the mass of the battery (kg) by

$$\text{energy} = (\text{specific energy}) \times (\text{mass}). \qquad (19.1)$$

While the specific energy is related to the mass of the active components of the battery (e.g., electrodes or electrolyte), for larger batteries the inactive components (e.g., case or terminals) typically contribute relatively little to the overall mass.

Example 19.1

Calculate the mass of Pb-acid batteries needed to supply 1 kWh of energy. Repeat for Li-ion batteries.

Solution

From Table 19.2, the energy density of a Pb-acid battery is 36 Wh/kg. From equation (19.1), the mass of a battery is related to its specific energy and the total energy available by

$$\text{mass} = \frac{\text{energy}}{\text{specific energy}}.$$

Thus

$$\text{mass} = \frac{1 \text{ kWh}}{0.036 \text{ kWh/kg}} = 27.8 \text{ kg}.$$

For a Li-ion battery, the specific energy is 160 Wh/kg, so the mass is

$$\text{mass} = \frac{1 \text{ kWh}}{0.160 \text{ kWh/kg}} = 6.25 \text{ kg}.$$

For a given vehicle, the driving range of the vehicle is related (more or less) to the total energy available. By comparison with a conventional gasoline-powered vehicle, the energy content of the batteries is analogous to the volume of the gas tank. The rate at which energy can be extracted from the batteries is also important because this is analogous to the power available from the engine of a traditional vehicle and

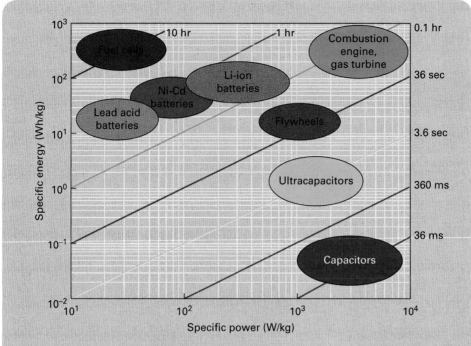

Figure 19.1: Relationship of specific energy and specific power for secondary battery types (Ragone plot). Other energy storage methods relevant to vehicles are shown for comparison.

Based on Moura, J. B. Siegel, D. J. Siegel, H. K. Fathy, A. G. Stefanopoulou "Education on Vehicle Electrification: Battery Systems, Fuel Cells, and Hydrogen," Proceedings of the 2010 IEEE Vehicle Power and Propulsion Conference, Lille, France, 2010

determines the vehicle performance (e.g., acceleration). Thus, in choosing a battery type for the design of an electric vehicle, it is important to maximize both the specific energy and the specific power or power density (i.e., the power per unit mass) as much as possible. The choice of a given battery chemistry and a given mass of batteries determines the total energy and power available and establishes the driving characteristics of the vehicle. In the next section, it will be shown how battery properties can put limitations on the vehicle's performance and range. Table 19.2 certainly suggests that, of the types listed, Li-ion batteries are the most desirable for electric vehicle use. The relationship of specific energy and specific power for major secondary battery chemistries are shown in Figure 19.1. This graph is referred to as a *Ragone plot* (pronounced ra-GOH-nee). It is also important to consider other factors such as safety, cost, and availability of materials. These considerations are now discussed.

19.3 BEV Requirements and Design

A comparison with traditional gasoline-powered vehicles is inevitable when considering the viability of an electric vehicle. Drivers are accustomed to the operating characteristics of these vehicles and will inevitably use them as a benchmark for

Table 19.3: Examples of some typical classes of nonhybrid production vehicles and specifications (2017 models). Fuel consumption (combined city/highway) values are from www.fueleconomy.gov. The specified energy used is the primary energy content of the gasoline. The energy utilized at the wheels is this value multiplied by a factor of 0.17, as discussed in the text.

class	typical example	mass kg	power kW	power/mass kW/kg	fuel consumption L/100 km	energy used MJ/km	range km	base price $(10^3$US$)$
subcompact	Honda Fit	1140	97	0.085	6.5	2.3	382	16
compact	Subaru Impreza	1354	113	0.083	7.3	2.5	422	18
family	Volkswagen Passat V6	1620	209	0.129	10.2	3.5	426	29
luxury	BMW 750i	2042	332	0.163	12.4	4.3	391	95
sport	Aston Martin V12 Vantage	1665	421	0.253	19.6	6.8	253	186
compact SUV	Toyota RAV4	1567	131	0.084	9.4	3.3	398	25
full-sized SUV	Lexus LX570	2963	286	0.130	15.7	5.5	369	89

comparison. Of course, all gasoline-powered vehicles do not have the same characteristics. Table 19.3 gives the properties of some typical classes of gasoline vehicles designed for the consumer market.

Fuel consumption is the average for city and highway and is given in liters of gasoline per 100 km (L/100 km). Energy used per kilometer is based on the energy content of gasoline (34.8 MJ/L) by

$$0.348 \cdot (L/100 \text{ km}) = \text{MJ/km}. \tag{19.2}$$

An overview of the automobile market suggests that vehicles like compact cars, family cars, and compact SUVs are viewed as the most practical and economical by the general public. Certainly from a practical standpoint, an electric vehicle that fits into this range of vehicles would be the most marketable. On the basis of the vehicle characteristics shown in Table 19.3, Table 19.4 gives the specifications of an ideal electric vehicle (in the context of market considerations for gasoline vehicles).

Current battery technologies dictate the properties of electric vehicles, and on the basis of the information in Table 19.2, the degree to which an actual electric vehicle can achieve the ideal design can be determined. The efficiency of an internal combustion vehicle is determined by the ratio of the energy provided to the wheels to the energy content of the gasoline in the tank. Typically this is only about 15–20%. Electric vehicles are, by comparison, quite efficient; typically 80–90% for the ratio of energy to the wheels to the energy stored in the batteries. Table 19.3 shows that a practical family

Table 19.4: Specifications of a benchmark family electric vehicle.	
property	**goal**
mass	1500 kg
power	150 kW
range	600 km
cost	US$30,000

vehicle might use about 3.5 MJ/km. The energy required to actually drive the wheels based on the efficiency (say 17%) of the gasoline engine is about 0.6 MJ/km. Thus the energy that has to be supplied by the batteries in an electric vehicle (based on an efficiency of 85%) is (0.6 MJ/km)/0.85 = 0.7 MJ/km. While this value could vary from about 0.46 MJ/km to about 1.4 MJ/km for different classes of vehicles (Table 19.3), the value of 0.7 MJ/km is used as a typical value in most of the examples in this text. If Pb-acid batteries, which contain 0.13 MJ/kg, are used, a range of 600 km would require batteries with a mass of

$$(600 \text{ km}) \times \frac{0.7 \text{ MJ/km}}{0.13 \text{ MJ/kg}} = 3230 \text{ kg},\qquad\textbf{(19.3)}$$

which is well above the design goal. The available power provided by these batteries would be

$$(3230 \text{ kg}) \times (0.10 \text{ kW/kg}) \times (0.85) = 275 \text{ kW}.\qquad\textbf{(19.4)}$$

This substantially exceeds the power requirements. In fact, to satisfy the power needs would require only

$$\frac{150 \text{ kW}}{0.10 \text{ kW/kg} \times 0.85} = 1760 \text{ kg},\qquad\textbf{(19.5)}$$

of batteries, still a substantial amount relative to expectations for total vehicle mass. However, this exercise clearly indicates that range may be a more difficult criterion to satisfy than power. While Pb-acid batteries may be acceptable for hybrid vehicles where the electric motor only supplements a traditional gasoline engine (Section 17.8), a purely electric vehicle requires batteries with a higher energy density.

Using the values of Li-ion battery parameters shown in Table 19.2, the performance characteristics calculated for Li-ion-powered BEVs are shown in Table 19.5.

Table 19.5: Characteristics of electric vehicles using Li-ion batteries.		
battery mass	**power**	**range**
100 kg	30 kW	80 km
300 kg	90 kW	250 km
725 kg	217 kW	600 km

Example 19.2

Calculate the mass of Li-ion batteries needed to power a vehicle with a range of 600 km and the corresponding maximum power available.

Solution

Using an energy requirement from the batteries of a typical electric vehicle of 0.7 MJ/km and a specific energy of 0.58 MJ/kg for a Li-ion battery gives the required mass:

$$(600 \text{ km}) \times \frac{0.7 \text{ MJ/km}}{0.58 \text{ MJ/kg}} = 724 \text{ kg}.$$

The maximum power available for 724 kg of Li-ion batteries is determined from the specific power of 300 W/kg:

$$(724 \text{ kg}) \times (0.3 \text{ kW/kg}) = 217 \text{ kW}.$$

There is active research in the development of new batteries with improved characteristics, and batteries with improved performance will certainly be available in the future. Calculations have considered the case for a battery mass that satisfies the requirements for power and range, and the table suggests that a reasonable compromise between power and range can be achieved by a battery-operated electric vehicle utilizing Li-ion batteries. It should be noted that the maximum power to the wheels is, in general, limited by the specifications of the electric motor and may be much less than the capabilities of the batteries and that, during normal driving, the power utilized is much less than the maximum available. In fact, a quick calculation would show that, at maximum power output, a Li-ion battery would be drained in less than half an hour.

Overall, a comparison of different energy sources for vehicles can be illustrated by a Ragone plot (Figure 19.2). For a particular energy source, the trade-off between energy and power is represented (more or less) by a diagonal line with a slope of −1, for example; the yellow area represents gasoline internal combustion engines (ICE). However, all known battery chemistries fall below the region for gasoline on the Ragone plot. The plot also illustrates some other energy storage mechanisms that have been discussed previously (e.g., flywheels) and some that will be discussed later in this chapter (supercapacitors) or in the next chapter [hydrogen internal combustion engines (H-ICE) and hydrogen fuel cells].

An important motivation for the development of battery electric vehicles deals with the overall energy consumption of an electric vehicle. Electric vehicle manufacturers claim that the equivalent fuel consumption of an electric vehicle in comparison with gasoline-powered vehicles is very low. The reason is that the conversion of energy stored in a battery to energy available at a vehicle's wheels is much more efficient than the conversion of chemical energy stored in gasoline to energy at a vehicle's wheels. This is because no heat engine is involved in the electric vehicle. If a traditional gasoline-powered vehicle is 15–20% efficient and an electric vehicle is 80–90% efficient, then the electric vehicle might consume the equivalent of about one-fifth of the fuel needed for the gasoline vehicle, or perhaps 1.5 to 2.0 L/100 km for a small- to medium-size car. However, it is important to realize that gasoline is produced very efficiently from a primary energy source (crude oil). Electricity used to charge batteries may or

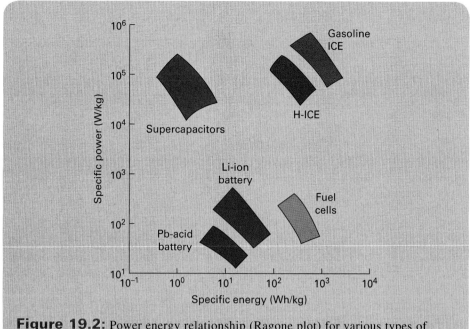

Figure 19.2: Power energy relationship (Ragone plot) for various types of batteries in comparison with other possible energy storage methods that might be used for transportation.

may not be produced very efficiently from a primary energy source. Hydroelectric power, for example, is very efficient and only slightly decreases the net efficiency between primary energy and energy at the wheels. However, if the electricity is produced by burning coal, then the efficiency of producing electricity is only about 40%, and the net efficiency of converting primary energy to energy at the wheels is typically $85\% \times 40\% = 35\%$, or about twice that of burning the fossil fuel directly. The increase in efficiency represents a net decrease of about 50% in carbon emissions, even if electricity comes exclusively from fossil fuels (and a greater reduction if the electricity comes from greener sources). Energy Extra 20.3 presents a more detailed analysis of the carbon footprint of various transportation technologies. An additional factor (in the favor of electric vehicles) is that it is easier to control the release of pollutants and to sequester carbon from a small number of coal-fired electric generating stations than from a large number of individual vehicles.

However, the production of a marketable electric vehicle requires the consideration of several important points.

The first consideration is economics: Li-ion batteries are quite expensive. An electric vehicle competing with a $25,000 internal combustion vehicle might consist of a $20,000 basic vehicle with $15,000 of Li-ion batteries for a total new-vehicle cost of about $35,000. While battery technology has improved greatly in recent years, there is the potential concern that rechargeable batteries have a finite lifetime and that their replacement would represent an unreasonable maintenance expense.

Second is the question of recharging the batteries when they have depleted their charge. In the early days of electric vehicles, typical recharge times could be 8–12 hours or longer. Advanced Li-ion batteries can sometimes be recharged in 1–2 hours. This is still much longer than the time required to refill a gasoline tank (a couple of minutes).

ENERGY EXTRA 19.2
Flow batteries

Flow batteries are a type of rechargeable battery in which electricity is produced by certain chemical reactions between liquid electrolytes. The first flow battery was constructed in the 1880s, although it was only in the 1970s that serious interest in this technology arose. The general features of a flow battery are illustrated in the following figure. Electrical energy can be used to charge the battery by causing an oxidation reaction in one electrolyte and a reduction reaction in the other. The two electrolytes are separated by an ion exchange membrane that allows the ions involved in the reaction to pass through. On discharge, the reverse reaction occurs, allowing electrical energy to be extracted from the cell. The electrolytes are stored in tanks external to the cell itself and are circulated through the cell, where they undergo chemical reactions by pumps.

Flow batteries have several interesting features that distinguish them from traditional rechargeable batteries. First, the power rating of the cell is a function of the size and geometry of the electrodes, and the energy storage capacity is limited only by the size of the electrolyte storage tank. This feature allows for the construction of flow batteries with very large energy storage capabilities and has made flow batteries an attractive possibility for large-scale electrical storage.

Another important feature of flow batteries is the ability to rapidly recharge them by merely pumping out depleted electrolytes (for later reprocessing) and pumping in freshly charged electrolytes. This feature would enable a flow battery electric vehicle to fill up much like filling up a conventional internal combustion engine vehicle at the gas station, thus eliminating the lengthy recharge stops that are a potential inconvenience for Li-ion-battery vehicle users. A prototype-scale vehicle (shown in the next page figure), operating on a flow battery, has

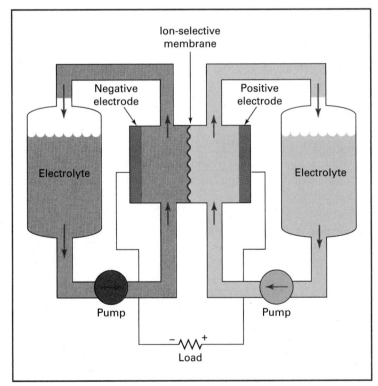

Diagram of a flow battery.

Continued on page 603

Energy Extra 19.2 continued

Courtesy of Robin Vanhaelst

Prototype vehicle running on a flow battery.

been constructed by researchers at the Fraunhofer Institute for Chemical Technology in Germany and is a step toward the possible future implementation of this technology.

Flow batteries are, however, not without their disadvantages. They are more complex and (at the present, at least) more costly and have lower energy densities than traditional rechargeable batteries.

Topic for Discussion

The highest energy density flow batteries currently available are based on zinc-bromine chemistry. These have energy densities in the range of 50–70 Wh/kg. Discuss how flow batteries compare with other battery technologies for use in BEVs. What are the expected vehicle ranges? What improvements would be desirable?

Vehicles utilized for short-range commuting that can be recharged at the user's home overnight are more practical than those that might be used for long trips.

There is also the need to establish an infrastructure of recharging stations, just as an infrastructure of gasoline stations was developed to make the use of gasoline-powered vehicles practical. It is also important to realize that as gasoline-powered vehicles are replaced with battery electric vehicles, the total electricity generating capacity must also be increased appropriately to provide power to charge the BEV batteries.

A final factor that requires consideration is the environmental impact of manufacturing batteries and the availability of starting materials for battery production. Recall the discussion in Chapter 7 of the limitations of fusion power based on the availability of lithium. The widespread use of Li-ion battery–powered electric vehicles would put additional demands on the world's Li resources. There are a number of uncertainties in the determination of the number of electric vehicles that could be produced from the

ENERGY EXTRA 19.3
Rechargeable sodium batteries

At present, the preferred battery choice for BEVs is Li-ion batteries because of their high specific power and high specific energy. Lithium is not the most desirable material for battery construction for at least two reasons: It is expensive, and if BEVs become widespread, the world lithium resources may not be sufficient to satisfy the need for this element. In this case, dwindling lithium supplies would only exacerbate problems of high lithium cost.

It is of interest to consider other battery chemistries that may make use of chemical reactions similar to those in lithium batteries. Lithium is the lightest of the alkali metals. A glance at the periodic table shows that the next heaviest alkali metal is sodium. In fact, sodium exhibits much of the same chemical behavior as lithium. Sodium has two huge advantages over lithium: It is inexpensive, and its supply is virtually unlimited, being half of all the atoms in common salt.

The earliest dedicated research effort on sodium-based batteries was conduced in the 1960s by Ford Motor Company. This work dealt with batteries utilizing the reaction between sodium and sulfur during discharge of the cell:

$$2Na + 4S \rightarrow Na_2S_4.$$

Because this reaction is reversible, the cell can be recharged. The design of a typical cell is illustrated in the figure and consists of a liquid sodium electrode and a liquid sulfur electrode separated by a solid electrolyte, through which sodium ions migrate to react with the sulfur. Na-S batteries have very high specific energy, and this feature would seem to make them attractive candidates for a number of portable power applications. However, Na-S batteries have two serious drawbacks: (1) The properties of the reaction and the behavior of sodium make it necessary to operate the cell at high temperature (typically around 350°C). (2) The sodium polysulfide formed during the discharge reaction is highly corrosive and represents a potential safety hazard.

Because of these features, Na-S batteries may be most suitable for large-scale stationary applications such as grid backup-power storage rather than transportation use. Recent progress has been made in the development of lower temperature (<100°C)

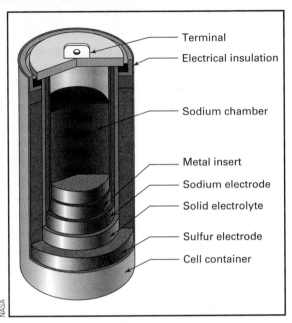

Construction of a sodium battery.

Na-S batteries where the electrodes remain solid. An ion-selective membrane that allows Na ions to pass is used in place of the solid electrolyte.

Sodium-ion batteries are another approach to using Na-based chemistry to produce rechargeable batteries that have been researched in recent years. Designs are similar to those used for Li-ion batteries and often use a solid carbon-based anode and a solid transition metal compound for the cathode. Major advantages of this design are the elimination of highly corrosive reaction by-products and the ability to operate the battery at or near room temperature. These features, along with a high specific energy, make Na-ion batteries an attractive possibility for vehicle use. Although research in this field is at its very early stages (compared with Li-ion battery research), Na-based batteries show promise as a viable future technology.

Topic for Discussion

If the world's supply of lithium were to be replaced by sodium (kilogram for kilogram) obtained from the oceans, estimate what fraction of the salt in the world's seawater would be utilized? Would this have undesirable environmental effects?

world's Li supplies. Some estimates put this number in the 500–600 million range, if all known Li is used for electric vehicle batteries. This is comparable to the number of vehicles currently on the road and suggests reasons for concern.

19.4 History of BEVs

The term *electric vehicles* actually applies to a broad range of devices that includes well-established technologies such as electric trains. Battery electric vehicles (BEVs) are devices that carry their own energy source in the form of batteries, and this requirement makes their technology much more complex. BEVs may include buses on one end of the size range and nonroad-going vehicles such as golf carts on the other. This discussion concentrates on vehicles for personal transportation (automobiles) that are suitable for road use. The history of BEVs predates that of gasoline-powered vehicles because the development of the electric motor predates the invention of the internal combustion engine. The first BEVs were constructed around 1830, but not until the 1880s were practical vehicles for human transportation available on the market. For some time, electric cars were more plentiful and popular than gasoline-powered automobiles. The reasons for this were that the BEV was

- Cleaner.
- Quieter.
- More reliable.
- Easier to start.
- More powerful.

The dominance of electric vehicles continued until around 1920, and by 1930 gasoline vehicles were in use almost exclusively. The reasons for the decline of the electric vehicle are:

- The invention of the electric starter in 1912 made starting gasoline vehicles much easier.
- The need for traveling longer distances made the limited range of the electric vehicle less convenient (particularly a factor in the United States).
- The development of mass production techniques (primarily a result of manufacturing techniques developed by Ford) for the manufacture of gasoline-powered vehicles made them much less expensive than they had been in the past.
- Technological developments made gasoline vehicles more reliable than they had previously been.

It was not until the 1970s and 1980s, with diminishing oil supplies, the increasing reliance (in the United States) on foreign energy, and concerns over pollution and greenhouse gas emissions, that the electric vehicle was reconsidered.

In the early days of automobile development, most speed records were held by electric vehicles because of their greater power output compared with gasoline-powered vehicles. The first automobile to exceed 100 km/h was the Jamais Contente (Figure 19.3 on the next page). On the more practical side, electric vehicles for urban use (Figure 19.4) were common in the early part of the last century. These typically had a range of 100–150 km and a top speed of around 30 km/h. This performance was sufficient to fulfill the needs of many drivers at that time. Some novel electric

Figure 19.3: The Jamais Contente (1899), designed and built by Camille Jenatzy.

Public Domain

Figure 19.4: A 1916 Detroit Electric vehicle.

Q-Images/Alamy Stock Photo

vehicles were produced during that era (Figure 19.5). The Bugatti Type 32 was a miniature electric sports car and was more of an expensive toy than an automobile.

In more recent years, the development of electric vehicles has been aimed at providing a clean replacement for gasoline vehicles. Perhaps the earliest of these recent vehicles that met with some success was the General Motors EV1 (Figure 19.6). The EV1 was produced from 1996 to 1999. Some of its specifications are given in Table 19.6 on page 608. Another vehicle that was produced in fairly substantial numbers during about the same time was the Toyota RAV4 EV (Figure 19.7; specifications in Table 19.6). By comparison with gasoline vehicles, these cars tended to be heavy, were somewhat underpowered, and had a relatively small range. This, coupled with recharge times of up to 8 hours, made their use limited. Cost was a factor that was difficult to analyze. Many of the electric vehicles produced during this era were available

Figure 19.5: A 1929 Bugatti Type 52 electric vehicle.

Dale Chappell

Figure 19.6: The GM EV1 (produced from 1996–1999) being charged.

Marmaduke St. John/Alamy Stock Photo

Figure 19.7: Toyota RAV4 EV produced 1997–2003.

IFCAR

Table 19.6: Specifications of the General Motors EV1 and the Toyota RAV4 EV. Because specifications (particularly of the EV1) changed over the years of its production, the information shown is for later year vehicles.

maker	general motors	toyota
model	EV1	RAV4 EV
years produced	1996–1999	1997–2003
number produced	1117	1249
battery type	NiMH	NiMH
mass	1320 kg	1565 kg
power output	102 kW	50 kW
top speed	128 km/h	126 km/h
range	260 km	160 km

to the general public, but not for sale; rather for lease only. In general, these vehicles were heavily subsidized and the EV1, for example, had a lease cost that was similar to a vehicle in the $30,000 to $40,000 range, whereas a detailed cost analysis suggested production costs that were closer to $80,000 per vehicle. General Motors destroyed virtually all EV1s after their leases expired.

The most successful (in terms of production numbers) electric vehicle in the early 2000s was the Reva (Figure 19.8). The vehicle was manufactured in India and sold worldwide. The United Kingdom was its major market, where it was called the G-Wiz. It was manufactured until 2012, and a total of about 4600 vehicles were sold. It was subsequently replaced by the Mahindra e2o (as discussed below). Early versions of the Reva used Pb-acid batteries, and the vehicle had a range of about 80 km and a top speed of about 80 km/h. From 2009–2012 it was renamed the Reva L-ion, and Li-ion batteries were used. These increased the range to about 120 km.

Figure 19.8: The 2001–2012 Reva electric vehicle made in India.

Premkudva

Figure 19.9: Tesla Roadster 2009–2012.

Olga Besnard/Shutterstock.com

The Tesla Roadster (Figure 19.9) was manufactured from 2009 to 2012 and represented a major step in the technical development of a viable electric vehicle. The Tesla Roadster was manufactured in the United States, although the components were fairly international in nature, with many of the (nonelectric) components of the car produced in collaboration with Lotus in the United Kingdom. The large amount of power available from batteries is an attractive feature in the design of a sports car. The Roadster was the first production electric vehicle that was highway-capable and had a range of over 300 km. Table 19.7 on the next page summarizes the specifications of the Tesla Roadster.

Example 19.3

Calculate the range of a 75 kW BEV utilizing NiMH batteries.

Solution

From Table 19.2, the density power of a NiMH battery is 0.8 kW/kg. Thus, 75 kW requires

$$\frac{75 \text{ kW}}{0.8 \text{ kW/kg}} = 94 \text{ kg}$$

of batteries. Using an energy density for NiMH from Table 19.2 of 0.28 MJ/kg gives a total energy available of

$$(94 \text{ kg}) \times (0.28 \text{ MJ/kg}) = 26 \text{ MJ}.$$

Using an estimated energy requirement of 0.7 MJ/km gives a total range of

$$\frac{26 \text{ MJ}}{0.7 \text{ MJ/km}} = 37 \text{ km}.$$

Table 19.7: Specifications of the Tesla Roadster electric vehicle.	
mass kg	1235
power kW	185
top speed km/h	210 [electronically limited]
range km	390
cost (US$)	109,000
charge time hours	3.5
battery type	Li-ion
years manufactured	2009–2012
total production	**2450**

Since about 2010 the BEV market has grown enormously, as evidenced by the number of electric vehicles registered in the United States (Figure 19.10). Table 19.8 gives the specifications of some of the notable battery electric vehicles currently available worldwide. While some are some dedicated electric vehicles (e.g., BMW i3, Tesla Model X), many are based on existing gasoline engine vehicle platforms (e.g., Honda Fit EV, Fiat 500e).

As Table 19.8 indicates, currently available battery electric vehicles range from small cars with relatively little power and small ranges that are designed primarily for city use through compact cars with more power and typically larger ranges, to fairly expensive high-performance vehicles. The Mitsubishi i-MiEV (Figure 19.11) is a very

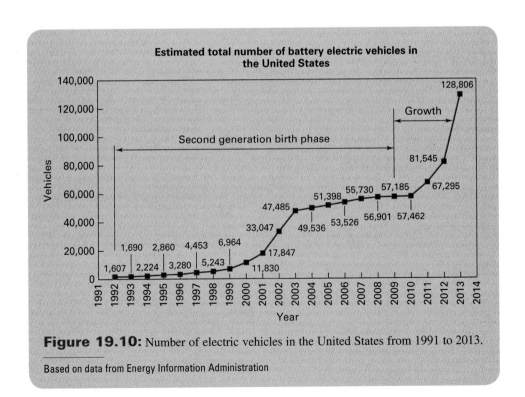

Figure 19.10: Number of electric vehicles in the United States from 1991 to 2013.

Based on data from Energy Information Administration

Table 19.8: Specifications of some currently available battery electric vehicles (for 2016 or 2017 year models). Some vehicles are available with optional larger-capacity batteries; however, specifications and pricing in the table are for currently available base models. Vehicles are ordered by power.

make	model	mass (kg)	power (kW)	power/ mass (kW/ kg)	top speed[1] (km/h)	range[2] (km)	price[3] (10³US$)	year[4]
Mahindra	e2o	830	19	0.023	90	120	12	2013
Mitsubishi	i-MiEV	1080	47	0.044	130	100	29	2009
Roewe	E50	1080	47	0.044	120	180	36	2012
smart	ed	900	55	0.061	125	145	26	2013
Renault	Zoe	1468	66	0.045	135	240	26	2012
Nissan	Leaf	1537	80	0.052	150	135	30	2010
Kia	Soul EV	1492	81	0.054	145	147	33	2014
Fiat	500e	1351	83	0.061	140	135	32	2013
VW	e-Golf	1538	86	0.056	138	190	30	2012
Honda	Fit EV	1478	92	0.062	144	130	37	2012
Chevrolet	Spark EV	1356	97	0.072	145	132	27	2013
Ford	Focus Electric	1643	107	0.065	135	122	30	2011
BMW	i3	1195	125	0.105	150	130	42	2013
Mercedes Benz	B250e	1725	132	0.077	160	200	42	2015
Tesla	Model X	2390	193	0.081	210	414	74	2015
Tesla	Model S	1961	225	0.115	210	340	66	2012

Notes:
(1) In many cases, top speed is limited electronically.
(2) For vehicles marketed in the United States, range figures are from the U.S. EPA. For vehicles that are not marketed in the United States., different testing protocols may be used, which make a direct comparison with EPA figures difficult.
(3) Pricing is before any government incentives and are exclusive of taxes. Prices vary depending on country, and U.S. prices for models sold in the United States are given when available. For vehicles that are not sold in the United States, prices are in the country of manufacture in equivalent US$.
(4) Year of first production model. In some cases, prototype, test, or limited edition models were sold or leased previously.

Figure 19.11: Mitsubishi i-MiEV

Fedor Selivanov/Shutterstock.com

Figure 19.12: Nissan Leaf.

Teddy Leung/Shutterstock.com

successful (in terms of total sales) example of the first category of electric vehicle. The Nissan Leaf (Figure 19.12) and the BMW i3 (Figure 17.35) are characteristic of the second group of vehicles. The Tesla S (Figure 19.13) has been a very successful high-end electric vehicle.

Two significant battery electric vehicles are the Chevrolet Bolt EV (Figure 19.14), which is becoming available for the 2017 model year, and the Tesla Model 3 (Figure 19.15), which is expected to become available in 2018. Specifications for these two vehicles are summarized in Table 19.9. A comparison of these specifications with the benchmark specifications (Table 19.4) shows that BEVs that are now becoming available to the consumer are very close to fulfilling these ideal criteria. Vehicle weight, power, and cost are becoming competitive with similarly sized gasoline vehicles. While range has improved greatly in recent years, charging time is still a serious concern, as is

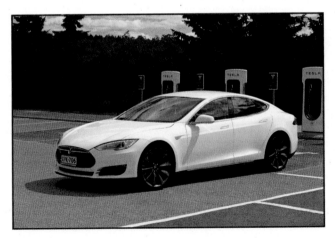

Figure 19.13: Tesla Model S.

Taina Sohlman/Shutterstock.com

Figure 19.14: 2017 Chevrolet Bolt EV.

Steve Lagreca/Shutterstock.com

Figure 19.15: Tesla Model 3

dpa picture alliance/Alamy Stock Photo

Table 19.9: Specifications of Chevrolet Bolt EV (2017 model) and Tesla Model 3 (advertised or estimated specifications for 2018 model). See notes for Table 19.8 for additional information.

make	model	mass (kg)	power (kW)	power/ mass (kW/kg)	top speed (km/h)	range (km)	price (10^3US$)	year
Chevrolet	Bolt EV	1624	150	0.092	146	383	37	2017
Tesla	Model 3	1500	225	0.150	200	345	35	2018

the infrastructure (or lack thereof) of charging stations that would be necessary for the use of BEVs for longer trips. Charging times depend not only on battery design but also on current supplied by the charging electronics. While charging from domestic power sources (i.e., home outlets) is still time consuming, overnight charging for short-range vehicle use is a viable option. Tesla has developed a "supercharger" that would allow adding 270 km of vehicle range for 30 minutes of charging, and the implementation of a network of such stations would go far to encouraging the more extensive use of BEVs. The development of battery-swapping stations is also a possible approach to facilitating the widespread use of electric vehicles. The practicality and economics of such a possibility require careful consideration.

19.5 Supercapacitors

A mechanism for storing electrical energy that may be of relevance to battery electric vehicles, as well as series hybrids (Chapter 17) and fuel cell vehicles (Chapter 20), is the *supercapacitor* (sometimes referred to as an *ultracapacitor*). The position of the supercapacitor on a Ragone plot is shown in Figures 19.1 and 19.2. The figures show clearly that the supercapacator stores relatively little energy (compared to batteries, for example) but that it can release this energy very quickly. Thus, this is a low-energy, high-power device. Its ability to release energy quickly also means that it can store energy quickly. The timescale for energy storage and release is illustrated on the Ragone plot in Figure 19.1. In principle, a supercapacitor is really just a normal capacitor that can store more energy and release more power. Thus, we begin with a discussion of the operation of capacitors in general.

Unlike a battery, which stores electrical energy using chemical reactions, a capacitor stores electrical energy in an electric field by separating positive and negative charges. The simple parallel-plate capacitor shown in Figure 19.16 consists of two conductive plates of area A separated by an insulating dielectric of thickness d. The capacitance of this device is

$$C = \frac{\varepsilon_r \varepsilon_0 A}{d}, \tag{19.6}$$

where ε_r is the *relative static permittivity* (sometimes in the past this has been referred to as the *dielectric constant*), and ε_0 is the permittivity of vacuum. In SI units, d is measured in meters, A is measured in square meters, ε_r is dimensionless, and ε_0 is 8.854×10^{-12} F/m (F = farad = J/V^2).

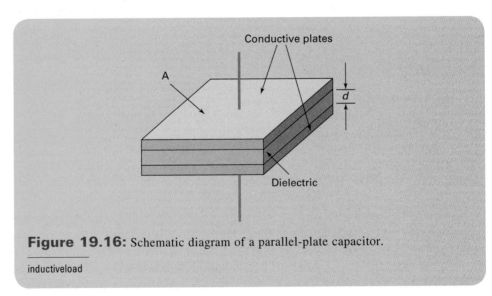

Figure 19.16: Schematic diagram of a parallel-plate capacitor.

inductiveload

Example 19.4

Calculate the capacitance (in Farads) of a parallel-plate capacitor utilizing a dielectric with a relative static permeability of $\varepsilon_r = 1$ and with an area of 10 cm × 10 cm and a distance between the plates of 0.1 mm.

Solution

Using all dimensions in meters in Equation (19.6), the capacitance is found to be

$$C = \frac{\varepsilon_r \varepsilon_0 A}{d} = \frac{(1) \times (8.854 \times 10^{-12}\,\text{F/m}) \times (0.1\,\text{m}) \times (0.1\,\text{m})}{0.0001\,\text{m}} = 8.85 \times 10^{-10}\,\text{F}.$$

If the capacitor is charged, a voltage difference, V, exists between the two plates. The energy stored in the capacitor under these conditions is

$$E = \frac{1}{2}CV^2. \tag{19.7}$$

It is easily seen that, for a capacitance in farads and a voltage in volts, the energy is given in joules. If a 1-F capacitor (compare with the capacitance calculated in Example 19.4) is charged to 10 V, then this represents 50 J of energy; this is enough to illuminate a 60-W lightbulb for just under a second, not a significant factor for electric vehicle use. The energy storage capabilities can be improved by increasing either V or C. Increasing V has a clear advantage because E is proportional to V^2. However, there is a limit for voltage, V_{max}, at which the dielectric breaks down and starts conducting, thereby shorting out the capacitor. The maximum voltage is related to the breakdown electric field of the dielectric material, E_b:

$$V_{max} = dE_b, \tag{19.8}$$

where E_b is in V/m. C can be increased by increasing either ε_r or A or by decreasing d. The simplest approach is merely to make the capacitor larger. However, there are certain limits to this approach if the capacitor is to be practical, and it is generally most desirable to maximize the energy per unit volume (or mass). Decreasing d beyond some limit is somewhat counterproductive. Although it increases C [equation (19.6)], it also decreases V_{max} [equation (19.8)]. An analysis of these equations shows that the maximum energy storage capacity is

$$E_{max} = \frac{1}{2}CV_{max}^2 = \frac{1}{2}\frac{\varepsilon_r \varepsilon_0 A}{d}(E_b d)^2 = \frac{1}{2}Ad\varepsilon_r \varepsilon_0 E_b^2. \tag{19.9}$$

The volume of the capacitor is Ad, and its mass is given in terms of its density, ρ, as $m = Ad\rho$. Most dielectric materials have densities around that of water. Thus, the maximum energy density per unit mass (in J/kg) from equation (19.9) is

$$\left(\frac{E}{m}\right)_{max} = \frac{1}{2}\left(\frac{\varepsilon_r \varepsilon_0}{\rho}\right)E_b^2. \tag{19.10}$$

Traditionally, capacitors have often been made by rolling up parallel plates with an insulating material between the layers. This does not increase the energy density, but it makes the geometry more compact. To actually maximize the energy density, it is

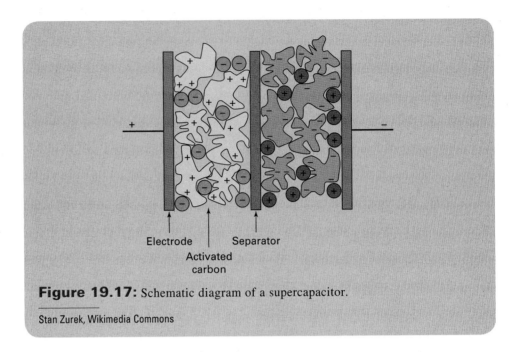

Figure 19.17: Schematic diagram of a supercapacitor.

important to consider ε_r and E_b. Although there is often a trade-off between these factors, maximizing E_b is obviously the more important.

Supercapacitors approach the energy density problem from two directions: maximizing E_b and reconsidering the geometry. Polyethylene terephthalate (PET) is the most commonly used dielectric. This material has values of $\varepsilon_r = 3$ and $E_b = 7.1 \times 10^8$ V/m and increases the capacitance by about a factor of four over traditional materials. It is electrode geometry, however, where supercapacitors benefit the most. As shown in Figure 19.17, nanoporous carbon is used as the electrode material. This material has numerous pores into which the dielectric penetrates, providing a substantially larger surface area onto which to accumulate charge than the macroscopic area of the surface. The improvement over normal capacitors is illustrated in Figure 19.1.

Example 19.5

Using a specific energy for a supercapacitor of 2 Wh/kg (Figure 19.1), calculate the range of a vehicle utilizing 1000 kg of supercapacitors.

Solution

The total energy content of 1000 kg of (fully charged) supercapacitors is

$$(2 \text{ Wh/kg}) \times 1000 \text{ kg} = 2 \text{ kWh}.$$

Converting to MJ gives

$$2 \text{ kWh} \times 3.6 \text{ MJ/kWh} = 7.2 \text{ MJ}.$$

Using a typical energy requirement for a vehicle of 0.7 MJ/km gives

$$\frac{7.2 \text{ MJ}}{0.7 \text{ MJ/km}} = 10.3 \text{ km}.$$

Even given the improvement of supercapacitors over normal capacitors, Figure 19.1 shows that the energy density is very low compared to batteries. An electric vehicle built by NASA using 1000 kg of supercapacitors rather than batteries had a range of about 7 km, consistent with the preceding example calculation. Clearly, this would suggest that it is unlikely that supercapacitors would compete with batteries as a sole source of energy for an electric vehicle. However, the place where supercapacitors excel is in power density. As Figure 19.1 shows, they can provide ten times or more the amount of power per unit mass as a Li-ion battery.

The advantages and disadvantages of supercapacitors over batteries can be summarized as follows:

Advantages

- High power density.
- High voltages available from a single unit.
- No chemical reactions.
- Up to 10^6 charge/discharge cycle life.
- Fast charge/discharge.
- No sophisticated charge/discharge electronics needed to avoid overcharging or damage.

Disadvantages

- Low energy density.
- Linear voltage decrease during discharge.
- High self-discharge rate compared with batteries.

These properties make them suitable for specific applications, including:

- Additional power for a BEV when there is a demand for excess power.
- Short-term energy storage for regenerative braking.
- Short-term power applications such as starter motors, where they could replace conventional batteries.

Also, nonvehicle applications can include short-term backup power for electronic devices.

19.6 Summary

Battery electric vehicles provide several very attractive features as alternatives to petroleum internal combustion engine vehicles. The chapter began with an overview of battery technology and showed that batteries are highly efficient storage devices for electrical energy and, combined with the high efficiency of an electric motor, provide a very energy-efficient vehicle. The Ragone plot has been introduced as a means of understanding the energy-power relationship for energy storage mechanisms. While the energy density of batteries is less than that of fossil fuels, Li-ion batteries, in particular, offer a good compromise between high specific energy and high specific power.

Electric vehicles themselves do not emit greenhouse gases during use, although an analysis of the net carbon footprint must take into account the source of the electricity used to charge the batteries. Even in the worst case (from a carbon standpoint) where electricity is generated from coal, the overall primary-energy-source-to-wheel efficiency of the BEV is very good, perhaps 30–35%, or about twice that for a conventional internal combustion engine. An added bonus is that, even if fossil fuels are used to generate electricity, it is much easier to sequester carbon from a few large coal-fired generating stations than from a very large number of individual vehicles.

The chapter reviewed the history of the development of the BEV and has shown that this was a popular technology in the early development of the passenger vehicle. In recent years, interest in BEVs has once again increased due to concerns over the future availability of petroleum-based fuels and their effects on the environment.

The major concerns for BEVs that utilize current battery technology are vehicle cost, range, and recharge time. Current vehicle characteristics that are suitable for typical urban vehicle use put constraints on the use of BEVs for longer-distance travel. Recent advances in range and the production of vehicles that are more price competitive with gasoline vehicles has made BEVs more attractive for consumer use.

As the availability and cost of lithium associated with the large-scale replacement of fossil fuel–powered vehicles with BEVs are uncertain, the development of new battery chemistries that do not require lithium is attractive. There is active research in the development of sodium-based batteries. Sodium shares many of the chemical properties of lithium (because of similarities in their electronic structure) but is inexpensive, and its availability is virtually unlimited (as a component of salt in the oceans).

This chapter has shown that the use of supercapacitors may help to improve some aspects of vehicle performance. However, because power is generally less of a concern for BEVs than range, supercapacitors do not resolve the principal concerns related to BEVs.

Problems

19-1 What are the ranges of typical (family-size) battery electric vehicles that utilize 250 kg of batteries if the batteries are (a) Pb-acid, (b) NiMH, and (c) Li-ion?

19-2 A solar photovoltaic installation generates an average of 20 MW of power during a 12-hour period of sunlight and no power during 12 hours of night. A Pb-acid battery system is designed to store one day of solar energy for use at night, and during periods of less than full sunlight. What is the mass of the battery system?

19-3 What is the range of a family-sized vehicle utilizing 100 kg of supercapacitors as an energy storage mechanism? What is the maximum power output (in watts)?

19-4 A family vehicle uses 100 kg of Li-ion batteries in conjunction with 200 kg of supercapacitors for energy storage. What is the maximum power output, and what is its range? Would it be advantageous to increase or decrease the relative mass of the batteries compared to the supercapacitors in order to improve vehicle characteristics?

19-5 A typical Pb-acid automobile battery weighs 20 kg. Compare the amount of energy it stores when fully charged to the energy content of 1 L of gasoline.

19-6 (a) Consider a Pb-acid battery backup system for a residence. Using information in Chapter 2, estimate the mass of batteries needed to provide the average electric power for a home (exclusive of heating).

(b) Using the energy density of Pb-acid batteries, how long could backup power be provided at this level?

19-7 A gasoline-powered automobile burns 7.0 L of fuel per 100 km while traveling on the highway at a constant velocity. If the engine is 19% efficient, how much power is actually being used to move the vehicle?

19-8 Assume that the batteries in an electric vehicle are designed to provide the same energy to the vehicle's wheels as 75 L of fuel in the tank of a gasoline-powered vehicle. The batteries are charged by connecting them to a DC charger that can supply 30 A at 220 V (this is typical of the AC circuit used to provide electricity for a kitchen stove if differences between DC and AC power are not considered). How long will it take to charge the batteries? Be sure to take the relative efficiencies of gasoline-powered vehicles and BEVs into account.

19-9 (a) How high would a Pb-acid battery have to be lifted so that its gravitational potential energy was equal to its chemical energy storage capacity?

(b) Repeat part (a) for a Li-ion battery.

19-10 (a) If primary alkaline D cells are used to power an electric (family-size) vehicle at a cost of $0.50 per cell, what is the cost of driving 1 km?

(b) What would be the mass and approximate volume of a battery pack consisting of alkaline D cells that could provide a range of 100 km for a family vehicle?

19-11 Estimate the mass of a disk-shaped steel flywheel that would be required to provide power for a family vehicle that has a range of 500 km.

19-12 For the vehicle considered in Example 19.5, estimate the available power.

19-13 An owner of an electric vehicle with a range of 200 km plans to drive the vehicle only at night and to charge the batteries using an array of solar photovoltaic cells during the daytime. What is the area of photovoltaic cells that is required? State whatever assumptions you make concerning insolation, photovoltaic characteristics, and geometry, as well as any relevant assumptions concerning vehicle design or performance.

19-14 An electric vehicle owner drives 25,000 km per year and uses electricity from a grid that is supplied by coal-fired generating stations. What is the annual savings in CO_2 emissions compared to a vehicle powered by a conventional internal combustion engine? State whatever assumptions you make concerning efficiencies, etc.

19-15 The Tesla Model S (performance version) has a total battery capacity of 85 kWh. The battery pack is composed of a large number of individual Li-ion cells. The cells are 3100 mAh 18650 cells that produce 3.6 V. The designation

18650 refers to the size of the cylindrical battery, 18 mm in diameter and 65.0 mm in length. Calculate the number of individual 18650 cells in the Tesla battery pack.

19-16 Supercapacitors benefit from large effective surface areas and small electrode separation distances. A typical small supercapacitor (produced by Maxwell Industries, San Diego, CA) has a capacitance of 100 F. It is a cylindrical device with a length of 4.5 cm and a diameter of 2.2 cm. What are the dimensions of a 100-F traditional parallel-plate capacitor (with a square geometry) that uses an electrolyte with a dielectric constant $\varepsilon_r = 2.0$ and has a plate separation of 0.01 mm?

19-17 If the number of electric vehicles in the United States continues to increase at the same rate as it did between 2012 and 2013, when will the number of electric vehicles reach 100 million?

19-18 (a) Using the specified maximum output power for the NiMH battery-powered vehicles in Table 19.6, calculate the battery mass on the basis of the characteristics of NiMH batteries given in the chapter and estimate the total energy content of the batteries.

(b) From the result in part (a) calculate the energy (in MJ) required per km.

19-19 Estimate the mass of supercapacitors necessary to provide a vehicle with the same range as an internal combustion gasoline engine vehicle with a 70-L fuel tank.

19-20 Describe the trends in specific energy as a function of battery size for Zn-C and alkaline cells. Show that the values given for large batteries in the text are reasonable.

Bibliography

J. O. Besenhard (Ed.). *Handbook of Battery Materials.* Wiley VCH, Weinheim, Germany (1999).

D. R. Blackmore and A. Thomas. *Fuel Economy of the Gasoline Engine.* Macmillan, London (1977).

H. Braess and U. Seiffert (Eds.). *Handbook of Automotive Engineering.* Society of Automotive Engineers, Warrendale, PA (2004).

J. A. Kraushaar and R. A. Ristinen. *Energy and Problems of a Technical Society,* (2nd ed.) Wiley, New York (1993).

J. MacKenzie. *The Keys to the Car: Electric and Hydrogen Vehicles for the 21st Century.* World Resources Institute, Washington, DC (1994).

D. A. J. Rand, R. Woods, and R. M. Dell. *Batteries for Electric Vehicles.* Research Studies Press, Somerset, United Kingdom (1998).

R. Stone and J. Ball. *Automotive Engineering Fundamentals.* Society of Automotive Engineers, Warrendale, PA (2004).

F. M. Vanek and L. D. Albright. *Energy Systems Engineering: Evaluation and Implementation.* McGraw Hill, New York (2008).

Hydrogen

Learning Objectives: After reading the material in Chapter 20, you should understand:

- The basic properties of hydrogen as a gas and as a liquid.
- Methods for hydrogen production.
- The factors that must be considered when storing gaseous or liquid hydrogen.
- The use of hydrogen as a fuel in internal combustion engines.

- The properties and types of fuel cells.
- The design of fuel cell vehicles.
- The viability of hydrogen as a fuel and efficiency considerations for its use.

20.1 Introduction

In previous chapters, two possible technologies for future nonfossil-fuel transportation have been discussed: biofuels and battery electric vehicles. Both of these have been the subject of intense development in recent years, and both have seen commercial availability in North America in the form of flex fuel vehicles and electric vehicles. However, neither of these technologies is without its own drawbacks. An alternative transportation technology that has also been promoted is hydrogen-powered vehicles.

Hydrogen, like a battery, is an energy storage mechanism. It is not a primary energy source, like oil or coal, because no significant sources of naturally occurring hydrogen exist in nature; although hydrogen-generation schemes based on organic processes may be viewed somewhat like biofuels. Hydrogen may be produced from electricity through electrolysis and can be used to provide thermal or mechanical energy by burning, or by conversion back into electricity by means of a fuel cell. The oxidation of hydrogen yields water as its only by-product (although nitrogen compounds can be created from atmospheric nitrogen in any combustion process that occurs at an elevated temperature). Like all methods of producing usable energy, the use of hydrogen must be viewed in detail, from production to end use, in order to fully understand its viability in a future energy economy.

Jim West/Alamy Stock Photo

20.2 Properties of Hydrogen

Energy is produced by combining the hydrogen with oxygen to produce water through the process

$$2H_2 + O_2 \rightarrow 2H_2O. \tag{20.1}$$

The energy content of hydrogen, that is, the energy gain during oxidation, is quite substantial compared with other fuels (Figure 20.1). Although this figure suggests the advantages of hydrogen as a fuel, there are some important considerations for its use. Firstly, hydrogen is a gas at standard temperature and pressure (STP = room temperature and 101 kPa) and occupies a substantial volume. Table 20.1 shows a comparison between liquid gasoline and gaseous hydrogen at STP. Clearly, the volume occupied by hydrogen at STP, as a result of its low density (0.0899 kg/m^3), is a drawback that would necessitate some means of reducing the volume of the fuel. This is true even for stationary applications, but is particularly the case for transportation use. The hydrogen equivalent of 60 L of gasoline (typical for an automobile gasoline tank) at STP would be about 200 m^3, or a cube 6 m on a side. One option would be to compress hydrogen gas while keeping it at room temperature. Another option would be to liquefy hydrogen

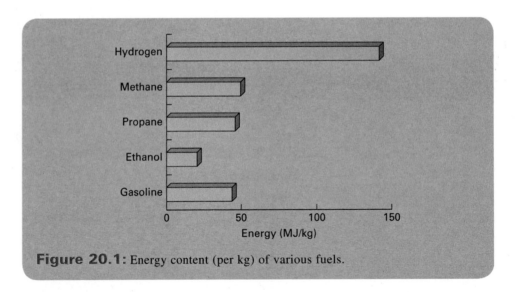

Figure 20.1: Energy content (per kg) of various fuels.

Table 20.1: Specific and volumetric energy densities for fuels shown in Figure 20.1. Gases are at standard temperature and pressure. Values are higher heating values as defined in Chapter 1.

fuel	energy per kg (MJ)	energy per m³ (MJ)
hydrogen	142	11.8
methane	55.5	36.4
propane	50.35	191.2
ethanol	29.8	23,512
gasoline	44.5	34,800

to make a liquid fuel and to maintain it at low temperature. The energy density of liquid hydrogen is 8520 MJ per m³; this is less than gasoline by about a factor of four. Although the energy density per kilogram is greater for hydrogen, its density is less than that of gasoline (71 kg/m³ vs. 720 kg/m³). The possibilities for storing hydrogen in a convenient form are discussed further in Section 20.4. The various mechanisms for producing hydrogen are considered first.

20.3 Hydrogen Production Methods

There are three basic methods for producing hydrogen:

1. Chemical reactions.
2. Electrolysis.
3. Thermal decomposition of water.

The basic features of each of these is discussed below.

20.3a Chemical Reactions

Steam reforming is a process where hydrogen is produced during a high-temperature reaction between methane and water:

$$CH_4 + H_2O \rightarrow CO + 3H_2. \tag{20.2}$$

The heat necessary to drive this reaction can be provided by the combustion of about 20% of the CH_4 that is input into the system.

This reaction is typically followed by the reaction of the carbon monoxide with water:

$$CO + H_2O \rightarrow CO_2 + H_2. \tag{20.3}$$

This is the water–gas shift reaction.

Other chemical reactions can also release hydrogen. Another common reaction used for hydrogen production is the high-temperature reaction of water with carbon:

$$H_2O + C \rightarrow H_2 + CO. \tag{20.4}$$

This can be followed by the reaction in equation (20.3). These chemical reactions can be an important source of hydrogen for industrial purposes. However, they may not suitable for producing hydrogen in the context of an environmentally conscientious energy technology because they produce CO_2.

20.3b Electrolysis

Electrolysis is the best known method of producing hydrogen. It is the separation of the oxygen and hydrogen atoms of water molecules using electricity. A basic electrolysis cell is shown in Figure 20.2 on the next page. The overall reaction is

$$2H_2O + energy \rightarrow O_2 + 2H_2, \tag{20.5}$$

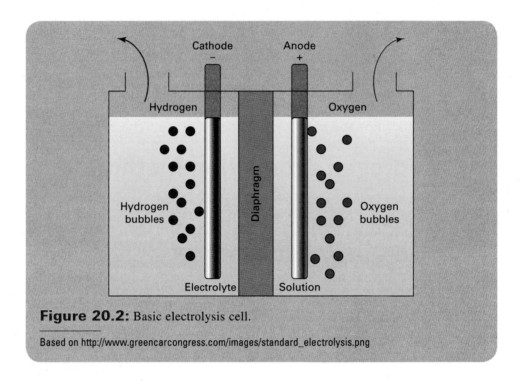

Figure 20.2: Basic electrolysis cell.

Based on http://www.greencarcongress.com/images/standard_electrolysis.png

where the energy is supplied in the form of electrical energy. It is convenient to consider the two half reactions that take place at the electrodes in the cell. Hydrogen is formed by decomposition of water at the cathode (negative electrode). In pure water this reaction is

$$2H_2O \rightarrow H_2 + 2OH^-, \qquad \textbf{(20.6)}$$

where the negative charge is supplied by an electron coming from the power supply. The negatively charged OH^- ion is repelled by the cathode and travels through the water to the anode where it undergoes the reaction

$$2OH^- \rightarrow (1/2)O_2 + H_2O. \qquad \textbf{(20.7)}$$

The negative charge travels back through the external circuit, in the form of an electron, to the power supply. The net result of this process is half of the reaction given in equation (20.5). The electrical conductivity of the water is increased with charged ions produced by dissolving an ionic compound such as H_2SO_4, KOH, or the like.

Electrical energy from the power supply is stored as chemical energy in the hydrogen. Another way in which this process can be viewed is to consider that the energy supplied by the circuit is used to reduce (that is the opposite of oxidize) the water. Energy stored in the hydrogen can be recovered by oxidizing the hydrogen either in a combustion process (Section 20.5) or in a fuel cell (Section 20.6). It is important to realize that this process, like all energy conversion processes, is not 100% efficient. In fact, it is typically about 70% efficient, and this must be considered in an evaluation of any process that involves the storage of energy in the form of hydrogen.

ENERGY EXTRA 20.1
Alternative hydrogen production methods

One interesting approach to the production of hydrogen, sometimes referred to as solar hydrogen, involves the possibility of producing hydrogen directly from the reaction of certain materials with sunlight. These materials may be organic materials that produce hydrogen as a result of photosynthetic processes, or they may be semiconducting materials that produce hydrogen by direct electrolysis with surrounding water molecules. One might envision a simple approach where sunlight is incident on a piece of silicon immersed in water, and electron-hole (e^--h^+) pairs are formed by the interaction of photons with the charges in the material (Chapter 9). These electron-hole pairs can then flow through the water, generating hydrogen and oxygen. The reaction is

$$2h^+ + H_2O \rightarrow 2H^+ + \tfrac{1}{2}O_2$$

and

$$2e^- + 2H^+ \rightarrow H_2.$$

There are several problems with this simple approach:

- The electron-hole pairs tend to recombine in the silicon rather than interact with the water molecules.

- The efficiency for converting photons into hydrogen is very low.
- The silicon is chemically reactive with water and degrades due to corrosion.

Peidong Yang at the University of California, Berkeley, has developed materials that are effective at dealing with these difficulties. By creating nano-sized wires of silicon with a coating of semiconducting TiO_2 on their surface (see the photograph), he has been able to resolve many of these issues. TiO_2 is much less reactive with water than Si, and the coating protects the Si from adverse chemical reactions with water. The junction between the two semiconducting materials helps to prevent electron-hole recombination. Finally, the large surface area of the nanowires presents a large cross section for absorption of photons and interaction with the water.

These materials have shown promise for the direct conversion of sunlight into hydrogen, which can then be used for creating heat by combustion or electricity by means of a fuel cell. It is hoped that this approach can help to alleviate problems related to the inefficiencies and expense of the traditional hydrogen production route.

Lawrence Berkeley National Laboratory

Highly dense array of 20-μm-long nanowires made from TiO_2-coated Si.

Continued on page 626

Energy Extra 20.1 continued

Topic for Discussion

The preceding description is only one of quite a few possible mechanisms for obtaining hydrogen by methods other than the electrolysis of water. Coal is not really pure carbon, but it contains various other elements, including hydrogen. It is possible to extract the hydrogen from coal using catalytic reactions. One possibility is the use of sulfur to react with the carbon and hydrogen in the coal:

$$4CH_{0.5} + S \rightarrow 4C + H_2S.$$

The hydrogen sulfide can then be decomposed to yield hydrogen,

$$H_2S \rightarrow H_2 + S,$$

and the sulfur can be recovered for further use. It has been suggested that soft coal may contain enough hydrogen that the hydrogen can be extracted for use as a carbon-free fuel and the carbon may be left sequestered in the coal and not burned. Discuss the merits of this approach.

20.3c Thermal Decomposition of Water

The thermal decomposition of water corresponds to the reaction given in equation (20.5), except that the energy is supplied in the form of thermal energy rather than electrical energy. Unfortunately, this process is not as simple as it may seem because the temperatures required to dissociate water are unrealistically high. It is, however, possible to use a catalyst to make the process easier. One common approach is the sulfur-iodine thermochemical cycle. This begins with heating sulfuric acid, which causes the reaction

$$\text{heat} + H_2SO_4 \rightarrow (1/2)O_2 + SO_2 + H_2O. \tag{20.8}$$

The sulfur dioxide is combined with iodine and water and heated further to yield

$$\text{heat} + SO_2 + I_2 + 2H_2O \rightarrow 2HI + H_2SO_4. \tag{20.9}$$

The sulfuric acid that is produced is used to fuel reaction (20.8), and the hydrogen iodide is further heated to give

$$\text{heat} + 2HI \rightarrow H_2 + I_2. \tag{20.10}$$

The iodine is used to fuel reaction (20.9), and net hydrogen is produced. The overall reaction results from the addition of one water molecule in equation (20.9) and the production of oxygen in equation (20.8) and hydrogen in equation (20.10) and corresponds to half of the reaction in equation (20.5).

A possible approach to producing hydrogen by the thermal decomposition of water uses heat produced by generation IV nuclear fission reactors (see Energy Extra 6.3). Certain types of generation IV reactors (very high temperature reactors, VHTR) operate at temperatures between about 500°C and 1000°C (compared to temperatures around 300°C for conventional water-moderated thermal neutron reactors). These VHTR may be thermal neutron reactors or fast breeder reactors. The high operating temperature is sufficient for the production of hydrogen by the sulfur-iodine thermochemical cycle. Reactors may be dedicated for the purpose of producing hydrogen, or they may produce electricity and use excess heat to produce hydrogen. The hydrogen produced by nuclear reactors may be used as a transportation fuel or as an energy storage mechanism for stationary fuel cell systems, as described below.

20.4 Storage and Transportation of Hydrogen

Once produced, hydrogen must be transported to its place of distribution and/or use, and it must be stored in an appropriate form. Because of the large volume occupied by hydrogen gas at STP, the hydrogen must be stored in a more compact form. There are three possible options for this:

1. Compressed hydrogen gas (CHG).
2. Liquid hydrogen (LH$_2$).
3. Metal hydrides.

To understand storage as CHG, it is important to look at the properties of hydrogen under pressure. Figure 20.3 shows the relationship between density and pressure for hydrogen.

At relatively low pressures, the density-pressure curve follows the ideal gas relationship reasonably well (i.e.):

$$PV = nRT. \tag{20.11}$$

Here P is pressure, V is volume, n is the number of moles of gas, R is the universal gas constant ($R = 8.315 \ \text{J} \cdot \text{K}^{-1} \cdot \text{mol}^{-1}$), and T is the absolute temperature. This expression can be rewritten in terms of the density, ρ:

$$\rho = \frac{MP}{RT}, \tag{20.12}$$

where M is the molecular mass. This expression shows that there is a linear relationship between density and pressure, and this is indicated in Figure 20.3. As the

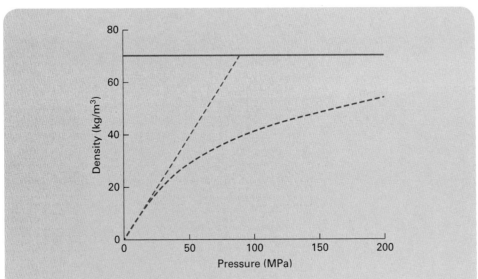

Figure 20.3: Density-pressure relationship for hydrogen gas showing the density of liquid hydrogen (red line) and the ideal gas law relationship (blue line). The green line shows the density of liquid hydrogen.

Example 20.1

A 1-m^3 tank of hydrogen at 100 MPa is used to generate electricity by combustion in a heat engine that drives an electric generator with an overall efficiency of 15%. What is the total electrical energy available in kilowatt-hours?

Solution

From Figure 20.3, the density of hydrogen gas at 100 MPa is about 40 kg/m^3. From Table 20.1, the energy content of hydrogen is 142 MJ/kg. Thus 1 m^3 of hydrogen at 100 MPa will yield

$$(40 \text{ kg}) \times (142 \text{ MJ/kg}) \times (0.278 \text{ kWh/MJ}) = 1580 \text{ kWh}.$$

At an efficiency of 15%, the total electrical energy generated is

$$(1580 \text{ kWh}) \times (0.15) = 237 \text{ kWh}.$$

pressure increases, the density-pressure relationship deviates from the ideal gas law, which ignores the presence of intermolecular interactions. As the pressure is increased and the gas becomes denser, the molecules come closer together, and these interactions become important. As a result, the ideal gas law becomes less applicable as the pressure increases, resulting in the behavior shown in Figure 20.3. This figure also shows the density of liquid hydrogen, which, since it is in the liquid state, is more or less incompressible and sets an upper limit to the density that can be achieved in the gaseous state. Although it is desirable to compress the hydrogen as much as possible to maximize the energy stored per unit volume, there are limits to what is practical. Higher pressures require stronger containment vessels in order to store the hydrogen safely. This means that, as the pressure increases, the cost of the container increases and the potential dangers increase. This is particularly a concern for use of hydrogen as a fuel in a vehicle where weight is a consideration and where the possible consequences of a collision must be considered. From a practical standpoint, hydrogen pressures that have been used in storage tanks for vehicles are typically in the range of about 35–70 MPa corresponding to densities in the range of about 20–35 kg/m^3, or about one-third to one-half the density of liquid hydrogen. A factor to consider for this and any other method for storing hydrogen is its efficiency; that is, how much energy is needed to store the hydrogen? In the case of CHG, the energy needed is that required to run a compressor to compress the gas. This energy is lost and is not recovered when the hydrogen is utilized and typically amounts to 10–20% of the energy content of the hydrogen, making this process 80–90% efficient.

From Figure 20.3, it is obvious that the energy density (MJ/m^3) of liquid hydrogen is greater than for CHG at any pressure. At atmospheric pressure, the boiling point of hydrogen is 20.3 K. Gaseous hydrogen must be cooled down below this temperature in order to convert it into a liquid. Two important factors need to be considered: how much energy is needed to liquefy the hydrogen, and how can it be contained to prevent it (as much as possible) from boiling away. Typically, the energy required for liquefaction compared to the energy content of the hydrogen is about 30%, making the process about 70% efficient.

The effective and efficient liquefaction of hydrogen is somewhat more complex than merely lowering the temperature. The hydrogen molecule consists of two hydrogen atoms bonded together. Each hydrogen atom consists of a nucleus (a proton) and

Example 20.2

If hydrogen behaves as an ideal gas, calculate the room temperature density at a pressure of 50 MPa.

Solution

We use equation (20.12), $\rho = \dfrac{MP}{RT}$, where the molecular mass for hydrogen is 2 g/mol, $R = 8.315 \text{ J} \cdot \text{K}^{-1} \cdot \text{mol}^{-1}$, and $T = 293 \text{ K}$. Note that in basic SI units, the joule is $\text{kg} \cdot \text{m}^2 \cdot \text{s}^{-2}$, and the pascal is $\text{Pa} = \text{kg} \cdot \text{m}^{-1} \cdot \text{s}^{-2}$. Thus equation (20.12) may be written as

$$\rho = \frac{(0.002 \text{ kg} \cdot \text{mol}^{-1}) \times (50 \times 10^6 \text{ kg} \cdot \text{m}^{-1} \cdot \text{s}^{-2})}{(8.315 \text{ kg} \cdot \text{m}^2 \cdot \text{s}^{-2} \cdot \text{K}^{-1} \cdot \text{mol}^{-1}) \times (293 \text{ K})}$$

$$= 41 \text{ kg/m}^3,$$

consistent with the value shown for the straight (ideal gas) line in Figure 20.3 for a pressure of about 50 MPa.

an electron. In the hydrogen molecule, the magnetic moments (spins) of the protons of the two atoms interact and can align either antiparallel or parallel (Figure 20.4). These two spin arrangements are referred to as parahydrogen and orthohydrogen, respectively. The ortho form is at a higher energy than the para form and can be considered an excited state of the parahydrogen molecule ground state. At room temperature, thermal energy causes many of the parahydrogen molecules to acquire enough energy to become orthohydrogen. The equilibrium state at room temperature corresponds to about 25% parahydrogen and 75% orthohydrogen. As the temperature is lowered, the equilibrium parahydrogen fraction increases at the expense of the orthohydrogen fraction. In the liquid state, the temperature is low enough that the equilibrium concentration of orthohydrogen is very small. However, if hydrogen is cooled and liquefied fairly quickly, then the liquid contains excess orthohydrogen. This slowly converts to parahydrogen by flipping the spin of one of the protons. Since the ortho–to-para conversion is an exothermic process (i.e., a transition from an excited state to the ground state), heat is released, causing some of the liquid hydrogen to boil. In practice, catalysts such as iron oxide or carbon, which convert orthohydrogen to parahydrogen, need to be used during the cooling process to minimize the formation of a nonequilibrium mixture of ortho and parahydrogen.

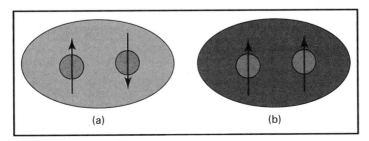

Figure 20.4: (a) Parahydrogen (ground state) and (b) orthohydrogen (excited state), showing the alignment of proton spins.

Table 20.2: Loss rates for typical liquid hydrogen storage tanks of different volumes.	
volume (m³)	loss rate (%/day)
0.1	2
50	0.4
20,000	0.06

Even given these considerations, liquid hydrogen must be effectively insulated from its surroundings to reduce heat transfer into the liquid, causing it to boil. Specially designed containers have been developed for this purpose. The loss rate depends on the quality of the insulation but depends on another important factor; the size of the container. As the heat capacity of a liquid depends on its volume (or mass) and the heat transfer depends on the surface area, the loss rate depends on the surface-to-volume ratio of the container, which is a function of the container volume. Some examples of loss rates (in percent per day) for hydrogen are given in Table 20.2. Although containers at distribution or storage facilities can be quite large, the fuel tank in a vehicle is limited in size. This means that loss of fuel may be a factor if the vehicle is unused for any period of time.

A final point concerns the transfer of liquid hydrogen. Transfer tubes must be well sealed to avoid contact between the liquid hydrogen and air. This is because air (consisting primarily of nitrogen and oxygen) has a higher boiling point than hydrogen, and contact between the two causes air to condense and contaminate the hydrogen.

The final method of storing hydrogen is in the form of metal hydrides. This method relies on reactions of hydrogen with certain metals of the form

$$2M + H_2 \rightarrow MH_2. \tag{20.13}$$

This reaction is exothermic and releases excess heat. One of the commonly used metals for this purpose is titanium, which forms the hydride TiH_2.

Example 20.3

Calculate the mass of hydrogen that can be stored in 1 m³ of titanium. *Note:* The density of titanium is 4510 kg/m³.

Solution

In the TiH_2 phase, two-thirds of the atoms are hydrogen. Using the approximate atomic masses of 1 g/mol for hydrogen and 48 g/mol for titanium, the ratio of the weight of hydrogen to titanium is

$$\frac{2 \text{ g/mol}}{48 \text{ g/mol}} = 0.042.$$

Thus, 1 m³ or 4510 kg of Ti contains

$$(4510 \text{ kg}) \times (0.042) = 188 \text{ kg of hydrogen.}$$

ENERGY EXTRA 20.2
Hydrogen storage in fullerenes

Hydrogen bonds readily with carbon, and this characteristic forms the basis for the use of carbon-containing materials for hydrogen storage. Elemental carbon is known to form a large number of crystal structures, referred to as allotropes. Three allotropes that form readily in nature are graphite, diamond, and amorphous carbon. Other allotropes can be formed under unusual conditions, and several of these based on the graphite structure may be useful materials for hydrogen storage.

The graphite structure consists of layers of carbon atoms arranged in strongly bonded hexagonal rings (analogous to the carbon ring in a benzene molecule), and these layers are loosely bonded together. This loose bonding between layers allows the hexagonal planes to slide with respect to one another, giving graphite its lubricating properties.

A single hexagonal layer of graphite may be removed (sometimes as simply as peeling it off with a piece of adhesive tape) and is referred to as graphene. Graphene was first identified and studied by Andre Geim and Konstantin Novoselov, who won the 2010 Nobel Prize in Physics for their work.

An interesting class of structures that can be derived from graphene are the fullerenes. The most common of these is buckminsterfullerene (named after the American architect Richard Buckminster Fuller who developed the geodesic dome, which has the same geometry as the buckminsterfullerene molecule or buckyball). If we begin with a sheet of graphene where the carbon atoms form hexagonal cells and periodically remove some of the carbon atoms, then we will end up with a structure in which some of the carbon atoms form pentagons rather than hexagons. The missing carbon bonds will cause the sheet to curl, and if we remove the right number of carbon atoms, then we will produce a spherical structure as shown in the figure. The sphere consists of 60 carbon atoms (hence its designation as a C_{60} molecule) with an arrangement of hexagonal and pentagonal faces that is the same as that found on a soccer ball.

Buckyballs and other fullerene molecules have empty space inside where other species of atoms (called endohedral atoms) can be trapped.

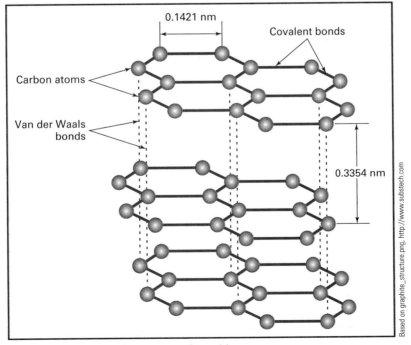

The layered hexagonal structure of graphite.

Based on graphite_structure.png, http://www.substech.com

Continued on page 632

Energy Extra 20.2 continued

Two-dimensional graphene structure.

Since carbon likes to bond to hydrogen, these are ideal structures for hydrogen storage. Theoretical studies of hydrogen atoms inside C_{60} molecules show that up to 58 endohedral hydrogen atoms can be placed inside the molecule [1]. The problem with this approach is that the bonds between the carbon

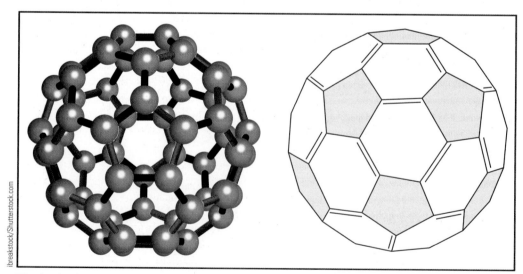

The buckminsterfullerenne C_{60} molecule and a soccer ball.

Continued on page 633

Energy Extra 20.2 continued

atoms in the C_{60} molecule are very strong, so it is very difficult to get the hydrogen atoms inside (and out of) the molecule.

Recent theoretical work [2] has described an approach for dealing with the problem of getting the hydrogen in and out of the C_{60} molecule. It is known that silicon, which is chemically similar to carbon because of its outer electron structure, can substitute for carbon in the C_{60} molecule. If six of the carbon atoms in C_{60} are replaced by silicon, yielding $C_{54}Si_6$, then the silicon atoms can form one of the hexagonal faces of the molecule. Because silicon atoms are larger than carbon atoms, then a hexagonal hole is formed that will allow the hydrogen atoms to pass in and out of the molecule more easily.

Since hydrogen likes to bond to carbon (and silicon), hydrogen atoms will tend to stick to the outside of the $C_{54}Si_6$ molecule rather than going inside. This problem is easily dealt with by covering the molecule with hydrogen atoms, one for each carbon and silicon atom, yielding a molecule of $H_{60}C_{54}Si_6$ as shown in the figure, so that additional hydrogen will preferentially go inside the structure. The theoretical work by Yong et al. [2] has shown that up to 42 hydrogen atoms (in the form of 21 H_2 molecules) can be stored (and recovered) from a single $H_{60}C_{54}Si_6$ molecule. This corresponds to 11.1% hydrogen by weight

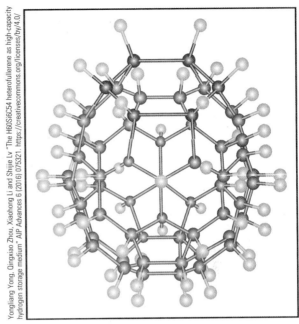

The structure of $H_{60}Si_6C_{54}$. Black atoms = carbon, yellow atoms = silicon, white atoms = hydrogen.

that can be stored reversibly in the material. These recent results are encouraging and are being followed up by experimental investigations to determine the viability of these materials for hydrogen storage.

Topic for Discussion

What would be the mass of a hydrogen storage tank utilizing $H_{60}C_{54}Si_6$ as a hydrogen storage material that could provide fuel to a fuel cell vehicle with a range of 500 km? The United States Department of Energy has placed a benchmark of 6% by weight of hydrogen in an acceptable hydrogen storage material. How does pure C_{60} compare with the U.S. DOE benchmark? Many metals form hydrides of the form MH_2. Which hydrides of this formula meet the U.S. DOE standard? Are there some that are potentially better than the fullerenes?

References

1. O. V. Pupysheva, A. A. Farajian, and B. I. Yakobson, "Fullerene Nanocage Capacity for Hydrogen Storage," *Nano Letters* **8** (2008) 767–774.

2. Y. Yong, Q. Zhou, X. Li, and S. Lv, "The $H_{60}Si_6C_{54}$ heterofullerene as high-capacity hydrogen storage medium," *AIP Advances* **6** (2016) 075321.

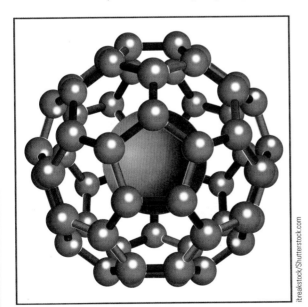

Endohedral buckminsterfullerene molecule.

Table 20.3: Storage capabilities for a 0.1-m³ volume for hydrogen in different forms and a comparison with gasoline. Total masses include the mass of a suitable storage container.

fuel	fuel mass (kg)	total mass, typical (kg)	energy per volume (MJ/0.1m³)	energy per mass (MJ/kg)
CHG (35 MPa)	2.0	100	280	2.8
CHG (70 MPa)	3.5	150	500	3.3
LH_2	7.2	100	1000	10
TiH_2	18	450	2550	5.7
gasoline	72	85	3480	44.5

A comparison of hydrogen storage capabilities of a reasonably sized fuel tank for vehicle use (e.g., 100 L, or 0.1 m³) is given for CHG, LH_2, and metal hydrides in Table 20.3. A comparison with gasoline is also shown.

While the total energy stored in 0.1 m³ of TiH_2 is attractive (in comparison with hydrogen stored by other methods) and begins to approach that of gasoline, there are two major difficulties with this approach. First, the mass of the titanium is quite large and adds substantially to the total mass of the fuel system. Secondly, it is not a simple matter to form the metal hydrides and to subsequently release the hydrogen for use. The process in equation (20.13) is exothermic, and this means that energy is required to remove the hydrogen from the metal hydride. Appropriate technology must be developed to enable effective charging and discharging of the metal hydride.

The transport of hydrogen also presents some challenges. In some ways, this can benefit from the methods and infrastructure associated with the use of natural gas. Hydrogen can be transported across land in pipelines or as CHG in tanker trucks. It may also be transported in appropriately insulated trucks in the form of LH_2. Across oceans, LH_2 ships are the most logical method. It may seem unclear why the transportation of hydrogen across oceans may be of interest because it is an energy storage medium that can be produced from electricity on the continent where it is to be used. However, hydrogen is a potential energy storage mechanism for transporting energy that is produced far offshore to users on land. This may be most relevant to possible future energy technologies such as OTEC, ocean currents, and possibly waves (if not close enough to shore for power transmission cables). Of course, the question of efficiency of energy conversions is very important, especially given the already marginal (at best) efficiency of methods like OTEC.

20.5 Hydrogen Internal Combustion Vehicles

The world's first automobile is shown in Figure 20.5. This vehicle utilized an internal combustion engine (ICE) that ran on hydrogen, and, although not practical for general use, it demonstrated the operational principle of a motorized vehicle. Other vehicle technologies (e.g., gasoline and even battery) dominated the development of the automobile for many years. In recent years, however, there has been renewed interest in the

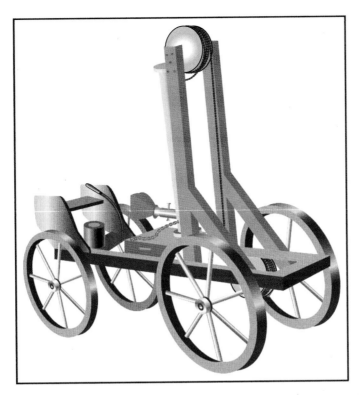

Figure 20.5: The first automobile was built in 1807 by Francois Isaac de Rivaz of Switzerland. It utilized compressed hydrogen gas as a fuel and used an electric spark produced by means of a battery for ignition.

Based on http://www.quantium.plus.com/derivaz/isaac/rivaz.jpg

use of hydrogen as a fuel. The remainder of this chapter is devoted to a consideration of hydrogen as a fuel, primarily for transportation. However, possible uses extend beyond the transportation sector and some of these are discussed here as well.

Current technologies for the use of hydrogen as a transportation fuel fall into two categories: internal combustion engines (discussed in this section) and fuel cells (discussed in the next section).

Hydrogen ICE vehicles are based on production gasoline-powered vehicles. The gasoline internal combustion engine will run on hydrogen with relatively minor modifications. The most significant differences between gasoline and hydrogen ICE vehicles are in the fuel supply system. Thus, the manufacturing infrastructure for the production of hydrogen ICE vehicles is largely in place. The use of hydrogen as a fuel in an ICE has clear advantages. Energy is produced by the reaction in equation (20.1) with the by-product being water. No carbon emissions are produced because there is no carbon in the fuel. Some NO_x compounds are produced becasue this inevitably occurs when anything is burned in air, but these emissions are relatively minor. The efficiency of a hydrogen ICE is similar to that of a gasoline ICE and is ultimately limited by the theoretical Carnot efficiency. For hydrogen ICEs developed for automobiles, this means an efficiency of about 15–20%. According to Table 20.3, it is clear that a major problem with hydrogen vehicles is carrying enough hydrogen in a compact enough space to provide a usable range. For this reason, at present, virtually all hydrogen vehicles that are under serious development are hybrids, that is, vehicles that utilize two energy sources. This is true both of hydrogen ICE vehicles, as well as hydrogen fuel cell vehicles (discussed in the next section). In the case of hydrogen ICE vehicles, the logical alternative fuel is gasoline because the engine can be designed as a two-fuel engine running on either hydrogen or gasoline, and fuel systems delivering both fuels can be incorporated into the vehicle.

Example 20.4

Hydrogen is produced by electrolysis using electricity produced by a natural gas–fired generating station. It is then compressed to 35 MPa for use in a H_2 ICE vehicle. Estimate the ratio of primary energy to energy at the wheels of the vehicle.

Solution

We need to consider the efficiency of each energy conversion in the process as follows:

- The efficiency of the natural gas–fired generating station is limited by the Carnot efficiency and might be around 40%.
- The efficiency of the electrolysis process is about 70%.
- The efficiency of the compression process is about 70%.
- The efficiency of the H_2 ICE is limited by the Carnot efficiency and might be around 17%.

Therefore, the overall efficiency of the process will be

$$100 \times (0.4) \times (0.7) \times (0.8) \times (0.17) = 4\%.$$

So the ratio of primary energy to energy at the wheels will be

$$(0.035)^{-1} = 25 \text{ to } 1.$$

Two companies that have been particularly active in pursuing hydrogen ICEs are BMW and Mazda. BMW's vehicle may be close to limited commercialization, and Mazda's vehicle is well along in its development. The BMW Hydrogen 7 (Figure 20.6) is a luxury vehicle derived from the 760Li gasoline-powered BMW that uses a 6.0-L V-12 internal combustion engine modified to run on either hydrogen or gasoline. It is perhaps unique in that it uses liquid hydrogen (LH_2) rather than compressed hydrogen gas (CHG) as a fuel. This feature has required the development of a specialized insulated hydrogen fuel tank with a suitable filling mechanism. Some of the specifications of this vehicle are given in Table 20.4.

Mazda has modified the rotary engine RX-8 for use as a two-fuel vehicle running on either hydrogen or gasoline. They claim that the details of the combustion cycle in the rotary (Wankel) engine make it particularly suited to burning hydrogen. The Mazda RX-8 RE (Figure 20.7) uses compressed hydrogen gas as a fuel. This is a common feature of virtually all (except the BMW) hydrogen-powered vehicles and has required the development of safe and light high-pressure gas cylinders. Some of the specifications of the Mazda RX-8 RE are shown in Table 20.4.

Table 20.4: Specifications of two hydrogen internal combustion engine (ICE) vehicles.

manufacturer	model	fuels	hydrogen range [km]	gasoline range [km]
BMW	Hydrogen 7	LH_2/gasoline	200	480
Mazda	RX-8 RE	CHG/gasoline	100	530

Figure 20.6: BMW Hydrogen 7, powered by a two-fuel internal combustion engine that runs on gasoline or hydrogen (LH_2).

ZUMA Press, Inc./Alamy Stock Photo

Figure 20.7: Mazda RX-8 RE, powered by a two-fuel internal combustion engine that runs on gasoline or hydrogen (CHG).

BG Motorsports/Alamy Stock Photo

Clearly, from the information in the table, both vehicles are primarily gasoline-powered vehicles with supplementary hydrogen power. This fact is indicated by the relative ranges of the vehicles on these two fuel sources and is a direct consequence of the difficulty in designing an internal combustion vehicle with sufficient usable range utilizing hydrogen as a sole fuel source. Hydrogen may be best utilized as a vehicle fuel in fuel cells, as discussed in the next section.

Example 20.5

In the context of the result in Example 20.4, estimate the CO_2 emissions per km for an H_2 ICE vehicle and compare it with a typical gasoline-powered vehicle.

Solution

From Chapter 19, the required energy to the wheels of a family sedan is about 0.6 MJ/km. Thus for the H_2 ICE vehicle, the primary (natural gas) energy required will be

$$25 \times 0.6 \text{ MJ/km} = 15 \text{ MJ/km}.$$

From equation (1.15), the energy content of methane is 55.5 MJ/kg, so the requirement for the vehicle will be

$$(15 \text{ MJ/km})/(55.5 \text{ MJ/kg}) = 0.27 \text{ kg/km of methane}.$$

The combustion of 1 mole (16 g) of methane produces 1 mole (44 g) of CO_2, so that the net CO_2 emissions for the vehicle will be

$$(0.27 \text{ kg/km}) \times (44/16) = 0.74 \text{ kg } CO_2 \text{ per km}.$$

From Table 19.3, the fuel consumption of a gasoline-powered family vehicle is about 10.7 L/100 km or 0.107 L/km. The density of gasoline is about 720 kg/m^3, so the fuel consumption will be

$$(0.107 \text{ L/km}) \times (0.72 \text{ kg/L}) = 0.077 \text{ kg/km}.$$

Approximating gasoline as octane, from equation (1.17) we see that 1 mole of octane (144 g) will produce 8 moles (44 $\times$ 8 = 352 g) of CO_2. Thus, the CO_2 emission from the gasoline vehicle will be

$$(0.077 \text{ kg/km}) \times (352/114) = 0.24 \text{ kg } CO_2 \text{ per km}.$$

This type of comparison is considered in more detail in Section 20.9.

Example 20.6

The Mazda RX-8 RE uses a CHG tank with a volume of 110 L at a pressure of 35 MPa.

(a) What is the total energy content of the fuel?
(b) Assuming an efficiency of 17% for conversion of chemical energy to energy to the wheels, what is the energy delivered to the wheels per km?

Solution

(a) From Figure 20.3, the density at 35 MPa (350 atm) will be about 25 kg/m^3. Thus a 110-L tank will contain

$$(0.11 \text{ m}^3) \times (25 \text{ kg/m}^3) = 2.8 \text{ kg}.$$

Continued on page 639

Example 20.6 continued

Since the energy content of hydrogen from Table 20.1 is 142 MJ/kg, then the energy content of the hydrogen in the fuel tank will be

$$(2.8 \text{ kg}) \times (142 \text{ MJ/kg}) = 400 \text{ MJ}.$$

(b) If the efficiency of converting chemical energy of the fuel to energy at the wheels is 17%, then the energy available at the wheels for a tank of fuel will be

$$(0.17) \times (400 \text{ MJ}) = 68 \text{ MJ}.$$

Since the range of the RX-8 RE in the hydrogen mode is 100 km, then the energy at the wheels per km will be

$$(68 \text{ MJ})/(100 \text{ km}) = 0.68 \text{ MJ/km},$$

in good agreement with the estimate in the previous chapter for vehicle energy requirements.

20.6 Fuel Cells

A fuel cell is a device that catalyzes the reaction in equation (20.1) and thereby produces energy by combining hydrogen and oxygen to produce water. Figure 20.8 illustrates a generic fuel cell. Typically, the fuel is in the form of hydrogen gas or a hydrogen-containing compound such as a hydrocarbon (e.g., methane). Oxygen is typically provided from air.

The concept of the fuel cell was first proposed by the German physicist C. F. Schönbein in 1838. The first operational fuel cell was constructed in 1843 by the Welsh scientist W. R. Grove; however, it was not until the 1950s that practical fuel cells were developed. In 1959, the farm equipment company Allis-Chalmers constructed a tractor

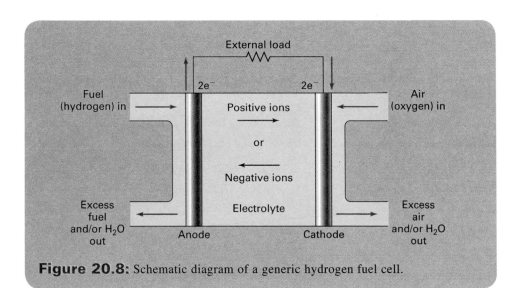

Figure 20.8: Schematic diagram of a generic hydrogen fuel cell.

Table 20.5: Typical properties of some common fuel cells. The efficiency is the typical system efficiency without cogeneration.

type	electrolyte	ion transferred	operating temperature (°C)	typical power output (kW)	efficiency (%)
proton exchange membrane	polymer membrane	H^+	50–100	100	50
alkaline	KOH	OH^-	80	10	65
phosphoric acid	H_3PO_4	H^+	150–200	100	50
molten carbonate	K_2CO_3	CO_3^{2-}	600–650	2000	50
solid oxide	ZrO_2	O^{2-}	800–1000	1000	60

that utilized a 15-kW fuel cell. Over the years, a variety of different types of fuel cells have been developed, and this section reviews the major types, their potential applications, and their advantages and disadvantages. Table 20.5 gives properties of the common varieties of fuel cells. These differ primarily in the electrolyte used. Most individual fuel cells produce an output voltage of about 0.8 V. Combinations of cells are arranged in series and/or parallel to provide the voltage and current needed for a particular application. The different types of fuel cells are discussed below.

20.6a Proton Exchange Membrane Fuel Cells

Proton Exchange Membrane (or PEM) fuel cells are probably the best choice for automobiles and are used in most prototype and limited-production fuel cell vehicles. The reasons for this include the fact that they can produce the power necessary for vehicle propulsion, they operate at fairly low temperature, and they have a high power density, so they are comparatively light.

Other features that make them attractive for vehicle use include the facts that they can operate at low temperatures, that they can respond quickly to a varying load, and that they have a high efficiency. One disadvantage is that they require fairly high-purity hydrogen. Another interesting feature of the PEM fuel cell is that it is reversible; that is, when hydrogen and oxygen are supplied as a fuel, it produces electricity and water. When water and electricity are supplied, then the system functions as an electrolysis cell and produces hydrogen and oxygen. This feature makes them attractive for remote power systems when they can produce and store electricity reversibly. This ability may be appropriate for use in conjunction with solar photovoltaic or wind-generated electricity. Ballard Power Systems in British Columbia, Canada, has been one of the leaders in the development of PEM fuel cells for vehicles and other uses. They provide cells for a number of vehicle manufacturers, although some vehicle manufacturers utilize in-house–produced cells. An example of a fuel cell installed in a prototype vehicle is shown in Figure 20.9.

20.6b Alkaline Fuel Cells

Alkaline fuel cells have been utilized by NASA as a source of energy for space vehicles. Although the need for a platinum catalyst makes these fuel cells expensive, cost is a secondary consideration for these applications. Also, these fuel cells are best suited to relatively low-power applications, which, again, is acceptable for spacecraft

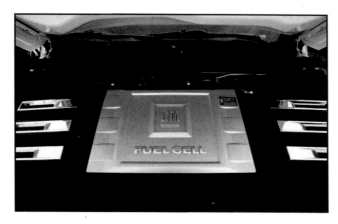

Figure 20.9: A fuel cell installed in the engine compartment of a prototype passenger vehicle.

Agencja Fotograficzna Caro/Alamy Stock Photo

use. One of the most attractive features of these fuel cells is that the water produced is of sufficiently high purity that it is drinkable. Thus, the fuel cell provides a source of both electricity and drinking water for the astronauts. The ability to operate over a wide range of low temperatures is also an advantage for space applications.

20.6c Phosphoric Acid Fuel Cells

The earliest fuel cells produced in the 1800s were similar in design to today's phosphoric acid fuel cells. They tend to be somewhat expensive, as are most fuel cells, because they use platinum as the catalyst. Recent research on this and other types of fuel cells is aimed at developing new materials with the aim of reducing their cost. Phosphoric acid fuel cells are available commercially and have been used for a number of applications. Because of the relatively low power density, these devices tend to be fairly large and heavy. For this reason, they are most applicable for stationary applications, such as backup electrical power for hospitals, police stations, and the like. One advantage of phosphoric acid fuel cells is that they are fairly insensitive to impurities in the hydrogen fuel. For this reason, hydrogen can be produced by less expensive processes (than electrolysis). Like many fuel cells, phosphoric acid fuel cells operate at an elevated temperature. This has certain disadvantages because it is necessary to utilize some energy to heat the cell in order to warm it up before it can be used. This feature may be most disadvantageous for applications, such as automobiles, where quick-starting capabilities are desirable. However, one advantage of the elevated temperature operation of this and some other fuel cells is that the water produced is actually steam. In some applications, the energy content of this steam may be utilized either directly as a source of heat or for the production of additional electricity by traditional generation methods. If the heat content of the steam is fully utilized, the overall efficiency of a phosphoric acid fuel cell can be as high as 85%. This is referred to as cogeneration, analogous to the discussion in Chapter 17.

20.6d Molten Carbonate Fuel Cells

As seen in the table, these fuel cells can be quite large and can provide considerable power. For this reason, they may be quite suitable for stationary backup-power applications. An additional advantage (as well as a disadvantage) is that they can utilize hydrogen from gaseous fossil fuels. This means that some natural fuels (i.e., fossil fuel products such as methane) can be used, as well as hydrogen produced by cheap-and-dirty methods. However, it also means that the by-products of the cell are hydrocarbons as well as water. Thus, these fuel cells contribute to the release of greenhouse gases. Because they operate at elevated temperatures, cogeneration can be used, and efficiency can be increased to about 85%.

20.6e Solid Oxide Fuel Cells

Solid oxide fuel cells typically use a ZrO_2 electrolyte that is stabilized by the addition of about 8% Y_2O_3 (yttrium oxide). As Table 20.5 indicates, hydrogen and oxygen are combined by the transport of doubly ionized oxygen ions across the electrolyte. High temperatures are needed for the efficient operation of the solid oxide fuel cell because the ionic conductivity of the ZrO_2 increases with increasing temperature, and temperatures in excess of 500°C are necessary to allow the O^{2-} ions to be transported between the cathode and the anode. Solid oxide fuel cells have the advantage of being able to use a variety of light hydrocarbons, such as methane, propane and butane, as fuel. For many applications, the high operating temperature creates difficulties because of the long start-up time required, typically about an hour. This makes solid oxide fuel cells unsuitable for vehicle applications. However, the large power output from these devices makes them ideal for many stationary applications, such as power for remote areas. For such applications, the high operating temperature has the advantage of facilitating cogeneration, where the excess heat can be used in a heat engine or for direct thermal applications.

Example 20.7

If tanks of compressed hydrogen gas at 70 MPa are used to store energy for grid back-up and are used to produce electricity using molten carbonate fuel cells, estimate the total tank volume needed to provide the same total energy storage as the Bath County pumped hydroelectric facility.

Solution

Energy Extra 18.1 indicates that the Bath County facility can provide 2772 MW_e for up to 11 hours. This is a total energy storage capacity of

$$(2772 \text{ MW}_e) \times (11 \text{ h}) \times (3.6 \times 10^3 \text{ MJ/MWh}) = 1.1 \times 10^8 \text{ MJ}.$$

The energy capacity of hydrogen is 142 MJ/kg, and for a molten carbonate fuel cell of 60% efficiency as shown in Table 20.5, the net energy from hydrogen will be

$$(0.6) \times (142 \text{ MJ/kg}) = 85 \text{ MJ/kg}.$$

So the energy storage requirement will need

$$(1.1 \times 10^8 \text{ MJ})/(85 \text{ MJ/kg}) = 1.29 \times 10^6 \text{ kg of hydrogen}.$$

Continued on page 643

Example 20.7 continued

From Figure 20.3, the density of hydrogen at 70 MPa ($\sim$700 atm) is 35 kg/m^3, so the required mass of hydrogen will occupy a volume, V, of

$$V = (1.29 \times 10^6 \text{ kg})/(35 \text{ kg/m}^3) = 3.69 \times 10^4 \text{ m}^3.$$

If this is stored in a single spherical tank (probably not a practical situation), the diameter, d, will be

$$d = [6V/\pi]^{1/3} = [6 \times 3.69 \times 10^4 \text{ m}^3/3.14]^{1/3} = 41 \text{ m}.$$

20.7 Fuel Cell Vehicles

Many major automobile manufacturers have programs for the development of fuel cell vehicles. The electricity produced by the fuel cell is used to power the vehicle using an electric motor. These vehicles are virtually all hybrid hydrogen/electric vehicles. Because the electric drive system is in the vehicle, it is convenient to utilize a supplementary electric storage mechanism to increase the vehicle's range and efficiency. This storage mechanism may consist of batteries or supercapacitors that may be charged externally and/or charged through regenerative braking.

A number of automobile manufacturers have experimented with PEM fuel cell vehicles since the mid-1960s. The first fuel cell vehicle approved for road use was the Honda FCX (see Figure 20.10). This vehicle was produced from 2002 to 2007 and offered only for lease for US$11,500 per month. Initially the vehicle was available only for corporate or government lease, but beginning in 2005, some were leased to noncommercial customers. The Honda FCX electric motor produced 80 kW, and the 1680 kg vehicle had a range of about 310 km from two tanks of

Figure 20.10: Honda FCX fuel cell vehicle available for lease from 2002 to 2007.

Hatsukari715

Figure 20.11: The Ford Focus FCV showing the fuel cell system in the engine compartment. NiMH batteries were used as a secondary energy source.

Jim West/Alamy Stock Photo

compressed hydrogen gas with a total capacity of 4 kg. Supercapacitors were used for supplementary electrical storage.

Between 2003 and 2014 several automobile manufacturers, including Ford (see Figure 20.11), Nissan, Mercedes Benz, and Chevrolet, leased a limited number of fuel cell vehicles. Many of these fuel cell vehicles are built on existing gasoline vehicle platforms, and all utilize compressed hydrogen gas as a storage mechanism.

As of 2017 three fuel cell vehicles are available to purchase or lease in different markets around the world: the Hyundai ix35 FCEV (available since 2014, Figure 20.12), the Toyota Marai (available since 2015, Figure 20.13), and the Honda Clarity Fuel Cell (available since 2016, Figure 20.14). The Hyundai sells for the equivalent of US$77,300 in South Korea and is available for lease in the

Figure 20.12: Hyundai ix35 FCEV.

VanderWolf Images/Shutterstock.com

Figure 20.13: Toyota Mirai.

EvrenKalinbacak/Shutterstock.com

Figure 20.14: Honda Clarity Fuel Cell.

Ed Aldridge/Shutterstock.com

United States. The Toyota and Honda sell for US$58,500 and US$67,300, respectively, in the United States. Vehicle specifications are shown in Table 20.6. All vehicles use compressed hydrogen gas at 70 MPa pressure and are fuel cell/battery electric hybrids. While the Toyota, for example, uses NiMH batteries with energy storage of 1.6 kWh, the Hyundai is closer to a true hybrid, using Li-ion batteries with energy storage of 24 kWh.

Table 20.6: Specifications of production fuel cell vehicles available as of 2017.

make	model	hydrogen storage at 70 MPa	mass (kg)	maximum power at wheels (kW)	battery	range (km)
Hyundai	ix35 FCEV	5.64 kg	2290	100	Li-ion	594
Toyota	Mirai	5 kg	1850	113	NiMH	502
Honda	Clarity Fuel Cell	5.4 kg	1890	100	Li-ion	589

20.8 Hydrogen: Present and Future

Table 20.6 allows for an analysis of the characteristics of fuel cell vehicles. A comparison with traditional gasoline vehicles (Table 19.3) and battery electric vehicles (Tables 19.6 and 19.7) is informative. In comparison to similarly sized gasoline vehicles, fuel cell vehicles are heavier but compare fairly favorably in power and range. Compared to battery electric vehicles, fuel cell vehicles tend to be a bit heavier and have similar power but substantially better range. A major selling point in favor of hydrogen vehicles is the time for refueling and general claims of the nonpolluting character of hydrogen as a fuel. Progress in the production of hydrogen-powered vehicles is perhaps evidenced by several manufacturers' marketing, on a limited basis, of hydrogen ICE or fuel cell vehicles and a number who have plans for marketing them in the relatively near future.

Nevertheless, hydrogen for transportation is a new technology and there are a number of concerns that need to be addressed. Principal among these are safety, infrastructure, and cost.

From a safety standpoint, hydrogen's reputation is greatly influenced (in a negative way) by the Hindenburg disaster (Figure 20.15). Recent studies have shown that it was, in fact, a combustible coating on the airship's exterior that was primarily responsible for the disaster and not the hydrogen fuel itself. However, hydrogen is flammable, and appropriate safety measures must be taken when using hydrogen as a fuel, either in gaseous or liquid form. Safety standards must take into consideration the wide range of concentrations of hydrogen in air that are flammable: by volume from 4% hydrogen in air to 75% hydrogen in air.

The infrastructure for the use of hydrogen as a transportation fuel involves the establishment of hydrogen stations to supplement or eventually replace gasoline and diesel stations. In fact, hydrogen and gasoline filling facilities could be integrated into common stations in the same way that many gas stations

Figure 20.15: The German Zeppelin Hindenburg burned on May 6, 1937 in Lakehurst, New Jersey. Of the 97 people on board, 35 people and one person on the ground died in the incident.

U.S. Navy

currently sell propane. Movement in this direction is already taking place and hydrogen fueling stations are currently available in many parts of the world. So-called *hydrogen highways* have been constructed in several locations to facilitate the use of hydrogen-powered vehicles. California (Figure 20.16), Southwest

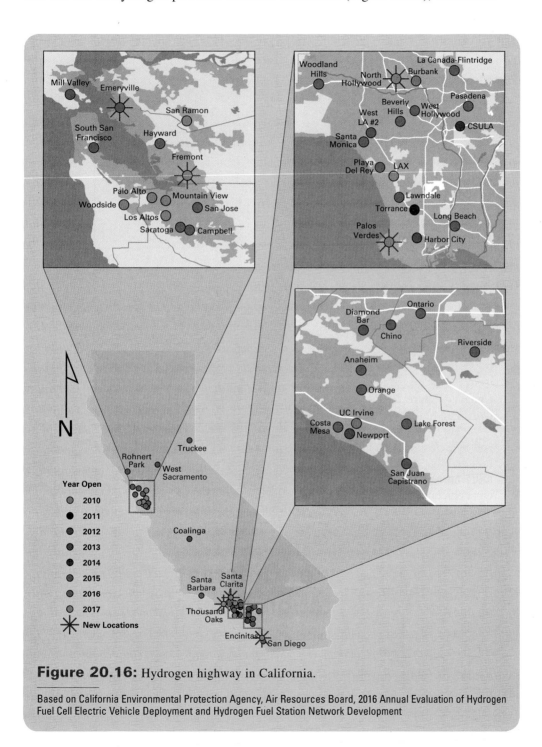

Figure 20.16: Hydrogen highway in California.

Based on California Environmental Protection Agency, Air Resources Board, 2016 Annual Evaluation of Hydrogen Fuel Cell Electric Vehicle Deployment and Hydrogen Fuel Station Network Development

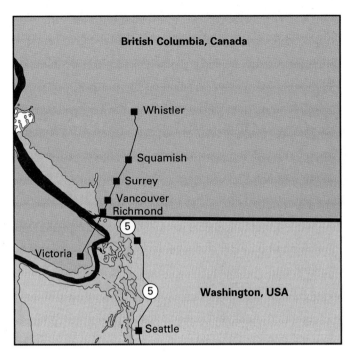

Figure 20.17: Hydrogen fueling station locations in British Columbia and the U.S. Northwest.

Based on data from www.hydrogencarsnow.com

Canada (Figure 20.17), Japan, and Europe (Figure 20.18) are principal locations of hydrogen highways. A typical hydrogen fueling station is shown in Figure 20.19.

This discussion may suggest that hydrogen ICE vehicles and hydrogen fuel cell vehicles are a technology that may provide environmentally friendly transportation and that could compete with battery electric vehicles (and in some ways may be more attractive). However, a careful consideration of the relative efficiency of various vehicle technologies, as discussed in the next section, is necessary for a complete understanding of the viability of hydrogen as a transportation fuel.

20.9 Efficiency of Different Transportation Technologies

One of the most important considerations for the use of any energy source is efficiency. In the case of a vehicle, it is the efficiency of converting a primary source energy such as oil, coal, solar, wind, and the like into mechanical energy delivered to the vehicle's wheels. Tables 20.7 through 20.10 show an efficiency analysis for some ways in which a primary fossil fuel energy source may be utilized to power a vehicle. These analyses include a traditional gasoline- or diesel fuel-powered vehicle, a hydrogen ICE vehicle, a hydrogen fuel cell vehicle, and a battery electric vehicle. The traditional gasoline

Figure 20.18: Hydrogen fueling station locations in Europe.

Volina/Shutterstock.com, data from www.hydrogencarsnow.com

Figure 20.19: Hydrogen fueling station. This facility in Reykjavík, Iceland opened in 2003 and was the world's first public hydrogen fueling station.

ARCTIC IMAGES/Alamy Stock Photo

Table 20.7: Efficiency analysis for gasoline-powered internal combustion engine vehicle showing net efficiency for conversion of primary energy (gasoline) to mechanical energy delivered to the vehicle's wheels.

process	efficiency (%)
fossil fuel → mechanical energy	17
net efficiency	17

Based on data from R. A. Dunlap, *Energy and Environment Research* "A simple and objective carbon footprint analysis for alternative transportation technologies" 3 (2013): 33–39.

vehicle (Table 20.7) has a relatively low efficiency because of the intrinsic limitation of the Carnot cycle in converting heat produced by the combustion of gasoline into mechanical energy. By comparison, burning a fossil fuel to produce electricity and using that electricity to charge batteries to run a battery electric vehicle is about twice as efficient (Table 20.8). This is because of the reasonably high efficiency of converting heat into electricity that results from the use of an effective cold reservoir at an electric generating station and the very high efficiency of converting electrical energy into mechanical energy. In addition to the efficiency factor, this approach has the advantage (assuming that fossil fuels continue to be used as a primary energy source) that it is much easier to sequester carbon from a few hundred large power plants than from millions of individual automobiles.

An analysis of the hydrogen ICE vehicle is shown in Table 20.9. Clearly, the number of processes necessary to convert fossil fuel energy to mechanical energy at the vehicle's wheels results in a very low overall efficiency for the process. The presence of two Carnot limited conversions—heat to electricity in the power plant and heat to mechanical energy in the vehicle—are detrimental to this approach. Thus, the cost in terms of primary source energy is unreasonable, and, although carbon sequestration at the power plant is easier than at the vehicle, the trade-off is not acceptable.

An analysis of the fuel cell vehicle (Table 20.10) shows that this is a somewhat better approach. However, the overall efficiency is barely comparable to that of the gasoline ICE vehicle and is less than half that of the battery electric vehicle.

Thus, if we continue to use fossil fuels as a primary energy source, the development of battery electric vehicles would seem to be a better approach than fuel cell or hydrogen ICE vehicles. This approach has the advantage of reducing carbon emissions by a factor of two, making carbon sequestration easier and allowing for the use of relatively plentiful coal for electricity generation.

Table 20.8: Efficiency analysis for battery electric vehicle showing net efficiency for conversion of primary energy (oil or coal) to mechanical energy delivered to the vehicle's wheels.

process	efficiency (%)
fossil fuel → electricity	40
electricity → mechanical energy	85
net efficiency	34

Based on data from R. A. Dunlap, *Energy and Environment Research* "A simple and objective carbon footprint analysis for alternative transportation technologies" 3 (2013): 33–39.

Table 20.9: Efficiency analysis for hydrogen-powered internal combustion engine vehicle showing net efficiency for conversion of primary energy (oil or coal) to mechanical energy delivered to the vehicle's wheels.

process	efficiency (%)
fossil fuel → electricity	40
electricity → hydrogen gas	70
hydrogen gas → CHG or LH_2	80
CHG or LH_2 → mechanical energy	17
net efficiency	4

Based on data from R. A. Dunlap, *Energy and Environment Research* "A simple and objective carbon footprint analysis for alternative transportation technologies" 3 (2013): 33–39.

Table 20.10: Efficiency analysis for hydrogen fuel cell-powered vehicle showing net efficiency for conversion of primary energy (oil or coal) to mechanical energy delivered to the vehicle's wheels.

process	efficiency (%)
fossil fuel → electricity	40
electricity → hydrogen gas	70
hydrogen gas → CHG	80
CHG → electricity	50
electricity → mechanical energy	90
net efficiency	10

Based on data from R. A. Dunlap, *Energy and Environment Research* "A simple and objective carbon footprint analysis for alternative transportation technologies" 3 (2013): 33–39.

If more environmentally friendly primary sources of energy (e.g., wind, solar, etc.) are used and the concern for carbon emissions is eliminated, then the comparison of efficiencies for battery electric vehicles and fuel cell or hydrogen ICE vehicles still shows that the net energy to the wheels, compared to the total primary energy input, favors batteries. This feature results from the fundamental fact that it is much more efficient to get energy into and out of batteries than into and out of hydrogen, and this basic principle is unlikely to change in the foreseeable future. However, the production of hydrogen by biological mechanisms may be an approach that can help to make hydrogen more attractive.

With respect to this type of assessment, *Technology Review* (March–April 2007) concluded (www.technologyreview.com/Energy/18301/):

In the context of the overall energy economy, a car like the BMW Hydrogen 7 would probably produce far more carbon dioxide emissions than gasoline-powered cars available today. And changing this calculation would take multiple breakthroughs—which study after study has predicted will take decades, if they arrive at all. In fact, the Hydrogen 7 and its hydrogen-fuel-cell cousins are, in many ways, simply flashy distractions produced by automakers who should be taking stronger immediate action to reduce the greenhouse gas emissions of their cars.

ENERGY EXTRA 20.3
Carbon footprint analysis for transportation technologies

The efficiencies that have been derived for different transportation technologies in this chapter certainly provide some guidance for the assessment of the environmental impact of these approaches. However, this analysis can be taken one step further by actually calculating the mass of carbon dioxide emitted per kilometer traveled. A fairly straightforward approach to calculating this quantity for the comparison of different types of vehicles is described here.[1] The amount of CO_2 emitted per km traveled is expressed as

$$\frac{kg(CO_2)}{km} = \frac{kg(CO_2)}{(MJ)_p} \times \frac{(MJ)_p}{(MJ)_w} \times \frac{(MJ)_w}{km},$$

where p is the primary energy, and w is the energy delivered to the vehicle's wheels. The first term on the right-hand side of the equation represents the amount of CO_2 generated per megajoule of primary energy consumed. The second term is the inverse of the efficiency from primary energy to wheel energy, as calculated in this chapter. The final term on the right-hand side is the average energy to the wheels needed to move the vehicle 1 km. This analysis does not consider the carbon footprint of nonfossil fuel–generated electricity or a life cycle analysis of vehicle materials. It also does not consider the possibility of sequestrating carbon from fossil fuel–generating stations. Because these effects tend to average out somewhat, this analysis is a reasonable comparison of different vehicle technologies.

For a gasoline internal combustion engine powered vehicle, gasoline is very close to a primary energy source, whereas for battery electric vehicles (BEVs) or hydrogen-powered vehicles the primary energy sources are first used to produce electricity.

Thus $kg(CO_2)/(MJ)_p$ depends on how electricity is produced. A rough breakdown of current electricity production in the United States follows:

fuel	% electricity
coal	44.9
natural gas	23.4
petroleum	1.0
nonfossil	30.7

The CO_2 emission per MJ for different fossil fuels is obtained from a simple analysis of their energy content and is shown in the following table. The CO_2 emissions from nuclear, hydroelectric, and alternative energy sources is considered to be zero.

fossil fuel	kg (CO_2)/MJ
coal (~pure carbon)	0.11
natural gas (methane, CH_4)	0.055
heavy hydrocarbons (>6 C/molecule)	0.069

Based on data from R.A. Dunlap, "A simple and objective carbon footprint analysis for alternative transportation technologies," *Energy and Environment Research* 3 (2013) 33–39.

A weighted average of these values according to how electricity is produced in the United States is 0.063 $kg(CO_2)/(MJ)_p$. For a typical family vehicle, the energy at the wheels per kilometer traveled was shown in the last chapter to be about 0.6 $(MJ)_w$/km. Combining all of this information, the CO_2 emission per kilometer is obtained.

technology	kg $(CO_2)/(MJ)_p$	$(MJ)_p/(MJ)_w$	$(MJ)_w$/km	kg (CO_2)/km
gasoline ICE	0.069	5.9	0.6	0.24
H_2 ICE	0.063	25	0.6	0.94
H_2 fuel cell	0.063	7.1	0.6	0.38
BEV	0.063	2.9	0.6	0.11

Continued on page 653

Energy Extra 20.3 continued

This analysis shows that the net environmental effect of BEVs is quite positive (compared with gasoline vehicles), whereas fuel cell vehicles are about neutral. Hydrogen ICE vehicles are seen to have a very negative environmental effect. Incremental improvements to the results for nonfossil fuel vehicles can be obtained by increasing energy conversion efficiencies (i.e., decreasing $(MJ)_p/(MJ)_w$) and by increasing vehicle efficiencies (i.e., decreasing $(MJ)_w/km$). However, it is through the shift in electricity-generating technology away from fossil fuels [i.e., the reduction of $kg(CO_2)/(MJ)_p$] that the most substantial improvements can be made. However, H_2 ICE vehicles would require that at least 85% of electricity would come from nonfossil fuel sources in order to be an environment improvement over gasoline vehicles (although from a financial standpoint this would not be an attractive alternative). A reduced value of $kg(CO_2)/(MJ)_p$ already exists, and puts alternative transportation technologies in a better light compared to gasoline-powered vehicles,

in countries like France and Canada, where a large fraction of electricity comes from nonfossil fuel sources.

———
[1]R. A. Dunlap, *Energy and Environment Research* **3** (2013): 33.

Topic for Discussion

Although the efficiency of energy use is an important consideration, it is also important to understand the financial aspects of any technological development and how results can be marketed. Consider, as an example, the fuel cost to the driver (in dollars per megajoule delivered to the wheels of a vehicle) for a gasoline-powered BMW 740Li. For comparison, estimate the cost (in U.S. dollars per megajoule) for a fossil-fuel-free BMW Hydrogen 7, running on hydrogen produced by the electrolysis of water using wind-generated electricity. Discuss the economic viability of this technology.

———
Based on R.A. Dunlap, "A simple and objective carbon footprint analysis for alternative transportation technologies," *Energy and Environment Research* **3** (2013) 33–39.

Although fueling times and ranges of hydrogen-powered vehicles make them attractive, the overall efficiency and, at present, the cost put them at a disadvantage compared to battery electric vehicles. There are, however, other potential applications where these factors are less critical. For example, hydrogen may have use as a nonfossil fuel for emergency backup power for installations such as hospitals in remote locations or as an aviation fuel where other nonfossil fuel alternatives are somewhat limited.

20.10 Summary

Hydrogen is not a primary energy source but merely an energy storage mechanism. Energy is required to produce hydrogen (typically from water), and this energy is recovered (at least in part) when the hydrogen is recombined with oxygen, either by combustion or in a fuel cell. This chapter reviewed the properties of hydrogen and the pros and cons of using it as a storage method.

Hydrogen is a gas at standard temperature and pressure. Its heat of combustion is 142 MJ/kg, compared with about 45 MJ/kg for gasoline. Unfortunately, its density is very low, leading to a very low energy per unit volume. It can be liquefied by cooling to about 20 K or compressed to increase its density. Even so, its energy per unit volume is less than one-third that of gasoline. This chapter reviewed the common methods of producing hydrogen.

Hydrogen is released in a variety of chemical reactions and by the decomposition of water into hydrogen and oxygen. Some reactions that yield hydrogen involve the release of CO_2 and are not desirable from an environmental standpoint. The electrolysis of water is a convenient and common method of hydrogen production that has no direct adverse environmental effects.

As described in this chapter, hydrogen is an attractive storage mechanism for energy for transportation use because its use to produce heat by combustion or to produce electricity by means of a fuel cell, yields no greenhouse gases. As noted, however, its low density requires that it be liquefied or compressed to high pressures for portable use. Liquid hydrogen must be stored in an insulated tank to minimize evaporation (although this cannot be eliminated entirely). Compressed hydrogen gas requires storage in sufficiently strong cylinders to withstand the high pressure. Either approach involves careful consideration in vehicle design.

This chapter has provided an overview of the operation of a fuel cell. This device combines hydrogen fuel with oxygen from the atmosphere to produce water and electricity. The chapter summarized the different types of fuel cells and showed that polymer electrolyte membrane fuel cells are probably the most suitable for vehicle use.

A description of a variety of prototype hydrogen ICE and fuel cell vehicles was presented in the chapter. The ICE vehicles are modified gasoline ICE vehicles and are typically hybrid vehicles because they carry gasoline to supplement the somewhat limited hydrogen range that results from the low energy per unit volume of hydrogen.

Fuel cell vehicles are also typically hybrid vehicles because they incorporate batteries or supercapacitors to increase the range and store electrical energy generated by regenerative braking.

A full analysis of hydrogen as a fuel must include how the hydrogen is produced and the overall efficiency of the energy production process from primary energy source to usable end-product energy. This has been considered in detail in this chapter. Clearly, the route by which hydrogen is produced, stored, and utilized is an important factor in determining the efficiency, as well as its environmental impact. If hydrogen is produced by chemical methods, the carbon footprint of this process is important in determining the overall greenhouse gas emissions. If hydrogen is produced by electrolysis, then the source of the electricity used for the process is important in determining the carbon emissions. The low efficiency for transportation using hydrogen ICE technology makes this approach of questionable economic viability, and any carbon emissions during the process become accentuated. Fuel cell technology is more promising and warrants further consideration as a future approach to transportation. The development of new technologies for the production of hydrogen would be an important step in making this technology competitive with battery electric vehicles, and improvements in fuel cell efficiency and cost could lead the way to commercial utilization.

Problems

20-1   Hydrogen gas is burned to provide heat for a typical North American home (see Chapter 2 for heating requirements). If the hydrogen is stored in a spherical tank at a pressure of 80 MPa, what would be the diameter of the tank needed to supply the average monthly heating requirement? Assume a typical furnace efficiency (Chapter 17).

20-2   Hydrogen gas is burned according to the reaction

$$2H_2 + O_2 \rightarrow 2H_2O$$

to provide heat for a typical North American home (see Chapter 2 for heating requirements). What would be the average daily water production in liters? Assume a typical furnace efficiency (Chapter 17).

20-3 If hydrogen followed an ideal gas law at all pressures, calculate the pressure at which the density of CHG would exceed that of LH_2 (71 kg/m^3).

20-4 Assume that the cost of hydrogen per unit energy is the same as gasoline ($0.03/MJ). What is the loss (in dollars per day) due to evaporation from a 100-L LH_2 fuel tank?

20-5 (a) Given a fuel cell efficiency of 75% and a required energy to the wheels of 0.6 MJ/km, calculate the mass of hydrogen fuel required to give a vehicle a range of 300 km.

   (b) Calculate the fuel volume if hydrogen is stored in the form of (i) liquid hydrogen, (ii) CHG at 50 MPa, and (iii) TiH_2.

20-6 A solid oxide fuel cell operates at 55% efficiency using methane as a fuel. Calculate the mass (in grams) of water and carbon dioxide produced for every kilowatt-hour electrical of output. Note the heat of combustion of methane is 50 MJ/kg (Chapter 3).

20-7 Calculate the percentage of improvement in the energy density of a CHG-fuel tank if the pressure is increased from 35 to 70 MPa.

20-8 For stationary applications, mass and volume considerations for a fuel source are much less important than they are for use in transportation. It therefore may be the most reasonable to sacrifice compactness to minimize requirements for expensive advanced technologies. Consider a plan to provide power to a small city with a population of 100,000, using hydrogen fuel cells with an efficiency of 60%. The average net per-capita power consumption is 6 kW for electricity, heat, and power for an electric vehicle.

   (a) Calculate the volume of hydrogen needed to provide the city's energy for one month if the hydrogen is stored at a (relatively low) pressure of 1 MPa.

   (b) If the city has a land area of 100 km^2, what fraction of this area would need to be devoted to hydrogen storage if the storage tanks were vertical cylinders 3 m in diameter and 5 m high and were packed tightly together in a square array?

20-9 The Mazda RX-8 RE hydrogen vehicle has a 110-L CHG tank that stores hydrogen at 35 MPa and a gasoline tank with a volume of 61 L. Calculate the relative efficiency for the engine when burning hydrogen compared to its efficiency when burning gasoline. Assume that the energy to the wheels per km is the same for both fuels.

20-10 One possible use for hydrogen as an energy storage mechanism is for large-scale grid storage instead of pumped hydroelectric or compressed air. Consider the possibility of using the cavern at the Huntorf facility to store hydrogen with the same maximum pressure that is now used to store compressed air. The hydrogen could then be burned in a hydrogen ICE to generate electricity. Estimate the relative electrical energy that could be produced by the two approaches.

20-11 MgH_2 has been considered as a possible hydrogen storage material. Calculate the mass and volume of Mg that would be required to store 100 kg of hydrogen.

20-12 The lightest element that is a solid at room temperature, lithium, absorbs hydrogen to form LiH. Compare the energy density of LiH (i.e., energy per kg) to the energy density of gasoline.

20-13 The U.S. Department of Energy has set a benchmark of 6% H by weight for an acceptable hydrogen storage material. Consider the following possible metals that form hydride phases: $NaAlH_4$, $LiAlH_4$, $LaNi_5H_6$, and $TiFeH_2$. Which meet the DOE standard?

20-14 Consider the possibility of extracting energy from 1 m^3 of water. One approach would be to lift the water to some elevation and then to generate electricity hydroelectrically (i.e., pumped hydroelectric storage). The second approach would be to produce hydrogen from the water by electrolysis and then generate electricity from the hydrogen in a proton exchange membrane fuel cell. Using typical efficiencies as given in Chapters 18 and 20, how high would the cubic meter of water have to be lifted to provide the same total electrical energy output as the fuel cell?

20-15 Consider the following three applications of fuel cells for producing electricity to power a vehicle.

(i) A proton exchange membrane fuel cell uses hydrogen produced by electrolysis using electricity generated by burning coal.

(ii) A proton exchange membrane fuel cell uses hydrogen produced by electrolysis using electricity generated by a photovoltaic array.

(iii) A solid oxide fuel cell uses methane as a fuel.

For each case discuss as quantitatively as possible

(a) The overall efficiency of converting primary energy to energy at the vehicle's wheels.

(b) The CO_2 emissions per MJ of energy at the vehicle's wheels.

20-16 Consider the situation where all 250,000,000 gasoline-powered vehicles in the United States, each of which is driven an average of 20,000 km per year and requires an average of 0.55 MJ/km from the batteries, would be replaced with PEM-fuel cell vehicles that use hydrogen produced by electrolysis using electricity generated by 1-GW_e nuclear power plants (operating at a 70% capacity factor). How many new nuclear power plants would need to be constructed to accommodate this increase in electricity demand?

20-17 Consider a simple model of the burning of butane as the oxidation of the carbon content and the oxidation of the hydrogen content separately. Does the combined heat of combustion of the components account for the heat of combustion of butane? Are there obvious reasons for any differences?

20-18 The BMW Hydrogen 7 consumes an average of 13.7 L/100 km of gasoline (when operating in the gasoline mode) and 50.0 L/100 km of liquid hydrogen (when operating in the hydrogen mode).

(a) If gasoline costs $0.90/L, what is the fuel cost per km in the gasoline mode?

(b) If the hydrogen is produced and processed using electricity that costs $0.12/kWh, what is the cost per km in the hydrogen mode?

20-19 Calculate the carbon footprint (i.e., $kg(CO_2)$/km) for a fuel cell vehicle requiring 0.6 MJ/km at the wheels if all of the electricity is generated by burning natural gas in thermal generating stations.

20-20 If the BMW Hydrogen 7 consumes an average of 13.7 L/100 km when burning gasoline and 50.0 L/100 km when burning liquid hydrogen, show that the engine efficiency is very nearly the same for both fuels.

Bibliography

A. I. Appleby and F. R. Foulkes. *Fuel Cell Handbook.* Krieger, Malabar (1993).

G. J. Aubrecht II. *Energy: Physical, Environmental, and Social Impact,* (3rd ed.) Pearson Prentice Hall, Upper Saddle River, NJ (2006).

D. R. Blackmore and A. Thomas. *Fuel Economy of the Gasoline Engine.* Macmillan, London (1977).

B. K. Hodge. *Alternative Energy Systems and Applications.* Wiley, Hoboken, NJ (2010).

P. Hoffmann. *Tomorrow's Energy: Hydrogen, Fuel Cells, and the Prospects for a Cleaner Planet.* MIT Press, Cambridge, MA (2001).

T. Koppel. *Powering the Future: The Ballard Fuel Cell and the Race to Change the World,* Wiley, New York (1999).

K. Kordesch and G. Simader. *Fuel Cells and Their Applications.* VCH Publishers, New York (1996).

P. Kruger. *Alternative Energy Resources: The Quest for Sustainable Energy.* Wiley, Hoboken (2006).

J. Larminie and J. Dicks. *Fuel Cell Systems Explained.* Wiley, Chichester, United Kingdom (2000).

J. MacKenzie. *The Keys to the Car: Electric and Hydrogen Vehicles for the 21st Century.* World Resources Institute, Washington, DC (1994).

C. Ngô and J. B. Natowitz. *Our Energy Future: Resources, Alternatives and the Environment.* Wiley, Hoboken, NJ (2009).

R. O'Hayre, S. W. Cha, W. Colella, and F. B. Prinz. *Fuel Cell Fundamentals.* Wiley, Hoboken, NJ (2006).

V. Quaschning. *Renewable Energy and Climate Change.* Wiley, West Sussex, United Kingdom (2010).

J. J. Romm. *The Hype About Hydrogen: Fact and Fiction in the Race to Save the Climate.* Island Press, Washington, DC (2005)

D. Sperling and J. Cannon (Eds.). *The Hydrogen Energy Transition: Moving Toward the Post-Petroleum Age.* Elsevier, Amsterdam (2004).

C. Spiegel. *Designing and Building Fuel Cells.* McGraw-Hill, New York (2007).

S. Srinivasan. *Fuel Cells.* Springer, New York (2006).

PART VI

The Future

Photovoltaic collectors with pumped hydroelectric storage in Geeshacht, Schleswig-Holstein, Germany.

Jörg Müller/Alamy Stock Photo

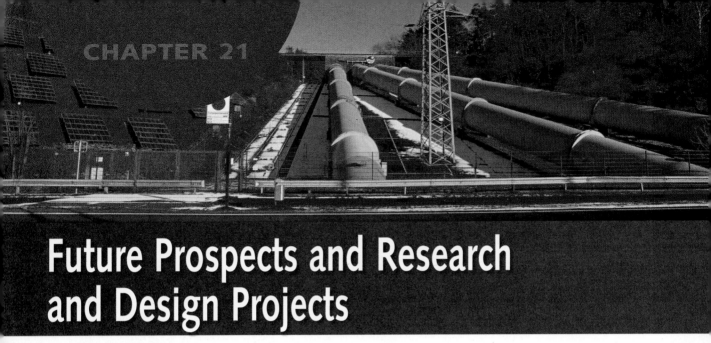
Jörg Müller/Alamy Stock Photo

CHAPTER 21

Future Prospects and Research and Design Projects

21.1 Introduction

Humanity has been dependent on fossil fuels to fulfill its energy needs for well over a century. The future of fossil fuel use is ultimately limited by the availability of oil, natural gas, coal, and other petroleum resources. Traditional oil resources are likely to provide energy for only a few more decades, and coal and natural gas produced by fracking and some alternative petroleum resources, such as oil sands, may last somewhat longer. There is, however, concern that the continued use of these resources will cause irreparable environmental damage. In addition to the clear relationship between greenhouse gas emissions and the quantity of energy produced by the combustion of fossil fuels, other adverse environmental consequences are associated with fossil fuel production and use, particularly for the many approaches to enhanced fuel recovery, such as shale oil extraction and natural gas well fracking. Thus, there are obvious environmental incentives to alternative energy development. Understanding future energy scenarios requires an understanding of future energy needs and possibilities for future energy production.

This concluding chapter provides an overview of the key points covered in this text in terms of possible technological approaches to future energy production, as well as methods for the efficient conservation, storage, and distribution of energy. The projects in this chapter provide an opportunity to take a broad look at various technologies and compare the advantages and disadvantages of different approaches in particular circumstances. In many cases, additional resources from the Internet, texts, and other sources are needed to provide the data necessary to undertake the project.

21.2 Approaches to Future Energy Production

Our energy needs over the years have increased, and this trend is expected to continue for some period into the future. Much work has been done on trying to predict future energy use, because knowing our future needs is beneficial in understanding which energy technologies are most viable for development. Figure 2.13 shows one prediction of future energy needs for the next couple of decades, and indicates a reasonably consistent increase on this timescale. However, predictions for the distant future can be quite variable, and some suggest that future energy use will actually decrease as a result of factors that include improved efficiency of equipment and increased awareness of energy conservation.

Predictions that extend very far into the future become progressively less certain because the effects of assumptions become compounded. Perhaps the most systematic recent approach to future energy prediction was presented in a report from the Intergovernmental Panel on Climate Change (IPCC) that was jointly established by the World Meteorological Organization (WMO) and the United Nations Environment Programme (UNEP). This study involved the prediction of future world energy use until the year 2100 on the basis of 40 different scenarios. Scenario assumptions dealt with factors such as predicted world population, economic growth, technological development, and the degree of cooperative energy management. Figure 21.1 shows the extremes of these predictions over the next century or so. The huge range of predictions for the year 2100 is apparent in the figure. In the best-case scenario, energy use in 2100 will be roughly the same as it is today. In the worst case, energy use will be about 5 times what it is today. The best case results from limited population growth, development of new energy-efficient technologies, and rapid economic growth. The worst case results from continuous population growth, poor economy, and lack of social equality.

The distribution of energy use from the different available sources must be considered in the context of overall energy needs, as suggested by Figure 21.1. Predictions

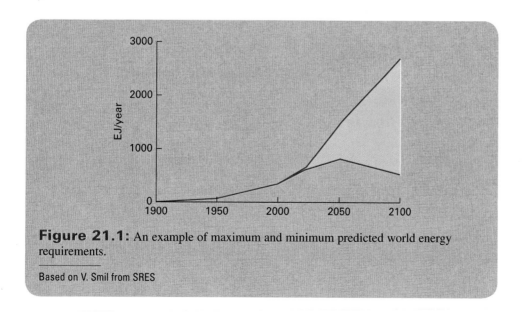

Figure 21.1: An example of maximum and minimum predicted world energy requirements.

Based on V. Smil from SRES

concerning the role of various energy sources in future energy production are perhaps even more uncertain than predictions for total energy needs. Certainly, the distribution of energy use can be no more accurate than the prediction of the total. This uncertainty may also be a feature of predictions of the relative proportions of the various contributions to energy use. Certain energy options—wind, for example, are not unlimited and may satisfy most of our energy needs if those needs do not exceed the availability. If the need is higher than the availability, the distribution of energy sources needs to be reevaluated. In any case, the evaluation of possible future energy economies is an interesting exercise that sheds some light on the viability of the options discussed in this text, but they should not be taken too literally.

Predictions for the future must, of course, start with what exists today. A typical current breakdown of energy sources (in the United States) is illustrated in Figure 21.2. It is inevitable that, at some point in the future, fossil fuel use will decrease. This decrease must be at the expense of the increased use of nuclear and/or renewable energy sources. Predictions concerning nuclear power that were made around 1970 were overly optimistic, and predictions concerning the use of OTEC have certainly not materialized so far. In one case, political factors played an important role in the development of nuclear energy, and in the other case, technological advances fell short of expectations. The significant implementation of alternative energy sources comes with a substantial economic cost, but the failure to implement these changes comes with a substantial environmental cost. There is a reluctance to change things that we are accustomed to, and, up to some point, there is a financial incentive to avoid the development of a new infrastructure.

The most reasonable options for a particular region are a function of a number of factors, including geography and climate. An appropriate approach to alternative energy in the U.S. Midwest would be quite different than it would be in Brazil or Scotland. Many regions have some commonality in resources; sunlight and wind are available in varying amounts almost everywhere. However, other regions may have more unique resources available, such as hydroelectric, tidal, or wave energy. Some

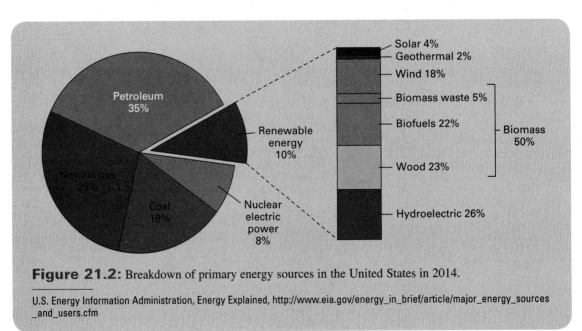

Figure 21.2: Breakdown of primary energy sources in the United States in 2014.

U.S. Energy Information Administration, Energy Explained, http://www.eia.gov/energy_in_brief/article/major_energy_sources_and_users.cfm

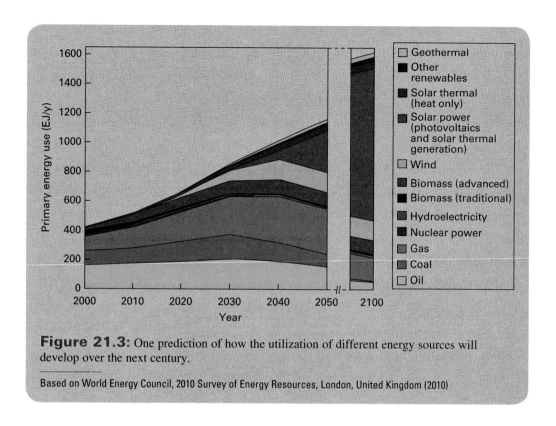

Figure 21.3: One prediction of how the utilization of different energy sources will develop over the next century.

Based on World Energy Council, 2010 Survey of Energy Resources, London, United Kingdom (2010)

energy options may not be mutually exclusive but may complement each other, as rows of photovoltaic collectors between the turbines of a wind farm.

It is likely that a combination of approaches will be taken in the foreseeable future and that our energy economy will be a mix of technologies. The exact nature of our energy economy is a complex matter that will develop over time. One prediction for a future energy economy is from the World Energy Council, shown in Figure 21.3. This prediction has some interesting features. The total energy use, 1600 EJ per year, falls almost exactly in the middle of the range of predictions from Figure 21.1. In this prediction, less than 20% of our energy comes from fossil fuels. The majority (more than 60%) comes from solar. Implementation of such a future energy mix will require ambitious efforts to develop new solar technologies, as well as a dedicated approach to freeing ourselves from the cheap and easy route of burning more coal for the next few hundred years.

21.3 Key Considerations

As pointed out in the introduction to Part IV of the text, an alternative energy technology must ideally be *c*lean, *u*nlimited, *r*enewable, *v*ersatile, and *e*conomical. It is clear, at this time at least, that there is no perfect solution to our future energy needs. Although different technologies fulfill different combinations of these key requirements, a comparison of their benefits is not always so obvious or straightforward.

Many predictions of future energy use, like that shown in Figure 21.3, conclude that by the end of the century the largest fraction of our energy use will be solar.

Certainly solar energy can be viewed very positively from the viewpoint of being unlimited and renewable. The degree to which solar and other energy sources satisfy other important criteria must be considered very carefully.

21.3a Clean

An objective rating of how an alternative energy technology satisfies the criterion of being clean is often difficult. As has been discussed in Part IV of this text, alternative energy sources can have a measurable or even significant environmental impact. In fact, such environmental influences inevitably include some quantity of greenhouse gas emissions. A thorough analysis of the environmental effects of each alternative energy technology is not generally straightforward. Quantitative assessments, such as that presented in Chapter 16 for biofuels, often yield results that are a major cause for concern. The environmental impact of so-called clean technologies for energy production is often surprisingly significant. In many cases, the reason for this is the rather low energy density of most alternative energy sources. The need for extensive manufacturing, transportation, maintenance, and related activities contributes to a net environmental impact per unit energy that can be greater than for traditional fossil fuel technologies. The analysis of this impact follows along the same lines as the risk assessment for energy sources presented in Chapter 6.

21.3b Unlimited

The power that is available from alternative energy resources depends on the nature and extent of the resources, and on the existence of a viable technology to utilize these

---

PROJECT 21.1
CO_2 sequestration

Consider the possibility of sequestering CO_2 from fossil fuel generating stations in depleted natural gas wells. Begin by summarizing the current production of electricity in the United States. Calculate the volume of CO_2 produced annually from the combustion of coal and natural gas in the United States for electric generation. Compare this volume to the volume of natural gas produced from wells in the United States and the volume of CO_2 from electric generating stations currently sequestered in this country.

(a) Are efforts to sequester CO_2 appropriate to the magnitude of the environmental problem of CO_2 emissions?

(b) In this electric generating scenario, is the availability of natural gas wells compatible with the needs for CO_2 sequestration?

(c) Would the volume of depleted oil wells in the United States contribute in a substantial way to CO_2 sequestration capacity?

(d) Investigate the potential technical difficulties in transporting CO_2 to the necessary locations for injection into depleted wells. Make a rough analysis of the distribution of fossil fuel–fired plants in the United States and the distribution of wells. It might be simplest to concentrate on coal-fired plants and natural gas wells. Make an estimate of the infrastructure that would be needed to sequester CO_2 by this method.

Table 21.1: Power available from different renewable energy sources. Power listed is that which is technologically feasible and economically viable on the basis of technical and scientific capabilities at present or in the foreseeable future.	
energy source	**power available (TW)**
solar	>1000
biomass	~6
wind	~4
tidal/waves/currents	~2
hydroelectric	~1
geothermal	<1
OTEC/salinity gradients	<1

energy sources. In some instances, such as with wind energy, a suitable technology exists that enables us to effectively make use of the resource, although this, in itself, does not ensure that development of the resource is viable economically and environmentally. In other cases, such as with OTEC or salinity gradient energy, further technological advances are clearly needed before an energy resource can be exploited in a practical way. Table 21.1 presents the amount of power available from different alternative energies on the basis of existing technologies or technologies that might conceivably be available in the foreseeable future. It is important to realize that the availability of power from a particular resource is limited by the physical availability of the resource itself, as well as by the availability of the technology to make use of the energy resource. This situation might, for example, arise as a result of the limits to the world's supplies of indium for the production of photovoltaic cells or limits to the supplies of the rare earth elements that are used for magnets in wind turbine generators.

The values in Table 21.1 must be viewed in the context of future energy needs as discussed in Section 21.2. As illustrated in Figure 21.1, annual energy requirements in the range of 1000 to 2000 EJ are reasonable expectations over the next century or so. This requirement amounts to an average power consumption in the range of 30–60 TW. Predictions like those shown in Figure 21.3 can be viewed in the context of the resources given in Table 21.1. The table also shows that solar energy exceeds (by far) the total energy requirements of society and that this is the only alternative energy resource (nuclear not included) that can fulfill all of the world's energy needs. Even the utilization of a fairly small fraction of available solar energy at a relatively low efficiency is sufficient to satisfy humanity's requirements. Biomass and wind can, at least in principle, make a major contribution to possible future energy mixes. Other resources may be locally significant but can contribute a relatively small fraction of total energy needs on a global scale. Thus, on a relevant scale, only solar energy may be considered to be unlimited.

21.3c Renewable

All of the alternative energy sources listed in Table 21.1 are basically manifestations of solar energy, except for geothermal and tidal energy. Because the longevity of solar energy is, for all practical purposes, infinite, we would anticipate that the longevity of

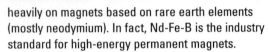

PROJECT 21.2
Rare earth resources

Many alternative energy technologies utilize materials (or particularly elements) with limited availability. While some cases have been made quite apparent in the text, such as lithium for rechargeable batteries or indium for photovoltaic cells, other cases are not so apparent. A somewhat controversial article by David Cohen ["Earth audit," *New Scientist* **194** (2007): 34], overviews the concerns for future resource availability. Although not specifically discussed in this article, the rare earth elements require careful consideration in the context of energy production and utilization, and, as their name implies, these elements are not common. Rare earth–containing materials are of importance largely because of their magnetic properties, and they form a critical component in many alternative energy technologies. Modern wind turbines and battery electric vehicles, for example, depend heavily on magnets based on rare earth elements (mostly neodymium). In fact, Nd-Fe-B is the industry standard for high-energy permanent magnets.

Research the ways in which rare earths are essential for the development of alternative energy technologies. Concentrate particularly on wind turbines and electric vehicles, but do not overlook other possible needs for these materials. How many kilograms of rare earth elements, such as Nd, are needed (for example) for a 1-MW wind turbine or a family-sized BEV? If all the world's energy needs were satisfied by electricity generated by conventional (nonsuperconducting) wind turbines and all passenger vehicles were BEVs, would enough rare earth elements be available? A good place to start researching the world's supplies of various elements is the website http://minerals.usgs.gov/minerals/pubs/commodity/ and the links therein.

these energy resources would also be infinite. This is only partly true for two reasons: (1) Changes that affect our ability to exploit the resource and (2) the need for non-renewable resources to convert primary energy into a usable form can both limit our ability to utilize a resource. In the first case, this might be the reduction in the viability of hydroelectric energy from a particular location due to sedimentation (Chapter 11). In the second case, this might necessitate the recycling of rare materials (Section 21.3b). For other possible energy technologies, these two factors may also be limitations. Some examples are the limited longevity of geothermal energy that results from the fact that thermal energy from geothermal deposits is typically extracted faster than it is replenished (Chapter 15) and limits to the resources of lithium used to breed fuel for a fusion reactor (Chapter 7).

PROJECT 21.3
The future use of lithium

Lithium may be a crucial element in future energy technologies. Summarize the ways in which lithium may be of importance. Discuss the availability of lithium, the requirements for its use in different energy technologies, the longevity of the world's lithium resources for different applications, and the possibility of recycling lithium from these technologies. Specifically contrast the need for lithium for current battery technology and the potential future need for fusion power.

21.3d **Versatile**

The goal of most alternative energy technologies is the generation of electricity. Although electricity is suitable for many of our energy needs, its utilization for transportation requires careful consideration. In fact, transportation places particular challenges on technology because it requires an energy storage mechanism that must, at a minimum, be portable and have a sufficiently high energy density. Added to this difficulty is the fact that the route from primary energy source to mechanical energy at a vehicle's wheels often involves a number of energy conversions that add up to an unacceptably low overall efficiency. To meet the needs of a suitable transportation technology, the method of energy production, as well as energy storage, must be versatile enough to provide acceptable overall efficiency. From a practical standpoint, this basically means that in the context of current technology, nonfossil fuel–based transportation will rely on either biofuels or involve some electricity storage mechanism.

Biofuels have been successful in Brazil but are unlikely to be an effective approach in more temperate countries unless new technologies are developed for utilizing the cellulose component of plant matter. Electric vehicles require an electricity storage mechanism, and at present this means either batteries or hydrogen. The efficiency analysis in Chapter 20 clearly favors batteries, and research into new battery chemistries has progressed in recent years. Lithium resources are somewhat limited. If fusion energy becomes a reality, then fusion reactors will compete with batteries for the world's lithium supply. From Chapter 7, the estimated Li resources worldwide are about 1.7×10^{10} kg. Based on the discussion in Chapter 19, a rough estimate of the amount of Li required for the Li-ion cells in a practical vehicle would be about 10 kg. Thus, the amount of Li available would provide batteries for about 1.7×10^9 vehicles, although improvement in vehicle efficiency and changes in our perception of vehicle requirements may increase this number. The current automobile population worldwide is about 6×10^8 vehicles, and that number is expected to double over the next few decades, as personal

---

PROJECT 21.4
The use of methane

The United States has substantial natural gas (methane) resources (compared with its oil resources). Natural gas has been promoted as a clean alternative to coal for producing electricity in thermal power plants. Consider the following options for use of methane in future transportation technologies:

- Combustion of methane in traditional ICE vehicles.
- The production of electricity by thermal generation by methane combustion and the use of electricity in battery-operated vehicles.
- The production of electricity by thermal generation using methane combustion and the subsequent use of electricity to produce hydrogen by electrolysis for fuel cell vehicles.
- The production of electricity by thermal generation by methane combustion and the use of electricity to produce hydrogen through electrolysis for hydrogen ICE vehicles.
- The direct use of methane in solid oxide fuel cell vehicles.

In each case, consider the environmental impact of the approach and its economy in terms of overall efficiency.

PROJECT 21.5
Dimethyl ether

Natural gas (primarily methane) is a fairly plentiful fossil fuel (Chapter 3), that has found a number of applications, e.g., electricity generation, home heating, and the like. A major factor that limits methane use for some applications (e.g., transportation) is the fact that it is a gas and therefore occupies a large volume. Because of its low boiling point, it is inconvenient to liquefy methane. Petroleum gas, which is about 85% propane combined with other gaseous hydrocarbons, has a much higher boiling point than methane and can be readily liquefied at ambient temperature under a modest pressure ($\sim$ 1.5 MPa) to produce liquefied petroleum gas (LPG). This product has advantages over methane for some applications.

There has been significant recent interest in the use of dimethyl ether (DME) as a fuel (by combustion) because it has properties similar to propane and can be produced from methane.

Investigate the use of DME as a fuel, including the following points:

(a) What are the properties of DME in comparison with methane and propane?

(b) What are the reactions and techniques by which DME can be produced from methane?

(c) What are the energetics of this production method? [Specifically, what is the net energy input or output (ideally) when methane is converted into DME?]

(d) What are the advantages (and disadvantages) of DME compared with methane and propane, and what applications would benefit most from this approach?

(e) If one considers the details of the process of producing DME from methane and the energy produced by combustion, are there environmental advantages to producing DME rather than merely using methane in terms of kg(CO_2) per MJ of energy produced?

motor vehicles become more common in developing countries. Thus, the viability of this approach as an alternative transportation technology may basically be a question of resource availability. The long-term sustainability of a BEV-based transportation system must consider this type of analysis and emphasizes the importance of understanding the global uses of Li and of effective Li recycling. Lithium-sulfur batteries are being developed (Figure 21.4) that have superior energy storage capabilities. Sodium shares common features with lithium in terms of its electronic structure and its subsequent chemical properties. Research into sodium-based batteries is underway and shows promise for batteries that rely on a plentiful element (contained in the salt of the oceans).

21.3e Economical

The economics of various approaches to energy production involve a number of factors. These include the cost of producing, operating, and maintaining the necessary infrastructure, as well as the overall efficiency of converting primary energy into a usable form. Alternative energy technologies are, for the most part, energy conversion technologies, with electricity, in many cases (although not all), as the end usable energy product. All alternative energy technologies must compete with fossil fuels. While coal, oil, and natural gas are readily available and (relatively) inexpensive, a major factor in promoting new energies is cost. It is also necessary to consider availability and environmental impact, as well as the availability of materials needed to utilize the resource. Figure 21.5 shows an estimate of cost of electricity in the United States for

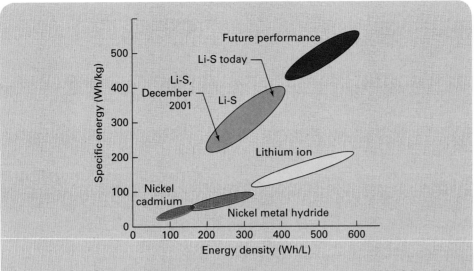

Figure 21.4: Ragone plot showing performance of newly developed Li-S batteries.

Based on Thom Mason, Oak Ridge National Laboratory

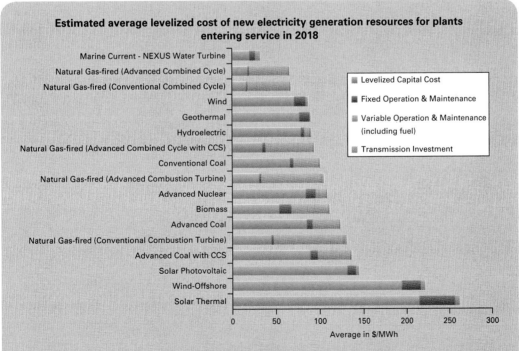

Figure 21.5: Estimated cost in US$ per MWh for new electricity generation facilities entering service in 2018, from the United States Energy Information Administration.

U.S. Energy Information Administration, Levelized Cost and Levelized Avoided Cost of New Generation Resources in the Annual Energy Outlook 2017, http://www.eia.gov/outlooks/aeo/pdf/electricity_generation.pdf

different technologies. These estimates are for new facilities entering service in 2018, showing the breakdown of different cost components (see Section 2.5c). The figure also shows the effects of the inclusion of carbon capture technology on the net cost of electricity for fossil fuel–fired power plants. While most well-established renewable sources (i.e., wind, geothermal, and hydroelectric) compete well from an economic standpoint with fossil fuels, solar (both thermal and photovoltaic) and offshore wind are not very cost effective. In particular, it is seen from the figure that infrastructure costs, as well as operational costs, are high for these technologies.

Although environmental concerns and convenience can offset economic factors to some extent, the cost per kilowatt-hour or megajoule of usable energy to the end user must somehow be competitive with other options. For a facility developed by, say, a public utility, development costs, infrastructure costs, maintenance and operating costs, as well as the longevity of the facility, are all important factors in determining the economic viability of an energy technology. Although long payback periods may be acceptable for well-established technologies such as coal-fired plants, nuclear power plants, or hydroelectric installations, new technologies carry more risk in terms of long-term reliability, maintenance costs, and operational costs, and a shorter payback period is certainly desirable. There may also be a tendency to delay investing in technologies, such as photovoltaics, pressure-retarded osmosis, or large-scale battery storage, that may become more economical in the future. For individuals investing in alternative energy resources for residential use, financial gains on a timescale of a few years is generally required. In any case, a consideration of energy options requires an analysis of a number of factors. A careful analysis of the environmental effects of alternative energies shows that their implementation often has undesirable and sometimes not so apparent consequences. The availability and longevity of many resources are less than we would hope. In fact, only solar energy can clearly supply all of our energy needs and is renewable indefinitely, but, at present, it is substantially more expensive than many other alternative options.

21.4 Overview of Future Energy Technologies

The need to develop energy technologies that are not based on fossil fuel combustion is clear. For this discussion, we will divide these technologies into two basic categories: nuclear energy and renewable energy. In the first case, nuclear energy, at least in the foreseeable future, refers to fission energy. In the second case, a number of options have been discussed in Part IV of this text. A brief summary of some of the available energy options is now presented.

21.4a Nuclear Energy

Traditional thermal neutron reactors are a well-established technology that contributes significantly to our present energy needs. The availability of ^{235}U resources is unlikely to extend very far into the future. However, the utilization of the energy content of non-fissile ^{238}U by more extensive fuel reprocessing and/or the use of fast breeder reactors will make this a viable option for quite some time. These possibilities are reasonably well-established technologically. The development of ^{232}Th as an energy source can extend these possibilities even further and has been promoted as an environmentally more attractive option. Fusion is unlikely to be technologically feasible in the foreseeable future.

21.4b **Solar Energy**

Solar energy can contribute to our energy needs in two ways: solar thermal and photovoltaic. Solar thermal energy is best suited to residential heating and can contribute to a small extent to our future energy use. It is based on a well-established and very mature technology. While there may be small technological advances that improve efficiency, no huge breakthrough is on the horizon for the use of solar thermal energy. Wide-scale use of solar thermal energy use, such as solar thermal electricity generation, is probably not a technology that can compete with other alternatives.

Solar photovoltaic electricity generation has received much interest in recent years, and there has been substantial research into improving the technology in this area. On the one hand, photovoltaics have been around for many years, while on the other hand, it is a fairly immature technology, and substantial developments are still likely. The low efficiency of solar photovoltaics is often seen as its major difficulty, and this is certainly an important consideration that has prompted much research into developing new photovoltaic materials with higher efficiency. The practical problem with photovoltaics is its relatively low efficiency combined with its relatively high cost. The significance of developing new materials must be weighed in the context of their cost and also their availability if photovoltaics are to fulfill a significant fraction of the world's energy needs. Also, resources such as the elements indium, cadmium, tellurium, and the like, which could play an important role in the development of high-efficiency photovoltaics, are limited. While it would be desirable to have photovoltaic efficiencies of 40%, we should consider whether 15–20% efficiency would be acceptable if the required materials are more plentiful and less expensive. Certainly, this situation would require more land area to fulfill our energy needs, but total land area may not be the most serious

---

PROJECT 21.6

Future photovoltaic applications

Evaluate the possibility of developing a sufficient solar energy infrastructure during the remainder of the twenty-first century to fulfill the predictions shown in Figure 21.3. You will need to estimate from the figure the solar energy per year (that is, W_e) as a function of year from now until 2100. What is the area of photovoltaic arrays needed to meet this requirement? On the basis of these needs, discuss the following questions:

(a) What is the most reasonable photovoltaic material based on current technology?

(b) What is the mass of materials (particularly critical elements) needed per year to satisfy this development? How does this compare with current photovoltaic cell production? Comment on the need for the development of additional photovoltaic manufacturing infrastructure.

(c) What is the anticipated lifetime of a facility, and will cells have to be replaced during the time period you are considering? How does this affect the quantity of cells that needs to be produced per year?

(d) Assuming that future developments increase the efficiency of photovoltaics to nearly the theoretical limit, how will this influence your analysis?

(f) If new photovoltaic technologies are developed from current research directions, are certain types of cells less desirable in the context of economics or resource availability?

PROJECT 21.7
Use of wind energy

Wind power is based on a well-established technology, its net cost per kWh$_e$ is competitive with other renewable sources, and it has minimal environmental impact (compared to other technologies). A convenient aspect of wind power is that it can be implemented on a vast range of scales, ranging from residential wind turbines of a few kW$_e$ capacity to wind farms of close to a GW$_e$ capacity. Not surprisingly, therefore, the utilization of wind energy has grown substantially in recent years. Five countries that have been very active in developing wind power in recent years are China, Denmark, Germany, India, and the United States. These countries have quite different energy needs and serve as good examples of how wind power can be implemented in different parts of the world.

(a) Locate information about wind energy use in these five countries. What fraction of the total electricity is produced from wind?

(b) Estimate the capacity factor for each country, realizing that the total energy produced per year (in Wh$_e$) is the rated capacity (in W$_e$) times the number of hours per year times the capacity factor.

(c) What fraction of its total electricity needs could each country expect to obtain from wind, assuming that 5% of its land area was utilized? Consider, for example, the diameter of a 2-MW$_e$-rated wind turbine and the optimal spacing of turbines.

(d) For each of the five countries prepare a brief report discussing the following points:

- How serious has the effort been thus far to develop wind power to meet the countries' electrical needs?
- Can a much greater capacity be expected from further development, and what fraction of total electrical needs would be reasonable to expect from wind?
- How important is the development of offshore wind farms?

limiting factor for the development of this technology. At the moment, the limiting factor seems to be cost, and, in the future, the limiting factor may be material availability. Lower efficiency may be a viable alternative if devices are inexpensive and utilize more common materials.

The availability of solar energy varies according to location around the earth. Clearly, latitude is significant, and locations far from the equator are less likely to benefit from solar energy. Also, local variations that depend on geography and climate influence the viability of this resource.

21.4c Wind Energy

Wind energy is based on a well-established technology that is unlikely to see any major changes in the foreseeable future. Wind energy is relatively inexpensive, its use has increased significantly in recent years, and it is likely to see continued increases in the future. Its possible use is widespread, although local geographic and climatic considerations are important. The major difficulty that limits the use of wind energy is the total power available worldwide. While wind is likely to be a factor in our future energy mix, it cannot satisfy all our energy needs.

PROJECT 21.8
Environmental effects of hydroelectric energy

Consider, as quantitatively as possible, the environmental effects of a moderately large high head hydroelectric facility with a capacity of about 1 GW_e compared with a typical coal-fired generating station of the same capacity. As a basis of comparison, calculate the CO_2 emissions per year for the coal station operating at full capacity. For the hydroelectric station determine the following:

(a) The typical area of a reservoir needed for such a facility. Look up specifications of similarly sized hydroelectric facilities to get an idea of reservoir size.

(b) If the facility is located in a temperate region with average forestation, calculate the loss of carbon sequestration capabilities that results from the replacement of forested areas with the reservoir.

(c) How would the calculated effect differ in tropical regions? Also, find information about methane production from reservoirs in tropical regions. What is the greenhouse gas effect of this methane production compared with the loss of CO_2 sequestration capabilities? How does the overall environmental impact of a hydroelectric facility in a temperate or in a tropical region compare with a fossil fuel plant of similar capacity?

21.4d Hydroelectric Energy

Hydroelectric energy, like wind, is based on a mature technology that has been utilized extensively in the past. There are relatively few opportunities for the expansion of large-scale hydroelectric facilities, particularly in North America. There are more possibilities for run-of-the-river hydroelectric development. Like wind energy, hydroelectricity will be a small factor in worldwide energy production. However, two considerations should be addressed: environmental impact and longevity. As discussed in Chapter 11, greenhouse gas emissions may be a factor for high head hydroelectric facilities, particularly in tropical areas, and many resources may not be renewable indefinitely due to sedimentation.

21.4e Tidal and Wave Energy

Energy from the movement of the oceans has been exploited in a few locations. Technology is developing in this area but is based on well-established scientific principles. Tidal energy has been harnessed in France and Canada, and it is possible that other locations may be developed in the future. The availability of locations where this source of energy may be utilized is fairly limited, and it is likely to be a local factor in certain areas rather than a substantial component of global energy use. Waves (and possibly ocean currents) are somewhat more widespread, but again the utilization of wave energy is likely to be only a small factor in the energy production for certain countries.

21.4f OTEC and Salinity Gradients

Ocean energy from gradients in temperature and salinity may be utilized. OTEC development has fallen well below one-time expectations. The low efficiency and technological difficulties associated with developing this resource are likely to limit its use for some time. Salinity gradient energy is a field that is still in the early stages of development, and time will tell whether it may be a small factor in our future energy mix.

21.4g Geothermal Energy

Geothermal energy has been developed extensively and relies, for the most part, on mature technologies. It is an important factor in energy production in specific geographical locations where appropriate underground resources are available. New technologies may be possible that will better utilize this resource, although the total power available is limited. Its longevity is perhaps questionable if thermal energy is extracted from a deposit more rapidly that it is replenished.

PROJECT 21.9
Comparison of different water-based energy technologies

A comparison of the advantages of various alternative energy technologies is complex. It must include an analysis of factors such as economics, resources availability, and environmental impact, many of which are difficult to quantify. A simple (and probably overly naïve) approach might be to look at energy produced per unit of resource. We should not try to overinterpret the meaning of such results because the analysis does not consider environmental and other factors, but this type of analysis can provide some insight into the relative desirability of different approaches. Consider, for example, the ways in which energy can be extracted from water, where we can define the quantitative measure of megajoules of energy per cubic meter of water. For the following water-based energy technologies, determine MJ/m^3 as indicated in the description.

- *Hydroelectric:* This technology converts the potential energy of the water into electricity. Calculate MJ/m^3 (for a freshwater density of 1000 kg/m^3) for a head of 100 m (typical for a high head hydroelectric facility) and a conversion efficiency of 90%.

- *Wave/tidal (current):* This technology coverts the kinetic energy of moving water into electricity. Assuming a seawater density of 1025 kg/m^3, a velocity of 5 m/s, and a conversion efficiency of 35%, calculate MJ/m^3.

- *OTEC:* This technology converts thermal energy in the oceans into electricity. For typical (tropical) conditions, calculate the ideal Carnot efficiency for conversion from thermal to mechanical energy by means of a heat engine and a conversion efficiency of 90% from mechanical to electrical. Calculate MJ/m^3.

- *Osmotic energy:* This method converts chemical energy in the seawater into electricity. Assume a seawater molarity of 500 mol/m^3. Calculate the effective head based on the osmotic energy, and assume that this can be converted into electricity by means of a turbine/generator with an efficiency of 80%.

Tabulate your results for each of these technologies and comment on the results. Which is the most efficient method of extracting electrical energy from water?

Does this method of assessment provide a reasonable view of which technologies are likely to be viable? To answer the more complex question of viability, consider an assessment of the following factors:

- Infrastructure needed per unit energy produced.
- Operating costs and difficulties.
- Energy distribution considerations.
- Overall efficiency of prime energy conversion to electricity for the user.

PROJECT 21.10
Hydrogen as an energy storage mechanism for OTEC

Research the following energy scenario:

A 10-MW$_e$ OTEC facility is located 500 km offshore in a very desirable ocean thermal environment. Electricity produced at the facility must be transported to shore. It is proposed that hydrogen will be produced by electrolysis and that the hydrogen will be liquefied and transported to shore by ship, where it will be converted back into electricity using fuel cell technology. To be environmentally friendly, the ship will utilize liquid hydrogen for propellant in a hydrogen ICE. Prepare a report summarizing the expectations for the overall economics and energy efficiency of this approach. Provide recommendations for investing in a program to develop this technology, and compare its viability to traditional alternative energy technologies, such as terrestrial-based wind energy.

PROJECT 21.11
A comparison of alternative energy resources

A major river runs through a productive agricultural region. The region has good insolation, good wind condition, and a reasonable elevation change. It is planned to develop alternative energy for the region using one of the following methods:

- Construct a nuclear power plant and use the river as a heat sink.
- Construct a 100-km^2 wind farm.
- Construct a 100-km^2 photovoltaic array.
- Convert 100 km^2 of food crops to corn for use for ethanol production.
- Dam the river and create a 100-km^2 reservoir for a hydroelectric generating station.

Discuss the pros and cons of each of these plans in terms of their energy productivity (i.e., in terms of energy per land area) and environmental impact. Be as quantitative as possible, and include an analysis of the following factors:

- Infrastructure cost, estimate cost per net MW$_e$ capacity.
- Anticipated longevity of resource.
- Operational cost and net cost per kWh$_e$.
- Environmental impact of the facility in terms of effects of infrastructure construction and environmental impact of operation.

21.4h Biomass Energy

The viability of biomass energy depends on the efficiency, that is, energy in (at the moment mostly from fossil fuels) vs. energy out, of converting biomatter into a usable fuel. Bioenergy could be a substantial component of future energy production, but, using present technology, its efficient use is limited to locations with suitable climate conditions (i.e., tropical regions). The development of efficient processes for using cellulose as a source for ethanol production would make biofuels more universally useful. Municipal waste is a minor factor in energy production. The use of municipal waste for energy production, however, is potentially beneficial from an environmental standpoint because any environmentally benign energy gain from waste material is preferable to most other methods of disposal.

21.5 Efficient Energy Utilization

While the efficient and economical production of energy from environmentally conscientious sources is a necessary step for a sustainable future, the efficient and effective use of this energy is also essential. The factors discussed in this text can be categorized in three ways:

1. Conservation of energy.
2. Efficient distribution of energy.
3. Storage of energy.

21.5a Conservation

While developing new energy technologies is crucial for a sustainable future, making the best use of energy that is produced helps to alleviate the demands on energy production. Thus, conservation is an important part of maintaining an adequate energy supply for the future and an essential part of dealing with environmental concerns that result from energy production and use. Because virtually all energy production and use have some environmental impact, reducing energy use reduces this impact.

As seen in Chapter 17, energy conservation can be approached at various levels, from national government energy policies and international cooperative efforts, to regional policies, to business and industrial practices, and finally to the actions of individuals. Energy conservation is an ideal example of a situation where individuals can be proactive in addressing future energy issues and where simple actions can have an important cumulative effect on world energy use. Although there is often reluctance to change effective practices and there are always economic concerns that need to be addressed, many conservation actions are effective at improving the overall comfort in our lives and have financial benefits on a quite reasonable timescale. Improved environmental quality and long-term energy security are added benefits of our conservation efforts.

PROJECT 21.12

Analysis of an integrated home-heating system

A homeowner plans to install an integrated central home-heating system consisting of solar thermal panels on the roof plus a natural gas-fired furnace that distributes heat through forced-hot-air ducts. The house is one story with a footprint of 15 m × 10 m and is located in a reasonably sunny location at a latitude of 40°N with one wall facing south. There are 3200 (°C) degree days per year. The heating requirements of the house are typical of those discussed in Chapter 8. For simplicity, consider the hot-water-supply system to be independent of the heating system. Assume that you work for an alternative heating consulting company that has been assigned the task of developing such a heating system. Prepare a proposal for this system. At a minimum, your proposal should show a diagram of the system illustrating how the various components would be integrated. Specify the parameters for the system, such as the area of the solar collectors, the size of an appropriate energy storage system, energy output requirements for the natural gas furnace, and any other relevant specifications. The homeowner is particularly interested in obtaining an estimate of the amount of natural gas (in cubic meters at STP) that will be used per year.

PROJECT 21.13

Analysis of an integrated smart-grid system

A region with a population of 320,000 and an average annual electricity requirement of 33 MWh$_e$ per capita would like to develop a self-contained electrical system with no substantial imports of electricity. It has been decided to develop a system consisting of a coal-fired thermal generator to cover base load capacity, a wind farm to provide additional power for peak periods, and a pumped hydroelectric storage system to store excess energy generated from wind during nonpeak periods. A smart-grid control system will manage the operation of the system. The region has good wind resources, a substantial water supply from a river, and variable terrain elevation. Prepare a proposal for such a system specifying the minimum requirements for each component. Make any reasonable assumptions that are needed about efficiencies, capacity factors, and natural resources availability. Use Figure 18.1 as a guideline for typical power-requirement variations. As a minimum, consider the following points:

- *Coal generating facility:* What is the proposed capacity in MW$_e$? How much coal on, say, a weekly basis needs to be used? What is the estimated cooling water requirement?

- *Wind farm:* What is the required capacity? Be sure to include a reasonable estimate of capacity factor. How many 1-MW$_e$ capacity turbines would be needed? What would be the land area required?

- *Pumped hydroelectric storage:* What total energy storage capacity (GWh) would be required? What would be reasonable dimensions (area, average depth) to achieve this? Propose minimum head and flow rates that would be compatible with the requirements.

- *Smart grid:* Describe how a smart-grid system might integrate the components of the system to most efficiently supply the community's electricity needs.

21.5b **Distribution**

The large-scale distribution of electricity is a major component of our energy use. The electric grid connects energy production facilities with users. In the past, electric generation was by means of fossil fuel-fired plants, nuclear power plants, and hydroelectric generating stations. Both fossil fuel and nuclear power stations have output that is controlled by operators. The timescale for changes to the output in response to changes in demand can be fairly short, as for combustion turbines, or much longer, as for coal-fired thermal or nuclear facilities. Apart from fairly predictable seasonal fluctuations, hydroelectric generating station output is readily controlled by the operator. Thus electricity supplied to the grid could be controlled to meet demands.

The implementation in recent years of alternative energy sources, such as wind, solar, and tidal (for example), has made the efficient distribution of electricity more complex. The need for storage mechanisms that results from the variations in output from most alternative generating methods, combined with the desire to utilize resources more efficiently by better matching supply and demand characteristics, has led to the development of smart-grid technologies. These technologies have been implemented in several locations worldwide, and development in this area will certainly continue in the future. This approach fulfills the need to integrate base-load supply, from sources such as nuclear energy, with fluctuating sources such as solar or wind and appropriate energy storage mechanisms, such as batteries or pumped hydroelectric, in order to most efficiently make use of the available resources. This approach, combined with real-time monitoring (or even control) of consumer energy use, will provide the most effective means of energy distribution.

21.5c Storage

Energy storage is an essential component of many viable energy systems because energy cannot always be produced where and/or when it is needed. This is particularly the case for alternative energy sources. The nature of the most suitable energy storage technology for a particular application depends on a number of factors:

1. *The form of the initial and final energy:* Energy storage often involves the storage of electrical energy with the need for electrical energy as a final product. This might be, for instance, the storage of electrical energy produced by wind turbines for later use on the electric grid when it is needed. In this case, pumped hydroelectric storage is a common approach. Another common situation is thermal-to-thermal energy storage. This might involve storing thermal energy from solar collectors for later use for domestic heating or hot water. In this situation, a tank of water or a container of rocks or sand might be appropriate, depending on the nature of the heat distribution system.

2. *Size and/or weight restrictions:* The size of an energy storage unit is most important for energy storage systems that need to be portable. This situation is most relevant to transportation where vehicle size and weight need to be within reasonable limits. Batteries for BEVs or hydrogen for fuel cell vehicles are the most common approaches to energy storage for transportation applications.

3. *Total energy storage capacity:* In cases (i.e., vehicles) where size is an important factor, high energy density is a requirement. Systems that store energy via chemical processes are the most suitable. These include batteries or hydrogen, and, although these mechanisms do not typically achieve the energy storage capacity of liquid fossil fuels, they are potentially adequate. Systems that store

---

PROJECT 21.14
Residential energy storage

A homeowner in a rural area would like to store enough energy to satisfy the electrical needs (without electric heating) for a period of 1 week in the event of an extended power outage (see Chapter 2 for typical electricity requirements for a single-family home in North America). This is a common situation that is typically satisfied by a gasoline (or sometimes diesel or propane) generator. In this case, the homeowner would like to consider nonfossil fuel alternatives for storing energy. Some options are:

- An ethanol-powered ICE generator.
- A CHG-fueled fuel cell.

- Pb-acid batteries.
- Li-ion batteries.
- A flywheel/generator system.
- A compressed air-powered generator.

Discuss the design of each of these systems. Consider, particularly, the quantity of fuel (if any) that would be required to supply 1 week of electricity, the space requirements for the system, and the actual environmental advantages over a gasoline-powered generator.

PROJECT 21.15

Batteries and supercapacitors for transportation use

A 1500-kg fuel cell vehicle requires an average of 0.7 MJ/km at the wheels. Hydrogen fuel is stored in a 0.1-m^3 CHG tank at 35 MPa, and electricity is produced by a 100-kW PEM fuel cell. It is desired to increase both power and range by adding an energy storage system, consisting of either Li-ion batteries or supercapacitors or both, which does not exceed 5% of the total vehicle weight. When the batteries and/or supercapacitors are fully charged externally, a 30% increase in both power to the wheels and range is required.

(a) Will either batteries or supercapacitors alone fulfill the requirements?

(b) If not, will some combination of batteries and supercapacitors suffice?

(c) If a viable design is feasible, what is the minimum increase in vehicle weight required? If a viable design is not feasible, specify the design of a battery/supercapacitor system that comes closest to satisfying the requirements and stays within the weight limitations.

mechanical energy (e.g., compressed air or pumped hydroelectric) have lower energy densities and are most appropriate where large quantities of energy need to be stored but size limitations are not a factor.

4. *Maximum power available:* The rate at which energy can be extracted from an energy storage system is generally an important consideration. For transportation applications, sufficient power is needed to provide acceptable vehicle performance. In the case of BEVs or fuel cell vehicles, this is generally less of a limitation than the total energy storage capacity. Supplementary power from, say, supercapacitors is an option when needed. Large-scale energy storage systems (for example, for connection to the grid) need to meet the demands of the users. System design, such as head, flow rate, and so on, for a pumped hydroelectric facility, needs to address these requirements.

21.6 Conclusions

Our future energy choices will be determined by a consideration of a number of factors:

- Science.
- Technology.
- Environment.
- Economics.
- Politics.

The basic science of many aspects of energy production still needs to be understood. For example, this would include understanding the physics of new materials for photovoltaic cells and batteries. Technologies need to be developed that make use of basic scientific principles—for example, establishing efficient methods for photovoltaic cell production or designing new devices for converting ocean wave motion into electricity. How energy production methods interact with our environment needs to be fully understood. It is clear from this discussion that so-called environmentally friendly or green energy production methods can have more of an environmental impact than might be initially perceived. Certainly, energy must be affordable to those who use it, and new technologies must, at some point, compete economically with traditional technologies; otherwise their development will be hindered. Political agendas often determine energy policy, and public perception of new energy technologies, such as hybrid vehicles or hydrogen vehicles, may not always be accurate. An informed public always makes the best decisions.

There are no obvious immediate solutions for our energy needs. Deciding the best route to take and developing the technology to follow that route will be a challenge and will require dedication and persistence. The effort required to deal with this challenge is enormous. In the Preface, the immensity of this task was suggested—one new major energy production facility every day for the next 50 years. A worldwide effort of this magnitude, with a clear vision of the development of facilities for sustainable and environmentally conscious energy is needed to secure our energy needs for the future.

Bibliography

J. Fanchi. *Energy: Technology and Directions for the Future.* Elsevier, Amsterdam (2004).

R. Heinberg. *Power Down: Options and Actions for a Post-Carbon World.* New Society, Gabriola Island, Canada (2004).

R. Hinrichs and M. Kleinbach. *Energy: Its Use and the Environment,* (5th ed.) Brooks-Cole, Belmont, CA (2012).

T. B. Johansson, H. Kelly, A. Reddy, and R. Williams (Eds.). *Renewable Energy: Sources for Fuels and Electricity.* Island Press, Washington, DC (1992).

D. J. C. MacKay. *Sustainable Energy—Without the Hot Air.* UIT Cambridge, Cambridge, MA (2009).

V. Smil. *Energy at the Crossroads—Global Perspectives and Uncertainties.* MIT Press, Cambridge, MA (2003)

B. Sorensen. *Renewable Energy: Its Physics, Engineering, Environmental Impacts, Economy, and Planning Aspects,* (2nd ed.) Academic Press, London. (2002).

Special Report on Emission Scenarios—Summary for Policymakers. WMO and UNEP, Geneva (2001), available at http/www.ipcc.ch/pub/SPM_SRES.pdf

C. Starr, M. F. Searl, and S. Alpert. "Energy sources: A realistic outlook." *Science* **256,** (1992): 981.

J. W. Tester, D. O. Wood, and N. A. Ferrari (Eds.). *Energy and the Environment in the 21st Century* Cambridge. MIT Press, Cambridge, MA (1991).

J. Tester, E. Drake, M. Driscoll, M. W. Golay, and W. A. Peters. *Sustainable Energy: Choosing Among Options.* MIT Press, Cambridge, MA (2006).

J. R. Wilson and G. Burgh. *Energizing Our Future: Rational Choices for the 21st Century.* Wiley, Hoboken, NJ (2008).

R. Wolfson. *Energy, Environment, and Climate,* (2nd ed.) W.W. Norton, New York (2011).

World Energy Council. *World Energy Resources 2016,* available online at https://www.worldenergy.org/publications/2016/world-energy-resources-2016/

APPENDICES

Powers of Ten

number	prefix	abbreviation
10^{18}	exa	E
10^{15}	peta	P
10^{12}	tera	T
10^{9}	giga	G
10^{6}	mega	M
10^{3}	kilo	k
10^{-3}	milli	m
10^{-6}	micro	μ
10^{-9}	nano	n
10^{-12}	pico	p
10^{-15}	femto	f
10^{-18}	atto	a

Physical Constants

quantity	symbol	value	units
alpha particle binding energy	B_α	28.296	MeV
alpha particle mass	m_α	4.00150618 3727.409	u MeV/c^2
atomic mass unit	u	$1.6605402 \times 10^{-27}$ 931.494	kg MeV/c^2
Avogadro's number	N_A	6.0221367×10^{23}	$mole^{-1}$
Boltzmann's constant	k_B	$1.3806488 \times 10^{-23}$ 8.6173324×10^{-5}	$J \cdot K^{-1}$ $eV \cdot K^{-1}$
Coulomb constant	$1/4\pi\varepsilon_0$ $e^2/4\pi\varepsilon_0$	8.987551×10^9 1.439976	$N \cdot m^2 \cdot C^2$ $MeV \cdot fm$
deuteron binding energy	B_d	2.224	MeV
deuteron mass	m_d	2.013553214 1875.628	u MeV/c^2
electron mass	m_e	$5.48579903 \times 10^{-4}$ 0.5109988	u MeV/c^2
electron volt	eV	$1.60217733 \times 10^{-19}$	J
electronic charge	e	$1.60217733 \times 10^{-19}$	C
gas constant	R	8.314510	$J \cdot K^{-1} \cdot mol^{-1}$
gravitational acceleration	g	9.80665	m/s^2
gravitational constant	G	6.67259×10^{-11}	$N \cdot m^2 \cdot kg^{-2}$
neutron mass	m_n	1.008664904 939.56531	u MeV/c^2
permittivity of vacuum	ε_0	8.854×10^{-12}	F/m
Planck's constant	h	$6.6260755 \times 10^{-34}$ 4.13570×10^{-15}	$J \cdot s$ $eV \cdot s$
	$\hbar$	$1.05457266 \times 10^{-34}$ 6.58217×10^{-16}	$J \cdot s$ $eV \cdot s$
	hc	1.240×10^3	$MeV \cdot fm$
	$\hbar c$	1.973×10^2	$MeV \cdot fm$
proton mass	m_p	1.007276470 938.2723	u MeV/c^2
speed of light	c	2.99792458×10^8	$m \cdot s^{-1}$
Stefan-Boltzmann constant	σ	5.67051×10^{-8}	$W \cdot m^{-2} \cdot K^{-4}$

Miscellaneous Conversion Factors

To convert from the unit on the left to the units on the right, multiply the value in left-hand units by the conversion factor. To convert from the units on the right to the units on the left, divide the value in right-hand units by the conversion factor.

convert from (convert to)	multiply by (divide by)	convert to (convert from)
astronomical units (au)	1.49×10^{11}	meters (m)
barrels (bbl)	158.97	liters (L)
becquerels (Bq)	1.0	decays per second (s^{-1})
electron volts (eV)	1.6×10^{-19}	joules (J)
hectares (ha)	10^4	square meters (m^2)
kilowatt hours (kWh)	3.6×10^6	joules (J)
million electron volts (eV)	1.6×10^{-13}	joules (J)
years (y)	3.15×10^7	seconds (s)

Energy Content of Fuels

fuel	quantity	equivalence in joules
crude oil	liter	3.85×10^7
	barrel (bbl = 42 U.S. gal)	6.12×10^9
gasoline	liter	3.48×10^7
bituminous coal	kilogram	3.10×10^7
natural gas (STP)	cubic meter	3.85×10^7
^{235}U (fission)	gram	8.28×10^{10}
deuterium (fusion)	gram	2.38×10^{11}
wood (maple, 20% water)	kilogram	1.4×10^7
ethanol	liter	2.35×10^7
municipal waste	kilogram	1.0×10^7

The Elements

element	symbol	atomic number	element	symbol	atomic number
actinium	Ac	89	gold	Au	79
aluminum	Al	13	hafnium	Hf	72
americium	Am	95	helium	He	2
antimony	Sb	51	holmium	Ho	67
argon	Ar	18	hydrogen	H	1
arsenic	As	33	indium	In	49
astatine	At	85	iodine	I	53
barium	Ba	56	iridium	Ir	77
berkelium	Bk	97	iron	Fe	26
beryllium	Be	4	krypton	Kr	36
bismuth	Bi	83	lanthanum	La	57
boron	B	5	lawrencium	Lr	103
bromine	Br	35	lead	Pb	82
cadmium	Cd	48	lithium	Li	3
cesium	Cs	55	lutetium	Lu	71
calcium	Ca	20	magnesium	Mg	12
californium	Cf	98	manganese	Mn	25
carbon	C	6	mendelevium	Md	101
cerium	Ce	58	mercury	Hg	80
chlorine	Cl	17	molybdenum	Mo	42
chromium	Cr	24	neodymium	Nd	60
cobalt	Co	27	neon	Ne	10
copper	Cu	29	neptunium	Np	93
curium	Cm	96	nickel	Ni	28
dysprosium	Dy	66	niobium	Nb	41
einsteinium	Es	99	nitrogen	N	7
erbium	Er	68	nobelium	No	102
europium	Eu	63	osmium	Os	76
fermium	Fm	100	oxygen	O	8
fluorine	F	9	palladium	Pd	46
francium	Fr	87	phosphorus	P	15
gadolinium	Gd	64	platinum	Pt	78
gallium	Ga	31	plutonium	Pu	94
germanium	Ge	32	polonium	Po	84

Continued on page A-7

element	symbol	atomic number	element	symbol	atomic number
potassium	K	19	tantalum	Ta	73
praseodymium	Pr	59	technetium	Tc	43
promethium	Pm	61	tellurium	Te	52
protactinium	Pa	91	terbium	Tb	65
radium	Ra	88	thallium	Tl	81
radon	Rn	86	thorium	Th	90
rhenium	Re	75	thulium	Tm	69
rhodium	Rh	45	tin	Sn	50
rubidium	Rb	37	titanium	Ti	22
ruthenium	Ru	44	tungsten	W	74
samarium	Sm	62	uranium	U	92
scandium	Sc	21	vanadium	V	23
selenium	Se	34	xenon	Xe	54
silicon	Si	14	ytterbium	Yb	70
silver	Ag	47	yttrium	Y	39
sodium	Na	11	zinc	Zn	30
strontium	Sr	38	zirconium	Zr	40
sulfur	S	16			

Table of Acronyms

AC	alternating current
ACEA	Association des Constructeurs European d'Automobile
AEEFM	Assessment of Energy Efficiency Finance Mechanism
ALR	adiabatic lapse rate
ASDEX	Axially Symmetric Divertor EXperiment
BEE	Bureau of Energy Efficiency
BEV	battery electric vehicle
BWR	boiling water reactor
CAFE	Corporate Average Fuel Economy
CANDU	Canadian Deuterium Uranium (Reactor)
CFL	compact fluorescent lamp
CHG	compressed hydrogen gas
CHP	combined heat and power
CIGS	copper-indium-gallium-selenide
CRF	capital recovery factor
CURVE	clean, unlimited, renewable, versatile, economic
DC	direct current
DOE	Department of Energy
DSSC	dye-sensitized solar cell
ec	electron capture
EIA	Energy Information Administration
Eliica	electric lithium ion car
EPA	Environmental Protection Agency
EU	European Union
FBR	fast breeder reactor
FCP	Fuel Consumption Program
GDP	gross domestic product
GNP	gross national product
HC	hydrocarbon
H-ICE	hydrogen internal combustion engine
HTSC	high-temperature superconductor
HVAC	heating, ventilation, and air conditioning
IAEA	International Atomic Energy Agency
ICE	internal combustion engine
IEE	Intelligent Energy Europe
INES	International Nuclear and Radiological Event Scale
IPEEC	International Partnership for Energy Efficiency Cooperation
IPEEI	Improving Policies Through Energy Efficiency Indicators
IR	infrared
ITER	International Thermonuclear Reactor

LED	light-emitting diode
LEED	Leadership in Energy and Environmental Design
LH$_2$	liquid hydrogen-2
LMFBR	liquid metal fast breeder reactor
LNG	liquid natural gas
LOCA	loss of coolant accident
MCT	marine current turbine
NASA	National Aeronautics and Space Administration
NCAR	National Center for Atmospheric Research
NELHA	National Energy Laboratory of Hawaii Authority
NIMBY	not in my backyard
NOAA	National Oceanic and Atmospheric Administration
NREL	National Renewable Energy Laboratory
OECD	Organisation for Economic Co-operation and Development
OTEC	ocean thermal energy conversion
OWC	oscillating water column
PCRA	Petroleum Construction Research Association
PCV	positive crankcase ventilation
PEM	polymer electrolyte membrane
PET	polyethylene terephthalate
PRO	pressure-retarded osmosis
PTO	power takeoff
PV	photovoltaic
PWR	pressurized water reactor
RBMK	Reactor Bolshoi Moschnosti Kanalynyi (Russian water-cooled graphite moderated reactor)
RDF	refuse-derived fuel
RED	reverse electrodialysis
rpm	revolutions per minute
SAVE	specific actions for vigorous energy efficiency
SBN	sustainable building network
SEGS	solar energy generating station
SET	Strategic Energy Technology (Program)
SI	System Internationale
SMES	superconducting magnetic energy storage
STEP	Solar Total Energy Project
STP	standard temperature and pressure
SVO	straight vegetable oil
TBL	triple bottom line
TEPCO	Tokyo Electric Power COmpany
TMI	Three Mile Island
TVA	Tennessee Valley Authority
US$	U.S. dollars
USDOE	United States Department of Energy
USEIA	United States Energy Information Administration
UV	ultraviolet
WEACT	Worldwide Energy Efficiency Action Through Capacity Building and Training
ZNE	zero net energy (building)

INDEX

G